THE
WOODBOOK

THE WOODBOOK

THE COMPLETE PLATES

The American Woods
Romeyn Beck Hough

TASCHEN

CONTENTS

It may seem surprising at first for a publisher of art books to bring out an edition of what is, from its subject matter, a book that can be classified under the headings of botanical geography, dendrology, forestry, or the biology and technology of timber. On the other hand, by virtue of its publication history, its unique design, with plates using genuine wood samples, and its extreme rarity, the original of Romeyn Beck Hough's *The American Woods* (1888–1913, 1928) provides historical, artistic, and aesthetic reason enough to document it once more in contemporary guise, in order to rouse enthusiasm and lead the newly won enthusiast to the wealth of specialist literature on the subject.

While the focus of attention has varied, wood has been a matter of interest and concern to human society for thousands of years. The word itself – and this seems true of many languages – belongs to the core vocabulary, and denotes both the material and the colonies of trees from which it comes. Our present-day symbolism preserves the memory of the time when primeval man used wood for tools and weapons even before he gave thought to what to wear or where to live.

The heraldic figure, handed down to us in the history of art and culture, of the "wild man," naked but for a leafy loincloth, yet leaning against a massive wooden club, represents this earliest human wood-user as an ancestor worthy of respect, and it is not for nothing that he still plays a part in our present-day rituals, in for example the carnival celebrations of south-western Germany. Goethe dedicated respectful lines to him in Part II of his *Faust*:

> *Giants*
> *Here are the Wild Men, that's our name,*
> *The Wild Men of Harz Mountains fame*

Natural-naked in full strength,
Each with his club a pinetree's length,
We come as giants big and tall
And thickly girdled one and all
With leaves and branches bound like thatch.
No Pope has bodyguards to match!

In the beginning, and "beginning" in this context, extends over long periods of varying duration as the world's population grew and spread, periods which among so-called primitive peoples continued until the recent past – people were undemanding, and made do with wood unworked but for a few knocks with some stone tool, or else managed simply with the branches or sticks as they found them. In contrast to this simplicity, the wood itself was highly versatile: it could be used as simple tools or weapons, as firewood, or for building. Apart from caves, mankind's earliest form of shelter consisted of windbreaks of woven branches, or tents of rods and sticks covered in animal skins.

A major advance of the Neolithic Age was the invention of the wooden handle for stone tools, for example a flint hand ax attached to a shaft of wood, which, by lengthening the lever, increased the force that could be applied. This tool represents the generalized form of "ax" as we know it. For all the design improvements and material refinements it has undergone, it has remained the standard tool for felling trees and coarsely processing the timber needed for huts, log cabins and pillars for raised houses – and once it was realized that wood floats, also for rafts, dug-out canoes, and boats and ships.

These cultural achievements represented the basic preconditions for the overseas expansion of Europe to the North American subcontinent, two-fifths of which was then covered by ancient forest.

It is difficult to imagine just how much timber was swallowed up in Europe for shipbuilding, and timber here means forest trees; we should remember the extent of deforestation that had already taken place in Scotland and Ireland by the time the colonization of America started. And, decisively, ships needed masts: in other words, single, tall, straight trunks with no branches. To this extent, America must have seemed an inexhaustible source of timber, and what is more, a source of particularly suitable varieties such as were not, or no longer, available in Europe. As a result, shipyards also began to spring up on the eastern seaboard of the New World.

Fight for Conservation
Immigration, and the colonization of North America, led in a very short time, however, to the decimation of the continent's forests. The early period of the United States of America was characterized by a concept of colonization based on occupying, using, and cultivating land, having first got rid of the forests and the native peoples. This mindset led to the country's historically unique rise to world status and world power – albeit at the price of the latent danger of self-destruction that would follow the total annihilation of nature as the basis of life. This danger was not realized or taken seriously until quite late, namely the second half of the 19th century, and then by only a few, leading to the creation of a conservation movement and a specific environmental policy, and it is in this context that we should view the endeavors of R. B. Hough.

The forests that less than a hundred years earlier had been regarded as inexhaustible had been reduced so drastically through exploitation, above all by giant logging companies such as Weyerhäuser, that calls were made to save the forests by planting trees. The scale of exploitation may be surmised from the fact that Frederick Weyerhäuser (1834–1914), who had come to America as a barefoot country boy from the Palatinate in Germany, was able to set up a veritable timber empire, and though he himself avoided the public gaze, it was said that he was richer than Rockefeller.

The publicly ordered reforestation marks the start of a new view of nature. One prominent advocate of the cause of the tree was Charles Sprague Sargent (1841–1927), Professor of Botany at Harvard University and from 1873 Director of its Arnold Arboretum, an open-air collection of American trees, a living museum, so to speak, of native forests.

The new nature-conservation movement found support at about the same time in the idea of the National Park, in other words the creation of nature reserves under the protection and supervision of the state. Yellowstone National Park, founded in 1872, became the pattern for these novel nature-conservation institutions in North America. As they needed the approval of Congress through legislation and the appropriation of public funds, they were not uncontroversial. A considerable degree of public education on the part of committed scientists was needed to persuade the representatives of the people that individual measures and well-meant private tree-protection initiatives were not sufficient to overcome the crisis facing the forests. On the contrary, the interaction of water, forest, and soil – whole ecosystems, as we would say today – first had to be surveyed, analyzed, and assessed nationwide. Only when cause and effect had been clarified would it be possible to progress beyond uncertain protection measures toward a regulated and sustainable forestry economy on a scientific basis, something that did not at that time exist in the United States. Thus the

goal was to establish a foundation for the life of the whole nation on a par with, and complementary to, agriculture.

On behalf of the Commissioners of Agriculture, Franklin B. Hough (1822–1885), our author's father, a physician and enthusiastic connoisseur and lover of nature, submitted his official *Report upon Forestry* as a first survey of this kind. The report, an incredible mass of information, appeared in two volumes in 1877 and 1880. Other reports followed. The topic was now well and truly on the agenda.

It was no less a person than Secretary of the Interior Carl Schurz (1829–1906), Weyerhäuser's compatriot and contemporary, who set up within his Department a Forest Agency, which later became the Forest Division of the Department of Agriculture when the latter was created. The administrative link between agriculture and forestry thus created can be seen as a taking up of a "Fight for Conservation" on the part of the government of the United States, not just to conserve forests while agriculture prospered, but also to revitalize them from the bottom up on a long-term basis.

As a pioneer in the fight for conservation, the nationally minded puritan Gifford Pinchot (1865–1946) was well aware of all this. He studied forestry in France, and on his return had the opportunity to exercise some influence in this area in leading advisory capacities. In 1901 his endeavors, which extended to the study of water supply and regulation, soil erosion, forest fires, opencast mining, and land policy, earned him the friendship of President Theodore Roosevelt (1858–1919), whose outlook largely coincided with that of Pinchot. In his book *The Fight for Conservation*, Pinchot summarized all his ideas and experiences, including, especially, those resulting from disputes involving powerful vested interests. He served two terms as Governor of Pennsylvania, and on his death in 1946 bequeathed to us all a legacy of exhortation to constant vigilance.

Sargent and his researches into tree and shrub flora in the 19th century

It was this era of rethinking and stocktaking as a preliminary to reform that saw the publication of Sargent's 14-volume *Silva of North America* (1889–1901), with its catalogued descriptions, arranged by species, of all the American forest trees then known, accompanied by splendid plates prepared by his illustrator Charles E. Faxon, which remain impressive when reproduced as engravings. The book was reprinted in 1947 and is still regarded as a standard reference work.

It was Sargent who first came up with a map of what were then still the primeval forests of North America (1884); the maps were reproduced by the German botanical geographer Adolf Engler (1844–1930) in *Petermanns Mitteilungen* in 1886. These, and a later map by John W. Harshberger (1869–1929), which appeared with a detailed monograph in Volume XIII of Engler & Drude's *Vegetation der Erde* (1911), are the only authentic contemporary maps on the subject. The collaboration between American and German botanists at this time is worthy of note.

A friend of Sargent's, the Bavarian forestry botanist Heinrich Mayr (1854–1911), twice traveled the whole of the North American subcontinent, from Vancouver to Florida, in 1886 and 1887. In the course of a trip around the world between 1902 and 1906 he visited North America once again. Mayr was perhaps the last of a long series of Europeans, including Peter Kalm, Michaux senior and junior, von Wangenheim, and finally Douglas "of the forests," who all came with the firm intention of finding fast-growing and useful tree species from the American forests to plant

and domesticate in their own native countries. Parks, botanical gardens, and commercial forests in Europe are now home to numerous such introduced species, some of them reintroductions, bearing in mind the shared flora of pre-Ice Age times. The history of plant life in the past explains many of the common features of North American and European trees, including for example the often identical genera of maple, beech, oak, elm, ash, lime, birch, willow, or pine, fir, and spruce. However, there are differences in the number and characteristics of individual species: differences not only between America and Europe, but also between the Atlantic seaboard and the Pacific northwest of the continent and their widely separated mountain ranges. In addition, there are major physical differences owing to latitude and the resulting climatic zones, with coniferous forests in the subarctic northern regions to evergreen hardwoods and palm in the subtropical south.

Mayr traveled through North America at a relatively late stage, a fact that moved him to make the following comment in 1906: "In America the second generation of trees is growing up on an area reduced by three-quarters; the forests are less dense, the number of species is much reduced and the trees themselves are in many cases of poorer quality. Unspoiled ancient forests are rapidly disappearing." He excluded from this assessment the mountain forests under government protection. More impressive than any statistics were two "reverse images" produced by Mayr to illustrate developments in the north-western states between 1890 and 1906. Viewed from a mountain in 1890, the settlements were "insignificant holes in the dark-green carpet." In 1906 in the same place he found nothing but "remnants of forest showing as dark-green dots" in the "light-green and brown landscape" now characterized by agriculture. Like many before and since, he nonetheless had the comfort of finding that the giant trees of the Sierra Nevada and the coastal mountains, Wellingtonias and Coast Redwoods growing to heights of 330 to 380 feet, were to some extent already protected, and in the case of the latter, even forming forests and representing an important commodity as northern California's chief export timber. The Coast Redwood has the property of sprouting from stump after being felled, a property inherited by the daughter trunks, so that in this sense the tree is literally immortal. Other forest trees of the conifer-rich north-west also grow very tall, incidentally: heights of 240 to 300 feet have been reported. It is also certain that the Weymouth Pine of the north-eastern seaboard grows, or at least once grew, very tall, as Britain's Royal Navy marked specimens with the broad arrow – the sign of government property – to reserve them for masts.

The preparation of wood specimens
For the 10th Census of the United States, Sargent undertook the responsibility for counting trees in his capacity as Director of Harvard University's Arnold Arboretum. Building on this experience, he organized on his own account a collection, in triplicate, of trees in the form of five-foot sections of trunk. The trunks are each processed in such a way as to show three polished surfaces, in order to reveal the anatomy of the wood for identification purposes: the cross section, the radial longitudinal section and the tangential longitudinal section. The bark is also retained. The items were provided with their scientific and vernacular names and briefly described. Their place of origin and range are noted on a map. One of these collections was acquired by the American Museum of Natural History in New York and named the Jesup Collection after its Director, the patron of

this expensive undertaking. It is now in the city of Independence, Iowa, where there are plans to display it in a new museum. The second collection went to Portland, Oregon, while the third, with the trunks appropriately reduced in length, is behind glass in the Arnold Arboretum itself.

A short article in the *Botanical Gazette*, Vol. XI, of February 1886, entitled "Sections of Native Woods" (p. 40) reports that Sargent had given carefully named pieces of trunk resulting from the compilation of his Jesup Collection to a well-known Boston-based manufacturer of veneers, Charles W. Spurr, who turned them into remarkable sets of thin sections, which were then allegedly being offered for sale. These sets were, it was said, unique. Each included some 200 native species of tree, and where possible all three directions of cut were included. The thickness of the sections varied, according to the properties of the wood concerned, from a hundredth to a two-hundred-and-fiftieth of an inch. Each of the sections was said to be mounted between two sheets of mica in a flexible wooden frame. The frame consisted of two layers of maple stiffened with strong paper and coated with shellac. In addition, it was printed with the name of the buyer or the institution that acquired it, as well as Sargent's Jesup Collection catalogue number, the scientific and vernacular name, details of the direction of cut, and the name of the craftsman who prepared the item.

The final paragraph of the article states that the sections were cut using "a three-ton veneer machine" and that "almost 18,000 individual sections" had been carefully prepared and were now being offered for sale at a low price. The article is unsigned, however, and no such sections can be traced.

R. B. Hough and his lifework –
The American Woods

Thomas Houseworth & Co.:
"The Domes from Merced River",
Yosemite Valley, ca. 1874
Courtesy George Eastman House

Two years later, in 1888, Romeyn B. Hough began to publish his work *The American Woods*, which extended to 15 volumes each with 25 plates, each consisting of three sections taken from a particular species. The plates accord with those described and advertised by, or on behalf of, Charles W. Spurr, the Boston veneer-manufacturer, in the *Botanical Gazette* of 1886. So do the preparations themselves (each approximately two by five inches).

Hough, who described himself in his introduction as "inspired," makes no reference either to the Jesup Collection or to Spurr. His *The American Woods* appeared between 1888 and 1913, and (posthumously) in 1928, without Volume 15; it was regarded then as his lifework, and still is. Sargent for his part never mentioned it. If we can trust the only account of Hough's life – namely that in *The National Cyclopaedia of American Biography* – he received many honors and awards for *The American Woods*. It is striking that the order of plates and volumes does not follow the system of botanical classification, nor is the statement in an auction catalogue that they are "arranged geographically" entirely accurate. But this hardly detracts from the information value, nor from the accompanying text (cf. Sargent), and not at all of course from the charm of the Victorian presentation.

A "curiosity" is the best way to describe the attempt by Marjorie Hough, one of R. B. Hough's daughters, to publish a new, likewise 15-volume, encyclopedia of American wood using preparations bequeathed to her, and with editorial assistance from Professor Harrar, the Dean of Forestry at Duke University in Durham, North Carolina. By 1981 the work, begun in 1957, had reached eight double volumes (text and plates) and four further volumes of plates only, but will presumably remain unfinished. Harrar claims in his introduction that Hough had

cut the sections with a machine he had invented himself "because there were not yet any available." In this he misjudged the veneer-machine manufacturers of Cincinnati and probably misled a number of readers.

Works using natural-wood preparations instead of illustrations are not as rare as is often assumed. In particular Hermann Nördlinger's *Querschnitte von hundert Holzarten*, Volumes I to XI, published in Stuttgart and Tübingen by the J. G. Cotta'scher Verlag from 1852 to 1888, ought to be mentioned here, as it might be thought that it was the source of Hough's inspiration – which was certainly not the case.

Each volume consists of a hundred paper-thin cross-sections of wood in an oval window mount between stiff folded paper, with one such preparation for each species of native or exotic wood, suitable for examining under a magnifying glass (and thin enough to be treated as a transparency). Each sample is provided with a label bearing its name, and they are arranged unbound in alphabetical order. Each volume is accompanied by a paperbound booklet containing a description, and treated as a continuation of the preceding volume; each comes in a cardboard box designed to imitate a book, with half-leather back andgilt title, and otherwise a polychrome pattern. The whole edition took 36 years to produce, and earned Nördlinger medals in the London exhibitions of 1851 and 1861.

Compared with other publications, the particular charm of Hough's plates lies in their unique presentation, so that the few remaining complete copies will always be rarities much sought-after by bibliophiles.

Romeyn Beck Hough
Amerikanische Hölzer

Klaus Ulrich Leistikow

Es mag zunächst überraschen, dass ein Kunstbuch-Verlag diese Edition vorlegt, die thematisch eher botanischen Bereichen, der Pflanzengeografie, der Forstbotanik und Dendrologie oder der Holzbiologie und -technologie zuzuordnen wäre. Demgegenüber bietet das Original von Romeyn Beck Houghs *The American Woods* (Amerikanische Hölzer, 1888–1913, 1928) allerdings durch seine Entstehungs- und Publikationsgeschichte sowie die Besonderheit seiner Gestaltung aus Tafeln mit echten Holz-Präparaten in Buchform historische, künstlerische und ästhetische Anreize genug, um – zumal bei seiner außergewöhnlichen Rarität – in zeitgemäßer Form dokumentiert zu werden, weitergehende Interessen zu wecken und so gewonnene Liebhaber an die entsprechende, reichlich vorhandene Spezialliteratur heranzuführen.

Das Thema Holz beschäftigt unsere Gesellschaft seit Jahrtausenden mit unterschiedlicher Gewichtung. Holz, Hölzer und Gehölze: Die Begriffe bezeichnen sowohl Bäume und Baumgesellschaften als auch das von ihnen stammende Material, sind also sprach- und sinnverwandt – eine etymologische Tatsache, die in den Wurzeln der Menschheitsentwicklung begründet liegt. In der Symbolik wirkt die Erinnerung fort, dass der Vor- oder Urmensch, noch bevor er an seine Bekleidung und Behausung dachte, bereits mit Holz als Werkzeug und Waffe umgegangen ist.

Die heraldische Figur des „Wildemannes", nackt bis auf einen Lendenschurz aus Blattwerk, aber gestützt auf eine mannsstarke Baumkeule, stellt, von der Kunst- und Kulturgeschichte überliefert, diesen frühesten menschlichen Holznutzer als einen respektablen Vorfahren dar, der etwa in der alemannischen Fastnacht nicht ohne Grund noch immer an unserer Gegenwart teilnimmt. Goethe hat ihm in *Faust II* die prachtvollen Verse gewidmet:

Die wilden Männer sind's genannt
Am Harzgebirge wohlbekannt
Natürlich nackt in aller Kraft
Sie kommen sämtlich riesenhaft
Den Fichtenstamm in rechter Hand
Und um den Leib ein wulstig Band
Den derbsten Schurz von Zweig und Blatt:
Leibwache, wie der Papst nicht hat.

Anfangs – und das "anfangs" betrifft hier lange, und zwar unterschiedlich lange Zeiträume des Anwachsens und der Verbreitung der Erdbevölkerung, Zeiträume, die bei den sogenannten Naturvölkern bis in die jüngste Vergangenheit hineingereicht haben – genügte unbearbeitetes, oder mit wenigen Handgriffen oder Steinhieben hergerichtetes, oft sogar nur gesammeltes Holz von Ästen und Zweigen den geringen Ansprüchen der Nutzer. Demgegenüber waren die Verwendungszwecke des Holzes jedoch vielfältig, sowohl als einfaches Werkzeug oder Waffe als auch als Brenn- und Baustoff. Abgesehen von Höhlen gehören Windschirme aus Gestecken von Zweigen und Zelte aus fellüberzogenen Stangen oder Ruten zu den ersten von Menschen errichteten Unterkünften.

Hoch einzuschätzen ist die jungsteinzeitliche Erfindung der Kombination eines Werkzeugs mit einer Handhabe, zum Beispiel die Schäftung eines Feuersteinkeils als Schneide am oberen Ende eines Schaftes aus Holz als Verlängerung und Verstärkung des Kraftarms. Dieses Werkzeug ist die Axt in ihrer allgemeinsten Form. Sie bleibt über alle Verbesserungen und Verfeinerungen des Materials und der Konstruktion hinweg bis in unsere Tage das Allerweltswerkzeug zum Fällen von Bäumen und zur rohen Bearbeitung des Stammholzes als dem Baustoff von Pfahlbauten, Hütten und Blockhäusern und

Albert Renger-Patzsch:
"Beech Forrest in November", 1954
Courtesy George Eastman House

des Holzes zum Bau von Flößen, Einbäumen, Booten und Schiffen.

Mit diesen Leistungen der Kultur waren die basalen Voraussetzungen für die überseeische Landnahme von Europäern auf dem zu vier Zehnteln seiner Oberfläche von Urwald bedeckten nordamerikanischen Subkontinent in der Tat gegeben. Man kann nicht umhin zu überlegen, welche Mengen an Holz, und das heißt ja an Waldbäumen in Europa, der Schiffbau verschlang und welches Ausmaß die Entwaldungen in Schottland und Irland um die Zeit der Eroberung Amerikas bereits angenommen hatten. Dabei kam hohen, geraden, unbeasteten Stämmen als Mastbäumen eine entscheidende Bedeutung zu. Insofern war Amerika, das anscheinend unerschöpfliche Waldland, als Lieferant besonders geeigneter Hölzer, wie sie in Europa nicht oder nicht mehr aufzutreiben waren, gerade zur rechten Zeit willkommen und betrieb an der Ostküste alsbald auch selber Schiffbau.

„Fight for Conservation"

Bedingt durch die Einwanderung und Kolonisation Nordamerikas kam es allerdings in kürzester Zeit zu einer drastischen Dezimierung des Waldbestandes. Das Konzept einer Kolonisation durch Bodengewinn, Bodenbearbeitung und Bodennutzung nach Beseitigung des Waldes und der Indianer prägte die Anfänge der Entstehung der Vereinigten Staaten von Amerika und führte den historisch einzigartigen Aufstieg zur Weltgeltung und Weltmacht herbei – freilich um den Preis der latenten Gefahr der Selbstzerstörung durch völlige Vernichtung der Natur als Lebensgrundlage. Diese Gefahr wurde erst spät, in der zweiten Hälfte des 19. Jahrhunderts von einigen erkannt oder ernst genommen und rief eine Erhaltungsbewegung und

Politik der Umweltkultur hervor, in die auch das Bestreben R. B. Houghs einzuordnen ist.

Die Waldbestände, noch vor weniger als hundert Jahren für unerschöpflich gehalten, waren durch die Raubnutzung vor allem der gigantischen Holzkonzerne wie dem der Weyerhäusers so dramatisch zurückgegangen, dass Aufrufe zur Erhaltung des Waldes durch Anpflanzung von Bäumen ergingen. Für das Maß der Raubnutzung mag gelten, dass Frederick Weyerhäuser (1834–1914), der als barfüßiger Dorfjunge aus der Pfalz eingewandert war, ein wahres Imperium der Holzindustrie errichten konnte, und ihm – der die Publizität scheute – nachgesagt wurde, reicher zu sein als Rockefeller.

Die öffentlich verordnete Aufforstung markiert den Anfang einer neuen Sicht der Natur. Einer der prominenten Wortführer der Sache des Baumes war Charles Sprague Sargent (1841–1927), Professor der Botanik an der Harvard University und seit 1873 Direktor am dortigen Arnold Arboretum, einer Freiland-Sammlung amerikanischer Gehölzarten, quasi das lebende Inventar der heimischen Wälder.

Unterstützung fand die neue Bewegung des Naturschutzes um die gleiche Zeit durch den National-park-Gedanken, das heißt, die Schaffung von Naturreservaten unter dem Schutz und der erhaltenden Aufsicht des Staates. Der Yellowstone-Nationalpark von 1872 wurde zum Muster für diese neuartigen Einrichtungen des Naturschutzes in Nordamerika. Da sie der Zustimmung des Kongresses, der Gesetzgebung und der Bereitstellung von Steuermitteln bedurften, waren sie nicht unumstritten. Es war erhebliche Aufklärungsarbeit engagierter Wissenschaftler nötig, um die Volksvertreter davon zu überzeugen, dass Einzelmaßnahmen und gutwillige Privatinitiativen zum Schutz der Bäume nicht ausreichten, um der Krise des Waldes Herr zu werden. Vielmehr musste das

Zusammenwirken von Wasser, Wald, Boden und Klima – ganzer Ökosysteme, wie wir heute sagen – zu allererst in Bestandsaufnahmen bundesweit erfasst, analysiert und bewertet werden. Erst nach Klärung von Ursachen und Folgen würde es möglich sein, über einen ungewissen, unsicheren Schutz hinaus zu einer geregelten, nachhaltigen Forstwirtschaft auf der Grundlage einer Forstwissenschaft zu kommen, die bis dahin in den Vereinigten Staaten noch nicht etabliert war. Das Ziel wäre somit eine der Agrikultur komplementäre Lebensgrundlage der ganzen Nation.

Im Auftrag des Commissioners of Agriculture legte Franklin B. Hough (1822–1885), Arzt und begeisterter Kenner und Freund der Natur, der Vater unseres Autors, seinen amtlichen *Report upon Forestry* als eine erste Bestandsaufnahme dieser Art von unglaublicher Datenfülle in zwei Bänden vor (1877 und 1880). Andere Berichte folgten. Das Thema blieb von nun an aktuell.

Kein Geringerer als Carl Schurz (1829–1906), Weyerhäusers Landsmann und Altersgenosse, richtete in seinem Innenministerium eine Forstagentur ein, die spätere Forest Division im eigens ausgegliederten Department of Agriculture. Die so hergestellte administrative Verbindung von Landwirtschaft und Forstwirtschaft kann als die Aufnahme eines „Fight for Conservation" durch die Regierung der Vereinigten Staaten angesehen werden, und dies nicht nur für die Erhaltung der Wälder bei prosperierender Landwirtschaft, sondern auch zum Zweck ihrer Erneuerung von Grund auf und auf lange Sicht.

Gifford Pinchot (1865–1946) war der Mann, dem eben das als einem Vorkämpfer im Fight for Conservation vollauf bewusst war. Er studierte in Frankreich Forstwissenschaft und erhielt nach seiner Rückkehr

schem Fachgebiet zu wirken. Seine Bemühungen, die das Studium der Wasserversorgung und -regulation, der Abschwemmung des Bodens, der Waldbrände, des Übertage-Bergbaus und der Landpolitik einschlossen, trugen ihm 1901 die Freundschaft des Präsidenten Theodore Roosevelt (1858–1919) ein, dessen Überzeugungen sich mit denen des national denkenden Puritaners Pinchot programmatisch deckten. In seinem Buch *The Fight for Conservation* fasste Pinchot alle seine Ideen und Erfahrungen, auch und gerade die aus zahllosen Kontroversen mit mächtigen Partikular-Interessen, zusammen. Er wurde zweimal Gouverneur von Pennsylvania und hinterließ 1946 sein Vermächtnis zur dauernden Mahnung.

Sargent und die Forschungen zur Gehölzflora des 19. Jahrhunderts

In diese Epoche des Umdenkens und der Bestandsaufnahmen zur Vorbereitung von Reformen fiel die Veröffentlichung von Sargents vierzehnbändiger *Silva of North America* (1889–1901) mit seinen nach Arten katalogisierten Beschreibungen aller bis dahin bekannten amerikanischen Waldbäume, begleitet von prächtigen Bildtafeln seines Zeichners Charles E. Faxon, die auch in ihrer Wiedergabe als Stahlstiche beeindrucken. Das Werk erfuhr eine Reprint-Edition im Jahre 1947 und gilt weiterhin als Standard-Referenz.

Von Sargent stammte auch die erste Kartierung der seinerzeit noch ursprünglichen nordamerikanischen Waldgebiete (1884), die der deutsche Pflanzengeograf Adolf Engler (1844–1930) in *Petermanns Mitteilungen* 1886 reproduzierte. Diese und eine spätere Karte John W. Harshbergers (1869–1929), die zusammen mit einer ausführlichen Monografie in erschien (1911), sind die einzigen authentischen zeitgenössischen Kartenbilder zum Thema. Bemerkenswert ist die damalige Zusammenarbeit amerikanischer und deutscher Botaniker auf diesem Gebiet.

Ein Freund Sargents, der bayerische Forstbotaniker Heinrich Mayr (1854–1911), durchquerte 1886 und 1887 den gesamten Subkontinent zweimal und bereiste ihn von Vancouver bis Florida. Auf einer Weltreise 1902 bis 1906 besuchte er Nordamerika noch einmal. Mayr war vielleicht der letzte europäische Fachmann in der langen Reihe von Vorgängern, darunter Peter Kalm, Vater und Sohn Michaux, von Wangenheim und schließlich Douglas "of the forests", die alle mit dem Vorsatz kamen, schnellwüchsige und nutzbringende Baumarten aus dem amerikanischen Wald für ihre Heimatländer ausfindig und in ihnen heimisch zu machen. Parks, Botanische Gärten und Nutzwälder in Europa beherbergen zahlreiche solcher Einführungen, manchmal Rückführungen, wenn man an die Gemeinsamkeit der Floren vor der Eiszeit denkt. Die vegetationsgeschichtliche Vergangenheit erklärt viele Übereinstimmungen der Sippen nordamerikanischer und europäischer Bäume, so zum Beispiel die vielfach gleichen Gattungen wie Eiche, Ahorn, Buche, Ulme, Esche, Linde, Birke, Weide, oder Kiefer, Tanne, Fichte. Bei den Artmerkmalen und bei der Anzahl der Arten zeigen sich allerdings Unterschiede: Unterschiede zu Europa und ebenso zwischen dem atlantischen Osten und dem pazifischen Westen Nordamerikas und ihren weit voneinander getrennten Gebirgen. Darüber hinaus treten die breitenparallelen Klimagürtel physiognomisch stark hervor: die Nadelwälder im subarktisch-borealen Norden, die immergrünen Hartlaubgehölze und Palmen im subtropischen Süden.

Mayr bereiste Nordamerika zu einem relativ späten Zeitpunkt, was ihn 1906 zu folgendem Kommentar bewegte: „In Amerika wächst die zweite Baumgeneration auf zu einem an Fläche um drei Viertel verkürzten, artenärmeren, durchlöcherten, vielfach minderwertigen Baumgemenge; unberührte, ursprüngliche Waldungen schwinden rasch dahin." Die Bergwälder unter dem Schutz der Regierung nahm er von seinem Urteil aus. Eindrucksvoller als alle Statistiken gab Mayr die Entwicklung zwischen 1890 und 1906 in zwei Umkehr-Bildern wieder, die sich ihm in den Nordwest-Staaten geboten hatten. Von einem Berg aus waren (1890) die Siedlungen „unscheinbare Löcher in dem dunkelgrünen Teppiche". 1906 fand er dort nur noch „Waldreste als dunkelgrüne Punkte" in der vom Feldbau bestimmten „hellgrünen und braunen Landschaft" vor.

Er hatte immerhin den Trost wie viele vor ihm und noch mancher nach ihm, die Baumriesen der Sierra Nevada und des Küstengebirges, den Riesen-Mammutbaum und den Küsten-Mammutbaum mit ihren Wuchshöhen von 100 und 120 Metern vorzufinden, teils bereits geschützt, teils – wie das Redwood des Küsten-Mammutbaums – sogar noch waldbildend und als Hauptausfuhrholz Nordkaliforniens ein bedeutendes Handelsgut. Der Küsten-Mammutbaum hat nämlich die Eigenschaft, nach einem Abholzen aus seinem Stubben Tochterstämme zu bilden, die alle auch diese Eigenschaft behalten, sodass der Baum auf diese Weise unsterblich ist. Hoher Wuchs kommt übrigens noch anderen Waldbäumen des nadelholzreichen Nordwestens zu, die Angaben schwanken zwischen 70 und 90 Metern. Es ist sicher, dass die Weißkiefer oder Strobe (Weymouth-Kiefer) im Norden der Ostküste auch zu den Riesenbäumen gehört oder doch gehört hat, als sie gekennzeichnet mit der Hiebmarke des Broad Arrow für die Mastbäume der Royal Navy reserviert wurde.

Für die Bestandsaufnahme des 10. Zensus der Vereinigten Staaten übernahm Sargent als Direktor des Arnold Arboretums der Harvard University den Bereich der Bäume. Darauf aufbauend organisierte er selbst eine Sammlung der Bäume in Form von Stammstücken (zu 5 Fuß) und in 3-facher Ausfertigung. Die Stämme sind so bearbeitet, dass sie jeweils drei polierte Schnittflächen zeigen, wie sie die Holzanatomie zur Identifizierung fordert: den Quer- oder Hirnschnitt, den radialen Längs- oder Spiegelschnitt und den parallel zur Tangente geführten, daher tangentialen Längs- oder Fladerschnitt. Die Borke bleibt im Übrigen erhalten. Die Stücke wurden mit ihren wissenschaftlichen und den Volks-Namen versehen und von Kurzbeschreibungen erläutert. Ihre Herkunft und ihr Verbreitungsgebiet sind in einer Karte vermerkt. Eine dieser Sammlungen fand Aufstellung im American Museum of Natural History in New York und wurde nach dessen Direktor und dem Gönner des kostspieligen Unternehmens Jesup Collection genannt. Sie befindet sich inzwischen in Independence in Iowa und soll dort in einem neuen Museum wieder gezeigt werden. Eine zweite Fertigung ist nach Portland in Oregon gegangen. Die dritte aber, mit entsprechend gekürzten Stammstücken, befindet sich in Vitrinen im Arnold Arboretum selbst.

Ein kurzer Artikel in der *Botanical Gazette*, Vol. XI vom Februar 1886 unter dem Titel „Sections of Native Woods" (S. 40) berichtet, Sargent habe sorgfältig benannte Stammstücke aus der Fertigung seiner Jesup Collection einem bekannten Fabrikanten von Furnieren in Boston, Charles W. Spurr, überlassen, der daraus bemerkenswerte Sets von Dünnschnitten hergestellt habe, die nun zum Verkauf angeboten würden. Die Sets seien in der Tat einzigartig. Jeder

repräsentiere etwa 200 Arten heimischer Bäume, wenn möglich mit allen drei Schnittebenen. Die Dicke der Schnitte variiere je nach der Beschaffenheit des jeweiligen Holzes zwischen einem hundertstel und einem zweihundertfünfzigstel Zoll (1 Zoll, inch = 2,54 cm). Jeder dieser Schnitte sei zwischen zwei Glimmerplättchen auf einem flexiblen Holzrahmen montiert. Der Rahmen bestehe aus zwei Lagen von Gekräuseltem Ahorn, verstärkt mit festem Papier und mit Schellack überzogen. Zudem sei er mit dem Namen des Käufers oder der erwerbenden Institution bedruckt, trage ferner die Nummer der Art in Sargents Ausstellungskatalog der Jesup Collection, den wissenschaftlichen und den Volks-Namen, die Angabe der Schnittrichtung und den Namen des Präparators.

Aus dem abschließenden Abschnitt ist zu entnehmen, dass die Schnitte „mit einer 3-Tonnen-Furniermaschine geschnitten wurden", und dass „fast 18 000 einzelne Schnitte" sorgfältig und aufwendig hergestellt worden seien und nun preiswert zum Verkauf stünden. Das Referat ist nicht gezeichnet. Die genannten Präparate haben sich nirgendwo nachweisen lassen.

R. B. Hough und sein Lebenswerk – *The American Woods*

Im Jahr 1888 beginnt Romeyn B. Hough mit der Veröffentlichung seines Werks *The American Woods*, angelegt auf 15 Bände zu je 25 Tafeln mit je drei Schnitten einer bestimmten Art. Die Tafeln entsprechen denjenigen, die in der *Botanical Gazette* von 1886 von oder für Charles W. Spurr, den Furnierhersteller aus Boston, beschrieben und angezeigt worden waren. So auch die Präparate (ca. 2 x 5 Zoll) selbst.

Hough, der sich selbst in seiner Einleitung als inspiriert" bezeichnet hat, nimmt auf die Jesup

Colection keinen Bezug und auch nicht auf Spur.

Houghs *The American Woods* erschien 1888 bis 1913 und (postum) 1928, ohne Band 15, und galt und gilt noch als sein Lebenswerk. Sargent hat es seinerseits nie erwähnt. Traut man der im Grunde einzigen Nachricht über Houghs Leben in *The National Cyclopaedia of American Biography*, dann ist Hough für *The American Woods* vielfach geehrt und ausgezeichnet worden. Auffallend ist, dass die Reihenfolge der Tafeln und der Bände nicht der Systematik folgt, auch die Angabe „arranged geographically" in einem Auktions-Katalog wird der Abfolge nicht durchweg gerecht. Das beeinträchtigt den Informationswert, auch der Begleittexte (vgl. Sargent), im Grunde wenig und natürlich auch den Reiz der viktorianischen Aufmachung nicht.

Unter Kuriosa fällt der Versuch Marjorie Houghs, einer Tochter Romeyn Becks, aus Präparaten des Nachlasses mit editorialer Hilfe Professor Harrars, des Dekans der Forstwissenschaft an der Duke University in Durham, North Carolina, eine neue, wieder auf 15 Bände angelegte Enzyklopädie der amerikanischen Hölzer zu veröffentlichen. Das Werk, 1957 begonnen, ist bis 1981 auf 8 Doppelbände (Text und Tafeln) und 4 weitere Tafelbände gediehen und wird vermutlich ein Torso bleiben. Harrar behauptet eingangs, Hough habe die Präparate mit einer von ihm selbst erfundenen Maschine hergestellt, „weil es noch keine gab". Darin verkennt er die Furniermaschinenbauer von Cincinnati gründlich und führt wahrscheinlich manchen Leser in die Irre.

Werke, die anstelle von Abbildungen natürliche Holzpräparate verwenden, sind nicht so selten, wie man gemeinhin annimmt. Vor allem Hermann Nördlingers *Querschnitte von hundert Holzarten*, I–XI, Stuttgart und Tübingen, J. G. Cotta'scher Verlag, 1852–1888 soll an dieser Stelle genannt werden, da

man vermuten könnte, Hough habe sich vielleicht gerade von diesem Werk inspirieren lassen, was, wie oben dargelegt, gewiss nicht der Fall war. Es handelt sich um je eine Centurie von hauchdünnen Holzquerschnitten in ovaler Fenstermontierung zwischen doppelseitig gefaltetem, steifem Papier, je ein Präparat für eine Art einheimischer und exotischer Hölzer, zur Lupenbetrachtung in Durch- und Aufsicht. Jede Probe ist mit einem Namensetikett versehen, lose und alphabetisch nach den Namen geordnet. In einem broschierten Heftchen wird jedem Hundert eine Beschreibung vorangestellt. Der nächste Band wird jeweils als Fortsetzung behandelt. Er ist ein Buchimitat aus einer steifen Pappschachtel mit Halbleder-Rücken und Goldtitel, im Schnitt vielfarbig gemustert. Herstellungszeit der ganzen Ausgabe: 36 Jahre. Nördlinger erhielt 1851 und 1861 je eine Preismedaille in London.

Im Vergleich mit anderen Publikationen liegt der besondere Reiz der Hough'schen Tafeln allerdings vor allem in ihrer einzigartigen Präsentation, sodass die wenigen komplett erhaltenen Ausgaben immer eine Liebhaber-Rarität bleiben werden.

The Jesup Wood Collection,
Western Forestry Center,
Portland, OR, 1980
Courtesy of the Arnold Arboretum Archives

Romeyn Beck Hough
Les Bois américains

Klaus Ulrich Leistikow

Au premier abord, il pourrait sembler étonnant qu'une maison d'édition spécialisée dans le livre d'art présente cet ouvrage dont le thème relève plutôt de la botanique et de la géographie végétale, qu'il s'agisse de sylviculture et de dendrologie ou de biologie et de technologie du bois. Il est vrai que l'original de Romeyn Beck Hough *The American Woods* (*Les Bois américains*, 1888–1913, 1928) de par l'histoire de sa création et de sa publication ainsi que par sa présentation particulière – plusieurs recueils de planches réunis sous forme de livre et montrant des préparations de bois véritable – offre suffisamment de stimulations historiques, artistiques et esthétiques, et ce d'autant plus qu'il est exceptionnellement rare, pour être documenté dans une version moderne, éveiller des intérêts qui seront approfondis, et familiariser ainsi les amateurs conquis avec la littérature spécifique, largement disponible aujourd'hui.

Depuis des milliers d'années, notre société s'intéresse au bois en accordant à ce thème une valeur variable. Le mot bois qui désigne aussi bien les arbres et la forêt que la matière dont ils sont faits, est ainsi un phénomène étymologique dont les racines plongent dans l'évolution de l'humanité. De son côté, la symbolique perpétue le souvenir du préhominien ou de l'homme préhistorique qui, bien avant de penser à se vêtir et s'abriter, manipulait déjà du bois, l'utilisant comme outil et comme arme. La figure héraldique du «sauvage» vêtu uniquement d'un pagne de feuillage mais prenant appui sur une massue de la taille d'un homme, telle que l'ont transmise l'histoire de l'art et l'histoire des civilisations, donne de ce précoce utilisateur de bois humain l'image d'un ancêtre respectable qui continue, non sans raison, de participer à notre vie actuelle, puisqu'on le retrouve par exemple au carnaval alémanique. Dans *Faust II*, Goethe lui a consacré ces vers superbes :

On les nomme les hommes sauvages
Ils sont bien connus au mont du Harz
Dans leur nudité naturelle, dans toute leur vigueur
Ils viennent ensemble, gigantesques
Dans la main droite un tronc de sapin,
Autour du corps une ceinture en forme de bourrelet ;
Le plus rude des tabliers, fait de branches et de feuilles ;
Garde du corps comme le pape lui-même n'en a pas.

Au début – et le «début» se réfère ici aux longues périodes de croissance et d'expansion des populations de la terre, des périodes de durées diverses qui, chez les peuples dits primitifs, se sont prolongées jusqu'au passé le plus récent – au début donc, le bois non travaillé ou apprêté en quelques manipulations ou à coups de pierre, souvent même sous forme de simples assemblages de rameaux et de branches, suffisait aux exigences minimes des utilisateurs. Pourtant, le bois offrait un vaste champ d'application, puisqu'il pouvait aussi bien servir d'outil ou d'arme élémentaire que de combustible ou de matériau de construction. Outre les cavernes, les paravents faits de branches liées et les tentes en perches ou en baguettes recouvertes de peaux de bête font partie des premiers abris érigés par les hommes.

Apprécions comme il convient l'invention au néolithique de la combinaison d'un outil avec une prise, par exemple une pointe de silex coupante fixée à la partie supérieure d'un manche de bois allongeant et renforçant le bras de force, ce qui donne la hache dans sa forme la plus courante. Son importance est capitale car au-delà de toutes les améliorations et affinements du matériau et de la construction, la hache reste à ce jour l'outil le plus commun pour abattre des arbres et pour traiter le bois brut du tronc, qui servira à construire des habitations sur pilotis, des huttes et des cabanes et, après la découverte de la faculté du bois à flotter naturellement sur l'eau, à fabriquer des radeaux, des pirogues, des barques ou encore des vaisseaux.

Ces acquis culturels créent les conditions de base qui permettront aux Européens de se lancer à la conquête des territoires d'outre-mer sur le sous-continent nord-américain dont les forêts primitives recouvrent alors quatre dixièmes de la surface. On ne peut s'empêcher de réfléchir à la quantité de bois, c'est-à-dire de forêts, engloutie en Europe par la construction navale, et aux dimensions déjà prises par le déboisement en Ecosse et en Irlande à l'époque de la conquête de l'Amérique, une importance décisive revenant aux troncs de grande taille, droits et dénudés qui servaient de mâts de navires. De ce fait, l'Amérique, dont on croyait les forêts inépuisables, fut découverte juste au bon moment puisqu'elle livrait des bois particulièrement adaptés, introuvables ou disparus en Europe, et elle aussi devait bientôt se lancer sur la côte Est dans la construction navale.

Combat pour la conservation

L'immigration et la colonisation de l'Amérique du Nord causèrent néanmoins en très peu de temps une déforestation drastique des forêts. Le concept d'une colonisation par la conquête, le travail et l'occupation du sol après l'élimination de la forêt et des Indiens marqua la naissance des Etats-Unis d'Amérique et généra leur ascension unique en son genre sur le plan historique, qui les verra s'imposer sur le plan international et devenir une puissance mondiale – le prix à payer étant à vrai dire le danger latent d'auto-destruction provoqué par l'anéantissement total de la nature considérée comme base vitale. Ce péril ne fut reconnu ou pris au sérieux par quelques-uns que

tardivement, dans la seconde moitié du 19ᵉ siècle, et suscita la formation d'un mouvement de préservation et une politique de la culture de l'environnement, au sein de laquelle il faut ranger aussi les aspirations de R. B. Hough.

Les forêts, jugées inépuisables moins d'un siècle auparavant, s'étaient éclaircies de manière si dramatique à cause de l'exploitation abusive par les gigantesques groupes de l'industrie du bois comme celui des Weyerhäuser, que des voix s'élevèrent pour demander de préserver la forêt par la plantation d'arbres. Pour se faire une idée de la dimension de cette exploitation, il suffit de savoir que Frederick Weyerhäuser (1834–1914), jeune va-nu-pieds immigré de son village du Palatinat, put construire un véritable empire de l'industrie du bois et que l'on disait de lui – il redoutait la publicité –, qu'il était plus riche que Rockefeller.

Le reboisement préconisé officiellement marque le début d'une nouvelle vision de la nature. L'un des porte-parole les plus célèbres des «affaires» arboricoles fut Charles Sprague Sargent (1841–1927), professeur de botanique à l'Université Harvard et, à partir de 1873, directeur de l'Arnold Arboretum installé sur ce site, une vaste collection d'espèces ligneuses, quasi l'inventaire vivant des forêts américaines.

Vers la même époque, le nouveau mouvement de protection de la nature trouva un soutien dans l'idée de parc national, c'est-à-dire la création de réserves naturelles placées sous la protection de l'Etat, qui les surveille et les entretient. Le parc national de Yellowstone (Wyoming), fondé en 1872, devint le modèle de ces nouvelles institutions vouées à la protection de la nature en Amérique du Nord, et controversées car elles nécessitaient l'accord du Congrès, de la législation et le dégagement de fonds fiscaux. Des scientifiques engagés durent fournir un travail sentants du peuple que des mesures isolées et des initiatives privées d'hommes de bonne volonté pour protéger les arbres ne suffisaient pas pour maitriser la crise. Il fallait bien plus que cela, et avant toute chose, il fallait procéder à un recensement à l'échelon national, analyser et valoriser l'action conjuguée de l'eau, de la forêt, du sol et du climat – des écosystèmes, comme nous disons aujourd'hui. Pour dépasser le stade d'une protection incertaine et illusoire, il ne serait possible d'arriver à une gestion forestière rationnelle et durable qu'après avoir étudié les causes et les conséquences et sur les bases d'une science sylvicole qui n'était pas encore établie aux Etats-Unis. L'objectif serait ainsi de créer une base vitale pour toute la nation et complémentaire de l'agriculture. Par ordre du Commissioners of Agriculture, Franklin B. Hough (1822–1885) – le père de l'auteur du présent ouvrage –, médecin, connaisseur enthousiaste et ami de la nature, soumit aux autorités son *Report upon Forestry* officiel, un des premiers inventaires de ce genre renfermant une foule incroyable de données dans ses deux volumes (1877 et 1880). D'autres rapports suivirent – le sujet restait maintenant à l'ordre du jour.

Carl Schurz (1829–1906) lui-même, allemand comme Weyerhäuser et du même âge que lui, institua dans son ministère de l'Intérieur une agence des forêts, la future Forest Division dans le Department of Agriculture séparé exprès. La liaison administrative ainsi établie entre l'agriculture et la sylviculture peut être considérée comme l'engagement du Fight for Conservation (combat pour la conservation) par le gouvernement des Etats-Unis, et ce non seulement pour la préservation des forêts en temps d'agriculture prospère mais aussi dans le but de leur renouvelle-

Pinchot (1865–1946), avait justement tout à fait conscience de cela. Elève des Eaux et Forêts à Paris, il eut l'occasion à son retour de devenir conseiller à des postes importants dans sa spécialité. Ses efforts, qui comprenaient l'étude de l'approvisionnement en eau et de la régulation de l'eau, de l'érosion des sols par l'eau, des incendies de forêts, de l'exploitation minière à ciel ouvert et de la politique agricole lui valurent en 1901 l'amitié du président Theodore Roosevelt (1858–1919), dont les convictions s'accordaient de manière programmatique à celles du puritain Pinchot. Dans son livre *The Fight for Conservation*, Pinchot récapitule toutes ses idées et ses expériences, dont celles nées de polémiques innombrables avec les puissants intérêts particuliers. Il fut deux fois gouverneur de Pennsylvanie et laissa en 1946 son testament à valeur d'avertissement durable.

Sargent et les recherches sur les essences au 19e siècle

C'est à cette époque, alors que l'on révise les conceptions et que l'on procède à des inventaires afin de préparer les réformes, que paraissent les quatorze volumes de la *Silva of North America* (1889–1901) de Sargent. L'ouvrage décrit tous les arbres connus des forêts américaines, classés selon les espèces, et les planches superbes dessinées par Charles E. Faxon impressionnent aussi en tant que gravures sur métal. Réédité en 1947, il est aujourd'hui encore considéré comme un ouvrage de référence.

Sargent est aussi l'auteur de la première carte des aires forestières d'Amérique du Nord encore originelles à son époque (1884), que le géographe botanique Adolf Engler (1844–1930) a reproduite en autre réalisée antérieurement par John W. Harshber ger (1869–1929) – elles parurent ensemble avec une monographie détaillée dans le volume XIII, *Vegetation der Erde*, d'Engler & Drudes (1911) –, sont les seules cartes authentiques de l'époque traitant de ce sujet. La collaboration entre les botanistes américains et allemands dans ce domaine en ce temps-là est remarquable.

Un ami de Sargent, le botaniste bavarois Heinrich Mayr (1854–1911) traversa en 1886 et en 1887 par deux fois l'Amérique du Nord dans sa totalité et voyagea de Vancouver à la Floride. Il y revint une fois encore pendant son tour du monde qui dura de 1902 à 1906. Mayr fut peut-être le dernier spécialiste européen des forêts d'une longue lignée comportant Peter Kalm, Michaux père et fils, von Wangenheim et finalement Douglas « of the forests », venus avec le dessein de trouver dans la forêt américaine des espèces utiles à croissance rapide et de les acclimater dans leur pays. Tous les parcs, les jardins botaniques et les forêts de rapport européens abritent nombre de ces essences importées, qui sont parfois des réintroductions, si l'on songe aux caractères communs des flores avant l'ère glaciaire. La longue histoire végétale explique de nombreuses affinités entre les tribus de plantes ligneuses d'Amérique du Nord et d'Europe, par exemple les genres semblables en bien des points comme le chêne, l'érable, le hêtre, l'orme, le frêne, le tilleul, le bouleau, le saule ou le pin, le sapin, l'épicéa. Néanmoins, des différences existent sur le plan des caractéristiques et du nombre des espèces. Des différences avec l'Europe et tout autant entre la façade atlantique et la façade pacifique nord-américaine et entre leurs montagnes très éloignées les unes des autres. Au-delà de ces considérations, l'aspect des deux ceintures climatiques parallèles en latitude est très accusé avec des peuplements de résineux dans les zones boréales et subarctiques, des forêts toujours vertes et des palmiers dans les régions subtropicales.

Le voyage de Mayr en Amérique du Nord eut lieu relativement tard, ce qui lui fit dire en 1906 : « En Amérique croît la deuxième génération d'arbres pour devenir un mélange plus pauvre en espèces, plein de brèches, de valeur inférieure à bien des égards et dont la surface est réduite des trois quarts ; les forêts vierges d'origine disparaissent rapidement. » De son constat étaient exclues les forêts de montagnes placées sous la protection du gouvernement.

A l'aide de deux images plus impressionnantes que toutes les statistiques et qui s'étaient offertes à lui dans les Etats du Nord-Ouest, Mayr rend compte de l'évolution entre 1890 et 1906. Vues d'une montagne (1890), les zones colonisées étaient des « brèches à peine visibles dans le tapis vert foncé ». En 1906, il n'y trouva plus que des « restes de forêts comme des points vert foncé » dans le « paysage vert clair et brun » défini par l'agriculture.

Il eut au moins la consolation comme beaucoup d'autres avant lui et quelques-uns encore après lui de pouvoir contempler les arbres géants de la Sierra Nevada et du massif côtier : le séquoia géant et le séquoia toujours vert ou Redwood hauts de 100 et de 120 mètres. Ces arbres sont en partie déjà protégés et, comme le séquoia toujours vert, forment même encore parfois des forêts ; de plus, étant le bois le plus exporté par la Californie du Nord, ce dernier est un produit commercial important. Il faut dire qu'il est pratiquement impérissable, vu que les souches des séquoias toujours verts abattus possèdent la faculté intéressante de former des tiges filles dotées des mêmes propriétés. D'autres arbres du Nord-Ouest riche en conifères sont également de taille élevée – les données vont de 70 à 90 mètres de hauteur. Il est certain que le pin blanc du Canada ou pin de partie des arbres géants ou en faisait partie, lorsqu'il fut marqué du signe d'abattage de la Broad Arrow réservé aux futurs mâts de la Royal Navy.

La préparation des bois

A l'occasion du 10e recensement des Etats-Unis, Sargent, alors directeur de l'Arnold Arboretum de l'Université Harvard, prit en charge le domaine des arbres. A partir de là, il organisa une collection des arbres eux-mêmes sous la forme de sections de troncs (d'environ 1,50 m) en trois exemplaires chacun. Les pièces sont traitées de telle sorte que chacune montre trois surfaces polies, nécessaires sur le plan anatomique si l'on veut identifier l'essence : la coupe transversale en bout, la coupe longitudinale radiale ou tranche sur maille et la coupe longitudinale tangentielle, parallèle à la tangente, dite coupe sur dosse. L'écorce reste intacte. Les pièces sont accompagnées de leurs noms latins et de leurs noms populaires et éclairées par de brèves descriptions. L'origine des arbres et les zones de peuplement sont notées sur une carte. Une de ces collections fut montée à l'American Museum of Natural History de New York et intitulée Jesup Collection en hommage au directeur du musée qui avait soutenu cette entreprise onéreuse. Elle se trouve aujourd'hui à Independence dans l'Etat d'Iowa et sera exposée dans un nouveau musée. Le second exemplaire a été envoyé à Portland dans l'Oregon ; quant au troisième, les pièces ayant été raccourcies de manière adéquate, il se trouve dans les cabinets vitrés de l'Arnold Arboretum lui-même.

Un bref article de la *Botanical Gazette*, vol. XI de février 1886, rapporte sous le titre «Sections of Native Woods» (p. 40) que Sargent aurait cédé des sections de troncs minutieusement désignés de sa Jesup Col-

lection à Charles W. Spurr, un fabricant de placages connu de Boston. Celui-ci aurait réalisé avec ces pièces de remarquables Sets de feuilles de bois qui seraient maintenant mis en vente. Les Sets seraient vraiment uniques en leur genre, chacun représentant environ 200 espèces d'arbres américains et montrant si possible les trois coupes mentionnées plus haut. L'épaisseur des tranches varierait selon la structure du bois entre un centième et un deux cent cinquantième de pouce (1 pouce = 2,54 cm). Chaque feuille serait montée entre deux petites plaques de mica sur un cadre de bois flexible. Celui-ci serait fait de deux couches d'érable frisé, renforcées avec du papier kraft et vernies. En outre, on pourrait y lire le nom de la personne ou de l'institution qui l'acquiert, le numéro de l'espèce telle que notée par Sargent dans le catalogue d'exposition de la Jesup Collection, le nom latin et populaire, la mention de la direction de coupe et le nom du préparateur.

La fin de l'article nous permet de conclure que les pièces «ont été tranchées à l'aide d'une machine à couper les placages pesant trois tonnes», et que «près de 18 000 tranches séparées» auraient été fabriquées avec beaucoup de soin et de sophistication et qu'elles seraient maintenant mises en vente à bas prix. L'article n'est pas signé. Les préparations mentionnées n'ont pu être mises en évidence où que ce soit.

R. B. Hough et l'œuvre de sa vie –
The American Woods

C'est en 1888 que débute la parution de l'ouvrage de Romeyn B. Hough, avec ses 15 volumes contenant chacun 25 planches qui montrent chacune trois coupes d'une espèce particulière. Les planches correspondent à celles qui sont décrites et annoncées dans la *Botanical Gazette* de 1886 par ou pour Charles

W. Spurr, fabricant de placages à Boston, et cela vaut aussi pour les préparations elles-mêmes (env. 2 x 5 pouces).

Hough qui s'est lui-même déclaré «inspiré» dans son introduction, ne se réfère pas à la Jesup Collection et ne mentionne pas non plus Spurr. *The American Woods* paru de 1888 à 1913 et de manière posthume en 1928, sans le volume 15, est encore considéré comme l'œuvre de sa vie. Sargent, de son côté, n'a jamais fait mention de cet ouvrage. Si l'on en croit l'unique information sur la vie de Hough telle qu'elle est contenue dans *The National Cyclopaedia of American Biography*, celui-ci fut honoré et décoré à de nombreuses reprises pour *The American Woods*. Il saute aux yeux que les planches et les volumes ne sont pas classés systématiquement, et la mention «arranged geographically» dans un catalogue de vente aux enchères n'est pas conforme d'un bout à l'autre à l'ordre fixé. Cela nuit peu à la valeur de l'information et, au fond, tout aussi peu aux textes d'accompagnement (cf. Sargent); quant au charme de la présentation victorienne, il n'en est naturellement pas amoindri.

Dans la rubrique curiosités, il faut mentionner la tentative de Marjorie Hough, une fille de Romeyn B. Hough, de réaliser une nouvelle Encyclopédie des bois américains, en 15 volumes elle aussi, en se basant sur des préparations qu'elle avait héritées et avec l'aide éditoriale du professeur Harrar, le doyen du département de recherche forestière à l'Université Duke de Durham, Caroline du Nord. L'œuvre commencée en 1957 et qui comptait en 1981 huit volumes doubles (texte et planches) et quatre autres volumes de planches, restera probablement tronquée. Harrar affirme dans l'introduction que Hough a fabriqué lui-même les préparations avec une trancheuse de son invention «parce qu'elle n'existait pas encore»,

méconnaissant complètement les fabricants de machines à placage de Cincinatti et induisant vraisemblablement plus d'un lecteur en erreur.

Les ouvrages utilisant des préparations de bois naturel au lieu d'illustrations sont moins rares qu'on ne le pense généralement. Nous ne manquerons pas de citer avant tout celui de Hermann Nördlinger, *Querschnitte von hundert Holzarten*, I–XI, Stuttgart et Tübingen, J. G. Cotta'scher Verlag, 1852–1888, car on pourrait supposer que Hough s'en est peut-être inspiré, ce qui ne fut certainement pas le cas, ainsi que nous l'avons exposé plus haut. Chaque volume rassemble une centaine de coupes de bois fines comme du papier, montées dans des fenêtres ovales entre deux pages de papier rigide et plié – une préparation pour chaque espèce de bois nord-américain ou exotique, à observer à la loupe normalement et par transparence. Les échantillons sont volants, pourvus d'une étiquette portant leur nom et rangés par ordre alphabétique. Une brochure est consacrée à chaque centaine d'échantillons précédés d'une description. Les volumes suivants sont traités chaque fois comme une continuation. Il s'agit d'une imitation de livre fait d'une boîte en carton rigide avec un dos mi-cuir et un titre en lettres dorées, la tranche présentant des motifs multicolores. Trente six années seront nécessaires pour éditer l'ensemble de l'œuvre. En 1851 et en 1861, une médaille fut décernée à Nördlinger à Londres.

Si on les compare à d'autres publications, le charme particulier des planches de Hough réside surtout dans leur présentation unique en son genre, ce qui fait que les éditions complètes, dont il reste peu d'exemplaires, resteront toujours des pièces rares réservées aux amateurs.

The present edition of the reprint of *American Woods* by R. B. Hough contains photographic illustrations of original prepared specimens of the wood of 354 tree species and varieties. For each tree species, the three diagnostically relevant sections (cross, radial, and tangential, in that order) are illustrated, and each tree is supplied with a brief descriptive "portrait."

The order of the plates here (unlike the original edition) is by **botanical family** and within each, by genus. Within each category, the order is alphabetical. For the naming, the Synonymized Checklist of the Vascular Flora of the United States, Canada, and Greenland (BONAP 1996) served as a guide.

For the drawing up of the "tree portrait," both Hough's original texts and Sargent's *Silva of North America* (1889–1901) were used as contemporary sources. At the beginning of each portrait, the **English family name** is given in the top line, and below it, the names of the individual species. There follow the **German and French family names** with the names of the individual species in the respective language, although it should be noted that the German and French names do not always reflect common usage. At the very bottom, the **botanical (Latin) family name** is given, and below it, the **scientific species name** of the tree. This often includes the (usually abbreviated) name of the scientist who first described the species in question.

In another line, following the abbreviation **syn**. (= synonym), the name used by Hough is listed if it differs from the nomenclature current today. A name is preceded by **"?"** if it can no longer be unambiguously determined from which tree Hough's specimens were taken.

In all the languages, there are numerous identical names which, however, denote different trees according to region. Only the botanical names are unambiguously codified and internationally valid. The rules, adherence to which is necessary for valid publication, can be found in the Internet as the International Code of Nomenclature for algae, fungi and plants (Melbourne Code) (http://www.iapt-taxon.org/nomen/main.php).

The brief descriptive portraits explain the shape (habit) of the tree, its habitat requirements, its distribution, and its use. For those species introduced to North America for cultivation, all the information given goes back to Hough's original texts. In individual cases where neither Hough nor Sargent contain information on the structure of the wood or its use, details are cited from other works.

The brief descriptions cannot be seen as aids to identification and are not intended thus. Rather, they serve the following ends: documentation of the trunk height, diameter, and form that could be reached in North America at the turn of the 19th and 20th centuries; illustration of the customary uses at that time, in particular of the timber, but also of other parts of the plant, in North America; and the conveying of a general impression of the appearance of the tree, not least with respect to its potential value for decorative planting.

Details of the species described are in most cases illustrated by a drawing by E. Faxon, as already published in Sargent's *Silva*. A correct identification of the trees after felling is only possible after thorough examination. For this purpose, supplementary herbarium material from the specimen in question would be needed, in the form for example of leaves, flowers and fruits. However, nothing is known of any Hough herbarium.

Die vorliegende Auflage des Reprints der *American Woods* von R. B. Hough enthält fotografische Darstellungen von Originalpräparaten des Holzes von 354 Baumarten und Varietäten. Für jede Baumart werden die drei diagnostisch relevanten Schnittebenen (in der Reihenfolge: Quer-, Radial- und Tangentialschnitt) abgebildet, und jeder Baum wird durch ein Kurzporträt beschrieben.

Die Anordnung der Tafeln erfolgt hier (anders als in der Originalausgabe) nach **Pflanzenfamilien**. Innerhalb jeder Familie erfolgt jeweils eine alphabetische Reihung. Bei der Benennung diente die Synonymized Checklist of the Vascular Flora of the United States, Canada, and Greenland (BONAP 1996) als Orientierung.

Für die Erstellung der Baumporträts wurden sowohl Houghs Originaltexte als auch Sargents *Silva of North America* (1889–1901) als zeitgenössische Quellen ausgewertet. Zu Beginn eines jeden Porträts wird in der obersten Zeile der **englische Familienname** und darunter die verschiedenen Namen der einzelnen Bäume genannt. Es folgen im weiteren die **deutschen und französischen Familiennamen** mit den Namen der einzelnen Bäume in der jeweiligen Landessprache, wobei diese nicht immer wirklich gebräuchliche Begriffe wiedergeben. Ganz unten wird zunächst der **lateinische Familienname** und darunter der **wissenschaftliche Artname** des Baumes genannt. Dazu gehört oft auch der meist abgekürzte Name desjenigen Wissenschaftlers, der die betreffende Baumart als erster beschrieben hat. In einer weiteren Zeile werden hinter der Bezeichnung **syn.** (= Synonyme) von Hough benutzte Namen aufgeführt, wenn sie von der heutigen Nomenklatur abweichen. Mit einem vorangestellten „**?**" werden Namen gekennzeichnet, wenn nicht mehr eindeutig geklärt werden konnte, von welcher Baumart Houghs Präparate stammen. In allen Sprachen finden sich zahlreiche gleich lautende Namen, die aber je nach Region unterschiedliche Bäume bezeichnen. Eindeutig kodifiziert und international gültig sind dagegen nur die wissenschaftlichen Namen. Die Regeln, deren Befolgung für die gültige Veröffentlichung erforderlich sind, können im Internet als International Code of Nomenclature for algae, fungi and plants (Melbourne Code) (http://www.iapt-taxon.org/nomen/main.php).

Die Kurzporträts erläutern die Gestalt des Baumes, seine Standortansprüche und Verbreitung sowie seine Nutzung. Bei den zur Kultur nach Nordamerika eingeführten Arten gehen alle Angaben auf die Originaltexte von Hough zurück. In einzelnen Fällen, in denen weder Hough noch Sargent Informationen zur Struktur des Holzes und seiner Nutzung enthalten, wurden Angaben aus weiteren Werken zitiert.

Die Kurzbeschreibungen können und sollen keine Bestimmungshilfen darstellen. Sie dienen vielmehr den folgenden Zielen:

Der Dokumentation der einst erreichbaren Dimensionen von Stammhöhe, Durchmesser und Wuchsform in Nordamerika an der Wende des 19. zum 20. Jahrhundert, der Illustration der zu dieser Zeit gebräuchlichen Nutzungen, vor allem des Holzes, aber auch anderer Teile der Pflanze in Nordamerika und der Vermittlung eines Gesamteindruckes vom Aussehen der Bäume, auch im Hinblick auf ihre potenziellen Zierwerte.

Details der beschriebenen Baumarten werden in den meisten Fällen durch eine Zeichnung von E. Faxon verbildlicht, wie sie bereits in Sargents *Silva* publiziert wurden.

Eine korrekte Bestimmung der gefällten Bäume ließe sich nur durch eingehendere Untersuchungen vornehmen. Dazu wäre auch ergänzendes Herbarmaterial dieser Exemplare nötig, zum Beispiel von Blättern, Blüten und Früchten. Über den Verbleib eines Herbars von Hough ist uns aber nichts bekannt.

Remarques

sur la préparation de la
présente édition

Cette édition de la réimpression de l'*American Woods* de R. B. Hough offre au lecteur des représentations photographiques des préparations originales du bois de 354 espèces de plantes ligneuses. Chaque essence est étudiée dans trois plans de l'espace, à savoir en coupes transversale, radiale et tangentielle et fait l'objet d'une courte description.

Les planches sont présentées ici par **familles** de ligneux (contrairement à ce que nous trouvons dans l'édition originale) et, à l'intérieur de celles-ci, par genre. Les noms sont classés par ordre alphabétique à l'intérieur de chaque catégorie. La nomenclature que nous avons choisie reprend celle de la Synonymized Checklist of the Vascular Flora of the United States, Canada, and Greenland (BONAP 1996).

Pour constituer les notices descriptives, nous nous sommes référés aussi bien aux textes de Hough qu'au livre de Sargent, *Silva of North America* (1889–1901) et qu'à des sources contemporaines. Chaque portrait commence par **le nom de famille en anglais** et en-dessous par les différents noms de chaque arbre. Ensuite apparaissent **les noms de famille en allemand et en français** et, en-dessous le nom commun de chaque arbre dans les diverses langues étant entendu que nombre d'appellations vernaculaires ne correspondent pas toujours aux noms réellement usités. En bas figurent **le nom de famille en latin** et le nom scientifique de l'arbre, dont fait partie souvent le nom, abrégé dans la plupart des cas, du savant qui a le premier décrit l'essence concernée.

À la dernière ligne, sous la dénomination **syn.** (= synonyme) sont indiqués les noms utilisés par Hough, lorsqu'ils ne correspondent pas à la nomenclature actuelle. Les noms sont précédés d'un «?» quand il a été impossible d'éclaircir avec certitude de quelle essence proviennent les préparations de Hough.

Il est courant, et ce dans toutes les langues, que le même nom désigne des arbres différents selon les régions. Seuls les noms scientifiques sont clairement codifiés et valables sur le plan international. Les règles d'écriture qu'il convient d'observer pour que la publication soit valide peuvent être examinées sur Internet dans l'International Code of Nomenclature for algae, fungi and plants (Melbourne Code) (http://www.iapt-taxon.org/nomen/main.php).

Les notices descriptives donnent des indications sur les caractères visuels des arbres, sur leur biotope et leur aire de répartition ainsi que sur leur utilisation. Les indications concernant les espèces introduites pour les cultiver en Amérique du Nord sont basées sur celles de Hough. Nous nous sommes cependant appuyés sur d'autres publications dans le cas où Hough et Sargent ne livraient aucune information sur la structure du bois ou son utilisation.

Précisons que les descriptions n'ont pas été compilées dans le but de déterminer des essences mais de faire connaître : les dimensions – hauteur et diamètre du tronc – et le type de croissance que pouvait présenter l'arbre en Amérique du Nord au tournant du 19e au 20e siècle ; d'illustrer une partie des usages que l'on faisait du bois et d'autres parties de la plante en Amérique du Nord, à cette époque ; de donner une idée de l'aspect général de l'arbre en tenant compte aussi de son caractère ornemental potentiel.

Des détails anatomiques des espèces décrites sont généralement illustrés par E. Faxon dont les dessins ont déjà été publiés dans la *Silva* de Sargent.

Seule une étude approfondie permettrait une détermination exacte des espèces traitées. Nous aurions besoin, pour ce faire, de connaître aussi leurs caractères évidents tels que feuilles, fleurs et fruits, mais nous ignorons si Hough possédait un herbier et, si tel était le cas, nous ne connaissons pas sa localisation.

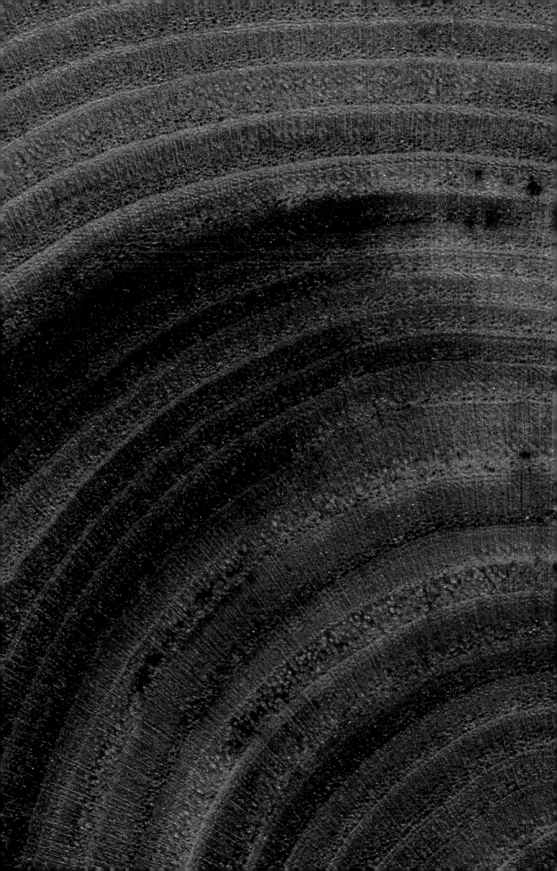

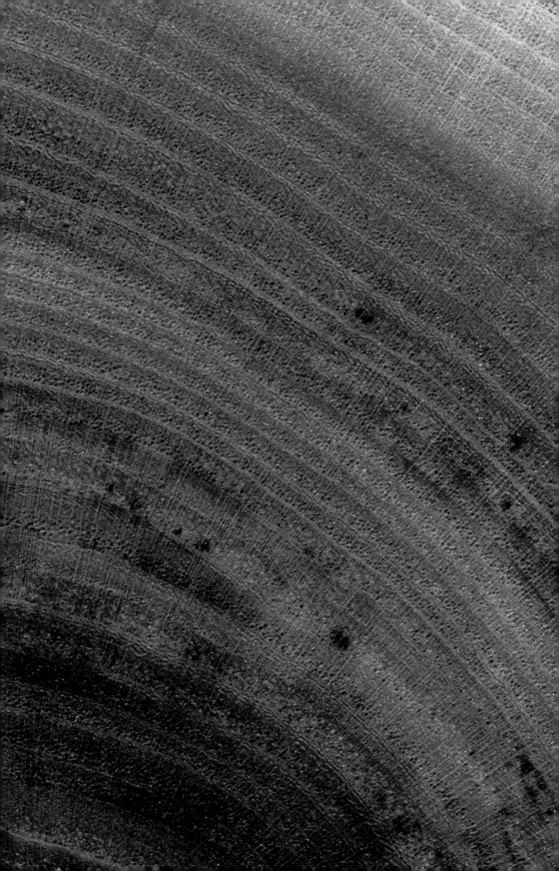

B | BAYBERRY FAMILY

Pacific Bayberry, Pacific Wax Myrtle

FAMILIE DER GAGELSTRAUCHGEWÄCHSE

Kalifornischer Wachsstrauch

FAMILLE DU PIMONT ROYAL

Myrique de Californie

MYRICACEAE

Myrica californica Cham.

Description: 40 ft. (13 m) high, 10 to 12 in. (0.28 to 0.3 m) in diameter, but usually smaller, and small and shrubby towards northern and southern limits of distribution. Bark smooth, compact, dark gray or pale brown on the surface, red-brown within.

Habitat: Sand dunes and damp hills in the coastal region of Pacific North America.

Wood: Heavy, very hard, very strong, brittle, and close-grained. Heartwood pale pink; sapwood thick and paler.

Use: Occasionally grown as a decorative shrub in California.

Discovered in San Francisco Bay by Chamisso de Boncourt on his circumnavigation of the world on the *Rurik*; described in 1831.

Beschreibung: 13 m hoch bei 0,28 bis 0,3 m Durchmesser, meist kleiner und gegen die Nord- und die Südgrenze des Verbreitungsgebiets jeweils als kleiner Strauch. Borke glatt, kompakt, dunkelgrau oder an der Oberfläche hellbraun, innen dunkel rotbraun.

Vorkommen: Sanddünen und feuchte Hügel in der Küstenregion des pazifischen Nordamerika.

Holz: Schwer, sehr hart, sehr fest, spröde und mit feiner Textur. Kern hell rosenfarben. Splint dick, heller.

Nutzung: In Kalifornien gelegentlich als Zierstrauch kultiviert.

Während seiner Weltumsegelung mit der „Rurik" von Chamisso de Boncourt in der Bucht von San Francisco entdeckt, 1831 beschrieben.

Description : 13 m de hauteur sur un diamètre de 0,28 à 0,3 m; d'ordinaire moins grand et sous forme de petit buisson sur les marges septentrionale et méridionale de son aire. Ecorce lisse, compacte, gris foncé ou brun clair en surface, brun rouge foncé dessous.

Habitat : dunes et collines humides des régions côtières de la façade pacifique de l'Amérique du Nord.

Bois : lourd, très dur et résistant, cassant et à grain fin. Bois de cœur rose clair; aubier épais, plus pâle.

Intérêts : quelquefois cultivé en ornement en Californie.

Découvert par Chamisso de Boncourt dans la baie de San Francisco au cours de son voyage autour du monde sur le «Rurik» décrit en 1831.

California Wax-Myrtle, Bayberry, Grease-wood.

TRANSVERSE SECTION.

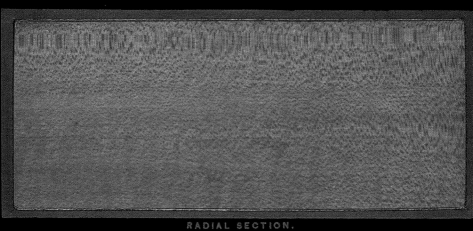

RADIAL SECTION.

TANGENTIAL SECTION

Ger. Californischer Kerzenbeer. *Fr.* Cirier de Californie.
Sp. Arrayan de California.

B | BAYBERRY FAMILY

Wax-myrtle, Candleberry, Tallow Shrub

FAMILIE DER GAGELSTRAUCHGEWÄCHSE
Wachsmyrte

FAMILLE DU PIMONT ROYAL
Arbre à cire, Myrique de Louisiane, Cirier de Louisiane

MYRICACEAE
Myrica cerifera L.

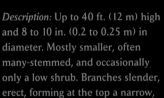
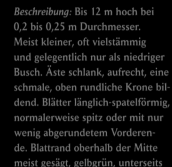

Description: Up to 40 ft. (12 m) high and 8 to 10 in. (0.2 to 0.25 m) in diameter. Mostly smaller, often many-stemmed, and occasionally only a low shrub. Branches slender, erect, forming at the top a narrow, roundish crown. Leaves oblong-spatulate, normally pointed or with only a slightly rounded tip. Leaf margin above the middle mostly serrate, yellow-green. Underside of leaf covered with distinct, pale orange-colored glands, later becoming darker. Dioecious. Flowers in short, oblong spikes. Fruit globular, light green coated with a thick layer of pale blue wax.

Habitat: Sandy swamps and ponds, growing as a shrub near the coast on uncultivated land with pines and on dry, sandy hills in southeastern North America.

Wood: Light, soft, brittle, and close-grained.

Use: The wax provides fuel for lamps, and was formerly processed in great quantities. The bark has an astringent and stimulating effect ("Thompsonian powder").

Beschreibung: Bis 12 m hoch bei 0,2 bis 0,25 m Durchmesser. Meist kleiner, oft vielstämmig und gelegentlich nur als niedriger Busch. Äste schlank, aufrecht, eine schmale, oben rundliche Krone bildend. Blätter länglich-spatelförmig, normalerweise spitz oder mit nur wenig abgerundetem Vorderende. Blattrand oberhalb der Mitte meist gesägt, gelbgrün, unterseits mit deutlichen, jung hell orangefarbenen, später dunklen Drüsen bedeckt. Zweihäusig. Blüten in kurzen länglichen Ähren. Frucht kugelig, hellgrün mit dickem blassblauen Wachsreif.

Vorkommen: Sandige Sümpfe und Teiche, als Strauch nahe der Küste auf Ödland mit Kiefern und auf trockenen, sandigen Hügeln im südöstlichen Nordamerika.

Holz: Leicht, weich, brüchig und mit feiner Textur.

Nutzung: Wachs als Feuerungsmittel für Lampen, früher in großem Maßstab verarbeitet. Rinde mit adstringierender und stimulierender Wirkung („Thompsonian powder").

Description : jusqu'à 12 m de hauteur sur un diamètre de 0,2 à 0,25 m. Plus petit en général, souvent à tronc multiple et quelquefois sous forme de buisson. Branches fines et dressées, constituant un houppier mince et rond dans sa partie supérieure. Feuilles oblongues, terminées en pointe le plus souvent ou à l'apex légèrement arrondi, vert jaune et dentées dans la moitié supérieure, munies à la face inférieure de glandes apparentes orange clair, devenant sombres avec l'âge. Espèce dioïque. Fleurs en courts épis allongés. Fruits globuleux, vert clair, recouverts d'une substance cireuse épaisse et bleu pâle.

Habitat : marécages et étangs sablonneux ; de type buisson près du littoral sur les terrains incultes en compagnie de pins, et sur les collines sablonneuses sèches dans le sud-est de l'Amérique du Nord.

Bois : léger, tendre, cassant et à grain fin.

Intérêts : la cire sert de combustible pour les lampes (autrefois à grande échelle). L'écorce a des propriétés astringentes et stimulantes (« Thompsonian powder »).

Wax Myrtle. Bayberry. Candleberry.

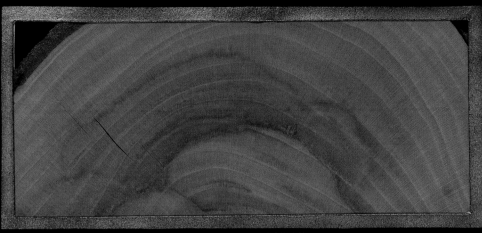

TRANSVERSE SECTION.

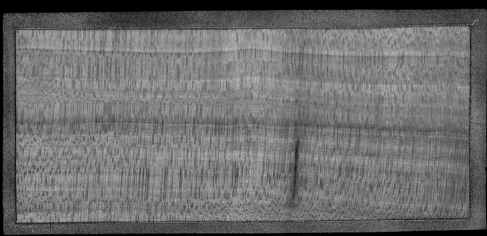

RADIAL SECTION.

TANGENTIAL SECTION.

B | BEECH FAMILY
American Beech

FAMILIE DER BUCHENGEWÄCHSE
Amerikanische Buche

FAMILLE DU CHÊNE ET DU HÊTRE
Hêtre d'Amérique, Hêtre rouge (Canada)

FAGACEAE
Fagus grandifolia Ehrh.
Syn.: *F. americana* Sweet, *F. ferruginea* Aiton

Description: Up to more than 100 ft. (30 m) high and 3 to 4 ft. (0.9 to 1.2 m) in diameter, with a straight, columnar trunk. In old age often with shoots sprouting from the base of the trunk, and forming thickets round the parent plant. This latter is replaced after its death by the strongest of its offspring. It has a strikingly smooth, blue-gray bark.

Habitat: Rich, well-drained uplands and slopes, in the north with Sugar Maple, birches, hornbeams, American Lime, Hemlock, etc. In the south also in the lowlands and the edges of swamps. Eastern North America.

Wood: Hard, tough, strong, and very close-grained; polishable; not durable in the ground and difficult to season, because it warps. Heartwood varying from dark to pale red; sapwood narrow, almost white.

Use: Chairs, shoe lasts, walking sticks, handles, firewood and charcoal. Beech nuts are food for many forest-dwellers, and are also collected and found at markets in the north.

Ornamental at any time of the year.

Beschreibung: Bis über 30 m hoch bei 0,9 bis 1,2 m Durchmesser. Gerader, säulenförmiger Stamm. Im Alter oft mit aufschießenden Trieben von der Stammbasis, Dickichte um den Mutterstamm bildend. Nach dessen Absterben ersetzt durch stärksten der Schößlinge. Auffallend glatte, blaugraue Borke.

Vorkommen: Reiche, gut drainierte Hochlagen und Hänge, im Norden mit Zucker-Ahorn, Birken, Hainbuche, Amerikanischer Linde, Hemlock u. a. Im Süden auch im Tiefland und an Sumpfrändern. Östliches Nordamerika.

Holz: Hart, zäh, fest und von sehr feiner Textur; polierfähig, nicht haltbar im Boden und schwer zu trocknen, weil es sich wirft. Kern dunkel- bis hellrot, variierend; Splint schmal, fast weiß.

Nutzung: Stühle, Schuhständer, Stöcke, Handgriffe, Brennholz und Holzkohle. Die Bucheckern finden sich auf den Märkten im Norden und eignen sich für viele Waldbewohner als Nahrungsmittel.

Ornamental zu jeder Jahreszeit.

Description: arbre qui atteint 30 m ou plus de hauteur, avec un tronc columnaire de 0,9 à 1,2 m de diamètre, drageonnant à la base avec l'âge et formant un taillis autour du tronc mère. Remplacé, après la mort de celui-ci, par les rejetons les plus vigoureux. Ecorce remarquable, demeurant lisse et grise.

Habitat: hautes terres et pentes aux sols riches, bien drainés. Dans le nord, en compagnie de l'érable à sucre, des bouleaux, du charme, du tilleul d'Amérique, de la pruche etc.; dans le sud, on le trouve aussi en plaine et au bord des marécages. Est de l'Amérique du Nord.

Bois: dur, résistant et à grain très fin, susceptible d'être poli; non durable sur pied et difficile à sécher car il gauchit. Bois de cœur rouge foncé à rouge clair, variable; aubier mince, presque blanc.

Intérêts: utilisé pour fabriquer des chaises, des porte-chaussures, des cannes et des poignées. Bois de chauffage et charbon de bois. Les faînes sont vendues sur les marchés des régions septentrionales.

Arbre ornemental en toutes saisons.

TRANSVERSE SECTION.

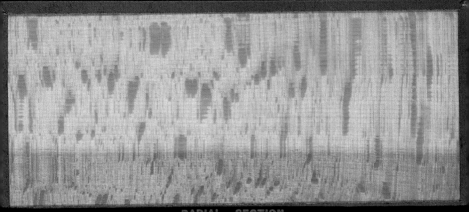

RADIAL SECTION.

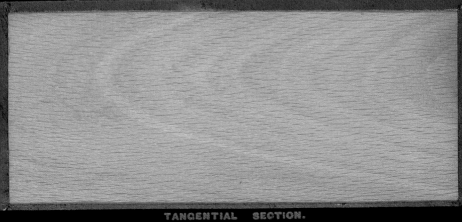

TANGENTIAL SECTION.

Ger. Amerikanische Buche. Sp. Haya Americana.

B | BEECH FAMILY

American Chestnut

FAMILIE DER BUCHENGEWÄCHSE
Amerikanische Edelkastanie

FAMILLE DU CHÊNE ET DU HÊTRE
Châtaignier d'Amérique, Châtaignier (Canada)

FAGACEAE
Castanea dentata (Marshall) Borkh.

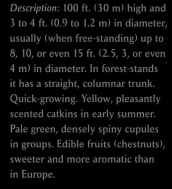

Description: 100 ft. (30 m) high and 3 to 4 ft. (0.9 to 1.2 m) in diameter, usually (when free-standing) up to 8, 10, or even 15 ft. (2.5, 3, or even 4 m) in diameter. In forest-stands it has a straight, columnar trunk. Quick-growing. Yellow, pleasantly scented catkins in early summer. Pale green, densely spiny cupules in groups. Edible fruits (chestnuts), sweeter and more aromatic than in Europe.

Habitat: Relatively rich soils; in the north on moraine soils; rarely, on chalk, dry and gravely hills in the interior of eastern North America.

Wood: Light, soft, not strong, and coarse-grained; very durable in the ground, warps on drying, and is easy to split. Heartwood reddish brown; sapwood thin and paler.

Use: Railroad sleepers, fence posts, railings, cheap furniture, interior construction. Durable owing to high tannin content, hence valuable. Fruits are sold at markets in eastern North America.

Beschreibung: 30 m hoch bei 0,9 bis 1,2 m Durchmesser, selten (im Freistand) bis zu 2,5 und 3, sogar 4 m Durchmesser. (Im Bestand des Waldes) gerader, säulenförmiger Stamm, schnellwüchsig. Gelbe, wohlriechende Kätzchen im Frühsommer. Fahlgrüne, igelstachelige Fruchtbecher in Gruppen. Nussfrüchte (Maronen) süßer und aromatischer als in Europa.

Vorkommen: Relativ reiche Böden, im Norden Moränenböden, selten auf Kalk, trocken-kiesige Hügel im Inneren des östlichen Nordamerika.

Holz: Leicht, weich, nicht fest und von grober Textur; im Boden sehr haltbar, wirft sich beim Trocknen und ist leicht zu spalten. Kern rotbraun; Splint dünn und heller.

Nutzung: Eisenbahnschwellen, Zaunpfähle, Geländer, billige Möbel, Innenausbau. Wertvoll infolge der Haltbarkeit wegen des hohen Gerbsäuregehalts. Früchte wild gesammelt auf allen Märkten im Osten Nordamerikas.

Description: arbre de 30 m de hauteur sur un diamètre qui est en général de 0,9 à 1,2 m, rarement 2,5 ou 3 m, voire 4 m (en situation isolée). Tronc droit, columnaire, à croissance rapide (en formation forestière). Chatons jaunes et odorants au début de l'été. Bogues épineuses groupées, d'un vert terne. Fruits comestibles («marrons»), plus doux et aromatiques qu'en Europe.

Habitat: sols plutôt riches; dans le nord, sols morainiques; rarement sur calcaires, collines sèches caillouteuses dans l'est de l'Amérique du Nord.

Bois: léger, tendre, pas résistant et à grain grossier; très durable sur pied. Gauchit au séchage et se fend facilement. Bois de cœur brun rouge; aubier mince et plus clair.

Intérêts: utilisé pour la fabrication de traverses de chemin de fer, de pieux de clôtures, de balustrades, de meubles bon marché et en menuiserie intérieure. Bois recherché pour sa durabilité provenant de sa haute teneur en tannin. Les fruits sont vendus sur tous les marchés de l'est du continent nord-américain.

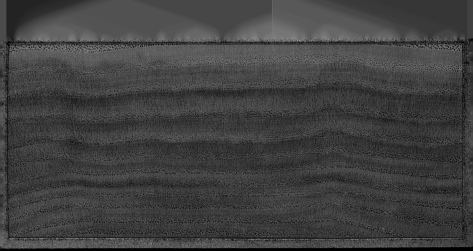

TRANSVERSE SECTION.

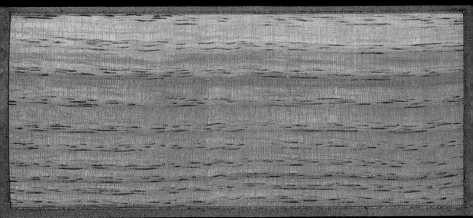

RADIAL SECTION.

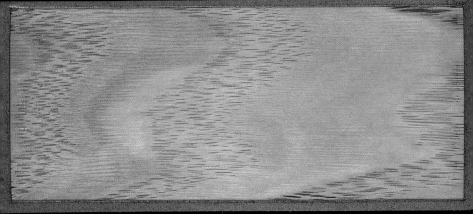

TANGENTIAL SECTION.

Ger. Kastanie.   Fr. Châtaignier.   Sp. Castaña.

B | BEECH FAMILY
Black Oak

FAMILIE DER BUCHENGEWÄCHSE
Emory-Eiche

FAMILLE DU CHÊNE ET DU HÊTRE
Chêne d'Emory

FAGACEAE
Quercus emoryi Torr.

Description: 30 to 40 ft. (9 to 12 m) high and 2 to 3 ft. (0.6 to 0.9 m) in diameter. Trunk short. Branches short, stout, forming a symmetrical round crown. Also grows as a shrub on exposed mountain slopes. Bark thick, rich dark brown to almost black, deeply fissured. Leaves oblong-lanceolate, pointed, with entire or toothed margins, leathery, dark green, very lustrous. Male flowers in long, pendent inflorescences, the female sessile or in short hanging clusters. Acorns sessile or with short stalks.

Habitat: The most important tree species in the canyons and on the mountain slopes of New Mexico and Arizona. Also occurs in western Texas.

Wood: Very heavy, not hard, strong, brittle, and close-grained. Heartwood dark brown to almost black; sapwood thick, brown with a reddish tinge. Bands of small cells, indicating the annual rings, and connected by narrow groups of similar cells, appear parallel to the medullary rays.

Use: The sweet-tasting acorns ("biotis" in Spanish) are much prized and are traded in great quantities.

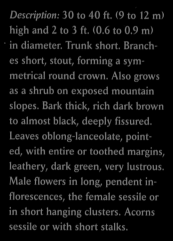

Beschreibung: 9 bis 12 m hoch bei 0,6 bis 0,9 m Durchmesser. Stamm kurz. Äste kurz, kräftig, eine runde, symmetrische Krone formend. Auf exponierten Berghängen auch als Strauch. Borke dick, stark dunkelbraun bis fast schwarz, tief rissig. Blätter länglich-lanzettlich, spitz, ganzrandig, oder gezähnelt, ledrig, dunkelgrün, stark glänzend. Männliche Blüten in lang herabhängenden Blütenständen, weibliche sitzend oder in kurzen, hängenden Blütenständen. Eicheln sitzend oder kurz gestielt.

Vorkommen: Wichtigste Baumart der Cañons und Berghänge Neumexikos und Arizonas. Ferner in West-Texas vorkommend.

Holz: Sehr schwer, nicht hart, fest, brüchig und mit feiner Textur. Kern dunkelbraun bis fast schwarz; Splint dick, braun, rötlich getönt. Bänder kleiner Gefäße, Jahresringe nachzeichnend und durch schmale Gruppen ähnlicher Gefäße verbunden, parallel zu den Markstrahlen.

Nutzung: Die süß schmeckenden Eicheln (spanisch: Biotis) werden in großen Mengen gehandelt.

Description : 9 à 12 m de hauteur sur un diamètre de 0,6 à 0,9 m. Tronc court, branches courtes, vigoureuses, formant une couronne ronde et symétrique. Sous forme de buisson sur les versants montagneux exposés. Ecorce épaisse, brun très sombre à presque noire, profondément fissurée. Feuilles allongées-lancéolées, à apex pointu, entières ou dentées, coriaces, vert foncé, très brillantes. Inflorescences mâles longues, pendantes, inflorescences femelles sessiles ou légèrement pendantes. Glands sessiles ou à court pédoncule.

Habitat : principale espèce arborescente des canyons et des versants montagneux du Nouveau-Mexique et de l'Arizona, aussi dans l'ouest du Texas.

Bois : très lourd, pas dur, résistant, cassant et à grain fin. Cœur brun sombre à noirâtre ; aubier épais, brun, teinté de rouge. Faisceaux vasculaires reliés à des groupes de vaisseaux similaires, effilés et parallèles aux rayons médullaires.

Intérêts : les glands doux (en espagnol, « biotis ») sont très appréciés et vendus en grande quantité dans le commerce.

Emory Oak, Arizona Black Oak.

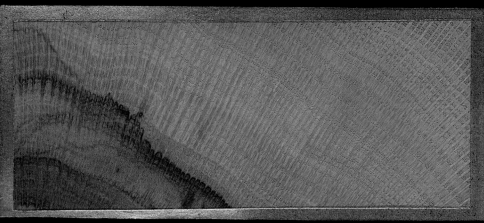

TRANSVERSE SECTION.

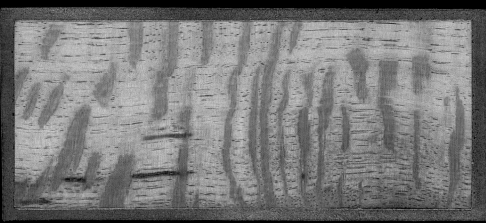

RADIAL SECTION.

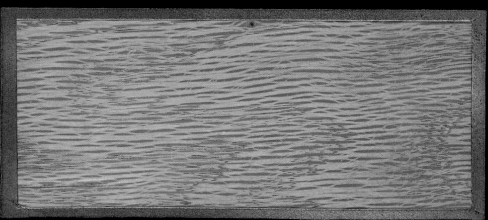

TANGENTIAL SECTION.

Ger. Eiche von Emory.   *Fr.* Chêne d'Emory.

Sp. Roble de Emory.

B | BEECH FAMILY

Black Oak, Yellow Oak, Quercitron Oak

FAMILIE DER BUCHENGEWÄCHSE
Färber-Eiche

FAMILLE DU CHÊNE ET DU HÊTRE
*Chêne des teinturiers, Chêne velouté,
Quercitron (Canada)*

FAGACEAE
Quercus velutina Lam.

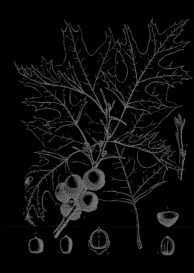

Description: 80 to 90 ft. (23 to 26 m), and quite often 100 ft. (30 m) or more in height and 3 to 4 ft. (0.9–1.2 m) in diameter. Bark deeply grooved and furrowed, dark brown or nearly black on the outer surface, inner bark distinctly yellow. Leaves strikingly large and shiny, especially on the lower branches and twigs. Fall coloring dull red to orange and brown.

Habitat: Found in considerable numbers in all the Atlantic oakwoods. West of the Allegheny Mountains it extends further south than the Scarlet Oak, a species to which it is quite closely related.

Wood: Heavy, hard, and strong.

Use: One of the most important timbers in the USA, as is the Red Oak. The bark is rich in tannin.

Beschreibung: 23 bis 26 m, nicht selten 30 m und mehr, bei 0,9 bis 1,2 m Durchmesser. Borke stark gerieft und gerillt, außen dunkelbraun oder schwärzlich, Rinde innen ausgeprägt gelb. Blätter, besonders an den unteren Ästen und Zweigen, auffallend groß und glänzend. Herbstfärbung trübrot bis orange und braun.

Vorkommen: Bewohnt in beträchtlichem Umfang alle atlantischen Eichenwälder. Westlich der Alleghenies weiter nach Süden als die Scharlach-Eiche, mit der sie als eine Art in Verbindung gebracht wird.

Holz: Schwer, hart und fest.

Nutzung: Eines der wichtigsten Nutzhölzer der USA, wie das der Rot-Eiche. Rinde gerbstoffreich.

Description: 23 à 26 m de hauteur, atteignant parfois 30 m ou plus, sur un diamètre de 0,9 à 1,2 m. Ecorce à texture profondément fissurée, brun foncé ou noirâtre en surface, d'un jaune prononcé en dessous. Feuilles très grandes et lustrées, tout particulièrement sur les branches et les rameaux inférieurs. Beau feuillage rougeâtre à orange et brun en automne.

Habitat: colonies importantes dans toutes les chênaies atlantiques. A l'ouest des Alleghanys, se rencontre plus au sud que le chêne écarlate, avec lequel on l'associe comme une même espèce.

Bois: lourd, dur et résistant.

Intérêts: l'un des bois utiles les plus importants des Etats-Unis avec celui du chêne rouge. Ecorce riche en tannins.

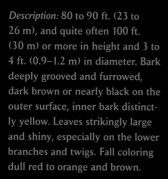

Yellow Oak, Yellow-bark Oak, Black Oak, Quercitron Oak.

TRANSVERSE SECTION.

RADIAL SECTION.

TANGENTIAL SECTION.

Ger. Färber Eiche. *Fr.* Chêne jaune. *Sp.* Roble amaril

B | BEECH FAMILY
Blue Oak, Mountain White Oak

FAMILIE DER BUCHENGEWÄCHSE
Douglas-Eiche

FAMILLE DU CHÊNE ET DU HÊTRE
Chêne bleu, Chêne de Douglas

FAGACEAE
Quercus douglasii Hook. & Arn.

Description: Rarely 85 to 90 ft. (24 to 27 m) high and 3 to 4 ft. (0.9 to 1.2 m) in diameter. More commonly only 20 to 30 ft. (6 to 9 m) high, and also as bush in the south. Branches short and sturdy, arising at right angles from the trunk and forming a rounded, symmetrical crown. Bark thick, pale, with brown or red tinge. Leaves oblong-ovate, entire or wavy-edged, blue-green and hairy. Flowers inconspicuous: male flowers in long, drooping inflorescences; female flowers in short, erect ones. Acorns sessile or on short stalks, singly or in pairs.

Habitat: Low, hilly country, dry mountain slopes and valleys in California.

Wood: Heavy, very hard, strong, and brittle; fades considerably when dried and turns almost black under exposure to light. Heartwood dark brown; sapwood thick, pale brown. Numerous medullary rays, scattered groups of smaller and rows of larger pores.

Use: Little used as building timber, but on the other hand excellent firewood.

Beschreibung: Selten 24 bis 27 m hoch bei 0,9 bis 1,2 m Durchmesser. Häufiger nur 6 bis 9 m hoch, im Süden auch als Busch. Äste kurz und kräftig, rechtwinklig vom Stamm abgehend und eine dichte, symmetrisch-runde Krone ausbildend. Borke dick, blass, mit brauner oder roter Tönung. Blätter oval-länglich, ganzrandig oder ausgerandet, blaugrün und behaart. Blüten unauffällig, die männlichen in hängenden, lang-zylindrischen, die weiblichen in kurzen, aufrechten Blütenständen. Eicheln sitzend oder kurz gestielt, einzeln oder zu zweien.

Vorkommen: Niedriges Hügelland, trockene Berghänge und -täler in Kalifornien.

Holz: Schwer, sehr hart, fest und brüchig; beim Trocknen stark schwindend und unter Lichteinfluss fast schwarz werdend. Kern dunkelbraun; Splint dick, hellbraun. Zahlreiche Markstrahlen, zerstreute Gruppen kleiner und Reihen größerer Gefäße.

Nutzung: Wenig geeignet als Bau- und Schnitzholz, dagegen exzellentes Feuerholz.

Description: arbre atteignant rarement 24 à 27 m de hauteur sur un diamètre de 0,9 à 1,2 m. Plus souvent 6 à 9 m de haut seulement, au sud aussi sous forme de buisson. Branches courtes et vigoureuses formant un houppier dense, symétrique et rond. Ecorce épaisse, pâle, teintée de brun ou de rouge. Feuilles vert bleu, ovales-allongées, entières ou à lobes peu marqués et pubescentes. Fleurs bisexuées, insignifiantes, les mâles en inflorescences allongées, cylindriques et pendantes, les femelles en inflorescences courtes et dressées. Glands sessiles ou courtement pédonculés, isolés ou par deux.

Habitat: régions de collines peu élevées, versants et vallées de montagnes secs de Californie.

Bois: lourd, très dur, résistant et cassant; fort retrait au séchage et noircissement à la lumière. Bois de cœur brun sombre; aubier épais, brun clair. Rayons médullaires nombreux; groupes de petits vaisseaux et alignements de plus gros vaisseaux çà et là dans le bois.

Intérêts: convient peu comme bois d'œuvre et de sciage, mais excellent bois de feu.

Blue Oak, California Rock Oak.

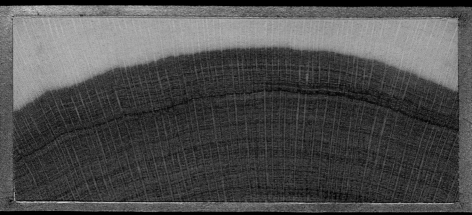

TRANSVERSE SECTION.

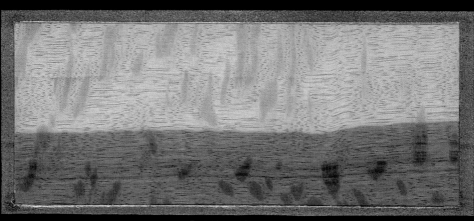

RADIAL SECTION.

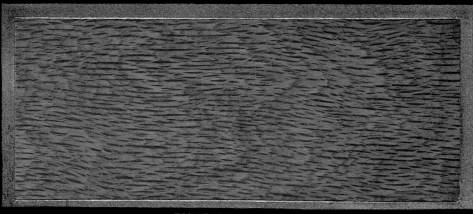

TANGENTIAL SECTION.

B | BEECH FAMILY
Burr Oak, Mossy Cup Oak

FAMILIE DER BUCHENGEWÄCHSE
Großfrüchtige Eiche

FAMILLE DU CHÊNE ET DU HÊTRE
Chêne à gros fruits, Chêne à gros glands (Canada)

FAGACEAE
Quercus macrocarpa Michx.

Description: At one time typically of gigantic size, but now only so in a few individuals, up to 200 ft. (55 m) high and 6 to 7 ft. (1.8 to 2.1 m) in diameter. Trunk up to 85 ft. (25 m) before branching. Quick-growing and ornamental.

Habitat: Almost exclusively on rich lowlands in association with Swamp White Oak, Black and Silver Maple, Shellbark Hickory, Nettle Tree, elms, etc., in eastern North America.

Wood: Heavy, hard, tough, strong, and close-grained; very durable in the ground, superior even to the White Oak in strength, and confused with this species in commerce. Heartwood light to rich brown or dark brown; sapwood much paler.

Use: Ship and boatbuilding, wooden constructions of all kinds, interior structures, cupboards, cooperage, wagons, agricultural implements, railroad sleepers, fences, baskets, and firewood.

Discovered for botany by André Michaux the Elder in the western Allegheny Mountains.

Beschreibung: Einst riesig, bis 55 m hoch bei 1,8 bis 2,1 m Durchmesser, heute nur noch in wenigen Exemplaren so hoch. Stamm bis 25 m ohne Äste. Schnellwüchsig und ornamental.

Vorkommen: Fast ausschließlich reiches Flachland in Gesellschaft von Sumpf-Weiß-Eiche, Schwarz- und Silber-Ahorn, Königsnuss, Zürgelbaum, Ulmen u. a. im östlichen Nordamerika.

Holz: Schwer, hart, zäh, fest und mit feiner Textur; sehr dauerhaft im Boden, übertrifft an Stärke noch die Weiß-Eiche und wird im Handel mit dieser verwechselt. Kern hell- bis kräftig braun, dunkelbraun; Splint viel heller.

Nutzung: Schiff- und Bootsbau, Holzkonstruktionen aller Art, Innenausbau, Schränke, Böttcherei, Wagen, Ackerbaugerät, Eisenbahnschwellen, Zäune, Körbe und Brennholz.

Botanisch entdeckt von André Michaux dem Älteren in den West-Alleghenies.

Description : essence autrefois de grande taille ; seuls quelques spécimens atteignent encore 55 m de hauteur sur un diamètre de 1,8 à 2,1 m. Tronc dénudé jusqu'à 25 m. Arbre à croissance rapide et ornemental.

Habitat : ne supporte pratiquement que les sols riches de plaine en association avec le chêne bicolore, les érables noir et argenté, le caryer lacinié, le micocoulier, les ormes etc. On le trouve dans l'est de l'Amérique du Nord.

Bois : lourd, dur, résistant et à grain fin ; très durable sur pied, plus fort encore que le frêne blanc et confondu avec celui-ci dans le commerce. Bois de cœur brun clair à brun vif ; aubier beaucoup plus clair.

Intérêts : utilisé dans la construction navale, en tonnellerie, en carrosserie, en menuiserie intérieure ; fabrication d'armoires, d'outils agricoles, de traverses de chemin de fer, de palissades, de paniers ; bois de chauffage.

Découvert botaniquement par André Michaux l'Ancien dans l'ouest des Alleghanys.

Burr Oak, Mossy-cup Oak, Over-cup Oak.

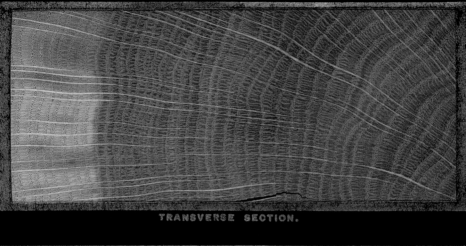

TRANSVERSE SECTION.

RADIAL SECTION.

TANGENTIAL SECTION.

Ger. Grossfrüchtige Eiche.   *Fr.* Chêne à gros gland.

Sp. Roble con bellotas musgosas.

B | BEECH FAMILY
California Black Oak

FAMILIE DER BUCHENGEWÄCHSE
Kalifornische Schwarz-Eiche

FAMILLE DU CHÊNE ET DU HÊTRE
Chêne de Kellogg, Chêne noir de Californie

FAGACEAE
Quercus kelloggii Newb.
Syn.: *Q. californica* (Torr.) Cooper

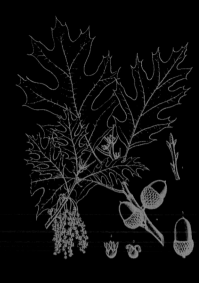

Description: Up to 100 ft. (30 m) high, 3 to 4 ft. (0.9 to 1.3 m) in diameter. Often much smaller, and shrubby at high altitudes. Bark dark brown, with a pale red tinge, or almost black. Base of old trunks divided by broad ridges. Otherwise broken into irregular thick plates covered with tiny scales.

Habitat: Scattered in coniferous forests in Pacific North America, sometimes forming extensive groves. Optimum in south-west Oregon and in the Sierra Nevada at around 6,600 ft. (2,000 m).

Wood: Heavy, hard, strong, and very brittle. Heartwood pale red; sapwood thin and paler.

Use: Little value as timber. Sometimes as firewood. Bark for tanning leather.

First discovered in 1846 near the Sonora by Karl Th. Hartweg.

Beschreibung: Gelegentlich 30 m hoch bei 0,9 bis 1,3 m Durchmesser. Oft viel weniger, und in großen Höhen nur noch als Strauch. Borke dunkelbraun, leicht rot getönt oder fast schwarz. An der Basis alter Stämme in breite Riefen geteilt. Darüber in unregelmäßige, dicke Platten gebrochen und mit winzigen anliegenden Schuppen bedeckt.

Vorkommen: Zerstreut in Koniferenwäldern im pazifischen Nordamerika, manchmal eigene Haine von beträchtlicher Ausdehnung bildend. Optimum in Südwest-Oregon und in der Sierra Nevada in etwa 2000 m Höhe.

Holz: Schwer, hart, fest und sehr spröde. Kern hellrot; Splint dünn und heller.

Nutzung: Von geringem Wert für den Holzbau. Manchmal als Brennholz. Rinde zum Gerben von Leder.

Erst 1846 von Karl Th. Hartweg nahe der Sonora entdeckt.

Description : atteint parfois 30 m de hauteur sur un diamètre de 0,9 à 1,3 m. Souvent beaucoup plus petit et, en haute altitude, à port de buisson. Ecorce brun sombre, légèrement teintée de rouge, ou presque noire, creusée de larges fissures à la base des vieux troncs et se craquelant en grosses plaques irrégulières, recouvertes de minuscules écailles adhérentes.

Habitat : disséminé dans les forêts de conifères de la façade atlantique de l'Amérique du Nord ; forme quelquefois des fourrés occupant de vastes étendues. Développement maximum dans le sud-ouest de l'Oregon et dans la Sierra Nevada à 2000 mètres au-dessus du niveau de la mer.

Bois : lourd, dur, résistant et très cassant. Bois de cœur rouge clair ; aubier mince et plus clair.

Intérêts : de qualité médiocre pour la construction. Utilisé parfois comme bois de chauffage ; écorce employée pour tanner le cuir.

Ne fut découvert qu'en 1846 par Karl Th. Hartweg, à proximité du Sonora.

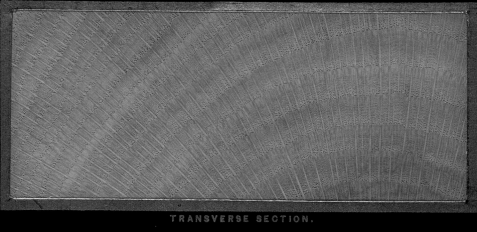

TRANSVERSE SECTION.

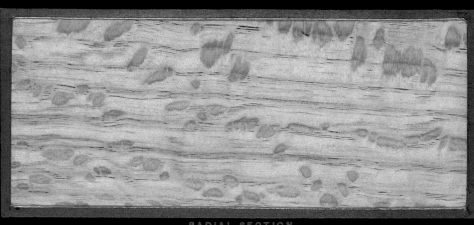

RADIAL SECTION.

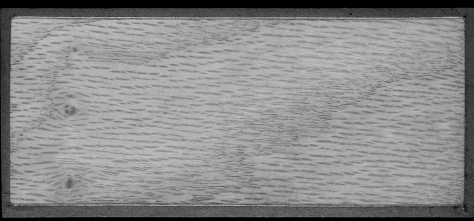

TANGENTIAL SECTION

Ger. Californische Schwarzeiche. *Fr.* Chêne noir de Californie.
Sp. Roble negro de California.

B | BEECH FAMILY
California Live Oak, Encina

FAMILIE DER BUCHENGEWÄCHSE
Kalifornische Steineiche

FAMILLE DU CHÊNE ET DU HÊTRE
Chêne de Californie, Chêne vert de Californie

FAGACEAE
Quercus agrifolia Née

Description: 85 to 100 ft. (25 to 30 m) high and 3 to 4 (6 to 7) ft. (0.9 to 1.2 (1.8 to 2.1)) in diameter. Trunk short, dividing into many subsidiary trunks near the base. These often lie on the ground, extending to 100 to 160 ft. (30 to 50 m). Also has upright trunks, and can be shrubby. Bark dark brown, pale red. Broad, rounded ridges with closely appressed scales. Dark brown or pale bluish-gray when young.

Habitat: Semi-prostrate on dunes in central California. Characteristic of area between the coastal mountains and the sea.

Wood: Heavy, hard, very brittle, close-grained. Heartwood pale brown or red-brown; sapwood thick and dark.

Use: Firewood, otherwise of little value. Acorns eaten by the Indians of Lower California.

Beschreibung: 25 bis 30 m hoch bei 0,9 bis 1,2 (1,8 bis 2,1) m Durchmesser. Kurzer Stamm. Nahe der Stammbasis zahlreiche Tochterstämme, die oft auf dem Boden aufliegen und 30 bis 50 m Querausdehnung annehmen. Daneben auch aufrechte Stämme, aber auch Strauchformen. Borke dunkelbraun, leicht rot. Breite, abgerundete Riefen mit dicht anliegenden Schuppen. Jung dunkelbraun oder blass bläulichgrau und geschlossen.

Vorkommen: Halb niederliegend auf Dünen Mittel-Kaliforniens. Charakteristisch zwischen Küstengebirge und Meer.

Holz: Schwer, hart, sehr spröde und mit feiner Textur. Kern hellbraun oder rötlichbraun; Splint dick und dunkler.

Nutzung: Als Brennholz, sonst wenig geschätzt. Die Eicheln als Nahrungsmittel der Indianer in Niederkalifornien.

Description : 25 à 30 m de hauteur sur un diamètre de 0,9 à 1,2 m (1,8 à 2,1 m). Tronc très variable : soit court et armé, à la base, de nombreuses tiges filles qui souvent reposent sur le sol, occupant ainsi une surface de 30 à 50 m de large (!), soit dressé. Se rencontre aussi sous forme de buisson. Ecorce brun foncé, légèrement rouge, se fissurant en larges crêtes arrondies et écailleuses ; brun sombre ou gris bleuâtre pâle, nette de craquelures à l'état juvénile.

Habitat : à port semi-prostré dans les dunes de la moyenne Californie. Typique de la zone littorale située entre les chaînes côtières et l'océan.

Bois : lourd, dur, très cassant et à grain fin. Bois de cœur brun clair ou brun rougeâtre ; aubier épais et plus foncé.

Intérêts : peu recherché sinon comme bois de chauffage. Les glands étaient consommés par les Indiens de la Basse-Californie.

Coast Live Oak, Holly-leaved Oak.

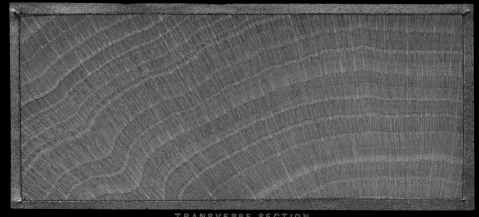

TRANSVERSE SECTION.

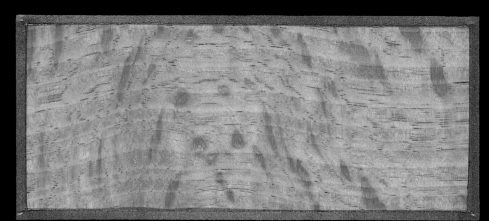

RADIAL SECTION.

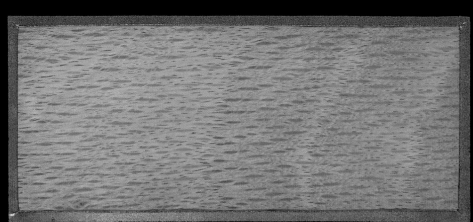

TANGENTIAL SECTION

Ger. Immergrüne Eiche von der Kuste.   *Fr.* Chêne vert de la côte

Sp. Encina.

B | BEECH FAMILY
Californian Oak, Valley Oak

FAMILIE DER BUCHENGEWÄCHSE
Kalifornische Weiß-Eiche

FAMILLE DU CHÊNE ET DU HÊTRE
Chêne lobé, Chêne blanc de Californie

FAGACEAE
Quercus lobata Née

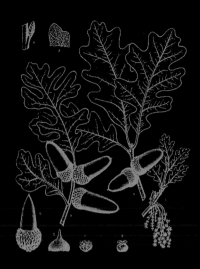

Description: 110 ft. (33 m) high, 3 to 4 (10) ft. (0.9 to 1.2 (3) m) in diameter. The largest and most attractive of the oaks in the American Pacific region. Spreads out into several branches close to the ground or at a height of 18 to 35 ft. (6 to 10 m) Crown diameter 100 ft. (30 m)! Bark pale gray. Scales lightly tinged orange. Towards the base of the trunk, shallow longitudinal furrows and broad ridges, transversely broken into short plates.

Habitat: Open areas in California, never in dense woodland. Alone, or with Blue Oak in open groves without undergrowth (resembling highly attractive parks; such areas have largely fallen victim to civilization).

Wood: Moderately hard, brittle, and close-grained; difficult to store.

Use: Few commercial uses. Firewood. Acorns used by Californian Indians by soaking in water and baking in primitive ovens dug into the sand dunes.

Not cultivated outside its native region.

Beschreibung: 33 m hoch bei 0,9 bis 1,2 (3!) m Durchmesser. Größte und schönste der Eichen im pazifischen Amerika. Nahe dem Grund oder in 6 bis 10 m Höhe in mehrere Teilstämme geteilt. Kronendurchmesser 30 m! Borke hellgrau. Schuppen leicht orangebraun getönt. An der Stammbasis flache Längsrillen und breite, flache Riefen, quer gebrochen in kurze Platten.

Vorkommen: Offenes Gelände in Kalifornien, niemals als dichter Waldbildner. Allein oder mit der Blau-Eiche in lichten Hainen ohne Unterwuchs, in der Art schönster Parklandschaften, meist der Zivilisation gewichen.

Holz: Mäßig hart, spröde und mit feiner Textur; schwierig zu lagern.

Nutzung: Wenig wirtschaftlicher Nutzen. Brennholz. Eicheln von den Indianern Kaliforniens gemörsert, mit Wasser angeteigt und in primitiven, ad hoc gegrabenen Öfen in den Sanddünen gebacken.

Außerhalb ihres natürlichen Wohngebiets nicht zu kultivieren!

Description: 33 m de hauteur sur un diamètre de 0,9 à 1,2 m (jusqu'à 3 m!). Le plus grand et le plus beau des chênes de la côte pacifique. Se divise en plusieurs tiges secondaires, près du sol ou à une hauteur de 6 à 10 m. Le houppier mesure 30 m de diamètre! Ecorce gris clair, avec des écailles légèrement brun orangé, parcourue à la base du tronc de sillons longitudinaux plats et de larges crevasses plates, se craquelant en courtes plaques à section transversale.

Habitat: terrains ouverts de Californie; ne constitue jamais de forêts denses. Seul ou avec le chêne bleu, en bosquets clairs sans sous-bois, composant ainsi à leur manière les paysages les plus beaux, presque toujours à l'écart de la civilisation.

Bois: moyennement dur, cassant et à grain fin; difficile à stocker.

Intérêts: sans grand rôle économique. Bois de chauffage. Les Indiens de Californie pilaient et délayaient les glands dans de l'eau pour les cuire dans des fours rudimentaires creusés dans les dunes.

Essence ne pouvant être cultivée en dehors de son aire naturelle!

Cal. White   Valley Oak.

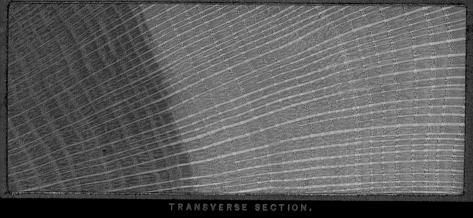

TRANSVERSE SECTION.

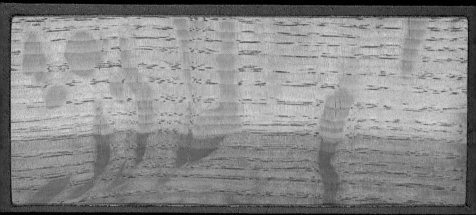

RADIAL SECTION.

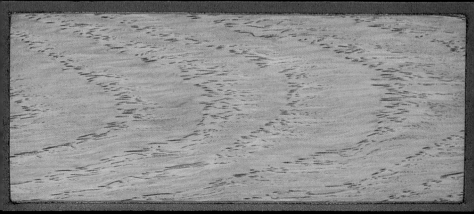

TANGENTIAL SECTION

Ger. Californische Weisseiche.   *Fr.* Chêne blanc de Californie

B | BEECH FAMILY
Canyon Live Oak, Maul Oak

FAMILIE DER BUCHENGEWÄCHSE
Goldschuppige Eiche

FAMILLE DU CHÊNE ET DU HÊTRE
Chêne des canyons, Chêne à écailles dorées

FAGACEAE
Quercus chrysolepis Liebm.

Description: 40 to 50 ft. (12 to 15 m) high, 3 to 5 ft. (0.9 to 1.5 m) in diameter. Crown and large horizontal branches stretch to 100 to 165 ft. (30 to 50 m) across. Branches and twigs droop down, brushing the ground. Trunks up to 8 to 9 ft. (2.4 to 2.7 m) in diameter in lowland and in protected canyons. On exposed mountain-sides forms dense thickets often only 2 to 10 ft. (0.6 to 3 m) high, and finally subalpine low shrubs. Bark pale or dark gray-brown, tinged pale red. Smooth surface sheds small plates.

Habitat: Western North America.

Wood: Heavy, hard, tough, very strong, close-grained; difficult to cut. Heartwood pale brown; sapwood thick and darker.

Use: Better timber than the other central Californian oaks. Otherwise for farming implements and carts.

Equaled only by Live Oak in the southern Atlantic and Gulf states for its majestic shape and massive strength.

Beschreibung: 12 bis 15 m hoch bei 0,9 bis 1,5 m Durchmesser. Große, horizontale Teilstämme überspannen mit der Baumkrone 30 bis 50 m Breite. Hängende Äste und Zweige streichen über den Grund. Stammstärken in geringer Meereshöhe und in geschützten Cañons 2,4 bis 2,7 m Durchmesser. Auf exponierten Bergseiten dicke Dickichte bildend, oft nur 0,6 bis 3 m hoch, schließlich subalpine Kriechsträucher. Borke hell oder dunkel graubraun, leicht rot getönt. Glatte Oberfläche schuppt anliegende Plättchen ab.

Vorkommen: Westliches Nordamerika.

Holz: Schwer, hart, zäh, sehr fest und mit feiner Textur; schwierig zu schneiden. Kern hellbraun; Splint dick und dunkler.

Nutzung: Als Bauholz besser als andere mittelkalifornische Eichen. Sonst für Ackergerät und Waggons.

An majestätischer Gestalt und massiver Stärke kommt ihr nur die Virginia-Eiche in den südlichen Atlantik- und den Golfstaaten gleich.

Description: 12 à 15 m de hauteur sur un diamètre de 0,9 à 1,5 m. Les grandes tiges secondaires horizontales s'étalent avec le houppier sur 30 à 50 m. Les branches et les rameaux retombants effleurent le sol. Le tronc atteint, à faible altitude et dans les canyons abrités, 2,4 à 2,7 m de diamètre. Sur les versants montagneux exposés, il forme d'épais fourrés n'atteignant souvent que 0,6 à 3 m de hauteur. Enfin, à l'étage subalpin, il forme des buissons à port rampant. Ecorce lisse, brun gris clair ou sombre, légèrement teintée de rouge, se desquamant en petites plaques.

Habitat: répandu dans l'ouest de l'Amérique du Nord.

Bois: lourd, dur, tenace, très résistant et à grain fin; difficile à couper. Bois de cœur brun clair; aubier épais et plus foncé!

Intérêts: meilleur bois de construction que les autres chênes de la moyenne Californie. Sinon instruments aratoires et wagons.

Seul le chêne de Virginie, dans le sud de la façade atlantique et le pourtour du Golfe, peut lui faire concurrence en majesté et en puissance.

Maul Oak, Thick-cup Live Oak, Hickory Oak, Cañon Live Oak.

TRANSVERSE SECTION

RADIAL SECTION.

TANGENTIAL SECTION

Ger. Schlagel-Eiche.   *Fr.* Chêne de maillet.

Sp. Roble de mazo.

B | BEECH FAMILY

Chestnut Oak, Rock Oak, Basket Oak

FAMILIE DER BUCHENGEWÄCHSE
Kastanien-Eiche

FAMILLE DU CHÊNE ET DU HÊTRE
Chêne des montagnes, Chêne prin (Canada)

FAGACEAE
Quercus prinus L.

Description: 70 to 85 (20 to 25), exceptionally to 100 ft. (30 m) high and 3 to 4 ft. (0.9 to 1.2 m) in diameter. Bark dark, thick, broadly furrowed.

Habitat: Well-drained slopes, upland soils and rocky ridges in association with Shagbark Hickory, Pignut Hickory, various species of oak, Tulip Tree, etc., in (north-) eastern North America. Avoids lime soils.

Wood: Heavy, and used like that of the White Oak.

Use: Bark rich in tannin, used in tanning leather.

A specimen 7 ft. (2.1 m) in diameter, and known historically as the "Washington Oak," stood or still stands on the east bank of the Hudson River in the vicinity of Fishkill. Its age, based on counting the annual rings of a neighboring tree after it was felled, was estimated at between 800 and 1,000 years.

Beschreibung: 20 bis 25, ausnahmsweise bis 30 m hoch bei 0,9 bis 1,2 m Durchmesser. Dunkle, kräftige, breit geriefte Borke.

Vorkommen: Gut drainierte Hänge, hoch gelegene Böden und Felsrücken in Gesellschaft von Schindelborkiger Hickory, Ferkelnuss-Hickory, verschiedenen Eichen, Tulpenbaum u. a. im (nord-) östlichen Nordamerika. Kalkmeidend.

Holz: Schwer und wie das der Weiß-Eiche genutzt.

Nutzung: Borke gerbstoffreich, zur Ledergerberei.

Ein Exemplar mit 2,1 m Durchmesser und geschichtlich als „Washington-Eiche" bekannt, steht noch am Ostufer des Hudson River in der Nähe von Fishkill. Das Alter konnte durch die Jahresringzählung bei einem gefällten Nachbarn bestimmt werden: 800 bis 1000 Jahre.

Description : 20 à 25 m de hauteur, exceptionnellement jusqu'à 30 m, sur un diamètre de 0,9 à 1,2 m. Ecorce de couleur sombre, épaisse et parcourue de larges sillons.

Habitat : pentes bien drainées, sols situés en altitude et sommets rocheux dans le (nord-)est de l'Amérique du Nord, en association avec le noyer blanc, le caryer ovale, divers chênes, le tulipier etc. Evite les sols calcaires.

Bois : lourd ; utilisé comme celui du chêne blanc.

Intérêts : l'écorce, riche en tannins, est employée dans le tannage des cuirs.

Un spécimen de 2,1 m de diamètre, connu sous le nom de « chêne de Washington », se trouvait, ou se trouve encore, sur la rive orientale de l'Hudson, non loin de Fishkill. On a pu évaluer son âge à partir du nombre de cernes annuels comptés dans le bois d'un arbre voisin : 800 à 1000 ans.

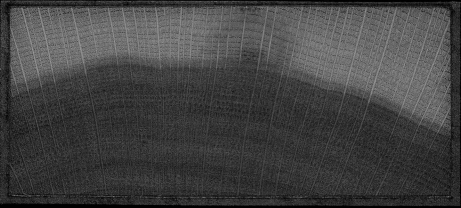

TRANSVERSE SECTION.

RADIAL SECTION.

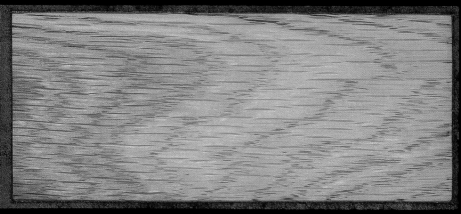

TANGENTIAL SECTION

Ger. Felsen-Eiche.   Sp. Roble de las rocas.

B | BEECH FAMILY
Chinquapin

FAMILIE DER BUCHENGEWÄCHSE
Zwergkastanie

FAMILLE DU CHÊNE ET DU HÊTRE
Châtaignier nain, Chicaquin, Chiquapin des Alléghanys (Canada)

FAGACEAE
Castanea pumila (L.) Mill.

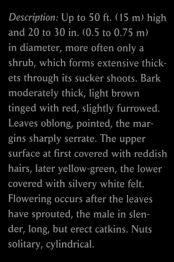

Description: Up to 50 ft. (15 m) high and 20 to 30 in. (0.5 to 0.75 m) in diameter, more often only a shrub, which forms extensive thickets through its sucker shoots. Bark moderately thick, light brown tinged with red, slightly furrowed. Leaves oblong, pointed, the margins sharply serrate. The upper surface at first covered with reddish hairs, later yellow-green, the lower covered with silvery white felt. Flowering occurs after the leaves have sprouted, the male in slender, long, but erect catkins. Nuts solitary, cylindrical.

Habitat: Dry sandy ridges, rich soils on hills, and swamp margins in eastern North America.

Wood: Light, hard, strong, and coarse-grained; very durable in the ground. Heartwood dark brown; sapwood of three or four annual rings, barely distinguishable. Numerous distinct medullary rays and bands consisting of a few rows of large, open pores along the annual rings.

Use: Fence posts, railings, railroad sleepers. The sweet-tasting nuts are sold in towns in the west and south of the USA.

Beschreibung: Bis zu 15 m hoch bei 0,5 bis 0,75 m Durchmesser, häufiger nur als Busch, der über Ausläufer ausgedehnte Dickichte bildet. Borke mäßig dick, hellbraun mit roter Tönung, leicht rissig. Blätter länglich, spitz, Ränder scharf gesägt. Oberseite zunächst rötlich behaart, später gelbgrün, unterseits silbrigweiß filzig behaart. Blüte nach dem Laubaustrieb, männliche in schlanken, langen und aufrechten Kätzchen. Nuss einzeln, zylindrisch.

Vorkommen: Trockene, sandige Kämme, reiche Böden auf Hügeln, Ränder von Sümpfen im östlichen Nordamerika.

Holz: Leicht, hart, fest und mit feiner Textur; im Boden sehr haltbar. Kern dunkelbraun; Splint 3 bis 4 Jahresringe, kaum unterscheidbar. Zahlreiche deutliche Markstrahlen und Bänder einiger Reihen großer offener Gefäße längs der Jahresringe.

Nutzung: Zaunpfosten, Geländer, Eisenbahnschwellen. Die süßen Nüsse werden gesammelt und in den Städten des Westens und Südens der USA verkauft.

Description: jusqu'à 15 m de hauteur sur un diamètre de 0,5 à 0,75 m, mais plus souvent arbrisseau touffu, constituant de vastes fourrés par son drageonnement vigoureux. Ecorce moyennement épaisse, brun clair teinté de rouge, légèrement fissurée. Feuilles allongées, pointues, bordées de dents aiguës, couvertes de poils roux puis vert jaune et glabres à la face supérieure, tomenteuses et blanc argenté à la face inférieure. Floraison après la feuillaison; fleurs mâles en longs chatons minces et dressés. Glands isolés, cylindriques.

Habitat: crêtes sablonneuses sèches, sols riches à l'étage collinéen, bords des marais dans l'est de l'Amérique du Nord.

Bois: léger, dur, résistant et à gros grain; très durable sur pied. Bois de cœur brun foncé et aubier à 3 ou 4 cernes annuels. Nombreux rayons médullaires distincts et bandes constituées de quelques grands faisceaux vasculaires ouverts le long des cernes annuels.

Intérêts: piquets de clôtures, balustrades. Les glands à chair douce sont vendus dans les villes de l'ouest et du sud des Etats-Unis.

Chinquapin. Chinkepin.

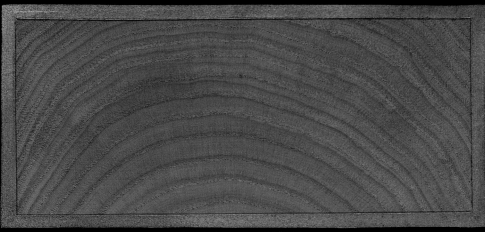

TRANSVERSE SECTION.

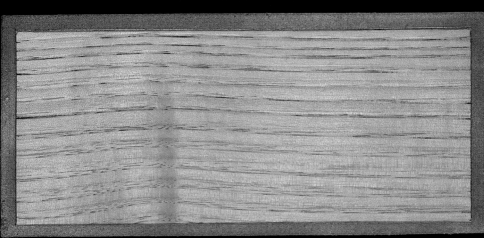

RADIAL SECTION.

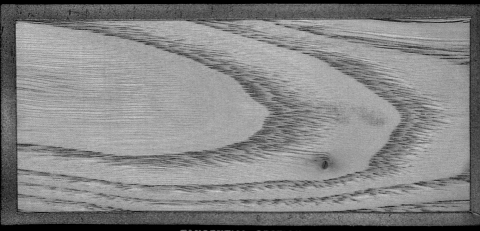

TANGENTIAL SECTION.

Ger. Kleine Kastanie. *Fr.* Chincapin. *Sp.* Castaño enano.

B | BEECH FAMILY

Cow Oak, Basket Oak

FAMILIE DER BUCHENGEWÄCHSE

Michaux-Eiche, Korb-Eiche

FAMILLE DU CHÊNE ET DU HÊTRE

Chêne de Michaux

FAGACEAE

Quercus michauxii Nutt.

Description: Over 100 ft. (30 m) high and 3 to 4 ft. (0.9 to 1.2 m) in diameter. Trunk columnar, with pale gray bark, furrowed, with narrow scales. Foliage dense. Leaves dark shiny green above, velvety white below, fluttering in the wind.

Habitat: Swampy soils, rich river valleys, and flooded riverbanks, with Water Hickory, Swamp Bay, Planer Tree, Water Oak, Laurel Oak, gums (*Nyssa* spp.), Red Maple, etc, in eastern North America.

Wood: Heavy, hard, tough, and strong; very durable in the ground.

Use: Much sought after for interior construction and furniture. Used for fence posts, firewood, and very good for baskets.

Beschreibung: Über 30 m hoch bei 0,9 bis 1,2 m Durchmesser. Säulenförmiger Stamm mit blassgrauer Borke, gerillt-gerieft, mit schmalen Schuppen. Dicht belaubt. Blätter oberseits glänzend dunkelgrün, unterseits samtig weiß. Wechselspiel im Wind.

Vorkommen: Sumpfige Böden und reiche Niederungen, Flussufer mit häufigen Überschwemmungen in Gesellschaft von Wasser-Hickory, Sumpf-Avocado, Wasser-Ulme, Wasser-Eiche, Lorbeer-Eiche, den Tupelos, Rot-Ahorn u. a. im östlichen Nordamerika.

Holz: Schwer, hart, zäh und fest; sehr dauerhaft im Boden.

Nutzung: Sehr gefragt für Innenausbau und Möbel. Zaunpfähle, Brennholz und sehr gutes Holz für Körbe.

Description: arbre de plus de 30 m de hauteur, avec un tronc columnaire de 0,9 à 1,2 m de diamètre. L'écorce gris terne se creuse de sillons formant d'étroites côtes écailleuses. Feuillage dense. Feuilles vert sombre et lustrées dessus, blanches et veloutées dessous, offrant un beau spectacle au moindre souffle de vent.

Habitat: sols marécageux et secteurs déprimés riches, berges souvent inondées de l'est de l'Amérique du Nord, en association avec le caryer aquatique, l'avocatier palustre, le planéra des marécages, le chêne noir, le chêne à feuilles de laurier, les nyssas, l'érable rouge etc.

Bois: lourd, dur, tenace et résistant très durable sur pied.

Intérêts: très recherché en menuiserie intérieure, en ameublement. Utilisé aussi pour fabriquer des pieux de clôtures et pour le chauffage. Excellent bois de vannerie.

Basket Oak, Cow Oak, Swamp Chestnut Oak

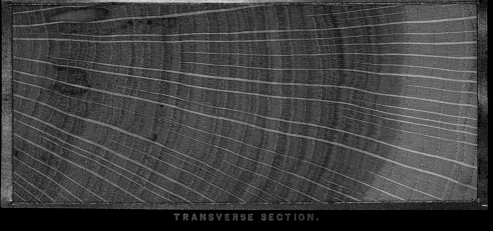

TRANSVERSE SECTION.

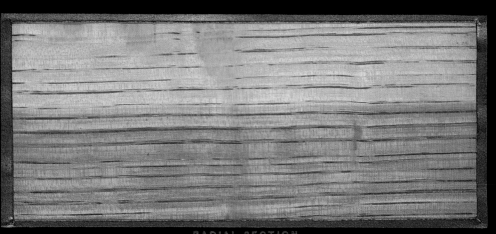

RADIAL SECTION.

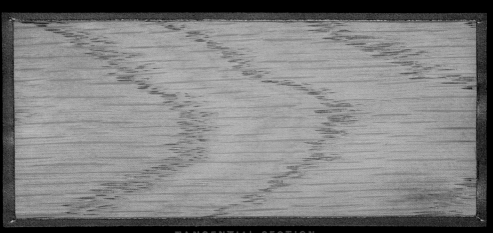

TANGENTIAL SECTION

ɪ. Korb-Eiche.   Fɪ. Chêne de panier.   Sp. Roble de canast

B | BEECH FAMILY
Evergreen White Oak

FAMILIE DER BUCHENGEWÄCHSE
Engelmanns Eiche

FAMILLE DU CHÊNE ET DU HÊTRE
Chêne d'Engelmann

FAGACEAE
Quercus engelmannii Greene

Description: 50 to 60 ft. (15 to 18 m) high and 2 to 3 ft. (0.6 to 0.9 m) in diameter. Branches sturdy, almost at right angles to trunk, forming a broad, rather irregular crown. Bark thick, pale gray with brown tints and deep furrows. Leaves oblong-ovate to lanceolate, entire or weakly wavy-edged, deeply serrated at tips of branches. Newly emerging leaves pale red, changing to dark blue-green when fully unfolded. Flowers inconspicuous: male inflorescences long and dangling; female short and erect. Acorns mostly clearly stalked, more rarely sessile.

Habitat: With *Quercus agrifolia* on low hills in the coastal mountain ranges of south-west California.

Wood: Very heavy, hard, strong, brittle, close-grained; hard to season. Heartwood dark brown; sapwood paler brown.

Use: Occasionally as firewood.

Beschreibung: 15 bis 18 m hoch bei 0,6 bis 0,9 m Durchmesser. Äste kräftig, fast rechtwinklig vom Stamm abgehend, eine breite, eher unregelmäßig geformte Krone bildend. Borke dick, hellgrau mit brauner Tönung und tiefen Rissen. Blätter ganzrandig oder schwach ausgerandet, an den Spitzen der Äste auch tief ausgerandet-gesägt, eiförmig-länglich bis lanzettlich. Zur Zeit des Laubaustriebes hellrot, ändert sich die Farbe im voll entfalteten Laub zu dunkel blaugrün. Blüten unauffällig, die männlichen in lang herabhängenden, die weiblichen in kurzen aufrechten Blütenständen. Eicheln meistens deutlich gestielt, seltener sitzend.

Vorkommen: Zusammen mit *Quercus agrifolia* auf niedrigen Hügeln der Küstengebirge Südwest-Kaliforniens.

Holz: Sehr schwer, hart, fest, brüchig und mit feiner Textur; problematisch zu trocknen. Kern dunkelbraun oder fast schwarz; Splint heller braun.

Nutzung: Gelegentlich als Feuerholz.

Description: arbre de 15 à 18 m de hauteur, avec un tronc de 0,6 à 0,9 m de diamètre. Angle de ramification proche de l'angle droit; branches vigoureuses, formant une couronne ample, plutôt irrégulière. Ecorce épaisse, d'un gris clair teinté de brun, creusée de profonds sillons. Feuilles entières ou à lobes très peu marqués, dentées et profondément lobées, ovoïdes à lancéolées à l'extrémité des rameaux, rouge clair au moment où elles se déploient, vert bleu sombre en pleine feuillaison. Fleurs peu marquantes, les mâles en longues inflorescences pendantes, les femelles en inflorescences courtes et dressées. Glands longuement pédonculés, rarement sessiles.

Habitat: basses collines des cordillères littorales du sud-ouest de la Californie en association avec *Quercus agrifolia*.

Bois: très lourd, dur, résistant, tendant à se fendre et à grain fin; très difficile à sécher. Bois de cœur brun foncé ou presque noir; aubier plus clair.

Intérêts: utilisé quelquefois comme bois de feu.

Engelmann Oak.

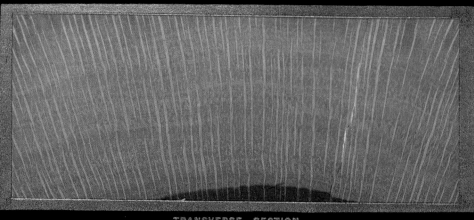

TRANSVERSE SECTION.

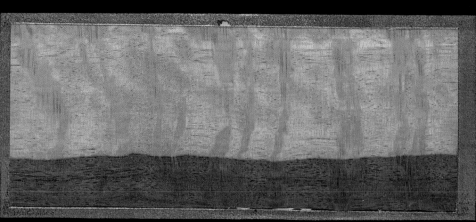

RADIAL SECTION.

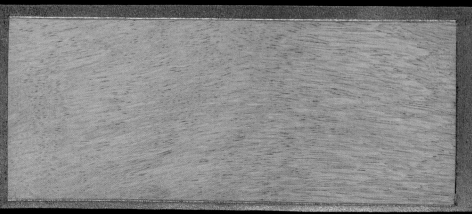

TANGENTIAL SECTION.

Ger. Eiche von Engelmann.   *Fr.* Chêne d'Engelmann.

Sp. Roble de Engelmann.

B | BEECH FAMILY
Golden Chestnut

FAMILIE DER BUCHENGEWÄCHSE
Goldblättrige Kastanie

FAMILLE DU CHÊNE ET DU HÊTRE
Chrysolepis doré

FAGACEAE
Castanopsis chrysophylla (Douglas ex Hook.) A. DC.

Description: 100 to 165 ft. (30 to 50 m) high, 4 (1.5 m) to more than 13 ft. (3 m) (!) in diameter. Trunk massive, buttressed. Often unbranched for first 85 ft. (25 m). Branches stocky. Measurements usually somewhat smaller. Shrubby growth at high altitudes or in the south. Bark deeply cut into rounded ridges, the width of two or three fingers, divided into thick plate-like scales, dark red-brown outside, pale red inside. Leaves shiny green on upperside, and with persistent golden-yellow scales beneath.

Habitat: Sporadic in coniferous woodland in the coastal states of Pacific North America.

Wood: Light, soft, not strong, close-grained. Heartwood tinged pale brown and red; sapwood thin and paler. Tannin-rich like the bark.

Use: Wood occasionally used for agricultural implements, plows, etc., in southern Oregon and northern California.

Seeds sweet and edible.

Beschreibung: 30 bis 50 m hoch bei 1,5 bis über 3 m (!) Durchmesser. Massiver Stamm, spannrückig. Häufig astfrei bis 25 m über Grund. Äste gedrungen. Maße im Allgemeinen erheblich kleiner. In größeren Höhen oder im Süden in Strauchform. Borke tief geteilt in zwei, drei Finger breite, abgerundete Riefen, quer gebrochen in dicke plattige Schuppen, außen dunkel rotbraun, innen hellrot. Blätter oberseits glänzend grün, unterseits mit bleibenden, goldgelben Schuppen besetzt.

Vorkommen: Als Einsprengsel in Nadelwäldern in den Küstenstaaten des pazifischen Nordamerika.

Holz: Leicht, weich, nicht fest und mit feiner Textur. Kern hellbraun und rot getönt; Splint dünn und heller. Wie die Rinde gerbstoffreich.

Nutzung: In Süd-Oregon und Nord-Kalifornien gelegentlich als Holz für Ackergerät, Pflüge u. a.

Samen süß und essbar.

Description : 30 à 50 m de hauteur, tronc massif, cannelé de 1,5 à 3 m (!) de diamètre, souvent non ramifié jusqu'à 25 m au-dessus du sol et aux branches trapues. Ses dimensions sont d'ordinaire bien inférieures ; il peut être plus grand ou pousser sous forme de buisson dans le sud. Ecorce brun rouge à fissures profondes et crêtes arrondies, de la largeur de deux ou trois doigts, qui se craquellent transversalement en épaisses écailles aplaties, révélant une couche inférieure rouge clair. Feuilles vert brillant dessus, couvertes d'écailles dorées persistantes dessous.

Habitat : disséminé dans les forêts de conifères des Etats côtiers de la côte pacifique nord-américaine.

Bois : léger, tendre, non résistant et à grain fin. Bois de cœur brun clair, teinté de rouge ; aubier mince et plus clair. Riche en tannins comme l'écorce.

Intérêts : utilisé parfois dans le sud de l'Oregon et le nord de la Californie pour fabriquer des instruments aratoires, des charrues etc.

Graines à chair douce et comestibles.

California Chinquapin, Evergreen Chestnut.

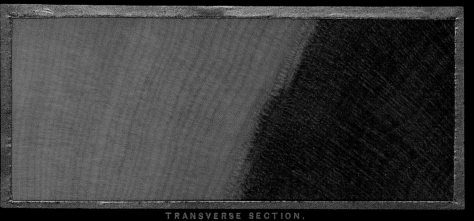

TRANSVERSE SECTION.

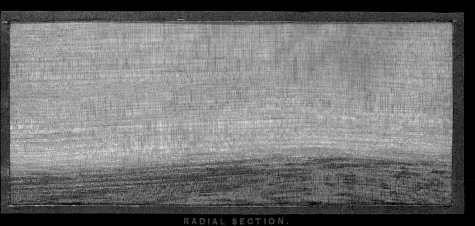

RADIAL SECTION.

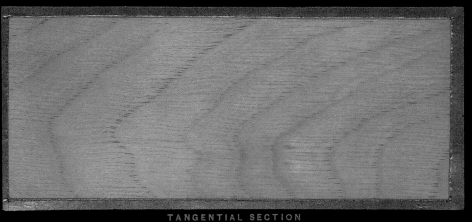

TANGENTIAL SECTION

Californianische Kastanie.   Fr. Châtaignier de Californie

Sp. Castaña de California.

B | BEECH FAMILY

Island Oak

FAMILIE DER BUCHENGEWÄCHSE
Insel-Stechpalme

FAMILLE DU CHÊNE ET DU HÊTRE
Chêne insulaire

FAGACEAE
Quercus tomentella Engelm.

Description: 35 to 45 (70) ft. (10 to 13 (20) m) high and 1 to 2 ft. (0.3 to 0.6 m) in diameter. Bark thin, red-brown, of large, closely appressed scales. Evergreen.

Habitat: Deep, narrow canyons and high, exposed slopes on the islands off the Californian coast.

Wood: Heavy, hard, silky-sheened, and close-grained. Heartwood pale yellow; sapwood paler.

Use: Unused owing to the inaccessibility of the habitat.

Beschreibung: 10 bis 13 (20) m hoch bei 0,3 bis 0,6 m Durchmesser. Borke dünn, rötlich braun, in großen, eng anliegenden Schuppen. Immergrün.

Vorkommen: Tiefe, enge Cañons und hoch gelegene, windexponierte Hänge auf Inseln vor der kalifornischen Küste.

Holz: Schwer, hart, seidig und mit feiner Textur. Kern blassgelb; Splint heller.

Nutzung: Aufgrund der Seltenheit und Schwerzugänglichkeit der Wuchsorte ungenutzt.

Description : arbre de 10 à 13 m de hauteur (20 m max.) sur un diamètre de 0,3 à 0,6 m. Ecorce mince, brun rougeâtre, recouverte de grandes écailles adhérentes. Essence sempervirente.

Habitat : canyons profonds et étroits, pentes élevées et exposées au vent sur les îles en face de la côte californienne.

Bois : lourd, dur, satiné et à grain fin. Bois de cœur jaune pâle ; aubier plus clair.

Intérêts : pas d'utilisation en raison de sa rareté et des terrains difficilement accessibles qu'il colonise.

Island Live Oak, Santa Catalina White Oak.

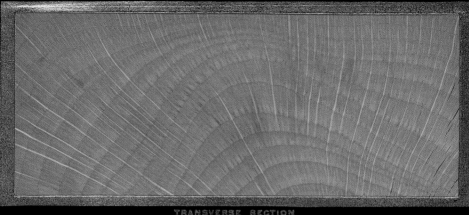

TRANSVERSE SECTION.

RADIAL SECTION.

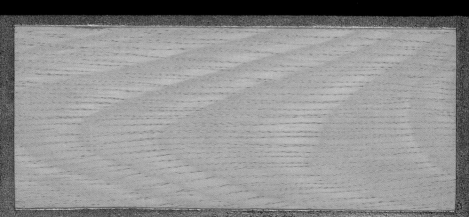

TANGENTIAL SECTION.

Ger. Eiländische Stechpalme. *Fr.* Chêne vert insulaire, *Sp.* Encina de isla.

B | BEECH FAMILY
Laurel Oak, Water Oak

FAMILIE DER BUCHENGEWÄCHSE
Lorbeer-Eiche

FAMILLE DU CHÊNE ET DU HÊTRE
Chêne à feuilles de laurier, Chêne laurier

FAGACEAE
Quercus laurifolia Michx.

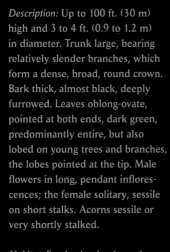

Description: Up to 100 ft. (30 m) high and 3 to 4 ft. (0.9 to 1.2 m) in diameter. Trunk large, bearing relatively slender branches, which form a dense, broad, round crown. Bark thick, almost black, deeply furrowed. Leaves oblong-ovate, pointed at both ends, dark green, predominantly entire, but also lobed on young trees and branches, the lobes pointed at the tip. Male flowers in long, pendant inflorescences; the female solitary, sessile on short stalks. Acorns sessile or very shortly stalked.

Habitat: Sandy riverbanks and swamps as well as rich hummocks in the neighborhood of the Atlantic and Gulf coasts of North America.

Wood: Heavy, very hard, very strong, and coarse-grained; perhaps unreliable during the seasoning process. Heartwood dark brown tinged with red; sapwood thick, paler. Broad, distinct medullary rays, as well as bands of small open pores along the annual rings.

Use: The wood is probably used only for firewood. As a street and park tree, it is one of the favorite woody plants in the towns along the coast.

Beschreibung: Bis 30 m hoch bei von 0,9 bis 1,2 m Durchmesser. Stamm groß und mit relativ schlanken Ästen, die eine dichte, breite, kugelige Krone bilden. Borke dick, fast schwarz, tief rissig. Blätter länglich oval, an beiden Enden zugespitzt, dunkelgrün, überwiegend ganzrandig, an jungen Bäumen und Ästen auch lappig mit zugespitzten Lappenenden. Männliche Blüten in lang herabhängenden Blütenständen, weibliche einzeln auf kurzen Stielchen sitzend. Eicheln sitzend oder sehr kurz gestielt.

Vorkommen: Sandige Flussufer und Sümpfe sowie reiche Hügel in der Nähe der Atlantik- und der Golfküste Nordamerikas.

Holz: Schwer, sehr hart, sehr fest und mit grober Textur; problematisch während der Trocknung. Kern dunkelbraun mit rotem Farbton; Splint dick, heller. Breite, deutliche Markstrahlen sowie Bänder kleiner, offener Gefäße längs der Jahresringe.

Nutzung: Das Holz vermutlich nur zur Feuerung. Als Straßen- und Parkbaum eines der beliebtesten Gehölze der Städte an den Küsten.

Description : jusqu'à 30 m de hauteur sur 0,9 à 1,2 m de diamètre. Branches relativement fines, constituant un houppier dense, ample et rond. Ecorce épaisse, presque noire, à texture profondément fissurée. Feuilles allongées-ovales, se terminant en pointe à la base et à l'apex, vert sombre, entières le plus souvent, mais également lobées et terminées en pointe sur les arbres et les branches juvéniles. Fleurs mâles en longs chatons pendants, fleurs femelles solitaires et portées par un court pédoncule. Glands sessiles ou brièvement pédonculés.

Habitat : berges sablonneuses et marécages ainsi que sur les collines riches à proximité de l'Atlantique et de la côte du golfe du Mexique.

Bois : lourd, très dur, très résistant et à grain fin ; difficile à sécher. Bois de cœur brun foncé teinté de rouge ; aubier épais, plus clair. Rayons médullaires larges et distincts, bandes de petits vaisseaux de type ouvert le long des cernes annuels.

Intérêts : vraisemblablement utilisé comme combustible. Un des arbres de parcs et d'alignement les plus appréciés dans les villes de la côte.

Laurel Oak.

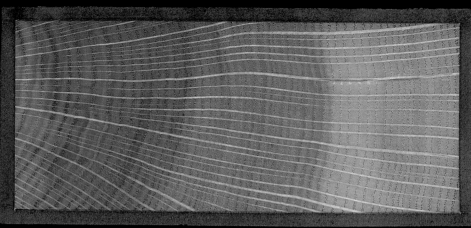

TRANSVERSE SECTION.

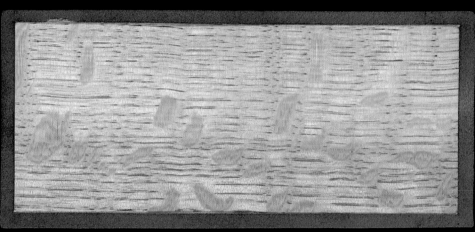

RADIAL SECTION.

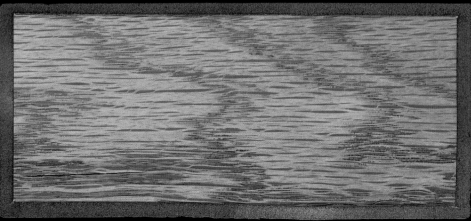

TANGENTIAL SECTION.

B | BEECH FAMILY
Live Oak

FAMILIE DER BUCHENGEWÄCHSE
Virginia-Eiche

FAMILLE DU CHÊNE ET DU HÊTRE
*Chêne de Virginie, Chêne de Caroline, Chêne des sables,
Chêne vert, Yeuse*

FAGACEAE
Quercus virginiana Mill.
Syn.: *Q. virens* Aiton

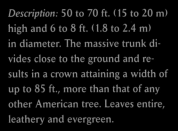

Description: 50 to 70 ft. (15 to 20 m) high and 6 to 8 ft. (1.8 to 2.4 m) in diameter. The massive trunk divides close to the ground and results in a crown attaining a width of up to 85 ft., more than that of any other American tree. Leaves entire, leathery and evergreen.

Habitat: Good soils near the coast in the south and east of North America. Often planted in the southern Atlantic states for its majestic shape. There it is often colonized by the lichen-like *Tillandsia usneoides* ("Spanish Moss").

Wood: Very heavy and very hard; difficult to work.

Use: Shipbuilding.

Once protected by Congress for shipbuilding.

Beschreibung: 15 bis 20 m hoch bei 1,8 bis 2,4 m Durchmesser. Der massive Stamm teilt sich dicht über dem Grund und gibt der Krone eine Breite von bis zu 25 m, mehr als sonst ein amerikanischer Baum aufweist. Blätter ganzrandig, ledrig und immergrün.

Vorkommen: Küstennahe, gute Böden im Süden und Osten Nordamerikas sowie auf Kuba. Häufig der majestätischen Wuchsform wegen in den atlantischen Südstaaten als Schmuckbaum gepflanzt. Dort auch häufig Träger der bartflechtenähnlichen *Tillandsia usneoides* („Spanish Moss").

Holz: Sehr schwer und sehr hart; schwer zu bearbeiten.

Nutzung: Schiffbau.

Einst für den Schiffbau durch den Kongress geschützt.

Description : arbre de 15 à 20 m de hauteur, avec un tronc massif de 1,8 à 2,4 m de diamètre, se divisant dès la base et constituant un houppier ample qui peut atteindre 25 m de largeur, plus que tout autre arbre américain. Feuilles entières, coriaces et persistantes.

Habitat : proximité des côtes, bons sols dans le sud et l'est de l'Amérique du Nord ainsi qu'à Cuba. Fréquemment planté en ornement pour son port majestueux dans les Etats du Sud du littoral atlantique. Il y est aussi souvent l'hôte d'une plante épiphyte, *Tillandsia usneoides*, la tillandsie usnoïde ou «mousse d'Espagne».

Bois : très lourd, et très dur ; difficile à travailler.

Intérêts : utilisé dans la construction navale.

Espèce autrefois protégée par le Congrès américain pour la construction navale.

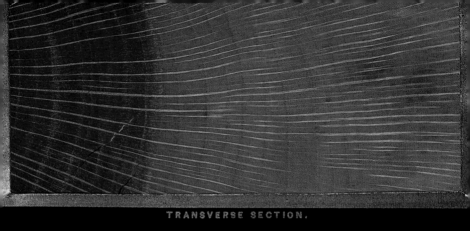

TRANSVERSE SECTION.

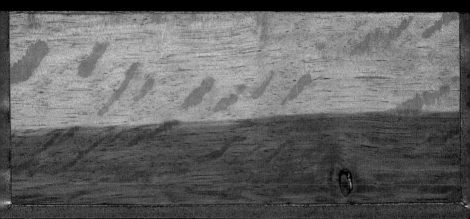

RADIAL SECTION.

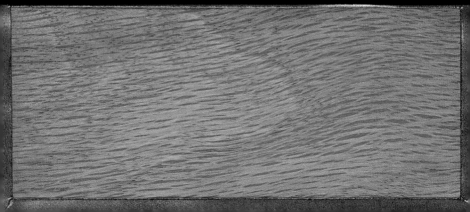

B | BEECH FAMILY
McDonald Oak

FAMILIE DER BUCHENGEWÄCHSE
MacDonald-Eiche

FAMILLE DU CHÊNE ET DU HÊTRE
Chêne broussailleux, Chêne broussailleux de Californie

FAGACEAE
Quercus dumosa Nutt.
Syn.: *Q. x macdonaldii* Greene

Description: Bush, 6 to 8 ft. (1.8 to 2.4 m) high with sturdy trunk. Often forms thickets. Occasionally (in sheltered canyons) as 25 to 35 ft. (7.5 to 9.0 m) high tree. Bark pale gray. Leaf shape very variable. Leaves ovate to oblong, entire, finely or coarsely serrated, hairy; mostly pale below. Flowers as leaves emerge: male inflorescences long and drooping; female short and erect. Acorns single, sessile, or short-stalked.

Habitat: Mountain slopes and canyons in California.

Wood: Little known. Heartwood pale brown to pale brown with pink tinge; sapwood whitish to brownish pink (Bärner 1962).

Use: Only as firewood owing to small diameter (Bärner, 1962).

Beschreibung: 1,8 bis 2,4 m hoher Busch mit kräftigen Stämmchen. Oft Dickichte bildend. Gelegentlich (in geschützt liegenden Cañons) als 7,5 bis 9 m hoher Baum. Borke blassgrau. Blattform extrem variabel. Blätter oval bis länglich, ganzrandig, fein oder grob gesägt, behaart, Unterseite meist blass. Blüte zur Zeit der Laubentfaltung, männliche Blüten in lang herabhängenden, weibliche in kurzen, aufrechten Blütenständen. Eicheln einzeln, sitzend oder kurz gestielt.

Vorkommen: Berghänge und Cañons in Kalifornien.

Holz: Kaum untersucht. Kern hellbräunlich bis blassbraun mit rosa Tönung; Splint weißlich bis bräunlich rosa (Bärner 1962).

Nutzung: Wegen des geringen Durchmessers nur als Brennholz (Bärner 1962).

Description : arbrisseau de 1,8 à 2,4 m de hauteur, avec un petit tronc vigoureux, constituant souvent des fourrés. Quelquefois (dans les canyons abrités par exemple) sous forme arborescente de 7,5 à 9,0 m de hauteur. Ecorce gris pâle. Feuilles extrêmement variables : ovales à allongées, entières, finement ou grossièrement dentées, pubescentes, face inférieure pâle le plus souvent. La floraison se réalise en même temps que les feuilles se déploient, les fleurs mâles en longues inflorescences pendantes, les femelles en inflorescences courtes et dressées. Glands isolés, sessiles ou courtement pédonculés.

Habitat : versants de montagnes et canyons de Californie.

Bois : peu étudié jusqu'ici. Bois de cœur brunâtre clair à brun clair, teinté de rose ; aubier blanchâtre à rose brunâtre (Bärner 1962).

Intérêts : utilisé uniquement comme bois de chauffage en raison de son faible diamètre (Bärner 1962).

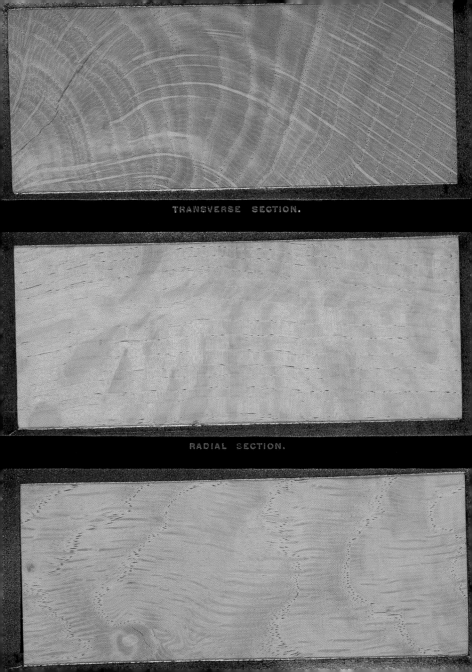

TRANSVERSE SECTION.

RADIAL SECTION.

B | BEECH FAMILY
Oregon White Oak

FAMILIE DER BUCHENGEWÄCHSE
Garrys Eiche

FAMILLE DU CHÊNE ET DU HÊTRE
Chêne de l'Oregon, Chêne blanc de Garry

FAGACEAE
Quercus garryana Douglas ex Hook.

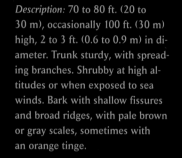

Description: 70 to 80 ft. (20 to 30 m), occasionally 100 ft. (30 m) high, 2 to 3 ft. (0.6 to 0.9 m) in diameter. Trunk sturdy, with spreading branches. Shrubby at high altitudes or when exposed to sea winds. Bark with shallow fissures and broad ridges, with pale brown or gray scales, sometimes with an orange tinge.

Habitat: Dry, gravely soils in valleys or slopes of low hills in Pacific North America.

Wood: Hard, often extremely tough, strong and close-grained. Heartwood pale brown or yellow; sapwood thin, almost white.

Use: Used for carts and wagons in Washington and Oregon; for cupboards, shipbuilding, and cooperage, and in large quantities as firewood.

The most important oak in Pacific North America.

Beschreibung: 20 bis 23 m, gelegentlich 30 m hoch bei 0,6 bis 0,9 m Durchmesser. Gedrungener Stamm mit ausgebreiteten Ästen. In Berghöhen oder unter Windwirkung vom Ozean nur als kleiner Strauch. Borke in flachen Rillen und breiten Riefen, mit hellbraunen oder grauen Schuppen, manchmal orangefarben getönt.

Vorkommen: Trocken-kiesige Böden in Tälern, Hänge kleiner Hügel im pazifischen Nordamerika.

Holz: Hart, häufig außerordentlich zäh, fest und mit feiner Textur. Kern hellbraun oder gelb; Splint dünn, nahezu weiß.

Nutzung: In Washington und Oregon für Kutschen und Waggons, für Schränke, Schiffbau und Böttcherei, in großem Umfang als Brennholz.

Die wichtigste Eiche im pazifischen Nordamerika.

Description: arbre de 20 à 23 m de hauteur, atteignant parfois 30 m, sur un diamètre de 0,6 à 0,9 m. Tronc trapu portant des branches étalées. Petit buisson en altitude ou sous l'action des vents marins. Ecorce creusée de sillons plats à larges crêtes écailleuses brun clair ou grises, quelquefois orangées.

Habitat: sols graveleux et secs en plaine, versants de petites collines sur la façade pacifique de l'Amérique du Nord.

Bois: dur, souvent extrêmement tenace, résistant et à grain fin. Bois de cœur brun clair ou jaune; aubier mince, quasiment blanc.

Intérêts: utilisé dans le Washington et l'Oregon en charronnage, en tonnellerie, dans la construction navale et pour la fabrication d'armoires, à grande échelle comme combustible

Le plus important des chênes de la façade pacifique de l'Amérique

Oregon Oak, Mountain White Oak.

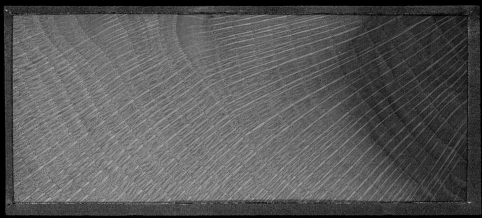

TRANSVERSE SECTION.

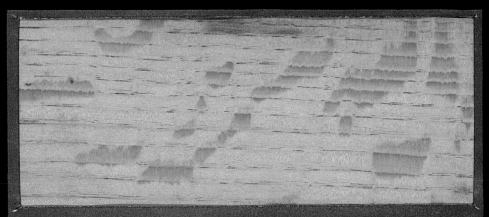

RADIAL SECTION.

TANGENTIAL SECTION

Ger. Oregonische Eiche.   Fr. Chêne de Oregon.

Sp. Roble de Oregon.

B | BEECH FAMILY
Overcup Oak, Swamp White Oak

FAMILIE DER BUCHENGEWÄCHSE
Leier-Eiche

FAMILLE DU CHÊNE ET DU HÊTRE
*Chêne à feuilles lyrées, Chêne à feuilles en lyre,
Chêné blanc aquatique*

FAGACEAE
Quercus lyrata Walt.

Description: Up to 100 ft. (30 m) high and 2 to 3 ft. (0.6 to 0.9 m) in diameter. Trunk mostly branching at a height of 15 to 20 ft. (4.5 to 6.0 m), forming a round, symmetrical crown or a broad, more irregular one. Bark thick, deeply furrowed, light gray, sometimes with a strong red tinge. Leaves ovate-oblong, deeply divided into five to nine lobes, at first bronze-green, later dark green, usually silvery white and hairy on the underside. Male inflorescences pendant; the female flowers solitary, surrounded by a dense felt of hairs. Acorns sessile or on short stalks.

Habitat: Swamps, hollows wet all year, on rich low-lying soils in eastern North America.

Wood: Heavy, hard, strong, and tough; very durable in the ground. Heartwood rich dark brown; sapwood thick, paler. Broad, distinct medullary rays, and bands of one to three rows of large, open pores along the annual rings.

Use: Shipbuilding, railroad sleepers, carriages, agricultural implements, cabinets, fences, firewood. Introduced into England in 1786, but

Beschreibung: Bis 30 m hoch bei 0,6 bis 0,9 m Durchmesser. Stamm verzweigt sich meist in 4,5 bis 6 m Höhe, eine symmetrisch-kugelige oder breitere, unregelmäßigere Krone bildend. Borke dick, tief rissig, hellgrau mit z. T. kräftiger roter Tönung. Blätter oval-länglich, tief geteilt in 5 bis 9 Lappen, anfangs bronzegrün, später dunkelgrün, Unterseite behaart, meist silbrig weiß. Männliche Blütenstände hängend, weibliche Blüten einzeln, von dichtem Haarfilz umgeben. Eicheln sitzend oder auf kurzen Stielen.

Vorkommen: Sümpfe, ganzjährig nasse Mulden in reichen Tieflandböden im östlichen Nordamerika.

Holz: Schwer, hart, fest und stark; im Boden sehr dauerhaft. Kern tief dunkelbraun; Splint dick, heller. Breite, deutliche Markstrahlen und Bänder von 1 bis 3 Reihen großer, offener Gefäße längs der Jahresringe.

Nutzung: Schiffbau, Bahnschwellen, Kutschen, Landwirtschaftsgeräte, Kabinette, Zäune, Feuerholz. 1786 nach England eingeführt, nur selten kultiviert.

Description : 30 m de hauteur maximum, avec un tronc de 0,6 à 0,9 m de diamètre, se ramifiant en général entre 4,5 et 6,0 m, et un houppier soit globuleux et symétrique, soit plus large et plus irrégulier. Ecorce épaisse, profondément fissurée, gris clair, teintée par endroit de rouge intense. Feuilles ovales-allongées, formées de 5 à 9 lobes très marqués, vert bronze jeunes puis vert foncé, à face inférieure duvetée et ordinairement blanc argenté. Inflorescences mâles pendantes, fleurs femelles solitaires, couvertes d'une épaisse pubescence. Glands sessiles ou brièvement pédonculés.

Habitat : marécages, dépressions mouillées dans des sols planitiaires riches (est de l'Amérique du Nord).

Bois : lourd, dur, résistant et fort ; très durable sur pied. Bois de cœur d'un brun profond ; aubier épais, plus clair. Rayons médullaires larges, distincts et alignements de 1 à 3 faisceaux vasculaires de type ouvert le long des cernes annuels.

Intérèts : construction navale, traverses de chemin de fer, calèches, outils, ébénisterie, clôtures et combustible. Introduit en Angleterre en

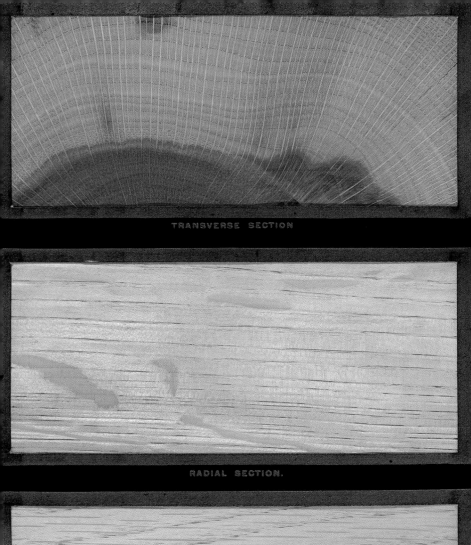

TRANSVERSE SECTION

RADIAL SECTION.

B | BEECH FAMILY

Pin Oak, Swamp Oak

FAMILIE DER BUCHENGEWÄCHSE
Sumpf-Eiche

FAMILLE DU CHÊNE ET DU HÊTRE
Chêne des marais, Chêne à épingles

FAGACEAE
Quercus palustris Münchh.

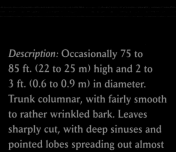

Description: Occasionally 75 to 85 ft. (22 to 25 m) high and 2 to 3 ft. (0.6 to 0.9 m) in diameter. Trunk columnar, with fairly smooth to rather wrinkled bark. Leaves sharply cut, with deep sinuses and pointed lobes spreading out almost horizontally. Fall color reddish.

Habitat: Deep, rich soils in low-lying ground, margins of ponds and swamps in association with Black Gum, Sweet Gum, Red Maple, Swamp Cottonwood, American Hornbeam, etc., but also thrives, when transplanted, on drier soils. Interior of eastern North America.

Wood: Heavy, hard, strong, coarse-grained. Heartwood light brown; sapwood thin, somewhat darker.

Use: For interior construction, roof shingles, staves of barrels, and other purposes in cooperage.

First described by Otto von Münchhausen in the last third of the 18th century (palace park at Schwöbber near Hamelin, Germany). Since then it has frequently been planted as a street tree.

Beschreibung: Gelegentlich 22 bis 25 m hoch bei 0,6 bis 0,9 m Durchmesser. Säulenförmiger Stamm mit ziemlich glatter bis eher runzliger Borke. Scharf geschnittene Blätter mit tiefen Buchten und fast waagerecht abstehenden, spitz endenden Lappen. Herbstfärbung rötlich.

Vorkommen: Tiefgründiger, reicher Boden im Tiefland, Teichufer und Ränder von Sümpfen in Gesellschaft mit Wald-Tupelo, Amberbaum, Rot-Ahorn, Sumpf-Pappel, Amerikanischer Hainbuche u. a., kommt aber auch, wenn verpflanzt, auf trockeneren Böden gut voran. Im Inneren des östlichen Nordamerika.

Holz: Schwer, hart, fest, mit grober Textur, hellbraun, Splint dünn, etwas dunkler.

Nutzung: Für Innenausbau, Dachschindeln, Fassdauben und sonst in der Böttcherei.

Zuerst im letzten Drittel des 18. Jh. von Otto von Münchhausen beschrieben (Schlosspark von Schwöbber bei Hameln). Seither auch häufiger Straßenbaum.

Description : arbre s'élevant quelquefois entre 22 et 25 m, avec un tronc columnaire de 0,6 à 0,9 m de diamètre. Ecorce plutôt lisse à ridée. Feuilles très découpées, à lobes mucronés, séparés par des sinus profonds. Beau feuillage rouge en automne.

Habitat : sols riches et profonds des plaines, berges des étangs et abords des marais, en association avec le nyssa sylvestre, le liquidambar, l'érable rouge, le peuplier des marais, le charme etc., mais, replanté, supporte aussi très bien des sols plus secs. On le trouve dans les régions intérieures de l'est de l'Amérique du Nord.

Bois : lourd, dur, résistant, à grain grossier. Bois de cœur brun clair ; aubier mince, un peu plus foncé.

Intérêts : utilisé en menuiserie intérieure, pour fabriquer des bardeaux, des douves et autres articles de tonnellerie.

Décrit pour la première fois par Otto von Münchhausen vers la fin du 18e siècle (parc du château de Schwöbber près de Hamelin en Allemagne). Arbre de plantations routières depuis lors.

Pin Oak, Swamp Spanish Oak, Water Oak.

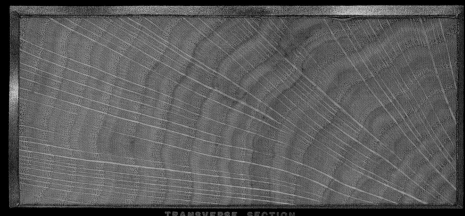

TRANSVERSE SECTION.

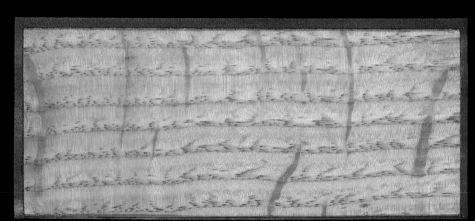

RADIAL SECTION.

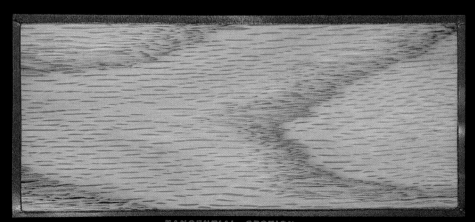

TANGENTIAL SECTION.

Ger. Sumpf-Eiche. Fr. Chêne marécageaux. Sp. Roble pantor

B | BEECH FAMILY
Post Oak, Iron Oak

FAMILIE DER BUCHENGEWÄCHSE
Pfahl-Eiche

FAMILLE DU CHÊNE ET DU HÊTRE
Chêne étoilé, Chêne à lobes obtus

FAGACEAE
Quercus stellata Wangenh.
Syn.: *Q. minor* (Marshall) Sarg.

Description: 50 to 70 ft. (15 to 20 m), in the Ohio basin apparently up to 100 ft. (30 m) high, and about 3 ft. (0.9 m) in diameter. Leaves shiny dark green and rough above, gray-green with stellate hairs beneath; the lobes, few in number, and broad towards the top, and the deep, rounded sinuses between them cause the leaves to resemble a cross. At the ends of the twigs the leaves are arranged in star-shaped clusters.

Habitat: On limestone uplands and sandy plains in eastern North America in association with Black Jack, Red, White, and other oaks, Sassafras, tupelos, Flowering Dogwood, Eastern Red Cedar, etc.

Wood: Heavy, hard, and durable.

Use: For agricultural implements, barrels, and furniture, also used on a large scale for railroad sleepers and fence posts, etc. Excellent as firewood.

Beschreibung: 15 bis 20 m, im Ohio-Becken angeblich bis 30 m hoch bei etwa 0,9 m Durchmesser. Blätter oberseits glänzend dunkelgrün, dabei rau, unterseits graugrün mit Sternhaaren. Durch wenige tiefe, abgerundete Buchten und wenige, vorn stumpf verbreiterte Lappen etwa kreuzförmig. An den Zweigenden in Sternform gedrängt.

Vorkommen: Höhenzüge aus Kalkstein und sandige Ebenen im östlichen Nordamerika, in Gesellschaft von Maryland-Eiche („Black Jack"), Rot-, Weiß- und anderen Eichen, Sassafras, Tupelos, Blumen, Hartriegel, Rotzeder u. a.

Holz: Schwer, hart und dauerhaft.

Nutzung: Für landwirtschaftliches Gerät, Fässer, für Möbel sowie in großem Umfang für Eisenbahnschwellen und Zaunpfähle usw. Ausgezeichnetes Brennholz.

Description : arbre de 15 à 20 m de hauteur, qui atteindrait dans le bassin de l'Ohio 30 m de haut sur un diamètre de 0,9 m environ. Feuillage vert foncé, luisant et rugueux à la face supérieure, vert gris et recouvert d'un duvet étoilé au revers. Feuilles découpées en lobes grossiers atténués en coin à la base, à sinus peu profonds et arrondis, leur donnant un aspect cruciforme, et disposées en étoile à l'extrémité des rameaux.

Habitat : chaînes de montagnes calcaires et plaines sableuses de l'est de l'Amérique du Nord, en association avec le chêne de Maryland («Black Jack»), divers chênes comme le rouge et le blanc, le sassafras, les nyssas, le cornouiller à fleurs et le cèdre rouge etc.

Bois : lourd, dur et durable.

Intérêts : employé en tonnellerie, en ameublement, pour fabriquer des outils agricoles et en grande partie pour les traverses de chemin de fer et les piquets etc. Excellent bois de chauffage.

Post Oak, Iron Oak.

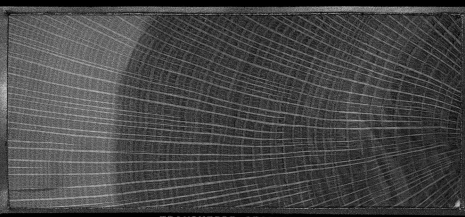

TRANSVERSE SECTION.

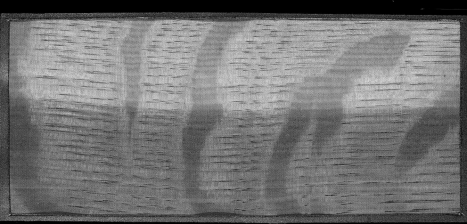

RADIAL SECTION.

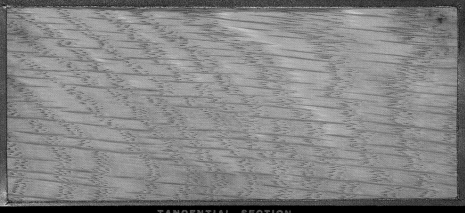

TANGENTIAL SECTION.

B | BEECH FAMILY
Red Oak

FAMILIE DER BUCHENGEWÄCHSE
Rot-Eiche

FAMILLE DU CHÊNE ET DU HÊTRE
Chêne rouge, Chêne rouge d'Amérique

FAGACEAE
Quercus rubra L.

Description: About 100 ft. (30 m), sometimes to 175 ft. (50 m), high and 3 to 4 ft. (0.9 to 1.2 m) in diameter, or even more. Trunk columnar, quick-growing.

Habitat: Rich uplands, well-drained slopes and riverbanks in association with White Pine and Red Pine, Aspen, Balsam Poplar, Red Maple, etc., in (north-)eastern North America.

Wood: Heavy, hard, strong, and coarse-grained; shrinks as it dries. Of lower quality than the wood of the White Oak. Heartwood pale or reddish brown; sapwood darker.

Use: Wooden buildings, interior construction, furniture.

The hardest and most adaptable of all the species of oak. The commonest of all the American oaks, with a wide distribution to the north.

Beschreibung: Etwa 30 m, aber manchmal bis zu 50 m hoch bei 0,9 bis 1,2 m Durchmesser, auch mehr. Säulenförmiger Stamm, schnellwüchsig.

Vorkommen: Reiche Hochlagen, gut drainierte Hänge und Flussufer in Gesellschaft von Weiß-Kiefer und Rot-Kiefer, Espe, Balsam-Pappel, Rot-Ahorn u. a. im (nord-) östlichen Nordamerika.

Holz: Schwer, hart, fest und von grober Textur; schwindet beim Trocknen. Qualitativ geringer als das Holz der Weiß-Eiche. Kern hell- oder rötlichbraun; Splint dunkler.

Nutzung: Holzbauten, Innenausbau, Möbel.

Härteste und anpassungsfähigste aller Eichenarten. Häufigste aller amerikanischen Eichen mit weiter Verbreitung nach Norden.

Description : arbre d'environ 30 m de hauteur, mais pouvant atteindre 50 m sur un diamètre de 0,9 à 1,2 m, parfois davantage. Tronc columnaire, à croissance rapide.

Habitat : sols riches des hautes terres, pentes bien drainées et bords des cours d'eau, en compagnie des pins blanc et rouge, du tremble, du peuplier baumier, de l'érable rouge etc. dans le (nord-) est de l'Amérique du Nord.

Bois : lourd, dur, résistant et à grain grossier ; retrait au séchage. De qualité inférieure à celle du chêne blanc. Bois de cœur rougeâtre ; aubier plus foncé.

Intérêts : utilisé pour fabriquer des charpentes, en menuiserie intérieure et en ameublement.

Le plus rustique et le plus plastique de tous les chênes. C'est aussi le plus répandu des chênes américains et celui qui se rencontre le plus au nord.

Red Oak.

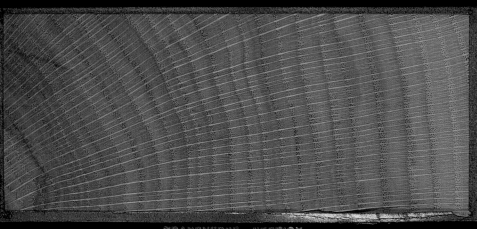

TRANSVERSE SECTION.

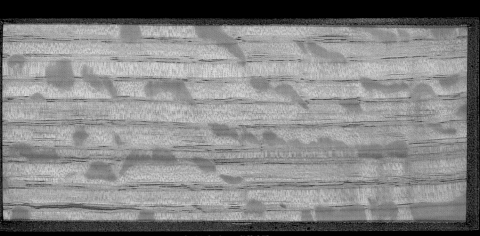

RADIAL SECTION.

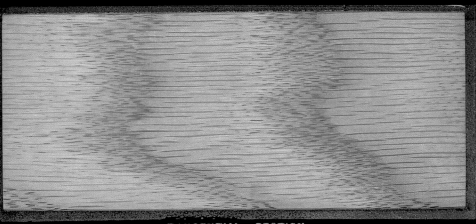

TANGENTIAL SECTION.

Ger. Roth Eiche. Fr. Chêne rouge. Sp. Roble rojo.

B | BEECH FAMILY
Scarlet Oak

FAMILIE DER BUCHENGEWÄCHSE
Scharlach-Eiche

FAMILLE DU CHÊNE ET DU HÊTRE
Chêne écarlate, Chêne cocciné

FAGACEAE
Quercus coccinea Münchh.

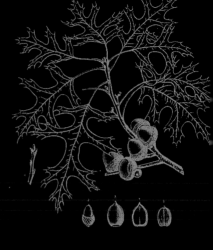

Description: 75 to 85 ft. (23 to 25 m) high and 2 to 3 ft. (0.6 to 0.9 m) in diameter. Bark dark brown divided up into flat ridges. Inner side of bark reddish. Leaves with distinctly deep sinuses between the three or four lateral, pointed lobes. Famous for its brilliant autumn color of scarlet and other shades of red, which lasts for a long time.

Habitat: An abundant tree. Grows best on damp, sandy soils and gravely slopes in association with Red, Black and other oaks, the Pignut Hickory and the Shagbark Hickory, Sweet Birch, Eastern Red Cedar, etc., in the interior of (north-)eastern North America.

Wood: Not differentiated in commerce and not differently used from the Red Oak, but it has fewer branches.

Use: As for the Red Oak.

A very valuable park tree. Considered to be the most beautiful of the red oaks.

Beschreibung: 23 bis 25 m hoch bei 0,6 bis 0,9 m Durchmesser. Borke dunkelbraun mit flachen Riefen. Innenseite der Rinde rötlich. Laub mit auffallend tiefen Buchten und 3 oder 4 seitlich abstehenden, zugespitzten Lappen. Berühmt wegen der brillanten Herbstfärbung in Scharlachrot und anderen Tönungen von Rot, die lange anhält.

Vorkommen: Als häufiger Baum gern auf feuchten Sandböden und kiesigen Hängen in Gesellschaft von Rot-, Schwarz- und anderen Eichen, der Ferkelnuss-Hickory und der Schindelborkigen Hickory, der Zucker-Birke, der Rot-Zeder u. a. im Inneren des (nord-) östlichen Nordamerika.

Holz: Von dem der Rot-Eiche im Handel und in der Nutzung nicht unterschieden, jedoch weniger Äste.

Nutzung: Wie bei der Rot-Eiche.

Sehr wertvoller Parkbaum. Gilt als schönste der Rot-Eichen.

Description : 23 à 25 m de hauteur sur un diamètre de 0,6 à 0,9 m. Ecorce brun foncé, se fissurant en crêtes aplaties qui laissent apparaitre dessous une écorce rouge. Feuilles profondément découpées en 3 ou 4 lobes parallèles à dents aiguës. Célèbre pour ses couleurs somptueuses en automne : son feuillage prend des tons de rouge et d'écarlate, qui persistent longtemps.

Habitat : arbre commun, ne craignant pas les sols sablonneux humides et les pentes caillouteuses des régions intérieures du (nord-) est de l'Amérique du Nord, en compagnie des chênes rouge et noir et d'autres essences de chêne, du caryer ovale, du caryer glabre, du noyer blanc, du bouleau flexible et du cèdre rouge etc.

Bois : aucune différence dans le commerce avec celui du chêne rouge, son emploi est le même mais il a moins de branches.

Intérêts : identiques à ceux du chêne rouge.

Arbre de parcs très précieux. Considéré comme le plus beau des chênes rouges.

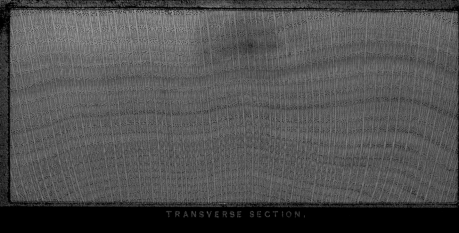

TRANSVERSE SECTION.

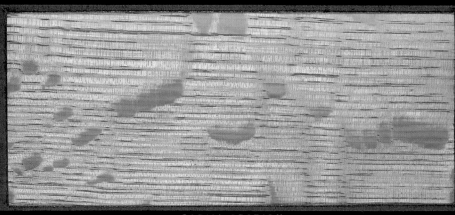

RADIAL SECTION.

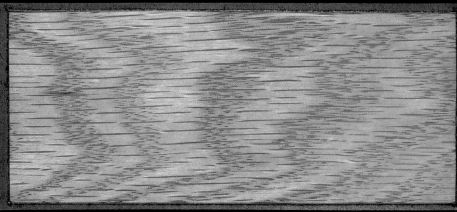

TANGENTIAL SECTION

Ger. Scharlach Eiche. Sp. Roble colorado.

B | BEECH FAMILY
Shingle Oak, Laurel Oak

FAMILIE DER BUCHENGEWÄCHSE
Schindeleiche

FAMILLE DU CHÊNE ET DU HÊTRE
Chêne imbriqué, Chêne à feuilles de laurier du nord, Chêne à lattes

FAGACEAE
Quercus imbricaria Michx.

Description: 50 to 60 ft. (15 to 18 m) high and 3 ft. (0.9 m) in diameter. Of broadly pyramidal growth when young, later with a narrow, round, open crown. Bark of young trees thin, light brown and smooth, on older trees thick and deeply furrowed, brown with a reddish tinge. Leaves oblong-lanceolate or oblong-ovate, usually entire, hairy beneath. When they unfold they are bright red, later becoming yellow-green and finally dark green. In fall the foliage is very colorful. Flowers insignificant, the male in long, pendant inflorescences, the female erect on small stalks. Acorns almost globular in shape, with a short stalk, attached singly or in pairs.

Habitat: Rich uplands in the interior of eastern North America, occasionally fertile river plains.

Wood: Heavy, hard, and rather coarse-grained; perhaps unreliable during the seasoning process. Heartwood light brown tinged with red; sapwood thin, paler.

Use: One of the most beautiful North American oaks, and cultivated in a large number of parks and gardens in North America.

Beschreibung: 15 bis 18 m hoch bei 0,9 m Durchmesser. Jung von breit pyramidalem Wuchs, später mit schmaler, runder, offener Krone. Borke junger Bäume dünn, hellbraun und glatt, die älterer dick und tief rissig, braun mit rötlicher Tönung. Blätter länglich-lanzettlich oder länglich oval, meist ganzrandig, Unterseite behaart. Bei Laubentfaltung hellrot, später gelbgrün und schließlich dunkelgrün. Im Herbst mit äußerst buntem Laub. Blüten unauffällig, männliche in lang herabhängenden Blütenständen, weibliche aufrecht auf kleinen Stielchen. Eicheln von fast kugeliger Form, mit kurzem Stiel, einzeln oder zu zweit.

Vorkommen: Reiche Hochländer, im Inneren des östlichen Nordamerika, gelegentlich fruchtbare Tiefebenen längs der Flüsse.

Holz: Schwer, hart und mit eher grober Textur; problematisch während der Trocknung. Kern hellbraun, rot getönt; Splint dünn, heller.

Nutzung: Eine der schönsten Eichen Nordamerikas, in zahlreichen Parks und Gärten Nordamerikas kultiviert.

Description: arbre de 15 à 18 m de hauteur, avec un tronc de 0,9 m de diamètre. Port pyramidal à l'état juvénile, constituant ensuite une couronne étroite, ronde et ouverte. Ecorce mince, brun clair et lisse sur les jeunes sujets, épaisse, profondément fissurée et brun rougeâtre sur les arbres plus âgés. Feuilles lancéolées-allongées ou ovales-allongées, entières le plus souvent, pubescentes au revers, rouge clair à leur naissance, puis vert jaune et enfin vert foncé. Feuillage aux coloris très variés en automne. Fleurs insignifiantes, les mâles en chatons pendants, les femelles dressées et portées par des pédicelles. Glands presque sphériques, à court pédoncule, isolés ou par deux.

Habitat: hautes terres riches dans l'est à l'intérieur de l'Amérique du Nord, quelquefois aussi plaines basses fertiles le long des rivières.

Bois: lourd, dur et à grain plutôt grossier; difficile à sécher. Bois de cœur brun clair teinté de rouge; aubier mince, plus clair.

Intérêts: un des plus beaux chênes d'Amérique du Nord, cultivé dans de nombreux parcs et jardins.

Shingle Oak.

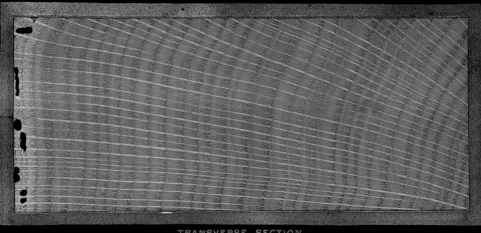

TRANSVERSE SECTION.

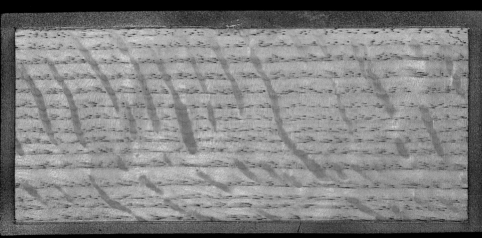

RADIAL SECTION.

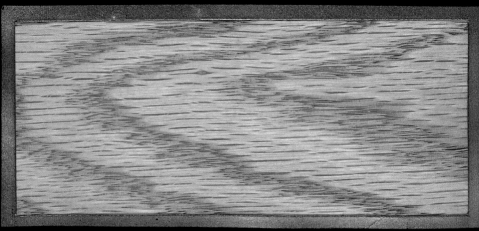

TANGENTIAL SECTION.

Ger. Schindel-Eiche. *Fr.* Chêne de bardeau. *Sp.* Roble de ripa

B | BEECH FAMILY
Sierra Live Oak

FAMILIE DER BUCHENGEWÄCHSE
Wislicenus-Eiche

FAMILLE DU CHÊNE ET DU HÊTRE
Chêne de l'intérieur

FAGACEAE
Quercus wislizenii A. DC.

Description: 75 to 85 ft. (23 to 26 m) high and 4 to 6 ft. (1.2 to 1.8 m) in diameter. Trunk short and stocky, with spreading branches, sometimes smaller, shrubby. Bark with broad, rounded, often interconnected ridges with small, thick, closely appressed, dark brown scales, lightly tinged red. Young bark thin, smooth and paler. Leaves dark green and shiny.

Habitat: The bottoms of slopes, or, as a shrub, the walls of canyons. Rather rare, sometimes found together with the shrubby oaks of the canyons and deserts.

Wood: Heavy, very hard and close-grained. Heartwood pale brown with red tinge; sapwood thick, paler.

Use: Firewood, here not distinguished from *Quercus agrifolia.*

Beschreibung: 23 bis 26 m hoch bei 1,2 bis 1,8 m Durchmesser. Kurzer, gedrungener Stamm mit ausgebreiteten Ästen, zuweilen kleiner, strauchig. Borke aus breiten, abgerundeten, oft verbundenen Riefen mit kleinen, dicken, dicht anliegenden, dunkelbraunen Schuppen, leicht rot getönt. Junge Borke dünn, glatt und heller. Blätter dunkelgrün und glänzend.

Vorkommen: Hangfüße oder (Strauchform) Wände der Cañons in Kalifornien. Eher selten, teilweise mit der Strauch-Eiche der Cañons und der Wüsten.

Holz: Schwer, sehr hart und mit feiner Textur. Kern hellbraun mit rot; Splint dick, heller.

Nutzung: Als Brennholz, hier nicht unterschieden von *Quercus agrifolia.*

Description : arbre de 23 à 26 m de hauteur, avec un tronc court, trapu de 1,2 à 1,8 m de diamètre, portant des branches étalées ; parfois plus petit, voire buissonnant. Ecorce à fissures et à larges crêtes arrondies, souvent entrelacées et recouvertes de petites écailles épaisses, adhérentes, brun rougeâtre foncé. Feuillage vert sombre et brillant.

Habitat : bas de pentes ou versants de canyons (forme buissonnante) de Californie. Plutôt rare, associé quelquefois au chêne des canyons et des déserts.

Bois : lourd, très dur et à grain fin. Bois de cœur brun clair lavé de rouge ; aubier épais et plus clair.

Intérêts : bois de chauffage ; aucune différence avec celui de *Quercus agrifolia.*

Highland Live Oak.

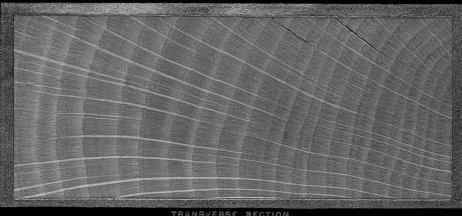

TRANSVERSE SECTION

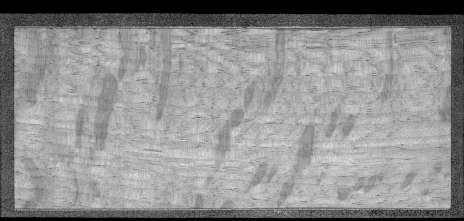

RADIAL SECTION.

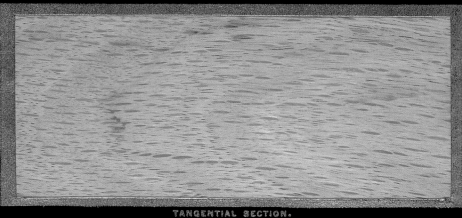

TANGENTIAL SECTION.

Ger. Hochländische Stechpalme. *Fr.* Chêne vert montagneux.

Sp. Encina montañesa.

B | BEECH FAMILY
Spanish Oak, Southern Red Oak

FAMILIE DER BUCHENGEWÄCHSE
Spanische Eiche

FAMILLE DU CHÊNE ET DU HÊTRE
Chêne digité, Chêne à feuilles en faucille, Chêne d'Espagne

FAGACEAE
Quercus falcata Michx.
Syn.: *Q. digitata* Sudw.

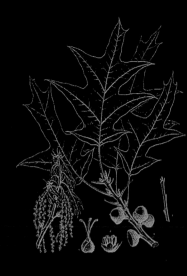

Description: 75 to 85 (-100) ft. (21 to 24 (-30) m) high and 2 to 3 (5) ft. (0.6 to 0.9 (-1.5) m) in diameter. Branches stout, forming a broad, open, round crown. Bark thick, dark brown tinged red or pale, with shallow fissures. Leaves oblong to ovate, three to five-lobed, the lobes (especially those nearer the tip) elongated, of very varied shape, often broadened towards the apex, with pale hairs on the lower surface. Flowering simultaneously with the unfolding of the leaves, the male flowers in long, pendent inflorescences, the female solitary, sessile.

Habitat: Dry, rocky uplands, and sandy, uncultivated ground, but occasionally also on the rich, flooded land of the lower reaches of rivers. Eastern North America.

Wood: Hard, strong, and close-grained; not reliable during the seasoning process and not durable in the ground. Heartwood light red; sapwood paler.

Use: Used to a great extent as firewood, also occasionally as building timber. A very popular shade-tree for houses and streets.

Beschreibung: 21 bis 24 (bis 30) m hoch bei 0,6 bis 0,9 (bis 1,5) m Durchmesser. Äste kräftig, aufrecht, eine breite, offene und runde Krone formend. Borke dick, dunkelbraun mit roter Tönung oder blass, flachrissig. Blätter länglich bis oval, drei- bis fünflappig, die Lappen (besonders der vordere) verlängert, von sehr variabler Form, vorne oft verbreitert, blass behaart auf der Unterseite. Blüte zur Zeit der Laubentfaltung, männliche Blüten in lang herabhängenden Blütenständen, weibliche einzeln sitzend.

Vorkommen: Trockene, steinige Hochländer und sandige Ödländer, gelegentlich aber auch auf den reichen Überschwemmungsböden der Flussunterläufe. Im östlichen Nordamerika.

Holz: Hart, fest und mit feiner Textur; problematisch während der Trocknung und im Boden nicht dauerhaft. Kern hellrot; Splint heller.

Nutzung: In großem Umfang als Feuerholz, gelegentlich auch als Bauholz. Sehr beliebter Schattenbaum für Häuser und Straßen.

Description: arbre de 21 à 24 m de hauteur (30 m max.), avec un tronc de 0,6 à 0,9 m de diamètre (1,5 m max.), armé de branches vigoureuses et dressées, constituan un houppier large, ouvert et rond. Ecorce épaisse, brun sombre teinté de rouge ou pâle, à fissures plates. Feuilles allongées à ovales, découpées en 3 à 5 lobes prolongés (surtout le lobe antérieur), de forme très variable, souvent élargies à la base, avec une pubescence pâle au revers. Floraison au moment de la feuillaison ; fleurs mâles en chatons pendants, fleurs femelles solitaires.

Habitat: hautes terres caillouteuses et sèches, terrains incultes sablonneux, mais parfois aussi sols inondés et riches du cours inférieur des fleuves. Croît dans l'est de l'Amérique du Nord.

Bois: dur, résistant et à grain fin ; difficile à sécher et non durable sur pied. Bois de cœur rouge clair ; aubier plus clair.

Intérêts: largement employé comme bois de chauffage, quelquefois aussi comme bois d'œuvre. Arbre d'ombrage très apprécié pour les maisons et en alignement.

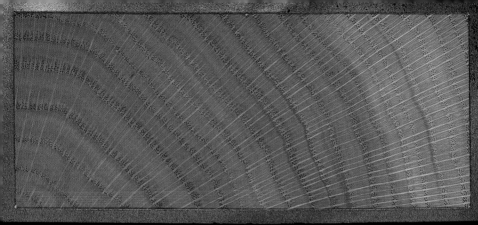

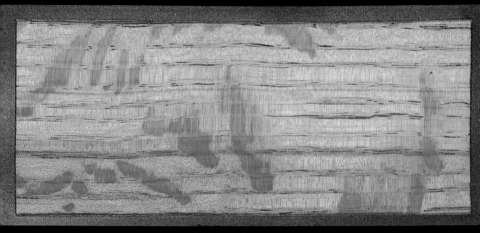

TRANSVERSE SECTION.

RADIAL SECTION.

B | BEECH FAMILY
Swamp White Oak

FAMILIE DER BUCHENGEWÄCHSE
Sumpf-Weiß-Eiche

FAMILLE DU CHÊNE ET DU HÊTRE
Chêne bicolore, Chêne bleu (Canada)

FAGACEAE
Quercus bicolor Willd.

Description: 70 to 85 ft. (20 to 25 m) high and 2 to 3 ft. (0.6 to 0.9 m) in diameter. More than 100 ft. (30 m) in height when growing in forest and surrounded by other trees. On the other hand, trunk when free-standing likely to be 6 to 8 ft. (1.8 to 2.4 m) in diameter. Crown with many lower branches drooping. Upper part of trunk shrouded in hanging branches. Bark light, separating into scales. Leaves shiny dark green above, whitish and felted beneath.

Habitat: Moist and lime-free, low-lying ground in association with Red and Silver Maple, King-nut Hickory, Sweet Gum, Black Gum, Overcup Oak, Swamp Oak and Burr Oak, Green and Black Ash, etc., in (north-)eastern North America but excluding the states on the Gulf of Mexico.

Wood: Heavy, hard, and tough.

Use: Like the White Oak. The two are not differentiated commercially.

Beschreibung: 20 bis 25 m hoch bei 0,6 bis 0,9 m Durchmesser. Wenn im Wald von anderen Bäumen bedrängt, Höhe auch über 30 m. Durchmesser des Stammes im Freistand größer: 1,8 bis 2,4 m. Krone mit zahlreichen gedrehten Ästen. Das obere Stammende von hängenden Zweigen eingehüllt. Borke hell, schuppig ablösend. Blätter oberseits glänzend dunkelgrün, unterseits weißlich und filzig.

Vorkommen: Feuchte und kalkfreie Niederungen im Tiefland in Gesellschaft von Rot- und Silber-Ahorn, Königsnuss-Hickory, Amberbaum, Wald-Tupelo, Leierblättriger Eiche, Sumpf-Eiche und Klettenfrüchtiger Eiche, Grün- und Schwarz-Esche u. a. im (nord-) östlichen Nordamerika ohne die Golfstaaten.

Holz: Schwer, hart und zäh.

Nutzung: Wie die der Weißen Eiche. Im Handel nicht unterschieden.

Description : 20 à 25 m de hauteur sur un diamètre de 0,6 à 0,9 m. S'élève jusqu'à 30 m lorsqu'il se trouve en compétition avec d'autres arbres pour la lumière ; à l'état dispersé, le diamètre du tronc peut atteindre 1,8 à 2,4 m. Houppier ample, avec de nombreuses branches tortueuses. La partie terminale du tronc porte des rameaux retombants. Ecorce claire, se détachant en larges écailles. Feuilles lustrées, vert foncé dessus, pubescentes et blanchâtres dessous.

Habitat : zones basses et humides des plaines, sur substrat non calcaire, dans le (nord-) est de l'Amérique du Nord, sans tenir compte des Etats proches du Golfe, en association avec les érables rouge et argenté, le caryer lacinié, le liquidambar, le nyssa sylvestre, les chênes à feuilles lyrées, des marais et à gros fruits, les frênes noir et rouge etc.

Bois : lourd, dur et tenace.

Intérêts : identiques à ceux du chêne blanc. Aucune différence dans le commerce.

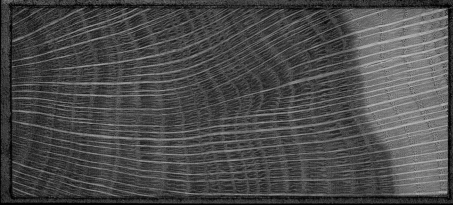

TRANSVERSE SECTION.

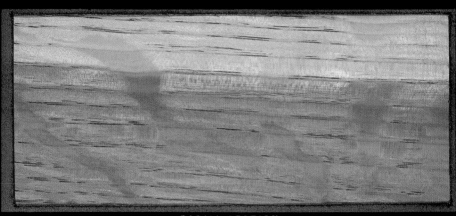

RADIAL SECTION.

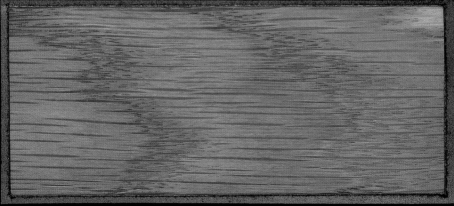

TANGENTIAL SECTION

Ger. Zweifarbige Eiche. Sp. Roble blanco de pantano.

Fr. Chêne de Marais.

B | BEECH FAMILY
Tanbark Oak

FAMILIE DER BUCHENGEWÄCHSE
Dichtblütige Eiche

FAMILLE DU CHÊNE ET DU HÊTRE
Lithocarpe de Californie

FAGACEAE
Lithocarpus densiflora (Hook. & Arn.)
Syn.: *Quercus densiflora* Hook. & Arn.

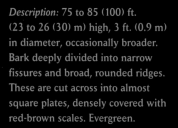

Description: 75 to 85 (100) ft.
(23 to 26 (30) m) high, 3 ft. (0.9 m)
in diameter, occasionally broader.
Bark deeply divided into narrow
fissures and broad, rounded ridges.
These are cut across into almost
square plates, densely covered with
red-brown scales. Evergreen.

Habitat: South-western North
America. Less common and smaller
further north and south.

Wood: Hard, strong, brittle, and
close-grained. Heartwood pale red-
brown; sapwood thick, dark brown.

Use: Little value as timber, mostly as firewood. Bark rich in tannin,
used for tanning leather, as are
all other oaks from Pacific North
America. Overharvested, but
sprouts well.

Attains a very attractive shape in
western North America. Interest-
ingly similar oaks occur in Asia.

Beschreibung: 23 bis 26 (30) m hoch
bei 0,9 m Durchmesser, vereinzelt
jedoch breiter. Borke tief geteilt in
enge Rillen und breite, abgerundete Riefen. Diese quer gebrochen in
fast quadratische Platten, mit dicht
anliegenden, rotbraunen Schuppen
bedeckt. Immergrün.

Vorkommen: (Süd-) Westliches
Nordamerika. Weiter nordwärts
und im Süden weniger häufig
und jeweils kleiner.

Holz: Hart, fest, spröde und mit feiner Textur. Kern hell rötlich braun.
Splint dick, dunkelbraun.

Nutzung: Meist Brennmaterial.
Rinde ist reicher an Gerbstoff
zum Gerben von Leder als alle
anderen Eichen im pazifischen
Nordamerika. Übernutzt, aber
Stockausschlag.

Bildet sehr pittoreske Wuchsform
im westlichen Nordamerika. Pflan-
zengeografisch interessant, da ähnliche Eichen in Asien vorkommen.

Description: 23 à 26 m (30 m) de
hauteur sur un diamètre de 0,9 m,
mais plus large en situation isolée.
Ecorce creusée d'étroits sillons
à larges côtes arrondies, cassant
transversalement en plaques
appliquées brun rouge et à section
presque rectangulaire. Feuilles
persistantes.

Habitat: (sud-) ouest de l'Amérique du Nord. Moins répandu, et
plus petit, vers le nord et au sud.

Bois: dur, résistant, fragile et
à grain fin. Bois de cœur brun
rougeâtre clair; aubier épais,
brun foncé.

Intérêts: bois de construction de
faible valeur, bois de chauffage
le plus souvent. Ecorce plus riche
en tannins, dont on se sert pour
tanner le cuir, que tous les autres
chênes de l'ouest de l'Amérique
du Nord. Exploitation excessive
mais il rejette bien de souche.

Présente une forme de croissance
très pittoresque dans l'ouest du
sous-continent nord-américain.
Espèce intéressante en géographie
botanique car on trouve des chênes
similaires en Asie.

Tan-bark Oak, Evergreen Chestnut Oak.

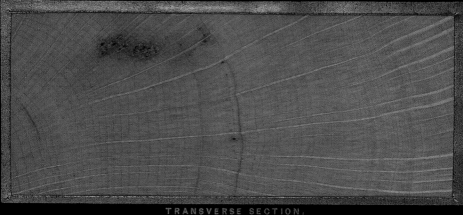

TRANSVERSE SECTION.

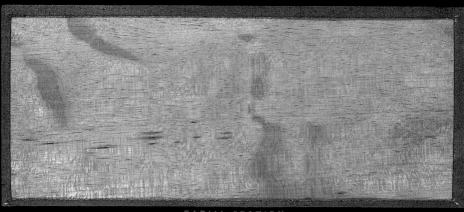

RADIAL SECTION.

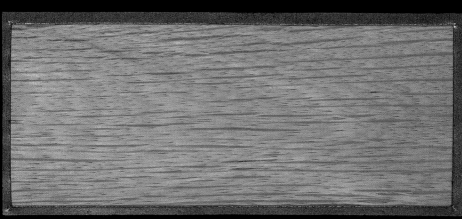

TANGENTIAL SECTION

Eiche mit ditchen Blumen.　　　　𝔉𝔷. Chêne à fleurs de

B | BEECH FAMILY
Texas Red Oak

FAMILIE DER BUCHENGEWÄCHSE
Südliche Roteiche

FAMILLE DU CHÊNE ET DU HÊTRE
Chêne de Nutall

FAGACEAE
Quercus texana Buckley

Description: Up to 200 ft. (60 m) high and 7 to 8 ft. (2.1 to 2.4 m) in diameter. Trunk thick, free of branches up to 90 ft. (27 m), with a greatly enlarged base. Branches relatively small, forming a narrow, open crown. Bark thick, light brown with a red tinge, deeply furrowed. Leaves ovate, deeply divided into seven to nine (rarely only five) lobes, with distinctly pointed tips, at first bright red, soon light green. Male inflorescences long, pendent; the female solitary on short stalks. Acorns solitary, sessile, or on short stalks.

Habitat: In the interior of eastern North America. On damp, low-lying soils near rivers, at times forming a considerable part of the lowland forest, but also on low limestone hills.

Wood: Heavy, hard, and close-grained. Heartwood light red-brown. A few distinct medullary rays, and bands of small pores along the annual rings. The description applies to a tree from Texas, but the wood from the Mississippi region is considered to be of higher quality.

Use: Building timber, also found as a street tree in New Orleans

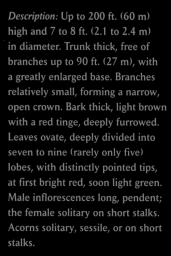

Beschreibung: Bis zu 60 m hoch bei 2,1 bis 2,4 m Durchmesser. Stamm dick, bis zu 27 m Höhe astfrei, mit stark verdickter Basis. Äste relativ klein, eine schmale, offene Krone bildend. Borke dick, hellbraun mit roter Tönung, tief rissig. Blätter oval, tief geteilt in 7 bis 9 (selten nur 5) Lappen mit deutlich zugespitzten Enden, anfangs hellrot, bald hellgrün. Männliche Blütenstände lang herabhängend, weibliche einzeln auf kurzen Stielen. Eicheln einzeln sitzend oder auf kurzen Stielen.

Vorkommen: Auf feuchten Tieflandböden in der Nähe von Flüssen, z.T. wesentliche Teile des Tieflandwaldes stellend, aber auch auf niedrigen Kalkhügeln. Im Inneren des östlichen Nordamerika.

Holz: Schwer, hart und mit feiner Textur. Kern hell rotbraun. Wenige deutliche Markstrahlen und Bänder kleiner Gefäße längs der Jahresringe. Die Beschreibung bezieht sich auf einen Baum aus Texas, das Holz aus dem Mississippigebiet soll von höherer Qualität sein.

Nutzung: Bauholz, in New Orleans auch als Straßenbaum.

Description : jusqu'à 60 m de hauteur sur un diamètre de 2,1 à 2,4 m de diamètre. Tronc épais, non ramifié jusqu'à 27 m de hauteur, base très renflée et branches assez petites, houppier rond et étroit. Ecorce épaisse, brun clair teinté de rouge, à texture profondément fissurée. Feuilles ovales, composées de 7 à 9 lobes profonds (rarement 5), terminées en pointe aiguë, rouge clair à leur naissance, devenant très vite vert clair. Fleurs mâles en longs chatons pendants, fleurs femelles solitaires et courtement pédonculées. Glands isolés, sessiles ou sur un court pédoncule

Habitat : plaines humides près des cours d'eau, parfois élément dominant de la forêt planitiaire; aussi petites collines calcaires. Intérieur de l'Amérique du Nord à l'est.

Bois : lourd, dur et à grain fin. Bois de cœur brun rouge clair. Quelque rayons médullaires distincts et alignements de petits vaisseaux le long des cernes annuels. Description basée sur celle d'un arbre du Texas ; le bois serait de meilleure qualité dans le Mississippi.

Intérêts : construction ; arbre de parcs à la Nouvelle Orléans

Southern Red Oak. Schneck's Oak.

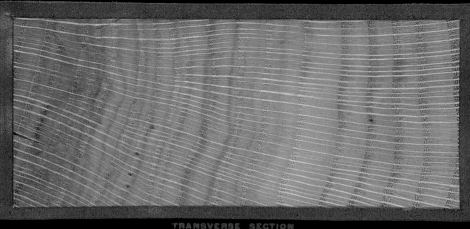

TRANSVERSE SECTION

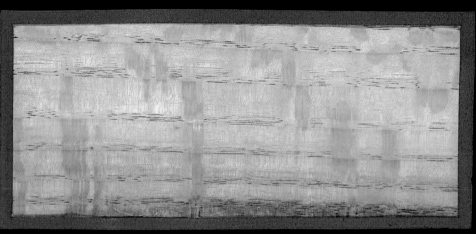

RADIAL SECTION.

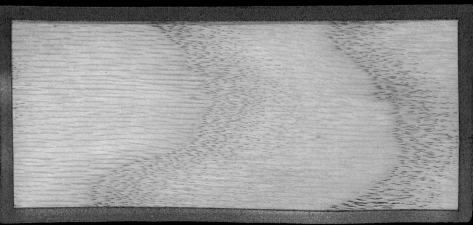

TANGENTIAL SECTION.

Ger. Südliche Rothe Eiche. *Fr.* Chêne rouge du sud.

B | BEECH FAMILY
Water Oak, Possum Oak, Duck Oak

FAMILIE DER BUCHENGEWÄCHSE
Wasser-Eiche

FAMILLE DU CHÊNE ET DU HÊTRE
Chêne noir, Chêne d'eau, Chêne possum

FAGACEAE
Quercus nigra L.

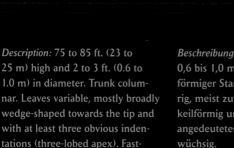

Description: 75 to 85 ft. (23 to 25 m) high and 2 to 3 ft. (0.6 to 1.0 m) in diameter. Trunk columnar. Leaves variable, mostly broadly wedge-shaped towards the tip and with at least three obvious indentations (three-lobed apex). Fast-growing.

Habitat: Damp valleys or nearby hills in (south-)eastern North America.

Wood: Heavy and hard.

Use: Little used except for firewood and charcoal.

Often planted as a shade tree on streets and in parks in the southern states. Easy to grow and transplant. In central Europe needs protection when young.

Beschreibung: 23 bis 25 m hoch bei 0,6 bis 1,0 m Durchmesser. Säulenförmiger Stamm. Verschiedenblättrig, meist zur Spitze hin verbreitert, keilförmig und mit wenigstens drei angedeuteten Buchten. Schnellwüchsig.

Vorkommen: Feuchte Niederungen und angrenzende Anhöhen im (süd-) östlichen Nordamerika.

Holz: Schwer und hart.

Nutzung: Wenig genutzt außer für Brennholz und Holzkohle.

In den Südstaaten häufig als Schattenbaum an Straßen und in Stadtparks. Gut anzupflanzen und leicht zu verpflanzen. In Mitteleuropa in der Jugend schutzbedürftig.

Description : arbre de 23 à 25 m de hauteur, avec un tronc columnaire de 0,6 à 1,0 m de diamètre. Feuille variables, généralement obovales, à base cunéiforme, avec trois lobes grossiers. Croissance rapide.

Habitat : zones basses humides ou hauteurs proches dans le (sud-) est de l'Amérique du Nord.

Bois : lourd et dur.

Intérêts : peu utilisé sauf comme bois de chauffage et bois de charbon.

Arbre d'ombrage courant dans les Etats du Sud le long des routes et dans les parcs. Se plante et se transplante facilement. Lorsque l'arbre est jeune, il nécessite une protection en Europe centrale.

TRANSVERSE SECTION.

RADIAL SECTION.

TANGENTIAL SECTION

_r._ Wasser-Eiche. _Fr._ Chêne aquatique. _Sp._ Roble acuati

B | BEECH FAMILY
White Oak

FAMILIE DER BUCHENGEWÄCHSE
Weiß-Eiche

FAMILLE DU CHÊNE ET DU HÊTRE
Chêne blanc

FAGACEAE
Quercus alba L.

Description: 85 to 100 ft. (25 to 30 m) high, under favorable conditions up to 170 ft. (50 m) in height, and 3 to 4 ft. (0.9 to 1.2 m) or even 5 ft. (1.5 m) in diameter. Trunk short and compact, with strong branches growing out horizontally. Bark strikingly light. A symbol of strength and steadfastness.

Habitat: On low-lying ground by rivers and in rich soils on higher ground, if not too damp, in association with other trees in eastern North America. Sometimes forms forests by itself. Was the first to be felled in the "great clearing" of the New England states.

Wood: Very heavy and hard, tough, strong, and close-grained; durable in the ground, but warps if not carefully seasoned.

Use: One of the most valuable commercial woods of North America, even being exported to Europe (staves for barrels). For wooden structures of all kinds: shipbuilding, carts and coaches, cooperage, agricultural implements, railroad sleepers, fences, interior construction, furniture, baskets, excellent as firewood.

Beschreibung: 25 bis 30 m, unter günstigen Bedingungen bis 50 m hoch bei 0,9 bis 1,2 m bzw. 1,5 m Durchmesser. Der Stamm wirkt kurz und gedrungen, starke, horizontal abgehende Äste. Borke auffallend hell. Symbol der Stärke und Festigkeit.

Vorkommen: In Flussniederungen wie auf reichen, höher gelegenen Böden, wenn nicht zu feucht, in Gesellschaft anderer Bäume im östlichen Nordamerika. Manchmal selbst waldbildend. Fiel in den Neuenglandstaaten dem „great clearing" zuerst zum Opfer.

Holz: Sehr schwer und hart, zäh, fest und mit feiner Textur; dauerhaft im Boden, aber verzieht sich, wenn nicht sorgfältig gelagert.

Nutzung: Eines der wertvollsten Nutzhölzer Nordamerikas, auch für den Export nach Europa (Fassdauben). Für Holzbauten aller Art, Schiffbau, Wagenbau, Böttcherei, Ackergerät, Eisenbahnschwellen, Zäune, Innenausbau, Möbel, Körbe, ausgezeichnetes Brennholz.

Description: 25 à 30 m de hauteur, en situation favorable jusqu'à 50 m sur un diamètre de 0,9 à 1,2 m, voire 1,5 m. Tronc paraissant court et trapu, avec de grosses branches horizontales. Ecorce claire. Symbol de puissance et de résistance.

Habitat: aussi bien lits de hautes eaux que sols riches situés plus en hauteur mais pas trop humides, en association avec d'autres arbres dans l'est de l'Amérique du Nord. Constitue parfois des peuplements forestiers. Première victime du « great clearing » en Nouvelle-Angleterre.

Bois: très lourd et dur, tenace, résistant et à grain fin; durable sur pied et gauchissant s'il n'est pas entreposé de manière appropriée.

Intérêts: un des bois utiles les plus précieux d'Amérique du Nord, exporté également vers l'Europe (douves de tonneaux). Usages multiples: construction navale, carrosserie, tonnellerie, instruments aratoires, traverses de chemin de fer, palissades, menuiserie intérieure, meubles, paniers; excellent bois de chauffage.

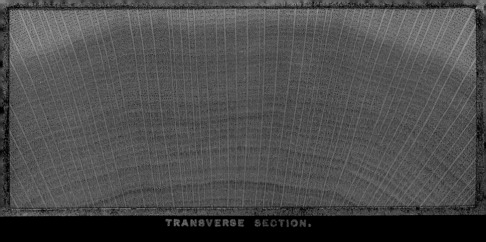

TRANSVERSE SECTION.

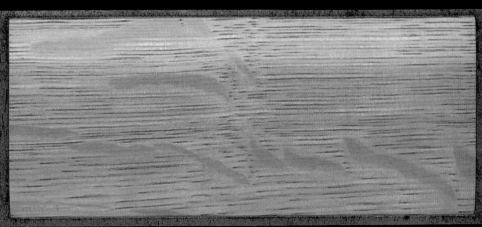

RADIAL SECTION.

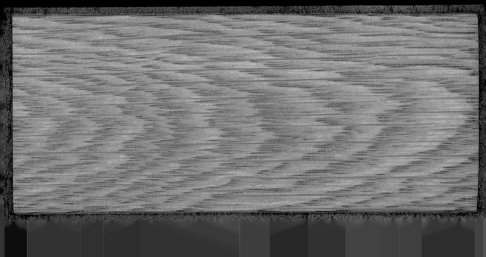

B | BEECH FAMILY
White-leaf Oak

FAMILIE DER BUCHENGEWÄCHSE
Mexikanische Weideneiche

FAMILLE DU CHÊNE ET DU HÊTRE
Chêne à feuilles argentées

FAGACEAE
Quercus hypoleucoides A. Camus
Syn.: *Q. hypoleuca* Engelm.

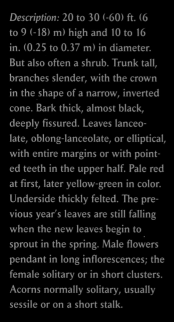

Description: 20 to 30 (-60) ft. (6 to 9 (-18) m) high and 10 to 16 in. (0.25 to 0.37 m) in diameter. But also often a shrub. Trunk tall, branches slender, with the crown in the shape of a narrow, inverted cone. Bark thick, almost black, deeply fissured. Leaves lanceolate, oblong-lanceolate, or elliptical, with entire margins or with pointed teeth in the upper half. Pale red at first, later yellow-green in color. Underside thickly felted. The previous year's leaves are still falling when the new leaves begin to sprout in the spring. Male flowers pendant in long inflorescences; the female solitary or in short clusters. Acorns normally solitary, usually sessile or on a short stalk.

Habitat: In pinewoods on the slopes of canyons and ridges in southwestern North America. On high land it grows between 5,500 and 6,500 ft. (1,800 to 2,100 m) above sea level, at lower altitudes it forms only a shrub.

Wood: Heavy, hard, very strong, and close-grained.

Use: None known, presumably occasionally used as fuel.

Beschreibung: 6 bis 9 (bis 18) m hoch bei 0,25 bis 0,38 m Durchmesser. Häufig aber auch als Strauch. Stamm hoch, Äste schlank, eine schmale, umgekehrt konische Krone bildend. Borke dick, fast schwarz, tief rissig. Blätter lanzettlich, länglich-lanzettlich oder elliptisch, ganzrandig oder in der vorderen Hälfte spitz gesägt. Anfangs hellrot, später gelbgrün gefärbt. Unterseite mit Haarfilz. Laubfall der Vorjahresblätter allmählich während des neuen Laubaustriebes im Frühjahr. Männliche Blüten in langen Blütenständen herabhängend, weibliche einzeln oder in kurzen Blütenständen. Eicheln normalerweise einzeln, meist sitzend oder auf kurzem Stiel.

Vorkommen: In Kiefernwäldern an den Hängen von Cañons und Graten im südwestlichen Nordamerika. In Höhen zwischen 1800 bis 2100 m, in tieferen Lagen nur als Busch.

Holz: Schwer, hart, sehr fest und mit feiner Textur.

Nutzung: Nicht bekannt, vermutlich gelegentlich als Feuerholz.

Description: 6 à 9 m de hauteur (18 m max.) sur un diamètre de 0,25 à 0,38 m, souvent aussi à port de buisson. Tronc élancé avec des branches minces, formant une couronne étroite s'inscrivant dans un cône renversé. Ecorce épaisse, presque noire, profondément fissurée. Feuilles lancéolées, allongées-elliptiques ou elliptiques, entières ou à dents aiguës sur la première moitié du limbe, rouge clair devenant vert jaune, pubescentes dessous, ne tombant progressivement qu'au cours de la feuillaison suivante. Inflorescences mâles longues, pendantes, fleurs femelles solitaires ou en inflorescences courtes. Glands isolés et sessiles en général ou courtement pédonculés.

Habitat: forêts de conifères sur les versants de canyons et les arêtes, entre 1800 et 2100 m dans le sud-ouest de l'Amérique du Nord; sous forme de buisson à une altitude inférieure.

Bois: lourd, dur, très résistant et à grain fin.

Intérêts: inconnus; probablement utilisé à l'occasion comme bois de chauffage.

White-leaf Oak.

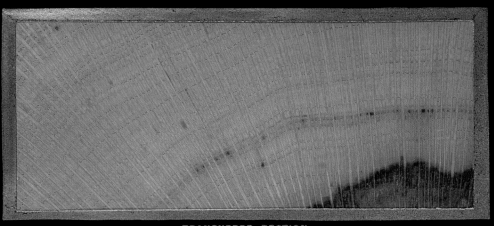

TRANSVERSE SECTION.

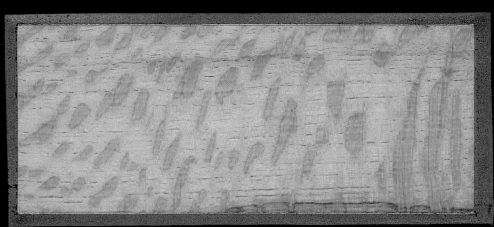

RADIAL SECTION.

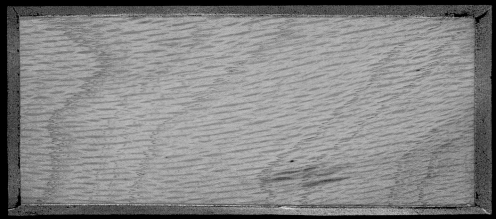

TANGENTIAL SECTION.

Ger. Weiszblätterige Eiche. *Fr.* Chêne à feuilles blanches.
Sp. Roble de hojas blancas.

B | BEECH FAMILY
Willow Oak

FAMILIE DER BUCHENGEWÄCHSE
Weideneiche

FAMILLE DU CHÊNE ET DU HÊTRE
Chêne saule, Chêne à feuilles de saule

FAGACEAE
Quercus phellos L.

Description: 75 to 85 ft. (21 to 24 m) high and 2 to 4 ft. (0.6 to 1.2 m) in diameter, often considerably smaller. Branches slender, forming a relatively narrow and open or dense and rounded crown. Bark thick, light reddish brown and smooth, but on old trees slightly fissured. Leaves linear-lanceolate, pointed at both ends, smooth, usually entire, yellow-green at first, later light green above and paler beneath. Male flowers in long, pendent inflorescences, the female solitary or in pairs erect on short stalks. Acorns shortly stalked or almost sessile, solitary or in pairs.

Habitat: Damp margins of swamps and rivers, and rich, sandy upland soils in the states on the Atlantic and Gulf coasts.

Wood: Heavy, not hard, strong, and rather coarse-grained. Heartwood light brown, tinged red; sapwood thin, paler.

Use: Cultivated in Europe since 1724, but only rarely planted since that time although it is a quick-growing and attractive tree.

Beschreibung: 21 bis 24 m hoch bei 0,6 bis 1,2 m Durchmesser, oft wesentlich kleiner. Äste schmal, eine relativ schmale und offene oder kugelig-geschlossene Krone bildend. Borke dick, hell rotbraun und glatt, nur im Alter schwach rissig. Blätter linealisch-lanzettlich, an beiden Enden zugespitzt, glatt, meist ganzrandig. Zunächst gelbgrün, später hellgrün auf der Oberseite, blasser auf der Unterseite. Männliche Blüten in lang herabhängenden Blütenständen, weibliche einzeln oder zu zweien aufrecht auf kurzen Stielchen. Eicheln kurz gestielt oder fast sitzend, einzeln oder in Paaren.

Vorkommen: Feuchte Ränder von Sümpfen und Flüssen, reiche sandige Hochlandböden in den Staaten an den Küsten des Atlantiks und des Golfes von Mexiko.

Holz: Schwer, nicht hart, fest und mit eher grober Textur. Kern hellbraun, rot getönt; Splint dünn, heller.

Nutzung: In Europa seit 1724 kultiviert, seither aber nur selten angepflanzt, obwohl es sich um einen raschwüchsigen, attraktiven Baum handelt.

Description: 21 à 24 m de hauteur sur un diamètre de 0,6 à 1,2 m, souvent beaucoup plus petit. Branches minces, formant un houppier relativement étroit et ouvert ou globuleux et fermé. Ecorce épaisse, brun rouge clair et lisse, se fissurant seulement avec l'âge. Feuilles linéaires-lancéolées, pointues aux deux extrémités, lisses, entières le plus souvent, vert jaune puis devenant vert clair dessus, plus pâles dessous. Fleurs mâles en chatons pendants, fleurs femelles solitaires ou par deux sur de courts pédoncules. Glands courtement pédonculés ou subsessiles, isolés ou par deux.

Habitat: bords humides des marais et des cours d'eau, hautes terres sablonneuses riches. Croît dans les Etats de la côte atlantique et du golfe du Mexique.

Bois: lourd, pas dur, résistant et à grain plutôt grossier. Bois de cœur brun clair teinté de rouge; aubier mince, plus clair.

Intérêts: introduit en Europe en 1724 mais très rarement planté depuis malgré sa belle apparence et sa croissance rapide.

Willow Oak.

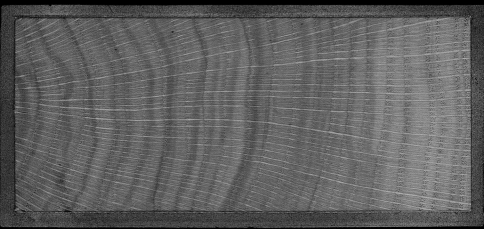

TRANSVERSE SECTION.

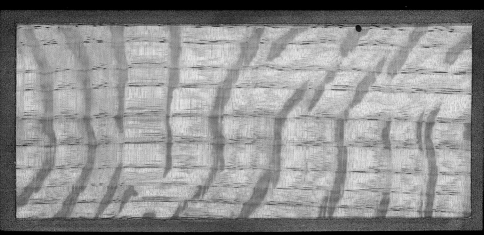

RADIAL SECTION.

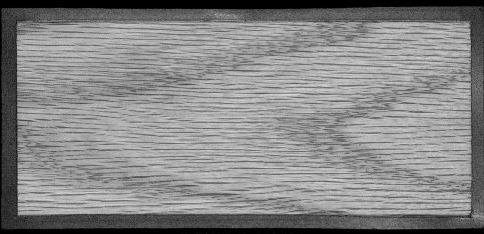

TANGENTIAL SECTION.

Ger. Weiden Eiche.   *Fr.* Chêne de Saule.   *Sp.* Roble de Sauce.

B | BEECH FAMILY

Yellow Oak, Chestnut Oak, Chinquapin Oak

FAMILIE DER BUCHENGEWÄCHSE
Gelb-Eiche

FAMILLE DU CHÊNE ET DU HÊTRE
Chêne jaune, Chêne à chinquapin

FAGACEAE
Quercus muehlenbergii Engelm.
Syn.: *Q. acuminata* (Michx.) Houba

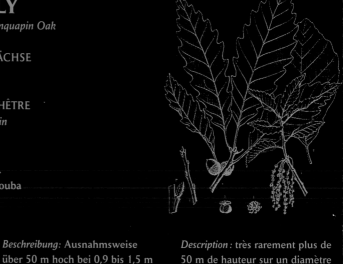

Description: In exceptional cases more than 165 ft. (50 m) high and 3 to 5 ft. (0.9 to 1.5 m) in diameter. In general, much smaller. Trunk straight, columnar, with large buttresses. Bark pale gray and scaly. Foliage very similar to that of the Sweet Chestnut. Fall color various shades of orange and red.

Habitat: The interior of eastern North America.

Wood: Heavy, hard, and strong.

Use: Extensively used in cooperage, for wooden structures and agricultural implements, furniture, posts, and railroad sleepers.

In central Europe grown as a park tree, mainly in older-established parks.

Beschreibung: Ausnahmsweise über 50 m hoch bei 0,9 bis 1,5 m Durchmesser. Im Allgemeinen viel kleiner. Gerader, säulenförmiger Stamm mit weiten Wurzelanläufen. Borke fahlgrau und schuppig. Laub dem der Esskastanie sehr ähnlich. Herbstfärbung: verschiedene Tönungen von Orange und Rot.

Vorkommen: Im Inneren des östlichen Nordamerika.

Holz: Schwer, hart und stark.

Nutzung: Ausgiebig gebraucht in der Böttcherei, für Holzkonstruktionen und Ackergerät, Möbel, Pfosten und Eisenbahnschwellen.

In Mitteleuropa als Parkbaum vorwiegend in älteren Anlagen.

Description: très rarement plus de 50 m de hauteur sur un diamètre de 0,9 à 1,5 m; de taille nettement inférieure en général. Tronc droit, columnaire, avec des contreforts étendus. Ecorce grisâtre et écailleuse. Feuillage ressemblant beaucoup à celui du châtaignier. Se pare en automne de divers tons d'orange et de rouge.

Habitat: dans les régions intérieures de l'est de l'Amérique du Nord.

Bois: lourd, dur et fort.

Intérêts: abondamment utilisé en tonnellerie, en construction et pour fabriquer des instruments aratoires, des meubles, des poteaux et des traverses de chemin de fer.

Arbre de parcs, surtout lorsqu'ils

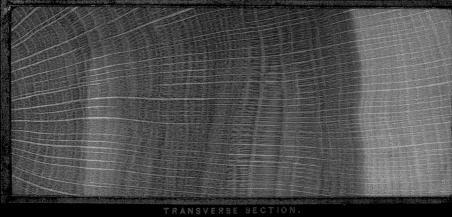

TRANSVERSE SECTION.

RADIAL SECTION.

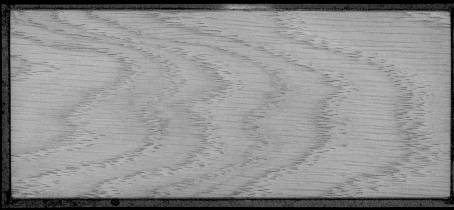

TANGENTIAL SECTION

Ger. Kostanien-Eiche. *Sp.* Roble amarillo.

Fr. Chêne jaune.

B | BEEFWOOD FAMILY, AUSTRALIAN PINE FAMILY

Beefwood, She Oak, Australian Pine

FAMILIE DER KASUARINENGEWÄCHSE

Eisenholz, Keulenbaum

FAMILLE DE LA CASUARINA

Casuarina à feuilles de prêle, Filao

CASUARINACEAE
Casuarina equisetifolia L.

Description: A large tree with upright growth and mottled, dark brown bark only slightly furrowed. The branches, covered with pointed, scale-like leaves, are reminiscent in form of the shoots of horsetails. The flowers are strongly reduced, very small and insignificant. The fruits are yellowish-brown and usually roundish in shape.

Habitat: Its native country is Australia. It has been introduced into large areas of the tropics and has also established itself in southern Florida, where it has formed colonies on the most varied types of soil.

Wood: The very hard and heavy wood is reddish when freshly cut, but later becomes deep red-brown in color.

Use: The tree is greatly valued as a street tree because of its erect growth and its undemanding nature. It has also been used successfully on occasions for controlling drifting dunes. The wood is employed for many purposes, and is especially suitable for masts because of its flexibility.

Beschreibung: Ein großer Baum mit geradem Wuchs und fleckig dunkelbrauner Borke, die nur leicht rissig ist. Die mit schuppenartigen, spitzen Blättern bedeckten Äste erinnern in ihrer Gestalt an die Sprosse von Schachtelhalmen. Die Blüten sind stark reduziert, sehr klein und unauffällig. Die Früchte sind gelblich braun und meist von rundlicher Form.

Vorkommen: Seine Heimat ist Australien. Er wurde in weiten Teilen der Tropen eingeführt und hat sich auch in Süd-Florida eingebürgert, wo er die verschiedensten Böden besiedelt.

Holz: Das sehr harte und schwere Holz ist frisch nach dem Schlagen noch rötlich, wird aber später tief rotbraun.

Nutzung: Der Baum ist wegen seines raschen und geraden Wuchses sowie seiner Anspruchslosigkeit sehr begehrt als Straßenbaum. Gelegentlich wurde er auch erfolgreich zur Festlegung von Wanderdünen verwendet. Das Holz ist vielfältig einsetzbar, wegen seiner Elastizität ist es besonders für Masten gut geeignet.

Description : grand arbre à port droit, avec une écorce tachée de brun sombre et parcourue de fissures peu profondes. Il doit son nom à son port équisétiforme, c'est-à-dire rappelant celui de la prêle : ses rameaux sont en effet articulés et ses feuilles verticillées réduites à des écailles. Les fleurs sont fortement réduites, très petite et insignifiantes. Les fruits sont brun jaunâtre et sphériques le plus souvent.

Habitat : originaire d'Australie. Il fut importé dans de nombreuses régions tropicales et s'est également naturalisé dans le sud de la Floride où il s'accommode de tous les sols.

Bois : très lourd et dur ; rougeâtre après la coupe, devenant ensuite d'un brun rouge profond.

Intérêts : très recherché comme arbre d'alignement en raison de sa croissance rapide et droite et de son adaptation à des conditions écologiques variées. Il a servi quelquefois à la fixation de dunes mouvantes. Le bois se prête à toutes sortes d'usages, en particulier à la fabrication de mâts car il est élastique.

TRANSVERSE SECTION.

RADIAL SECTION.

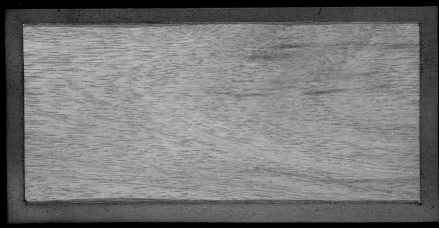

TANGENTIAL SECTION.

B | BIRCH FAMILY

American Hornbeam, Blue Beech, Water Beech, Ironwood

FAMILIE DER BIRKENGEWÄCHSE
Amerikanische Hainbuche

FAMILLE DU BOULEAU
*Charme de la Caroline, Charme américain,
Bois dur (Canada)*

BETULACEAE
Carpinus caroliniana Walt.

Description: 35 to 40 ft. (10 to 12 m) high and 15 in. to 2 ft. (0.4 to 0.6 m) in diameter. Trunk short, fluted ("buttressed"), with spreading branches drooping at the ends, hence its Indian name "Otantahrteweh" ("feeble tree"). Bark smooth blue-gray, often with lighter or darker markings. Foliage bluish green.

Habitat: Damp, low-lying ground in association with American Holly, Sassafras, Swamp Bay, Black Gum, Red Maple, Water Locust, Prickly Ash (Toothache Tree), etc. In eastern North America; more abundant and developing better in the southern states. In the north, often stunted or shrub-like.

Wood: Heavy, very hard, very strong, close-grained. Heartwood light brown; sapwood thick and almost white.

Use: Levers, handles of tools, firewood.

Beschreibung: 10 bis 12 m hoch bei 0,4 bis 0,6 m Durchmesser. Kurzer, gekehlter („spannrückiger") Stamm, daher Indianername „Otantahrteweh" („kraftloser Baum"). Borke blaugrau, oft heller oder dunkler gefleckt. Laub blaugrün.

Vorkommen: Tief gelegenes, feuchtes Flachland in Gesellschaft von Amerikanischer Stechhülse, Sassafras, Sumpf-Avocado, Tupelos, Rot-Ahorn, Wasser-Gleditschie, Eschenblättrigem Gelbholz („Zahnwehholz") u.a. im östlichen Nordamerika. Häufiger und besser ausgebildet in den Südstaaten. Im Norden oft krüppelhaft oder strauchig.

Holz: Schwer, sehr hart, sehr fest und feine Textur. Kern hellbraun; Splint dick und fast weiß.

Nutzung: Hebebäume, Werkzeuggriffe, Brennholz.

Description: de 10 à 12 m de hauteur sur un diamètre de 0,4 à 0,6 m. Tronc court, cannelé, d'où son nom indien «otantahrteweh» (arbre manquant de vigueur). Ecorce gris bleu, présentant souvent des taches plus claires ou plus foncées. Feuillage vert bleu.

Habitat: basses plaines humides dans l'est de l'Amérique du Nord en compagnie du houx, du sassafras, de l'avocatier palustre, du nyssa, de l'érable rouge, du févier aquatique, du clavalier etc. Prospère dans le sud; au nord, sou une forme souvent buissonnante ou rabougrie.

Bois: lourd, très dur et résistant, à grain fin. Bois de cœur marron clair; aubier épais et quasiment blanc.

Intérêts: utilisé pour fabriquer des leviers, des manches d'outils; bois de chauffage.

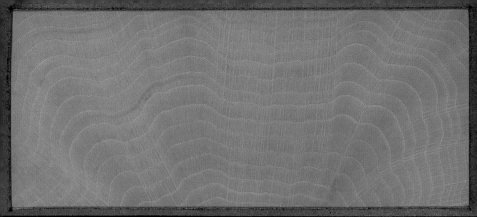

TRANSVERSE SECTION.

RADIAL SECTION.

TANGENTIAL SECTION.

Ger. Hainbuche.   Fr. Charme.   Sp. Ojaranzo.

B | BIRCH FAMILY

Eastern Hop Hornbeam, Ironwood

FAMILIE DER BIRKENGEWÄCHSE
Virginische Hopfenbuche

FAMILLE DU BOULEAU
Ostryer de Virginie, Bois à levier (Canada)

BETULACEAE
Ostrya virginiana (Mill.) K. Koch
Syn.: *O. virginica* Willd.

Description: Rarely more than 75 ft. (20 m) high and 2 ft. (0.6 m) in diameter. Trunk below the crown short. Bark rough, in narrow, oblong scales. Fruit-clusters pale green, pendent, reminiscent of hops.

Habitat: On even relatively poor and dry soils, but certainly on well-drained gravely slopes and ridges. Most frequent in the north and thriving best in association with Beech, Sugar Maple, Yellow Birch, White and Cork Elm, Butternut, White Ash, etc., in eastern North America.

Wood: Heavy, very hard ("Ironwood"), tough, very strong, and extremely close-grained; polishable, durable in the ground. Heartwood light brown tinged with red or almost white; sapwood thick and pale.

Use: Fence posts, levers, handles, mallets, also firewood. Bark rich in tannin.

Beschreibung: Selten über 20 m hoch bei 0,6 m Durchmesser. Stamm bis zur Krone kurz. Borke rau, in länglichen, schmalen Schuppen. Blassgrüne, hängende, an Hopfen erinnernde Fruchtstände.

Vorkommen: Auch relativ arme und trockene, jedenfalls gut drainierte, kiesige Hänge und Hangrücken. Im Norden am häufigsten und am besten gedeihend in Gesellschaft von Buche, Zucker-Ahorn, Gelb-Birke, Weiß- und Kork-Ulme, Butternuss, Weiß-Esche u. a. im östlichen Nordamerika.

Holz: Schwer, sehr hart ("Iron Wood"), zäh, sehr fest und mit überaus feiner Textur; polierfähig, im Boden haltbar. Kern hellbraun mit rot oder aber fast weiß; Splint dick und fahl.

Nutzung: Zaunpfähle, Hebebäume, Handgriffe, Holzhämmer, auch Brennholz. Rinde gerbstoffreich.

Description: rarement plus de 20 m de hauteur sur un diamètre de 0,6 m. Tronc court jusqu'au houppier. Ecorce à texture rugueuse, en écailles étroites et allongées. Chatons pendants vert pâle, rappelant ceux du houblon.

Habitat: croît aussi sur les versants et les sommets de versants graveleux, relativement pauvres et secs, bien drainés en tout cas. C'est au nord qu'il se développe le mieux et le plus fréquemment, en association avec le hêtre, l'érable à sucre, le bouleau jaune, l'orme blanc et l'orme liège, le noyer cendré, le frêne blanc etc. On le trouve dans l'est de l'Amérique du Nord.

Bois: lourd, très dur («Iron Wood»), tenace, très résistant et d'un grain extrêmement fin; pouvant se polir, durable sur pied. Bois de cœur brun clair teinté de rouge ou presque blanc; aubier épais et blafard.

Intérêts: utilisé pour fabriquer des pieux de clôtures, des leviers, des manches d'outils, des maillets; bois

Hop-Hornbeam, Iron-wood, Lever-wood.

TRANSVERSE SECTION.

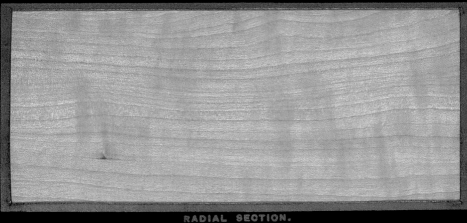

RADIAL SECTION.

TANGENTIAL SECTION.

Ger. Amerikanische Hopfenbuche.   Fr. Bois dur.

Sp. Ojaranzo de lupulo.

B | BIRCH FAMILY
Gray Birch, White Birch

FAMILIE DER BIRKENGEWÄCHSE
Grau-Birke, Pappelblättrige Birke

FAMILLE DU BOULEAU
Bouleau gris, Bouleau à feuilles de peuplier,
Bouleau rouge (Canada)

BETULACEAE
Betula populifolia Marshall

Description: Mostly only about 23 to 33 ft. (7 to 10 m) high, exceptionally to about 40 ft. (12 m) in height and sometimes 12 in. (0.35 m) in diameter. Trunk commonly surrounded by daughter stems arising from its base. Young bark dull white with dark, broadly triangular markings at the insertion of the branches, slowly separating into horizontal strips. In old age it becomes darker and rough from horizontal fissures. Leaves long-stalked, and continually moving in the wind, like those of the Aspen.

Habitat: Dry, often barren, sandy soils and wasteland in northeastern North America. Springs up in vast numbers after forest fires and then provides shade and shelter for seedlings of more valuable species until finally overwhelmed by them.

Wood: Moderately heavy.

Use: For reels, hoops of barrels, and other small articles. Excellent as firewood and for charcoal.

Beschreibung: Meist nur etwa 7 bis 10 m, ausnahmsweise etwa 12 m hoch bei manchmal 0,35 m Durchmesser. Der Stamm ist gewöhnlich von Tochterstämmen („Ästen") aus der Stammbasis eingeschlossen. Borke jung trübweiß mit dunklen, breit dreieckigen Marken an den Insertionsstellen der Zweige, löst sich träge in Querstreifen ab. Im Alter dunkler und rau von Querrissen. Blätter langgestielt, wie bei der Zitter-Pappel in ständiger Bewegung bei Wind.

Vorkommen: Trockene, oft sterile Sandböden und Ödflächen im nordöstlichen Nordamerika. Kommt massenhaft nach Waldbränden auf und dient dann als Schutz und Schattenbaum für aufkommende bessere Baumarten, bis sie von diesen überwachsen wird.

Holz: Mäßig schwer.

Nutzung: Für Spulen, Faßreifen und andere kleine Artikel. Ausgezeichnet als Brennholz und für Holzkohle.

Description : d'ordinaire 7 à 10 m environ de hauteur, exceptionnellement 12 m, sur un diamètre pouvant atteindre 0,35 m. Le tronc qui se divise généralement dans sa partie basale, est encerclé par des tiges secondaires. A l'état juvénile, l'écorce blanchâtre, ponctuée de grandes marques sombres et triangulaires aux endroits où s'insèrent les rameaux, se détache en lamelle transversales ; elle devient plus foncée avec l'âge et se creuse de fissures transversales. Feuilles à long pétiole, tremblant au moindre souffle de vent comme celles du peuplier tremble.

Habitat : sols sablonneux secs, souvent stériles et terrains incultes du nord-est de l'Amérique du Nord. Espèce colonisatrice très abondante après le passage d'un feu et protectrice d'espèces d'ombre plus vigoureuses, qui grandissent puis la dominent en hauteur.

Bois : moyennement lourd.

Intérêts : utilisé pour fabriquer des bobines, divers petits objets et en tonnellerie. Excellent bois de chauffage et charbon de bois.

White Birch, Poplar-leaved Birch, Old-field Birch

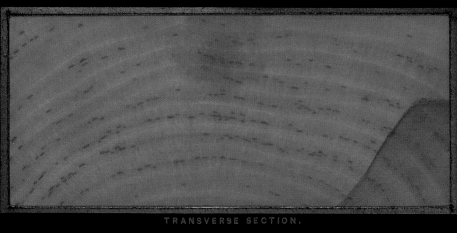

TRANSVERSE SECTION.

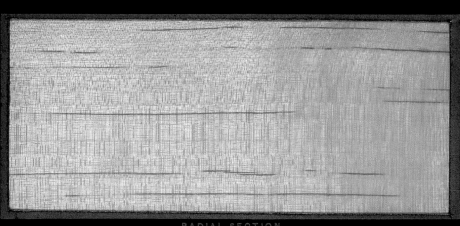

RADIAL SECTION.

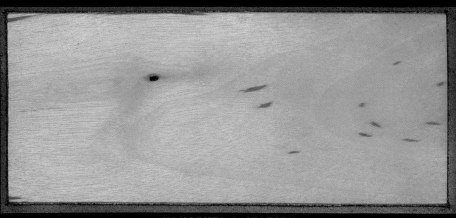

TANGENTIAL SECTION

Ger. Pappelblättrige Birke   *Sp.* Abedul blanco.

Fr. Bouleau blanc.

B | BIRCH FAMILY
Oregon Alder, Red Alder

FAMILIE DER BIRKENGEWÄCHSE
Oregon-Erle

FAMILLE DU BOULEAU
Aulne rouge, Aulne de l'Oregon

BETULACEAE
Alnus rubra Bong.
Syn.: *A. oregona* Nutt.

Description: 40 to 50 ft. (12 to 15 m) high and 6 to 24 in. (0.15 to 0.6 m) in diameter. Branches slender, somewhat drooping, forming a narrow, pyramidal crown. Bark fairly thick, smooth, pale gray to almost white, with small, warty outgrowths. Removal of thin outer bark reveals the paler inner bark. Leaves ovate to elliptic, finely toothed, dark green above, and with rusty-colored hairs beneath. Male flowers in long, narrow inflorescences; female flowers in erect, spindle-shaped ones. Infructescences ovoid.

Habitat: Common on riverbanks in the coastal region of Pacific North America.

Wood: Light, soft, not strong, brittle, silky-sheened, and close-grained; polishable and easy to work. Heartwood pale brown, with red tinge; sapwood thick, almost white. Medullary rays broad and distinct.

Use: In large amounts for furniture in Washington and Oregon. Hollowed out trunks used as canoes by Alaskan Indians.

Beschreibung: 12 bis 15 m hoch bei 0,15 bis 0,6 m Durchmesser. Äste schlank, etwas herabhängend, eine schmale, pyramidale Krone bildend. Borke mäßig dick, bis auf kleine warzenförmige Auswüchse glatt und hellgrau bis fast weiß. Die dünne äußere Rinde entblößt beim Abschälen die helle innere Rinde. Blätter oval bis elliptisch, fein gezähnelt, dunkelgrün auf der Ober- und mit rostfarbener Behaarung auf der Unterseite. Männliche Blüten in schmalen, lang herabhängenden, weibliche in aufrechten, spindelförmigen Blütenständen. Fruchtstände oval-eiförmig.

Vorkommen: Häufig an Flussufern in der Küstenregion des pazifischen Nordamerikas.

Holz: Leicht, weich, nicht fest, brüchig, seidig und mit feiner Textur; polierbar und leicht zu bearbeiten. Kern hellbraun mit roter Tönung; Splint dick, fast weiß. Breite, deutliche Markstrahlen.

Nutzung: In Washington und Oregon in großem Umfang für Möbel. Die Indianer Alaskas stellten aus den ausgehöhlten Stämmen Kanus her.

Description: 12 à 15 m de hauteur sur un diamètre de 0,15 à 0,6 m. Houppier étroit et quasi pyramidal, formé de fines branches retombantes. Ecorce gris clair à presque blanc, moyennement épaisse, légèrement verruqueuse, qui en s'exfoliant révèle un endoderme clair. Feuilles ovales à elliptiques, finement dentées, vert sombre dessus, garnies d'une pubescence rousse dessous. Inflorescences mâles étroites, longues et pendantes, inflorescences femelles fusiformes et dressées. Infrutescences en strobiles.

Habitat: fréquent sur les rives des cours d'eau des régions de la côte pacifique de l'Amérique du Nord.

Bois: léger, tendre, non résistant, cassant, satiné et à grain fin; susceptible d'être poli et facile à travailler. Bois de cœur brun clair, teinté de rouge; aubier épais, presque blanc. Rayons médullaires larges et distincts.

Intérêts: utilisé dans le Washington et l'Oregon en grande partie pour la fabrication de meubles. Les Indiens d'Alaska se servaient des troncs pour fabriquer des canots.

Oregon Alder, Red Alder.

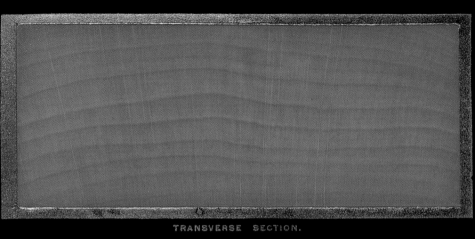

TRANSVERSE SECTION.

RADIAL SECTION

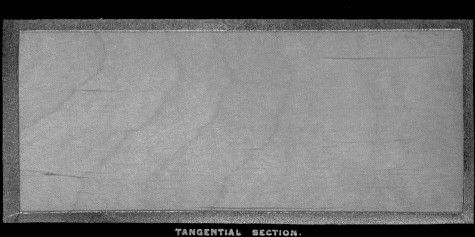

TANGENTIAL SECTION.

B | BIRCH FAMILY
Paper Birch, Canoe Birch

FAMILIE DER BIRKENGEWÄCHSE
Papier-Birke

FAMILLE DU BOULEAU
*Bouleau frontinal, Bouleau d'eau, Bouleau occidental,
Merisier rouge (Canada)*

BETULACEAE
Betula occidentalis Hook.

Description: 60 to 70 (-120) ft. (18 to 21 (-36) m) high and 2 to 3 ft. (0.6 to 0.9 m) in diameter. Also grows as a shrub in the mountains of New England. Freestanding specimens are well-branched, with a regular, pyramidal, dense crown, but in the forest they are only branched in the upper part, and have a rounded, open crown. Bark at the base of the trunk up to half an inch (1 cm) thick, dark brown, deeply fissured. Above it is paper-like, creamy white on the outside and orange on the inside. Branches thin. Leaves roundish, with irregularly serrate margins. The flowers are first of all in erect catkins, but later these become pendent.

Habitat: Riverbanks, the edges of lakes and swamps, and rich forest slopes in northern North America.

Wood: Light, hard, strong, and close-grained. Heartwood light brown with a tinge of red; sapwood thick, almost white. Numerous medullary rays.

Use: Spindles, paper, and as firewood. The Indians used the wood for paddles, snowshoes, and ax-handles, the bark for canoes, bags, beakers, etc.

Beschreibung: 18 bis 21 (bis 36) m hoch bei 0,6 bis 0,9 m Durchmesser. In den Gebirgen Neuenglands auch als Busch. Freistehende Exemplare tief beastet, mit regelmäßiger, pyramidaler, geschlossener Krone, im Bestand weit herauf astfrei mit rundlicher, offener Krone. Borke an der Stammbasis bis 1 cm dick, schwarzbraun, tief rissig. Oben papierartig, cremeweiß auf der Außenseite und orange auf der Innenseite. Äste dünn. Blätter rundlich, mit unregelmäßig gesägten Rändern. Blüten in zunächst aufrechten, später hängenden Kätzchen.

Vorkommen: An den Ufern von Flüssen, Seen und Sümpfen sowie reicher Waldhänge im nördlichen Nordamerika.

Holz: Leicht, hart, fest und von feiner Textur. Kern hellbraun mit roter Tönung; Splint dick, fast weiß. Zahlreiche Markstrahlen.

Nutzung: Spindeln, Papier und zur Feuerung. Von den Indianern für Paddel, Schneeschuhe und Beilgriffe, Rinde für Kanus, Taschen, Trinkbecher etc. genutzt.

Description : 18 à 21 m de hauteur (36 m max.) sur un diamètre de 0,6 à 0,9 m. Egalement sous forme de buisson dans les montagnes de la Nouvelle-Angleterre. Houppier fermé, régulier et pyramidal en situation isolée ; branches insérées beaucoup plus haut, formant une couronne ronde et ouverte en peuplement. Ecorce de 1 cm d'épaisseur, brun noir, profondément fissurée à la base ; papyracée, blanc crème dessus et orange dessous sur le reste du tronc. Branches fines. Feuilles rondes, irrégulièrement dentées. Inflorescences en chatons, dressés puis pendants.

Habitat : berges de cours d'eau, de lacs et de marais et pentes boisées du nord de l'Amérique du Nord.

Bois : léger, résistant et à grain fin. Bois de cœur brun clair, teinté de rouge ; aubier épais, presque blanc. Nombreux rayons médullaires.

Intérêts : fabrication de fuseaux, de papier, et combustible ; les Indiens fabriquent des pagaies, des chaussures de neige et des manches de haches avec le bois ; des canots, des sacs, des gobelets etc. avec l'écorce.

Western Birch, Puget Sound Birch.

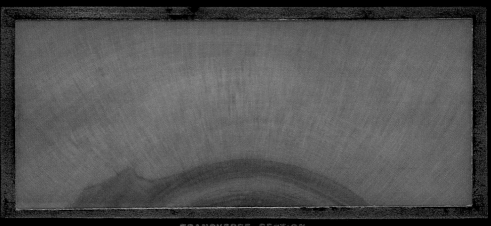

TRANSVERSE SECTION.

RADIAL SECTION.

TANGENTIAL SECTION.

B | BIRCH FAMILY
Paper Birch, Canoe Birch

FAMILIE DER BIRKENGEWÄCHSE
Papier-Birke

FAMILLE DU BOULEAU
*Bouleau blanc, Bouleau à papier occidental,
Bouleau à canot (Canada)*

BETULACEAE
Betula papyrifera Marshall
Syn.: *B. papyracea* Aiton

Description: Said to be 140 ft. (40 m) high and 3 to 4 ft. (0.9 to 1.2 m) in diameter west of the Rocky Mountains, elsewhere rarely more than 75 to 85 ft. (20 to 25 m) high, though with the same diameter. Often with several trunks. Young bark smooth, creamy white, and marked by long, horizontal lenticels, later rolling back and peeling away, finally in thick, irregular scales.

Habitat: Rich slopes, riverbanks, lake shores, margins of swamps, with Aspen, larches, poplars, pines, spruce, and fir in northern North America.

Wood: Light, hard, tough, and very close-grained. Heartwood light brown tinged with red; sapwood thick and almost white.

Use: Sledges and snow shoes of the Indians, firewood. Bark used by the Indians for building their ten to twelve-man canoes. Also used for drinking vessels and watertight containers. Used to cover the roofs of wigwams in the absence of animal skins. The sugary sap serves as a drink and syrup in the spring.

Beschreibung: Westlich der Rocky Mountains angeblich 40 m hoch bei 0,9 bis 1,2 m Durchmesser, anderswo selten mehr als 20, 25 m hoch bei gleichem Durchmesser. Oft mehrstämmig. Junge Borke glatt, cremeweiß und mit langen, horizontalen Lentizellen durchsetzt, später zurückgerollt und abschilfernd, zuletzt in dicken unregelmäßigen Schuppen.

Vorkommen: Reiche Hänge, Flussufer, Seeufer, Ränder von Sümpfen, mit Espe, Lärche, Pappel, Kiefer, Fichte und Tanne im nördlichen Nordamerika.

Holz: Leicht, hart, zäh und sehr feine Textur. Kern hellbraun und rot; Splint dick und fast weiß.

Nutzung: Schlitten und Schneeschuhe der Indianer, Brennholz. Borke von den Indianern zum Bau von Kanus für 10 bis 12 Mann benutzt. Auch als Trinkgefäße und wasserundurchlässige Gefäße. Wo Tierfelle fehlten, als Wigwambedachung. Zuckersaft im Frühjahr als Getränk und Sirup.

Description: atteindrait à l'ouest des Rocheuses 40 m de hauteur sur un diamètre de 0,9 à 1,2 m ; ailleurs, rarement plus de 20, 25 m sur le même diamètre. Espèce multicaule Ecorce lisse d'un blanc crème, couverte de longues lenticelles horizontales sur les jeunes arbres, se déchirant avec l'âge en minces pellicules enroulées et s'exfoliant finalement en épaisses écailles irrégulières.

Habitat : pentes riches, bords des cours d'eau, des lacs et des marécages dans le nord de l'Amérique du Nord, en association avec le mélèze, le tremble, le peuplier, le pin, l'épicéa et le sapin.

Bois : léger, dur et à grain très fin. Bois de cœur brun clair et rouge ; aubier épais et quasiment blanc.

Intérêts : traîneaux et chaussures de neige des Indiens ; bois de chauffage. Ecorce imperméable utilisée par les Indiens pour fabriquer des canots légers de 10 ou 12 personnes ainsi que divers récipients étanches et à recouvrir les habitations (« wigwam »), à défaut de fourrures. La sève sucrée était extraite au printemps pour être bu telle quelle ou épaissie en sirop.

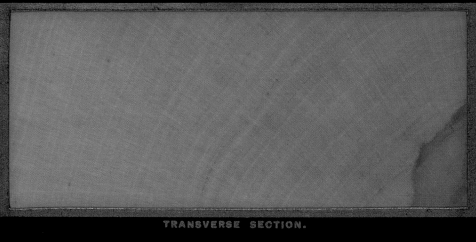

TRANSVERSE SECTION.

RADIAL SECTION.

TANGENTIAL SECTION.

B | BIRCH FAMILY
River Birch, Water Birch, Red Birch

FAMILIE DER BIRKENGEWÄCHSE
Schwarz-Birke

FAMILLE DU BOULEAU
Bouleau noir, Bouleau de rivière

BETULACEAE
Betula nigra L.

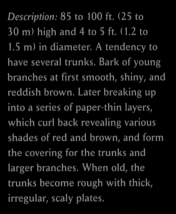

Description: 85 to 100 ft. (25 to 30 m) high and 4 to 5 ft. (1.2 to 1.5 m) in diameter. A tendency to have several trunks. Bark of young branches at first smooth, shiny, and reddish brown. Later breaking up into a series of paper-thin layers, which curl back revealing various shades of red and brown, and form the covering for the trunks and larger branches. When old, the trunks become rough with thick, irregular, scaly plates.

Habitat: Abundant on flooded river-banks from the coast of the Gulf of Mexico across the interior as far as the north-east of the USA.

Wood: Rather light.

Use: For all kinds of wooden articles.

Beschreibung: 25 bis 30 m hoch bei 1,2 bis 1,5 m Durchmesser. Mehr-stämmigkeit. Borke junger Äste zuerst glatt, glänzend rötlich braun. Später in eine papierdünne Lagen aufbrechend, die sich kräuseln und Schattierungen von Rot und Braun freigeben, an größeren Ästen und Stämmen eine regelrechte Matte bildend. Alte Stämme dann rau, mit dicken, unregelmäßig schuppi-gen Platten.

Vorkommen: Häufig an über-schwemmten Flussufern von der Golfküste durch das Landesinnere bis in den Nordosten der USA.

Holz: Eher leicht.

Nutzung: Für mannigfache Artikel aus Holz.

Description : 25 à 30 m de hauteur sur un diamètre de 1,2 à 1,5 m. Espèce multicaule. L'écorce des jeunes rameaux est d'abord brun rougeâtre, lisse et brillante avant de se fragmenter en feuillets qui se froissent et forment sur les plus grosses branches et le tronc un vé-ritable tapis aux nuances de rouge et de brun. Elle devient rugueuse sur les sujets âgés et se détache en grosses écailles irrégulières.

Habitat : souvent sur les berges inondées, depuis la côte du golfe du Mexique en passant par l'inté-rieur des terres et jusqu'au nord-es des Etats-Unis.

Bois : plutôt léger.

Intérêts : utilisé pour fabriquer divers objets en bois.

River Birch, Red Birch.

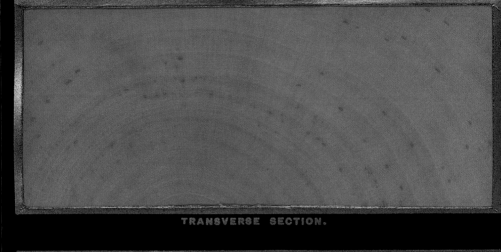

TRANSVERSE SECTION.

RADIAL SECTION.

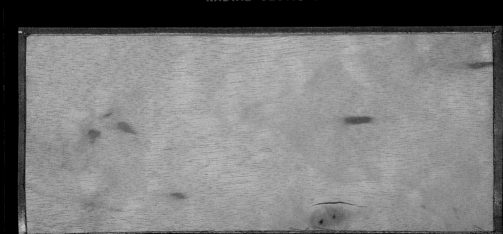

TANGENTIAL SECTION.

Ger. Schwarz-Birke.   Fr. Bouleau noire.   Sp. Abedul neg

B | BIRCH FAMILY

Sweet Birch, Black Birch, Cherry Birch

FAMILIE DER BIRKENGEWÄCHSE
Zucker-Birke

FAMILLE DU BOULEAU
*Bouleau flexible, Bouleau à sucre,
Merisier rouge (Canada)*

BETULACEAE
Betula lenta L.

Description: 85 to 100 ft. (25 to 30 m) high and 2 to 5 ft. (0.6 to 1.5 m) in diameter. The English names reflect qualities of the bark: aromatic and sweet when young, later with dark scales, similar to those of the Wild Black Cherry. Golden yellow color in the fall.

Habitat: Rich, well-drained upland soils in eastern North America.

Wood: Heavy, very hard, and strong, with a silky sheen, and close-grained; polishable. Heartwood dark brown tinged with red; sapwood thin, light brown or yellow.

Use: Ship and boatbuilding, furniture, firewood. Sugary sap fermented to make beer.

The most beautiful of the American birches.

Beschreibung: 25 bis 30 m hoch bei 0,6 bis 1,5 m Durchmesser. Die englischen Namen bezeichnen Eigenschaften der Rinde bzw. Borke: jung aromatisch duftend und süß, später dunkle Schuppenborke, die der der Wilden Schwarzkirsche (Späten Traubenkirsche) ähnelt. Herbstfärbung goldgelb.

Vorkommen: Höher gelegene, reiche, gut drainierte Böden im östlichen Nordamerika.

Holz: Schwer, sehr hart und fest, seidig und mit feiner Textur; polierfähig. Kern dunkelbraun mit rot; Splint dünn, hellbraun oder gelb.

Nutzung: Schiff- und Bootsbau, Möbel, Brennholz. Zuckersaft vergoren zu Bier.

Schönste der amerikanischen Birken.

Description: 20 à 30 m de hauteur sur un diamètre de 0,6 à 1,5 m. Le noms anglais (sweet birch, cherry birch) décrivent les particularités de l'écorce : elle est aromatique et douce sur les jeunes arbres puis elle se déchire avec l'âge en écailles sombres comme celle du merisier. Feuillage jaune intense en automne.

Habitat: sols riches, bien drainés sur les hauteurs dans l'est de l'Amérique du Nord.

Bois: lourd, très dur et résistant, satiné et à grain fin ; susceptible de prendre un beau poli. Bois de cœur brun foncé teinté de rouge ; aubier peu épais, brun clair ou jaune.

Intérêts: utilisé dans la construction navale et l'ameublement ; bois de chauffage. Fabrication de bière avec la sève fermentée.

Le plus beau des bouleaux américains.

TRANSVERSE SECTION.

RADIAL SECTION.

TANGENTIAL SECTION.

B | BIRCH FAMILY
White Alder

FAMILIE DER BIRKENGEWÄCHSE
Weiß-Erle, Rautenblättrige Erle

FAMILLE DU BOULEAU
Aulne de la sierra, Aulne blanc

BETULACEAE
Alnus rhombifolia Nutt.

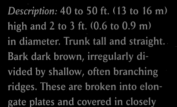

Description: 40 to 50 ft. (13 to 16 m) high and 2 to 3 ft. (0.6 to 0.9 m) in diameter. Trunk tall and straight. Bark dark brown, irregularly divided by shallow, often branching ridges. These are broken into elongate plates and covered in closely appressed scales.

Habitat: Banks of streams, especially in valleys of central California. Widespread in western North America from Idaho across the eastern slopes of the Cascade Mountains as far as California.

Wood: Light, soft, not strong, brittle, close-grained.

Use: Unknown.

Discovered by Thomas Nuttall near Monterey in 1835.

Beschreibung: 13 bis 16 m hoch bei 0,6 bis 0,9 m Durchmesser. Stamm hoch und gerade. Borke dunkelbraun, unregelmäßig geteilt in flache, oft verzweigte Riefen. Diese in längliche Platten gebrochen und mit kleinen, dicht anliegenden Schuppen besetzt.

Vorkommen: Bachufer, besonders der mittelkalifornischen Täler. Verbreitet im westlichen Nordamerika von Idaho über die Osthänge des Kaskadengebirge, bis nach Kalifornien.

Holz: Leicht, weich, nicht fest, spröde und mit feiner Textur.

Nutzung: Unbekannt.

1835 in der Nähe von Monterey von Thomas Nuttall entdeckt.

Description : arbre de 13 à 16 m de hauteur, avec un tronc droit et élancé de 0,6 à 0,9 m de diamètre. Ecorce brun foncé, creusée de fissures plates et irrégulières, souvent ramifiées, et se craquelant en plaques allongées aux petites écailles adhérentes.

Habitat : rives de ruisseaux, des vallées de la moyenne Californie en particulier. Répandu dans l'ouest de l'Amérique du Nord depuis l'Idaho jusqu'en Californie, en passant par les pentes est des montagnes Cascades.

Bois : léger, tendre, non résistant, cassant et à grain fin.

Intérêts : inconnus.

Découvert par Thomas Nuttall en 1835 dans les environs de Monterey.

California Alder, Mountain Alder, White Alder.

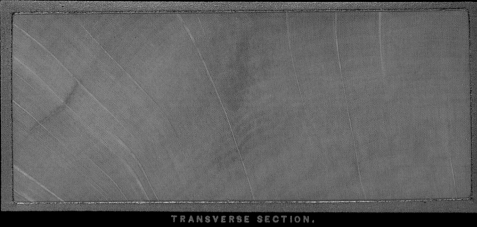

TRANSVERSE SECTION.

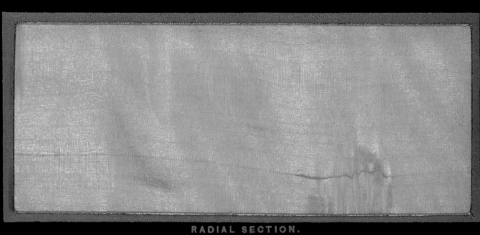

RADIAL SECTION.

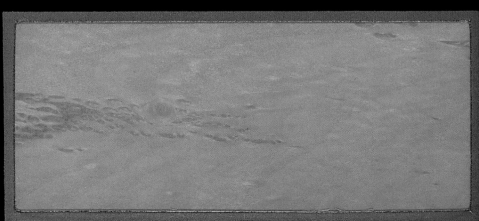

TANGENTIAL SECTION

Ger. Californische Erle.   Fr. Anue de Californie.
Sp. Aliso de California.

B | BIRCH FAMILY
Yellow Birch

FAMILIE DER BIRKENGEWÄCHSE
Gelb-Birke

FAMILLE DU BOULEAU
Bouleau jaune, Bouleau des Alleghanys (Canada)

BETULACEAE
Betula alleghanensis Britton var. *alleghanensis*
Syn.: *B. lutea* Michx. f.

Description: Sometimes up to about 100 ft. (30 m) high and 3 to 4 ft. (1 m) in diameter. Bark when young a shining gold or silver gray, later unmistakable with curled strips that rustle with every breath of wind. In old age rough and in irregular plates.

Habitat: Rich, moist soils on high ground, in the north in association with Beech, Sugar Maple, Red Maple, Black and White Ash, White Elm, Hop Hornbeam, etc., in northeastern North America.

Wood: Heavy, hard, strong, with a silky sheen and close-grained; polishable. Heartwood pale brown, slightly reddish; sapwood narrow, almost white. One of the most valuable woods in North America.

Use: Particularly valuable for furniture because of the occasional markings in the wood. Also used for agricultural implements. Within its range a very good firewood.

Beschreibung: Manchmal bis etwa 30 m hoch bei bis zu 1 m Durchmesser. Borke jung glänzend gold- oder silbergrau, später unverkennbar in zurückgerollten Streifen, die bei jedem Windhauch rascheln. Im Alter rau und unregelmäßig geplättet.

Vorkommen: Reiche, feuchte, höher gelegene Böden, im Norden in Gesellschaft von Buche, Zucker-Ahorn, Rot-Ahorn, Schwarz- und Weiß-Esche, Weiß-Ulme, Hopfenbuche u. a. im nordöstlichen Nordamerika.

Holz: Schwer, hart, fest, seidig glänzend und mit feiner Textur; polierfähig. Kern hellbraun, leicht rötlich; Splint schmal, fast weiß. Eines der wertvollsten Hölzer im Norden Amerikas.

Nutzung: Durch gelegentliche Zeichnungen im Holz besonders wertvoll für Möbel. Sonst für landwirtschaftliche Geräte. Sehr gutes Brennholz im Verbreitungsgebiet.

Description : parfois jusqu'à 30 m de hauteur sur un diamètre pouvant atteindre 1 m. L'écorce est gris argenté ou doré à l'état juvénile, puis en pellicules s'enroulant en bandelettes qui bruissent au vent. Elle est rugueuse sur les vieux sujets et se fragmente en plaques asymétriques.

Habitat : sols riches et humides des hautes terres ; au nord, en compagnie du bouleau, de l'érable à sucre, de l'érable rouge, des frênes blanc et noir, de l'orme d'Amérique, de l'ostryer de Virginie etc. On le trouve dans le nord-est de l'Amérique du Nord.

Bois : lourd, dur, résistant, satiné et à grain fin ; pouvant se polir. Bois de cœur brun clair, un peu rougeâtre ; aubier mince, presque blanc. Un des bois les plus précieux d'Amérique du Nord.

Intérêts : très recherché pour l'ébénisterie car il présente parfois des dessins. Sinon, utilisé pour fabriquer des outils agricoles.

Yellow Birch, Gray Birch.

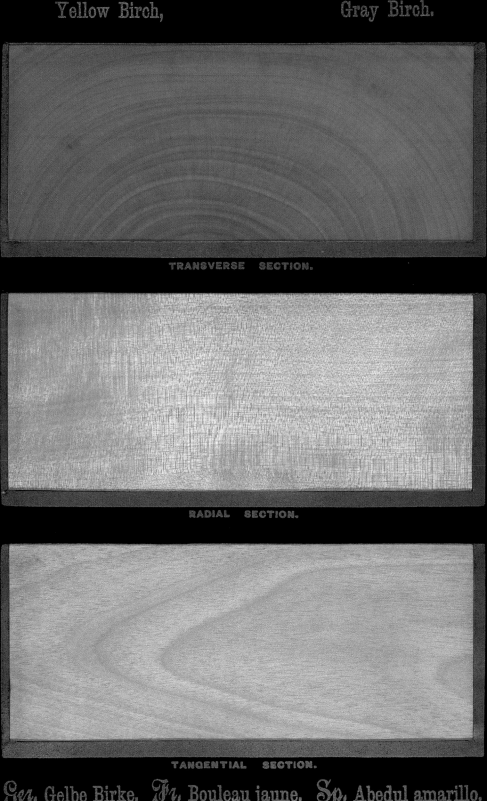

TRANSVERSE SECTION.

RADIAL SECTION.

TANGENTIAL SECTION.

Ger. Gelbe Birke. Fr. Bouleau jaune. Sp. Abedul amarillo.

B | BORAGE FAMILY
Strongbark, Strong-back

FAMILIE DER RAUBLATTGEWÄCHSE,
BORRETSCHGEWÄCHSE
Havanische Bourreria

FAMILLE DE LA BOURACHE
Bourrerier Havana

BORAGINACEAE
Bourreria ovata Miers
Syn.: *B. havanensis* Miers

Description: A shrubby tree, 40 to 50 ft. (12 to 15 m) high and 8 to 10 in. (0.2 to 0.25 m) in diameter, or more frequently much smaller and then a many-stemmed shrub. Bark thin, more or less fissured, light brown with a red tinge. Leaves leathery, ovate to oblong, dark green above, yellow-green beneath. Flowers with creamy white petals joined together, sepals also joined forming a usually five-toothed calyx. Fruit orange-red with a thick skin and thin dry flesh.

Habitat: Forests on the Bahamas and Antilles. In North America confined to southern Florida.

Wood: Hard, strong, with a silky sheen, and very close-grained; polishable. Heartwood brown, streaked with orange; sapwood thick, barely distinguishable. Numerous indistinct medullary rays.

Use: On account of its narrow diameter, presumably suitable only as firewood.

Beschreibung: Strauchartiger Baum, 12 bis 15 m hoch bei 0,2 bis 0,25 m Durchmesser, häufig viel kleiner als vielstämmiger Strauch. Borke dünn, mehr oder weniger rissig, hellbraun mit roter Tönung. Blätter ledrig, oval bis länglich, dunkelgrün auf der Oberseite, gelbgrün auf der Unterseite. Blüten mit verwachsenen, cremeweißen Kronblättern, Kelchblätter verwachsen, meist fünfspitzig. Frucht orangerot mit dicker Haut und dünnem, trockenem Fleisch.

Vorkommen: Wälder der Bahamas und der Antillen. Nordamerika nur in Süd-Florida erreichend.

Holz: Hart, fest, seidig und von sehr feiner Textur; polierfähig. Kern braun mit orangefarbenen Streifen; Splint dick, kaum unterscheidbar. Zahlreiche, undeutliche Markstrahlen.

Nutzung: Wegen des geringen Durchmessers vermutlich nur als Brennholz geeignet.

Description : arbre à port buissonnant de 12 à 15 m de hauteur sur un diamètre de 0,2 à 0,25 m ou plus souvent encore sous forme de buisson à tronc multiple. Ecorce mince, plus ou moins fissurée, brun clair teinté de rouge. Feuilles coriaces, ovales à allongées, vert sombre dessus, vert jaune dessous. Fleur à corolle monopétale, blanc crème et calice gamosépale, terminé en cinq pointes généralement. Fruit rouge orangé, à épicarpe épais et chair mince, sèche.

Habitat : forêts des Bahamas et des Antilles. En Amérique du Nord, seulement dans le sud de la Floride.

Bois : dur, résistant, satiné et à grain très fin ; susceptible d'être poli. Peu de contraste entre le bois de cœur brun à rubans orange et l'aubier épais. Nombreux rayons médullaires indistincts.

Intérêts : probablement convient-il tout au plus comme combustible, en raison de son faible diamètre.

STRONGBACK. STRONGBARK.

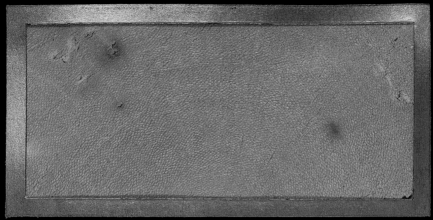

TRANSVERSE SECTION.

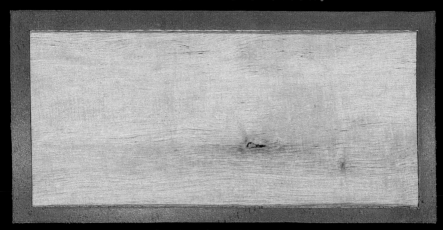

RADIAL SECTION.

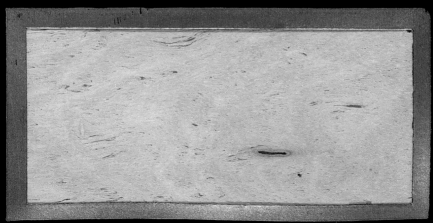

TANGENTIAL SECTION.

Ger. Havanische Bourreria. Fr. Bourrerier de Havana.

Sp. Ateje. Ircuma. Bureria.

B | BUCKTHORN FAMILY
Bearberry, Coffee-tree

FAMILIE DER KREUZDORNGEWÄCHSE
Faulbaum, Purshs Faulbaum

FAMILLE DE LA BOURDAINE
Frangula cascara, Ecorce sacrée, Nerprun de Pursh (Canada)

RHAMNACEAE
Frangula purshiana (DC.) Cooper
Syn.: *Rhamnus purshiana* DC.

Description: About 40 to 50 ft. (12 to 15 m) high and 14 to 16 in. (0.35 to 0.4 m) in diameter. Mostly branching 10 to 14 ft. (3 to 4 m) above the ground; also shrubby, in a creeping-cushion variety. Bark in older specimens dark to pale brown or gray with red. Thin, short scales.

Habitat: Shade-tolerant undergrowth in mountain forests of western North America, e.g. the Rocky Mountains or hills of southern California; creeping forms also occur near the coast.

Wood: Light, soft, or hard, not strong. Heartwood brown with red; sapwood thin and paler.

Use: Bark is strongly laxative and is used medicinally under the name "Cascara Sagrada."

Beschreibung: Etwa 12 bis 15 m hoch bei 0,35 bis 0,4 m Durchmesser. Meist schon 3 bis 4 m über dem Boden mehrstämmig, auch strauchig, sogar mit einer kriechend-polsterbildenden Varietät. Borke selbst bei alten Exemplaren dunkelbraun bis hellbraun oder grau mit rot. Dünne, kurze Schuppen.

Vorkommen: Als schattenverträglicher Unterwuchs in Bergwäldern des westlichen Nordamerika, z. B. den Rocky Mountains oder auf den Hügeln im südlichen Kalifornien, die Kriechformen auch küstennah vorkommend.

Holz: Leicht, weich oder hart und nicht fest. Kern braun mit rot; Splint dünn und heller.

Nutzung: Die Rinde stark abführend und als Drastikum unter dem Volksnamen Cascara Sagrada medizinisch genutzt.

Description : petit arbre d'environ 12 à 15 m de hauteur, avec un tronc de 0,35 à 0,4 m de diamètre, se divisant d'ordinaire à 3 ou 4 m au-dessus du sol. Se développe aussi sous forme de buisson ; il existe même une variété tapissante et rampante. Ecorce brun sombre à brun clair ou grise avec du rouge, même chez les spécimens âgés, couverte de courtes et minces écailles.

Habitat : espèce qui s'accommode de l'ombre et forme le sous-étage des forêts montagnardes de l'ouest de l'Amérique du Nord, dans les Rocheuses par exemple, ou sur les collines du sud de la Californie ; la forme à croissance prostrée se rencontre aussi à proximité du littoral.

Bois : léger, tendre ou dur et non résistant. Bois de cœur brun teinté de rouge ; aubier mince et plus clair.

Intérêts : écorce utilisée en médecine comme purgatif puissant sous le nom courant de « cascara ».

TRANSVERSE SECTION.

RADIAL SECTION.

TANGENTIAL SECTION

Ger. Kreuzdorn von Pursh. *Fr.* Nerprun de Pursh.

Sp. Cascara Sagrada.

B | BUCKTHORN FAMILY
Black Ironwood

FAMILIE DER KREUZDORNGEWÄCHSE
Schwarzes Eisenholz

FAMILLE DE LA BOURDAINE
Bois de fer noir, Bois de fer

RHAMNACEAE
Krugiodendron ferreum (Vahl) Urban

Description: The tree is frequently up to 35 ft. (10 m) in height, and the trunk has a maximum diameter of 20 in. (0.5 m). Usually, however, the trunk is considerably thinner. It has a distinctly furrowed, gray bark. The leathery, opposite leaves are ovate. Their upper surface is bright green, while the lower surface is paler. The flowers are yellowish green. The ovoid fruits have a thin, dark flesh surrounding the thin-walled seed.

Habitat: Very common in Florida.

Wood: The orange-brown heartwood is the heaviest of all the North American woods. Although it is hard, it is also brittle. After it is burned it leaves behind a large amount of ash.

Use: There is no information about the uses of this tree.

Beschreibung: Der Baum wird oft bis zu 10 m hoch und besitzt einen maximalen Stammdurchmesser von 0,5 m. Meist bleibt der Stamm jedoch erheblich dünner. Er besitzt eine deutlich rissige graue Borke. Die ledrigen, gegenständigen Blätter sind oval. Ihre Oberseite ist hellgrün, die Unterseite bleibt blasser. Die Blüten sind gelbgrün. Die ovalen Früchte besitzen ein dünnes schwarzes Fleisch, welches den dünnwandigen Samen umgibt.

Vorkommen: Sehr häufig in Florida.

Holz: Das orangebraune Kernholz ist das schwerste aller Hölzer Nordamerikas. Dennoch ist es zwar hart aber brüchig. Es hinterlässt bei der Verbrennung eine ungewöhnlich hohe Menge an Asche.

Nutzung: Über die Nutzung des Baumes liegen keine Informationen vor.

Description: arbre atteignant souvent 10 m de hauteur, avec un tronc dont le diamètre reste le plus souvent inférieur à 0,5 m. L'écorce est grise et à texture nettement fissurée. Les feuilles sont opposées, ovales et coriaces, vert clair à la face inférieure, plus pâles à la face inférieure. Les fleurs sont vert jaune. Les fruits sont noirs, ovales et peu charnus et renferment des graines protégées par un fin tégument.

Habitat: très répandu en Floride.

Bois: le plus lourd des bois de cœur d'Amérique du Nord, tendre certes mais aussi cassant; brun orange. Il laisse à la combustion une quantité inhabituelle de cendres.

Intérêts: aucun renseignement sur les utilisations éventuelles de cet arbre.

BLACK IRONWOOD.

RADIAL SECTION.

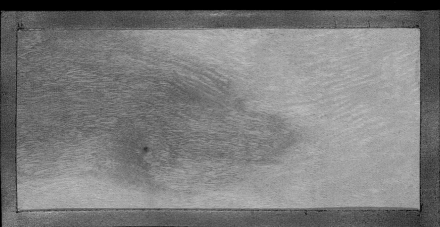

TANGENTIAL SECTION.

Fr. Petit bois-de-fer
Sp. Palo di hierro (Santo Domingo); Espejuelo (Porto Rico).

B | BUCKTHORN FAMILY
Blue Myrtle, California Lilac

FAMILIE DER KREUZDORNGEWÄCHSE
Kalifornischer Flieder

FAMILLE DE LA BOURDAINE
Céanothe thyrsiflore, Lilas bleu de Californie

RHAMNACEAE
Ceanothus thyrsiflorus Eschsch.

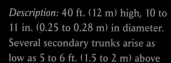

Description: 40 ft. (12 m) high, 10 to 11 in. (0.25 to 0.28 m) in diameter. Several secondary trunks arise as low as 5 to 6 ft. (1.5 to 2 m) above the ground; also shows shrubby growth. Bark thin, red-brown, with layered scales. Flowers in umbel-like clusters, blue or white, in early spring, scented.

Habitat: Shady hillsides, close to forest fringes or streams in western California. Largest specimens in the north of the region and in Redwood forests, often with Douglas Fir, bumelias, and willows and oaks. On wind-exposed banks only 1 to 2 ft. (0.3 to 0.6 m) high, but nevertheless flowering and fruiting.

Wood: Somewhat soft, close-grained. Heartwood pale brown; sapwood thin and darker.

Use: Introduced into England as early as 1837, but later displaced by flowering varieties and hybrids.

Discovered in 1816 by Escholtz during his circumnavigation of the world on the Rurik.

Beschreibung: 12 m hoch bei 0,25 bis 0,28 m Durchmesser. Schon 1,5 bis 2 m über dem Grund in mehrere Sekundärstämme geteilt, auch als Groß- oder Kleinstrauch. Borke dünn, rotbraun, in Schuppen geschichtet. Blüten in Scheindolden, blau oder weiß, im zeitigen Frühjahr, duftend.

Vorkommen: Schattige Hügel, nahe an Waldrändern oder Bächen Westkaliforniens. Die größten Exemplare im Norden des Gebiets und in den Wäldern des Küstenmammutbaums, oft mit Douglasien, Bumelien, verschiedenen Weiden und Eichen. An windgepeitschten Ufern nur 0,3 bis 0,6 m hoch, doch blühend und fruchtend.

Holz: Ziemlich weich und mit feiner Textur. Kern hellbraun; Splint dünn und dunkler.

Nutzung: Bereits 1837 nach England eingeführt, aber später von üppigen blühenden Varietäten und Hybriden verdrängt.

1816 auf der Weltumseglung der „Rurik" von Eschscholtz entdeckt.

Description: 12 m de hauteur, tronc de 0,25 à 0,28 m de diamètre, divisé en plusieurs tiges secondaires à 1,5 ou 2 m au-dessus du sol; aussi sous forme d'arbrisseau buissonnant ou de petit buisson. Fine écorce brun rouge en couches écailleuses. Fleurs bleues ou blanches groupées en corymbes, précoces et odorants.

Habitat: collines ombragées, à proximité de lisières forestières ou de ruisseaux en Californie. Les plus grands spécimens dans le nord de son aire et dans les forêts de séquoias côtiers, souvent en association avec le douglas, l'acoma, différents saules et chênes. Il ne fait que 0,3 à 0,6 m de hauteur sur les rives ventées, mais il fleurit et donne cependant des fruits.

Bois: assez tendre et à grain fin. Bois de cœur brun clair; aubier mince et plus sombre.

Intérêts: introduit en Angleterre en 1837 mais supplanté plus tard par d'autres variétés et hybrides à fleurs luxuriantes.

Blue Myrtle, Blue-blossom, California or Wild Lilac.

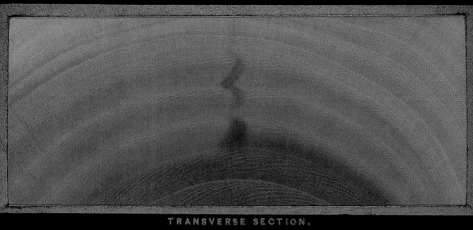

TRANSVERSE SECTION.

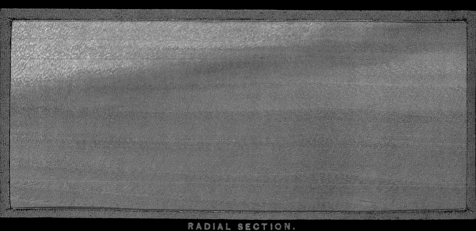

RADIAL SECTION.

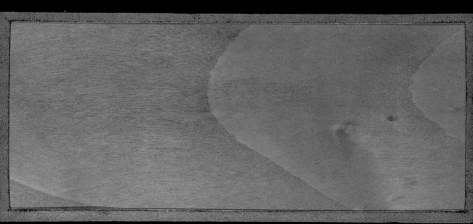

TANGENTIAL SECTION

Ger. Californischer Flieder. *Fr.* Lilas de Californie.
Sp. Lilas de California.

B | BUCKTHORN FAMILY
European Buckthorn, Waythorn

FAMILIE DER KREUZDORNGEWÄCHSE
Echter Kreuzdorn, Gemeiner Wegdorn

FAMILLE DE LA BOURDAINE
Nerprun purgatif, Bourge épine, Epine de cerf, Noirprun

RHAMNACEAE
Rhamnus cathartica L.

Description: Mostly a shrub, but also under favorable conditions a tree up to 30 ft. (9 m) high with a trunk 12 in. (0.3 m) in diameter, dark gray bark, and thorny branches. The leaves are ovate and opposite. The insignificant flowers arise from the axils of the leaves. The roundish black fruit has a very bitter taste.

Habitat: Native to Europe, and northern and western Asia. Since its introduction into North America, it has spread and now grows wild in many places.

Wood: Heavy, hard, and strong; very durable.

Use: The Buckthorn was introduced as a decorative shrub and a hedge plant.

Beschreibung: Meist als Strauch, unter günstigen Umständen auch als bis zu 9 m hoher Baum mit 0,3 m dickem Stamm, dunkelgrauer Borke und dornigen Ästen. Die ovalen Blätter sind gegenständig. Die unauffälligen kleinen Blüten stehen in den Achseln von Laubblättern. Die rundliche schwarze Frucht besitzt einen sehr bitteren Geschmack.

Vorkommen: Seine Heimat ist Europa, Nord- und Westasien. Nach seiner Einführung in Nordamerika verwilderte er vielerorts.

Holz: Schwer, hart und fest, sehr dauerhaft.

Nutzung: Wurde als Ziergehölz und für die Anlage von Hecken eingeführt.

Description: buisson le plus souven mais, en situation favorable, petit arbre pouvant atteindre 9 m de hauteur, avec un tronc de 0,3 m de diamètre, garni d'une écorce gris sombre et de branches épineuses. Feuilles ovales, opposées. Petites fleurs insignifiantes à l'aisselle des feuilles. Petits fruits noirs globuleux d'un goût très amer.

Habitat: originaire d'Europe, d'Asie occidentale et septentrionale. Introduit en Amérique du Nord et devenu subspontané dans de nombreux endroits.

Bois: lourd, dur et résistant; très durable.

Intérêts: introduit pour l'ornement et la plantation de haies.

Common or European Buckthorn. Waythorn.

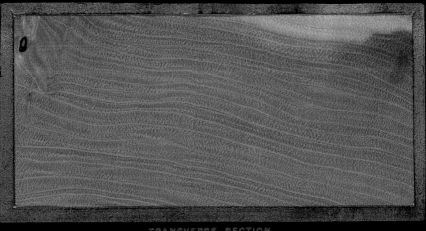

TRANSVERSE SECTION

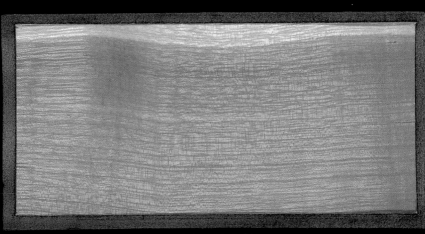

RADIAL SECTION.

TANGENTIAL SECTION.

Ger. Stechdorn. *Fr.* Nerprun.

B | BUCKTHORN FAMILY
Feltleaf Ceanothus

FAMILIE DER KREUZDORNGEWÄCHSE
Baum-Myrte

FAMILLE DE LA BOURDAINE
Céanothe arborescent, Céanothe en arbre

RHAMNACEAE
Ceanothus arboreus Greene

Description: A small tree, rarely higher than 25 ft. (7.5 m), with a trunk 12 in. (0.3 m) in diameter and a broad crown. In unfavorable circumstances it is often also a many-stemmed shrub. Leaves alternate, ovate to broadly elliptic, pointed at the tip and rounded at the base, the upper side dark green, the lower surface paler and covered with felted hairs. The flowers are pale blue in color, and grow in dense clusters near the ends of the young branches. The fruits are black.

Habitat: In the gorges of the islands off the Californian coast.

Wood: Very heavy and hard, with distinct annual rings. The reddish brown heartwood is surrounded by paler sapwood.

Use: The tree is not used, although its wood is suitable for firewood, and at flowering time it is a highly decorative plant.

Beschreibung: Der kleine Baum wird selten höher als 7,5 m mit 0,3 m dickem Stamm, der eine breite Krone trägt. Unter ungünstigeren Umständen auch oft als vielstämmiger Strauch. Blätter wechselständig, oval bis breit elliptisch, vorne spitz, an der Basis abgerundet, mit dunkelgrüner Ober- und filzig behaarter, blasser Unterseite. Die Blüten sind blassblau gefärbt und treten in üppigen Blütenständen nahe den Enden junger Äste auf. Die Früchte sind schwarz.

Vorkommen: In den Schluchten der Inseln vor der kalifornischen Küste.

Holz: Sehr schwer und hart, mit deutlichen Jahresringen. Das rotbraune Kernholz wird von hellerem Saftholz umgeben.

Nutzung: Der Baum wird nicht genutzt, obwohl sein Holz als Brennholz geeignet ist und er gerade zur Blütezeit ein lohnendes Ziergehölz darstellt.

Description : petit arbre atteignant rarement plus de 7,5 m de hauteu avec un tronc de 0,3 m de diamètre, surmonté d'une couronne ample. Se présente aussi sous forme de buisson à tronc multiple si les conditions sont plus défavorables. Feuilles alternes, ovales à larges-elliptiques, pointues à l'ape: arrondies à la base, vert foncé dessus, pâles et feutrées dessous. Fleurs bleu pâle, groupées en inflorescences touffues, terminales Petits fruits noirs.

Habitat : ravins dans les îles du littoral californien.

Bois : très lourd et dur, avec des cernes annuels distincts. Bois de cœur brun rouge ; aubier plus clai

Intérêts : l'arbre n'est pas utilisé, bien que son bois soit un bon com bustible et qu'il offre un beau spe tacle au moment de la floraison.

Tree Myrtle.

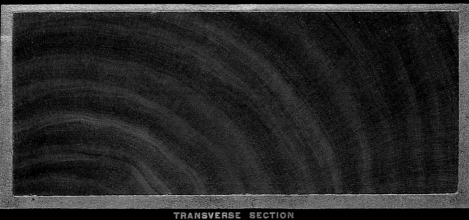

TRANSVERSE SECTION

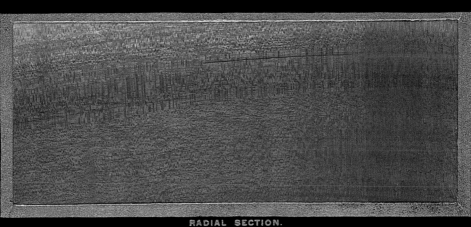

RADIAL SECTION.

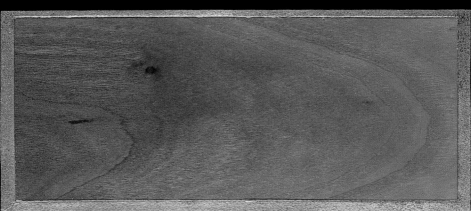

TANGENTIAL SECTION.

Ger. Baumische Myrte.   *Fr.* Myrte d'arbre.

Sp. Mirto de arbol.

B | BUCKTHORN FAMILY
Feltleaf Ceanothus

FAMILIE DER KREUZDORNGEWÄCHSE
Südlicher Flieder

FAMILLE DE LA BOURDAINE
Céanothe arborescent

RHAMNACEAE
Ceanothus oliganthus Nutt. subsp. *sorediatus*
(Hook. & Arn.) C. Schmidt
Syn.: *C. sorediatus* Hook. & Arn.

Description: 20 to 30 ft. (6 to 9 m) high and 6 to 10 in. (0.15 to 0.25 m) in diameter. Trunk straight, crown rounded. Branches slightly angled. Bark thin, dark brown, breaking off in small, square plates. Leaves alternate, broadly ovate, dark green above, densely felty below. Flowers in clusters at ends of young branches. Flowers pale blue. Fruit black when ripe.

Habitat: The islands of the Santa Barbara group off the Californian coast.

Wood: Heavy, hard, close-grained. Heartwood pale red-brown; sapwood with seven to eight annual rings, almost white. Thin, very clear medullary rays, annual rings clearly delineated by bands of small, open pores, with further irregular groups of pores between.

Use: Pretty tree, but apparently not yet widely grown.

Beschreibung: 6 bis 9 m hoch bei 0,15 bis 0,25 m Durchmesser. Stamm gerade mit runder Krone. Zweige leicht kantig. Borke dünn, dunkelbraun, in kleine, quadratische Plättchen zerfallend. Blätter wechselständig, breit oval, dunkelgrün auf der Oberseite, dicht filzig auf der Unterseite. Blütenstände basal mit Laubblättern. Blüten blassblau. Frucht reif von schwarzer Farbe.

Vorkommen: Inseln der Santa Barbara-Gruppe vor der Kalifornischen Küste.

Holz: Schwer, hart und mit feiner Textur. Kern hell rotbraun; Splint aus 7 bis 8 Jahresringen, fast weiß. Dünne, sehr deutliche Markstrahlen, Jahresringe deutlich abgegrenzt durch Bänder kleiner offener Gefäße, dazwischen unregelmäßige Gruppen weiterer Gefäße.

Nutzung: Ein hübscher Baum, bisher aber offenbar nicht in nennenswertem Umfang kultiviert.

Description : arbuste de 6 à 9 m de hauteur, avec un tronc droit de 0,15 à 0,25 m de diamètre et un houppier globuleux. Rameaux légèrement anguleux. Ecorce fine, brun sombre, se craquelant en petites plaques carrées. Feuilles alternes, à limbe ovale-large, vert foncé dessus, très pubescent dessous. Inflorescences bleu pâle, basales avec les feuilles. Fruits noi à maturité.

Habitat : groupe des îles de Santa Barbara en face de la côte californienne.

Bois : lourd, dur et à grain fin. Bois de cœur brun rouge clair ; aubier constitué de sept à huit cernes annuels, presque blanc. Rayons médullaires fins et très distincts, cernes annuels nettement circonscrits par des faisceaux vasculaires de type ouvert avec, entre les deux d'autres vaisseaux en groupes irréguliers.

Intérêts : bel arbuste ornemental mais peu cultivé jusqu'à maintenant.

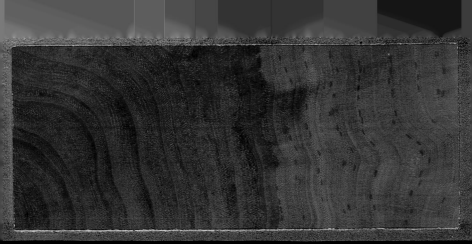

TRANSVERSE SECTION.

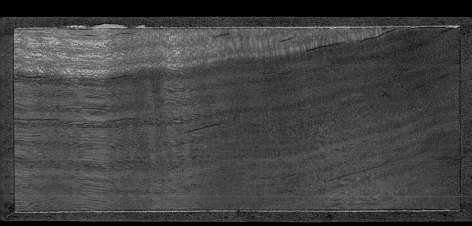

RADIAL SECTION.

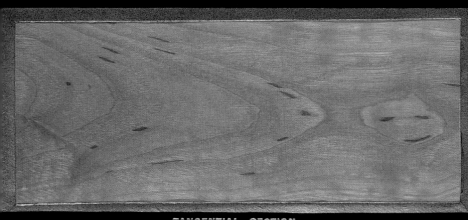

TANGENTIAL SECTION.

Ger. Süder Flieder   *Fr.* Lilas meridional

B | BUCKTHORN FAMILY
Island Redberry

FAMILIE DER KREUZDORNGEWÄCHSE
Insel-Kreuzdorn

FAMILLE DE LA BOURDAINE
Nerprun jaune des îles

RHAMNACEAE
Rhamnus crocea Nutt.

Description: This little tree is rarely more than 25 ft. (7 m) high, and has a trunk 12 in. (0.3 m) in diameter covered with a dark brown, irregularly fissured bark. The leathery leaves are pointed at both ends and their margins are finely toothed. The stipules soon fall off. The greenish flowers are in four parts, and are arranged in small clusters in the axils of the leaves. The red fruits are roundish.

Habitat: Slopes on the islands and coastal strips of southern and Lower California.

Wood: The rich reddish brown heartwood is very heavy, strong, and close-grained. Polishable, and surrounded by yellow sapwood that is only weakly developed.

Use: There are no details about its use. The bark is, however, like all buckthorns, suitable for use in tanning. The wood could be used as firewood.

Beschreibung: Der kleine Baum wird selten höher als 7 m mit einem 0,3 m dicken Stamm, der von einer dunkelbraunen, unregelmäßig rissigen Borke eingehüllt wird. Die ledrigen Blätter sind an beiden Enden zugespitzt und besitzen fein gezähnelte Ränder. Die Stipeln fallen früh ab. Die grünlichen Blüten sind vierzählig und sitzen büschelig in den Achseln der Laubblätter. Die roten Früchte sind rundlich.

Vorkommen: Abhänge auf den Inseln und Küstenstreifen Süd- und Niederkaliforniens.

Holz: Das kräftig rotbraune Kernholz ist sehr schwer, fest und feinkörnig. Es lässt sich gut polieren und wird vom nur schwach entwickelten gelben Saftholz umgeben.

Nutzung: Über eine Nutzung liegen keine Informationen vor. Die Rinde ist aber wie bei allen Kreuzdornen für die Verwendung in der Gerberei geeignet. Das Holz eignet sich zur Verfeuerung.

Description : arbuste atteignant rarement plus de 7 m de hauteur, avec un tronc de 0,3 m de diamèt et une écorce brun foncé, creusée de fissures irrégulières. Feuilles coriaces, terminées en pointe à la base et au sommet, finement dentées. Les stipules tombent très tôt. Fleurs tétramères vertes, disposée en touffes à l'aisselle des feuilles. Fruits globuleux rouges.

Habitat : pentes sur les îles et la frange côtière du sud et de la Basse-Californie.

Bois : très lourd, résistant et à grai fin ; se polissant bien. Bois de cœu d'un brun rouge intense ; aubier jaune, peu développé.

Intérêts : on ne dispose d'aucun renseignement sur les utilisations éventuelles de cette espèce. L'écorce se prêterait toutefois au tannage comme celle des autres nerpruns. Le bois pourrait aussi être utilisé comme combustible.

Island Buckthorn,   Island Bearwood.

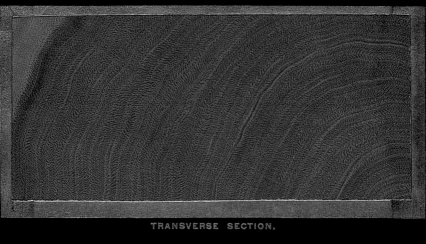

TRANSVERSE SECTION.

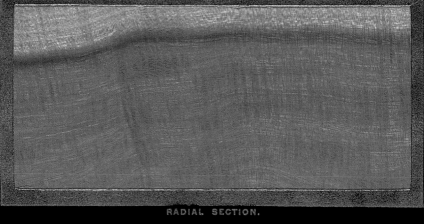

RADIAL SECTION.

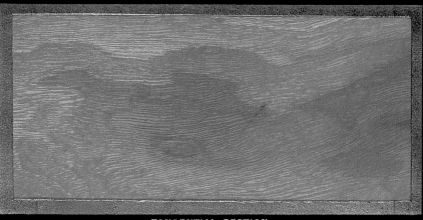

TANGENTIAL SECTION.

Ger. Eiländischer Kreuzdorn.   *Fr.* Nerprun insulaire.

Sp. Ramno de isla.

B | BUCKTHORN FAMILY
Lemonade Sumac, 'Mahogany'

FAMILIE DER KREUZDORNGEWÄCHSE
Sauerbeere

FAMILLE DE LA BOURDAINE
Sumac limonade

RHAMNACEAE
Rhus integrifolia (Nutt.) Benth. & Hook. f. ex H. Brewer &
S. Watson

Description: 35 ft. (10 m) high and 2 to 3 ft. (0.6 to 0.9 m) in diameter. Trunk short and stocky with many spreading branches or small creeping shrub. Bark a rich red-brown. Flaking in large scales. Evergreen. Fruit dark red, with resinous sap.

Habitat: Poor sandy soil by the sea and coastal cliffs in California; when exposed to coastal storms forming an impenetrable thicket just 1 to 2 ft. (0.3 to 0.6 m) high. Grows as a tree in protected sites. Optimum in the Todos Santos Bay in Lower California.

Wood: Heavy and hard. Heartwood a pretty, pure red; sapwood thin and pale.

Use: Good firewood and often used as such.

The oily white sap that exudes from the fruits is sometimes made into a refreshing drink in southern California.

Beschreibung: 10 m hoch bei 0,6 bis 0,9 m Durchmesser. Stamm kurz und gedrungen mit zahlreichen ausgebreiteten Ästen, oder kleiner Kriechstrauch. Borke kräftig rötlichbraun. Blättert in großen Schuppen ab. Immergrün. Frucht dunkelrot, mit harzigem Saft.

Vorkommen: Sterile Sandböden längs der Strände und Strandklippen in Kalifornien mit nur 0,3 bis 0,6 m hohen, dichten, undurchdringlichen Dickichten, wenn den Ozean-Stürmen ausgesetzt. An windgeschützten Stellen eher baumförmig. Optimum erst an der Todos Santos-Bucht in Niederkalifornien.

Holz: Schwer und hart. Kern hübsches, reines Rot; Splint dünn und blass.

Nutzung: Gutes Brennholz und viel dafür genutzt.

Der von den Früchten ausgeschwitzte ölige weiße Saft wird in Süd-Kalifornien gelegentlich zu Erfrischungsgetränken verarbeitet.

Description: 10 m de hauteur, avec un tronc court et trapu de 0,6 à 0,9 m de diamètre, armé de nombreuses branches étalées, ou petit buisson à port rampant. Ecorce d'un brun rougeâtre intens s'exfoliant en grandes écailles. Espèce sempervirente. Fruits roug foncé à suc résineux.

Habitat: sols sablonneux stériles le long des rivages et écueils de plag en Californie; ne formant que des fourrés impénétrables de 0,3 à 0,6 m de hauteur lorsqu'il est soumis aux tempêtes océaniques, ma à port arborescent dans les endroi abrités. Optimum de développement dans la baie de Todos Santos en Basse-Californie.

Bois: lourd et dur. Bois de cœur d'un beau rouge pur; aubier mince et pâle.

Intérêts: bois de chauffage de bonne qualité, très utilisé pour cette raison.

Dans le sud de la Californie, on prépare quelquefois une boisson rafraîchissante avec le suc blanc exsudé par les fruits.

Sour-berry, Sour-wood, Sour Oak, Mahogany.

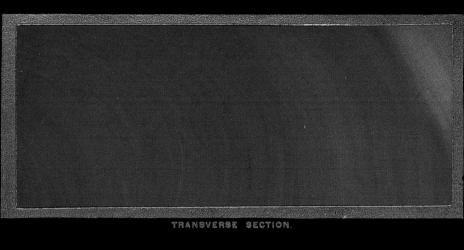

TRANSVERSE SECTION.

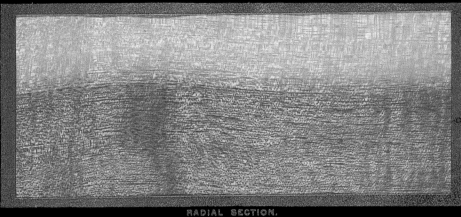

RADIAL SECTION.

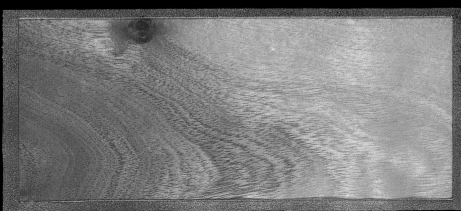

TANGENTIAL SECTION.

Ger. Sauerbeere. *Fr.* Baie aigre. *Sp.* Baya agria.

B | BUCKTHORN FAMILY
Naked Wood

FAMILIE DER KREUZDORNGEWÄCHSE
Westindisches Eisenholz

FAMILLE DE LA BOURDAINE
Colubrine elliptique, Bois mambi

RHAMNACEAE
Colubrina elliptica (Sw.) Briz. & Stern
Syn.: *C. reclinata* (L'Hér.) Brongn.

Description: 50 to 60 ft. (15 to 18 m) high and 3 to 4 ft. (0.9 to 1.2 m) in diameter. Trunk stout, later divided by numerous irregular furrows. Bark thin, orange-brown, separating into large, papery scales. Leaves elliptic, ovate or lanceolate, entire, yellow-green, persisting into the second year. Flowers in small groups, shorter than the flower stalks. Fruit globular, dry, dark orange, opening when ripe by splitting into three parts.

Habitat: Coasts and islands of the Caribbean and the Bahamas. In North America confined to the islands off the southern coast of Florida, forming a forest on Umbrella Key.

Wood: Heavy, hard, very strong, extremely brittle, with a silky sheen; polishable.

Use: On St. Croix and on other of the Virgin Islands the leaves are occasionally used to treat stomachache.

Beschreibung: 15 bis 18 m hoch bei 0,9 bis 1,2 m Durchmesser. Stamm kräftig, später durch zahlreiche unregelmäßige Furchen zerteilt. Borke dünn, orangebraun, sich in großen papierartigen Fetzen ablösend. Blätter elliptisch, oval oder lanzettlich, ganzrandig, gelbgrün, ausdauernd bis in das zweite Jahr. Blüten in kleinen Gruppen, kürzer als die Blütenstiele. Frucht kugelig, trocken, reif sich mit 3 Rissen öffnend, dunkelorange.

Vorkommen: Küsten und Inseln der Karibik sowie auf den Bahamas. In Nordamerika nur die Inseln vor Südflorida errreichend, auf Umbrella Key waldbildend.

Holz: Schwer, hart, sehr fest, äußerst brüchig, seidig; polierfähig.

Nutzung: Die Blätter werden auf St. Croix und auf den Virgin-Islands gelegentlich als Magenmittel genutzt.

Description : arbre de 15 à 18 m de hauteur, avec un tronc vigoureux de 0,9 à 1,2 m de diamètre, se scindant avec l'âge sous l'effet de nombreuses fissures irrégulières. Ecorce fine, d'un brun orangé, s'exfoliant en grands lambeaux papyracés. Feuilles elliptiques, ovales ou lancéolées, entières, vert jaune, demeurant sur l'arbre jusqu'à la deuxième année. Fleurs en petits groupes, plus courts que les pédoncules. Fruit sec, globuleux, s'ouvrant par trois petites fentes à maturité, orange foncé.

Habitat : côtes et îles des Caraïbes ainsi qu'aux Bahamas ; en Amérique du Nord, seulement sur les îles en face du sud de la Floride ; constitue des peuplements forestiers à Umbrella Key.

Bois : lourd, dur, très résistant, extrêmement cassant, satiné ; susceptible d'être poli.

Intérêts : à Sainte-Croix et sur les autres îles Vierges, les feuilles sont employées quelquefois pour soigner les maux d'estomac.

Naked-wood.   Naked-bark,   Soldier-wood

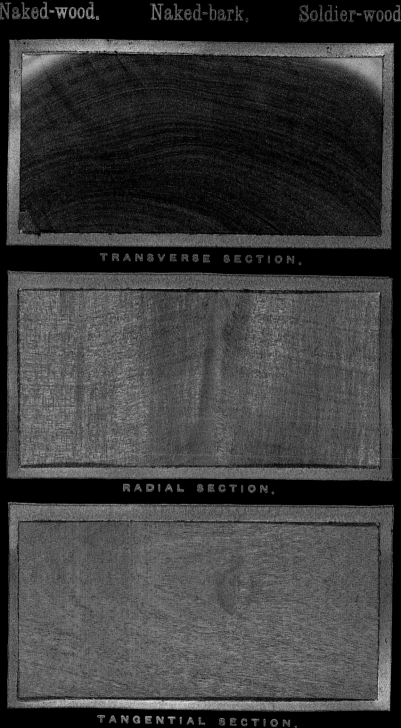

TRANSVERSE SECTION.

RADIAL SECTION.

TANGENTIAL SECTION.

B | BUCKTHORN FAMILY
Spiny Ceanothus, California Lilac

FAMILIE DER KREUZDORNGEWÄCHSE
Dornige Myrte

FAMILLE DE LA BOURDAINE
Céanothe épineux, Bois rouge de Californie

RHAMNACEAE
Ceanothus spinosus Nutt.

Description: Rare in the canyons of the San Rafael Mountains, 18 to 20 ft. (5.5 to 6.0 m) high and 6 to 8 in. (0.15 to 0.18 m) in diameter, with narrow, open crown. Usually shrubby. Branches angled and thorny. Bark dark red-brown. Leaves elliptic, leathery, rarely three-veined, long-lived. Inflorescence (thyrse) a composite of flowerheads, arising in the axils of leaves. Flowers pale to dark blue and very fragrant. Fruit subglobose, black.

Habitat: As thick undergrowth in woods, from mountain canyons down almost to sea level. Southern California.

Wood: Heavy, hard, strong, flexible. Heartwood pale yellow-brown to pale brown, partly tinged reddish; sapwood very thin, scarcely distinguishable (Begemann, 1990).

Use: Firewood. Bark and roots contain constituents that cause coagulation and stop bleeding; a tisane is prepared from the leaves (Begemann, 1990).

Beschreibung: Selten in den Cañons der San Rafael Mountains, 5,5 bis 6,0 m hoch bei 0,15 bis 0,18 m Durchmesser und mit schmaler, offener Krone. Normalerweise strauchförmig. Zweige kantig, bedornt. Borke dunkel rotbraun. Blätter elliptisch, ledrig, selten dreinervig, ausdauernd. Blütenstände aus Teilblütenständen zusammengesetzt (Thyrsen), aus den Achseln von Laublättern entspringend. Blüten hell- bis dunkelblau und sehr wohlriechend. Frucht abgeflacht kugelig, schwarz.

Vorkommen: Als dichter Unterwuchs in Wäldern, von Gebirgscañons bis fast auf Meereshöhe hinab. Süd-Kalifornien.

Holz: Schwer, hart, fest und elastisch. Kern hell gelblich braun bis hellbraun, z. T. mit rötlicher Tönung; Splint sehr dünn, kaum unterscheidbar (Begemann 1990).

Nutzung: Brennholz. Rinde und Wurzeln besitzen blutungsstillende und blutgerinnende Eigenschaften, aus den Blättern wird Tee bereitet (Begemann 1990).

Description: atteint dans les canyon des San Rafael Mountains raremen 5,5 à 6,0 m de hauteur sur un diamètre de 0,15 à 0,18 m. Couronne étroite et ouverte. Port de buisson le plus souvent. Rameaux anguleux et épineux. Ecorce brun rouge. Feuilles elliptiques, coriaces, rarement trinervées, persistantes. Inflorescences en grappes composées, axillaires (thyrses). Fleurs bleu foncé à bleu clair exhalant un odeur agréable. Fruits globuleux aplatis, noirs.

Habitat: constitue des fourrés denses dans les forêts, depuis les canyons d'altitude jusqu'au niveau de la mer pratiquement. Croît dans le sud de la Californie.

Bois: lourd, dur, résistant et élastique. Bois de cœur brun jaunâtre clair à brun clair, partiellement teinté de rouge; aubier très mince, à peine distinct (Begemann 1990).

Intérêts: bois de chauffage. L'écorce et les racines ont des propriétés hémostatiques et coagulantes; les feuilles entrent dans la compositior de tisanes (Begemann 1990).

Redwood Myrtle, Spiny Myrtle.

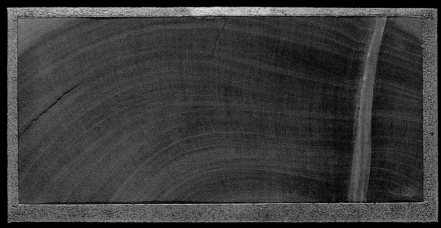

TRANSVERSE SECTION.

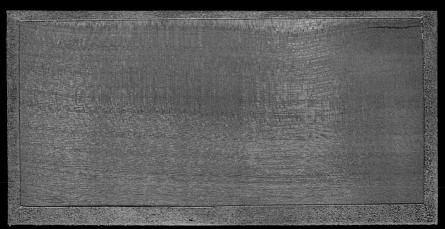

RADIAL SECTION.

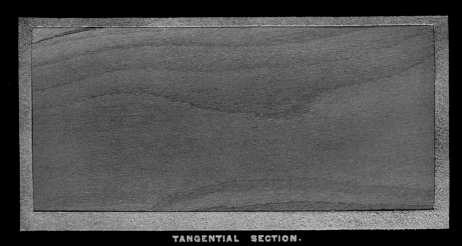

TANGENTIAL SECTION.

Ger. Dornige Myrte.　　　　*Fr.* Myrte espineuse.

Sp. Myrto espinosa.

B | BUCKWHEAT FAMILY, KNOT-WEED FAMILY, SMARTWEED FAMILY

Pigeon Plum, Cucubano

FAMILIE DER KNÖTERICHGEWÄCHSE
Coccaloba

FAMILLE DES RENOUÉES
Raisinier à feuilles variées, Maivisse, Raisinier des pigeons

POLYGONACEAE
? Coccolobis diversifolia Jacq.
Syn.: *C. laurifolia* Jacq.

Description: 60 to 70 ft. (18 to 21 m) high and 1 to 2 ft. (0.3 to 0.6 m) in diameter. Trunk straight, with spreading branches that form a dense, rounded crown. Bark thin, gray with a red tinge, broken on the surface into small, thin scales, which, when they fall, expose the dark purple inner bark. Leaves ovate to ovate-lanceolate, entire, thick, bright green above and pale beneath. Flowers in terminal clusters. Fruits ovoid, dark red with a rather acid flesh, distributed by raccoons, other mammals, and birds.

Habitat: Coastal strips in Florida, and also throughout the Caribbean.

Wood: Heavy, very hard, strong, brittle, and close-grained. Heartwood rich dark brown, tinged red; sapwood paler. Small, scattered pores, annual rings and medullary rays difficult to distinguish.

Use: Occasionally for cabinets. Introduced into Europe at the beginning of the 18th century, and flowering for the first time near Paris in 1820.

Beschreibung: 18 bis 21 m hoch bei 0,3 bis 0,6 m Durchmesser. Stamm gerade, mit ausladenden Ästen, die eine dichte, kugelige Krone bilden. Borke dünn, grau mit roter Tönung, an der Oberfläche in dünne Schüppchen zerfallend, die nach ihrem Herunterfallen die dunkelviolette innere Rinde freilegen. Blätter oval bis oval-lanzettlich, ganzrandig, dick, hellgrün auf der Ober- und blass auf der Unterseite. Blüten in endständigen Trauben. Früchte oval, dunkelrot mit säuerlichem Fleisch. Früchte werden von Waschbären, anderen Säugetieren und Vögeln verbreitet.

Vorkommen: Küstenstreifen in Florida, weiter im karibischen Raum verbreitet.

Holz: Schwer, sehr hart, fest, brüchig und mit feiner Textur. Kern tief dunkelbraun, rot getönt; Splint heller. Kleine zerstreute Gefäße, Jahresringe und Markstrahlen schwer unterscheidbar.

Nutzung: Gelegentlich für Kabinette. Anfang des 18. Jahrhunderts nach Europa eingeführt und erstmals 1820 nahe bei Paris blühend.

Description: 18 à 21 m de hauteur sur un diamètre de 0,3 à 0,6 m. Tronc droit, avec des branches étalées, formant un houppier globuleux et dense. Ecorce mince, grise teintée de rouge, à texture finement écailleuse, révélant une écorce interne violet foncé lorsque les écailles sont tombées. Feuilles ovales à ovales-lancéolées, entières, épaisses, vert clair dessus, pâles dessous. Feuilles en grappes terminales. Fruits ovales, rouge foncé, à chair acide, disséminés par les ratons laveurs, par d'autres mammifères et les oiseaux.

Habitat: zones côtières de Floride, répandu aussi dans la région des Caraïbes.

Bois: lourd, très dur, résistant, cassant et à grain fin. Bois de cœur brun profond teinté de rouge; aubier plus clair. Petits vaisseaux disséminés, cernes annuels et rayons médullaires difficiles à différencier.

Intérêts: utilisé de temps à autre en ébénisterie. Introduit en Europe au 18e siècle, et première floraison à Paris en 1820.

PIGEON PLUM.

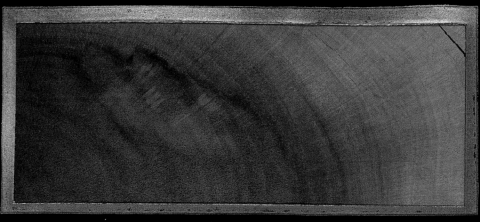

TRANSVERSE SECTION.

RADIAL SECTION.

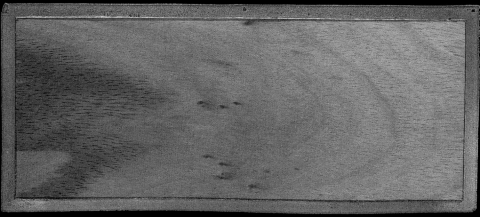

TANGENTIAL SECTION.

B | BUCKWHEAT FAMILY, KNOT-WEED FAMILY, SMARTWEED FAMILY

Sea Grape

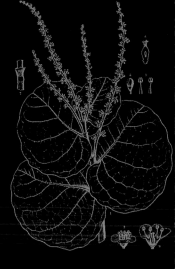

FAMILIE DER KNÖTERICHGEWÄCHSE
Seetraube

FAMILLE DES RENOUÉES
Raisinier bord-de-mer

POLYGONACEAE
Coccoloba uvifera (L.) L.

Description: Rarely higher than 15 ft. (4.6 m) or more than 3 to 4 ft. (0.9 to 1.2 m) in diameter. Trunk short, gnarled, and contorted, with short branches that form a compact, rounded crown. Often also a shrub with a creeping stem. Bark thin, smooth, light brown, with irregular pale blotches. Leaves broadly ovate or roundish, very thick and leathery, dark green in color, remaining two or three years on the branches. Flowers white, in terminal and lateral clusters, appearing almost throughout the whole year. Fruit roundish-ovoid, purple or greenish white, with a thin flesh.

Habitat: Saline beaches and coastal strips in Florida, and across the Caribbean as far as Brazil.

Wood: Very heavy, hard, and close-grained; polishable. Heartwood dark brown or purple; sapwood thick, paler.

Use: Occasionally for cabinets. Because of their astringent and styptic effect, the fruits are used medicinally in the West Indies.

Beschreibung: Selten höher als 4,6 m bei 0,9 bis 1,2 m Durchmesser. Mit kurzem, knorrig-verdrehtem Stamm und kurzen Ästen, eine runde, kompakte Krone bildend. Häufig auch als Strauch mit kriechendem Stamm. Borke dünn, glatt, hellbraun und mit unregelmäßigen blassen Flecken. Blätter breit oval, rundlich, sehr dick und ledrig, von dunkelgrüner Farbe, 2 bis 3 Jahre am Ast bleibend. Blüht fast über das ganze Jahr, Blüten weiß, in end- und seitenständigen Trauben. Frucht rundlich oval, violett oder grünlich weiß, mit dünnem Fleisch.

Vorkommen: Salzige Strände und Küstenstreifen in Florida, ferner über die Karibik bis nach Brasilien verbreitet.

Holz: Sehr schwer, hart und mit feiner Textur; polierfähig. Kern dunkelbraun oder violett; Splint dick, heller.

Nutzung: Gelegentlich für die Kunsttischlerei. Die Früchte werden aufgrund ihrer adstringierenden/blutstillenden Wirkung auf den Westindischen Inseln medizinisch genutzt.

Description: espèce atteignant rarement plus de 4,6 m de hauteur sur un diamètre de 0,9 à 1,2 m. Tronc court et noueux, avec des ramifications courtes, constituant un houppier rond et compact. Souvent aussi sous forme de buisson à port rampant. Ecorce mince, lisse, brun clair et ponctuée de taches pâles, irrégulières. Feuilles larges-ovales, arrondies, très épaisses et coriaces, vert sombre et demeurant sur l'arbre 2 à 3 ans. Floraison presque tout au long de l'année, en grappes blanches terminales ou latérales. Fruit arrondi-ovale, violet ou blanc verdâtre, à pulpe peu épaisse.

Habitat: plages et franges littorales salines de Floride, puis répandu depuis les Caraïbes jusqu'au Brésil.

Bois: très lourd, dur et à grain fin; susceptible d'être poli. Bois de cœur brun sombre ou violacé; aubier épais, plus clair.

Intérêts: utilisé quelquefois en ébénisterie. Les fruits sont employés en médecine aux Antilles en raison de leurs propriétés astringentes et hémostatiques.

SEA GRAPE.

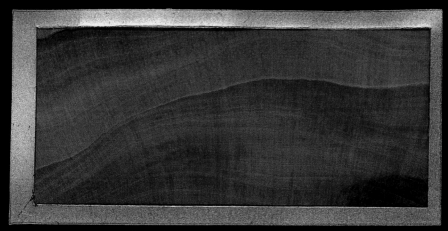

TRANSVERSE SECTION.

RADIAL SECTION.

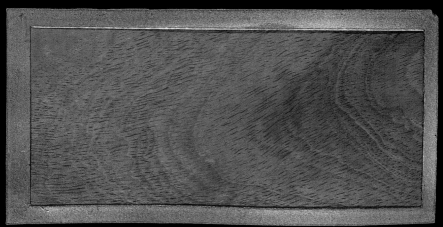

TANGENTIAL SECTION.

Ger. Coccaloba-holz.   Fr. Raisinier de mer.

Sp. Uvero. Uvifero. Uva del mar. Uva (Cuba).

C | CACAO FAMILY
Flannel Bush, California Slippery Elm

FAMILIE DER STERKULIENGEWÄCHSE
Fremontie

FAMILLE DES CACAOYERS
Fremontodendron de Californie

STERCULIACEAE
Fremontodendrom californicum (Torr.) Coville

Description: This little tree is rarely more than 30 ft. (9 m) high, with a trunk 12 in. (0.3 m) in diameter, and stout, erect branches. The deeply furrowed bark is very dark brown, and is broken into thick scales. The thick, usually weakly three-lobed, or sometimes five to seven-lobed leaves have short red-brown hairs. Flowers yellow, with felted hairs on the outside. The fruit forms a leathery capsule.

Habitat: Dry, stony soils at the foot of hills and mountains in California.

Wood: Hard, heavy, and strong. The heartwood is light red-brown at first, finally becoming dark chocolate-brown. The sapwood is a creamy white.

Use: The inner bark is very mucilaginous, and is used medicinally for poultices.

Beschreibung: Der kleine Baum wird selten höher als 9 m mit einem Stamm von 0,3 m Dicke und kräftigen, aufrechten Ästen. Die tief-rissige Borke ist sehr dunkel braun und löst sich in dicken Stücken ab. Die dicken und meist leicht dreilappigen, manchmal auch fünf- bis siebenlappigen Blätter tragen kurze, rotbraune Haare. Blüten gelb, außen filzig behaart. Die Frucht ist eine ledrige Kapsel.

Vorkommen: Trockene, steinige Böden am Fuße der Hügel und Berge Kaliforniens.

Holz: Hart, schwer und fest. Das Kernholz ist zuerst hell rotbraun und wird schließlich dunkel schokoladenbraun. Das Saftholz ist cremeweiß.

Nutzung: Die innere Rinde ist sehr schleimig und wird für medizinische Umschläge verwendet.

Description: arbuste atteignant rarement plus de 9 m de hauteur, avec un tronc de 0,3 m de diamètre et des branches dressées, vigoureuses. Ecorce d'un brun très sombre, profondément fissurée et se détachant en grosses plaques. Fleurs épaisses à trois, parfois aussi 5 à 7 lobes peu marqués, couvertes de petits poils brun rouge. Fleurs jaunes, feutrées sur la face supérieure. Fruits en capsules coriaces.

Habitat: sols caillouteux secs au pied des collines et des montagnes de Californie.

Bois: dur, lourd et résistant. Bois de cœur brun rouge clair puis couleur chocolat; aubier blanc crème.

Intérêts: plante médicinale dont l'écorce interne, très mucilagineuse est employée en compresses.

Fremontia, Cal. Slippery "Elm".

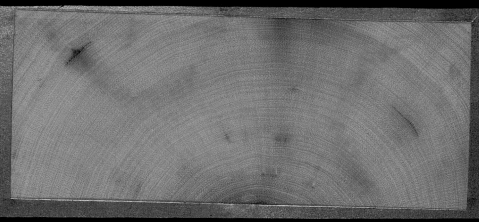

TRANSVERSE SECTION.

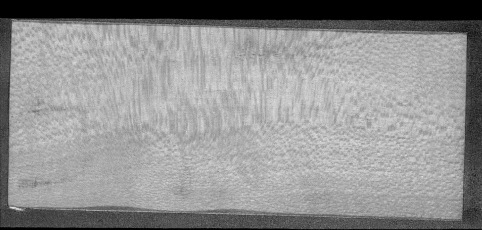

RADIAL SECTION.

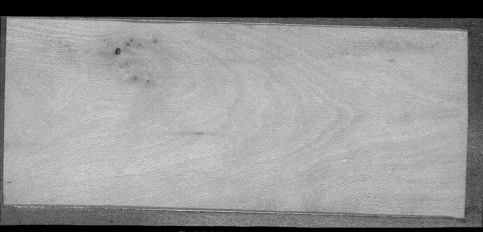

TANGENTIAL SECTION.

Ger. Fremontia. *Fr.* Fremontia.

Sp. Fremontia.

C | CACTUS FAMILY
Prickly Pear, Tuna

FAMILIE DER KAKTUSGEWÄCHSE
Opuntie

FAMILLE DES CACTUS
Opuntia figuier-de-barbarie

CACTACEAE
Opuntia tuna Mill.

Description: This cactus, which reaches a height of up to 15 ft. (4.5 m) and a diameter of more than 15 in. (0.4 m), is covered with relatively widely spaced tufts of yellow spines. Particularly long stems are formed in cactus thickets. The fleshy, carmine red fruits are armed with small spines.

Habitat: Native in the arid areas of South America and the islands of the West Indies. Introduced into the warm southern part of North America, occasionally escaping cultivation and growing wild.

Wood: Of a very loose, open texture, forming a lattice-like framework. After the cactus has died, this framework remains standing for a considerable time.

Use: Primarily as a hedge for protection purposes, and in earlier times even of military importance as a defense against attacks by Indians. The fruit is edible but troublesome to harvest on account of the fine spines. It is the host plant for a scale insect, from whose body an economically important red dye is obtained.

Beschreibung: Der bis zu 4,5 m hohe und einen Durchmesser von mehr als 0,4 m erreichende Kaktus ist mit relativ weit voneinander entfernt stehenden, gelben Stachelbüscheln bedeckt. Besonders lange Stämme werden in Dickichten des Kaktusses gebildet. Die fleischigen, karminroten Früchte sind mit kleinen Stacheln bewehrt.

Vorkommen: Heimisch in den Trockengebieten Südamerikas und auf den Westindischen Inseln. In den warmen Süden Nordamerikas eingeführt und z. T. verwildert.

Holz: Von sehr lockerer, offener Struktur, ein gitterartiges Rahmenwerk bildend. Nach dem Absterben der Kaktee bleibt das Gerüst noch einige Zeit stehen.

Nutzung: Vorwiegend als Schutzhecke und in früherer Zeit sogar von militärischer Bedeutung als Schutz gegen Indianerüberfälle. Die Frucht ist essbar, wegen der feinen Stacheln aber nur mühsam zu ernten. Nahrungspflanze einer Schildlaus, aus deren Körper eine wirtschaftlich bedeutsame rote Farbe gewonnen wird.

Description : cactus de 4,5 m de hauteur sur un diamètre de 0,4 m, couvert de touffes assez éparses de grandes épines jaunes. Des tiges très longues se développent dans la masse touffue du cactus. Les fruits sont des baies charnues, couleur carmin, armées de petits poils épineux.

Habitat : originaire des zones désertiques de l'Amérique du Sud et des Antilles. Introduit dans les régions chaudes du sud de l'Amérique du Nord, où il est retourné à l'état sauvage à certains endroits.

Bois : d'une structure très lâche, ouverte et formant un assemblage d'éléments entrecroisés, qui demeure encore quelque temps après la mort du cactus.

Intérêts : utilisé avant tout en haies protectrices, et même planté autrefois en haies défensives pour repousser les attaques des Indiens. Le fruit est comestible mais peu aisé à récolter en raison de ses petites épines. Espèce qui nourrit la cochenille, dont on extrait le carmin, une teinture d'un grand intérêt économique.

Mission Cactus, Indian Fig, Prickly Pear.

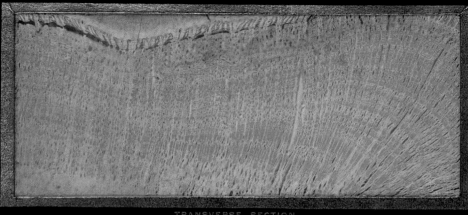

TRANSVERSE SECTION

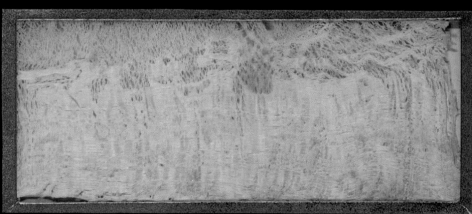

RADIAL SECTION.

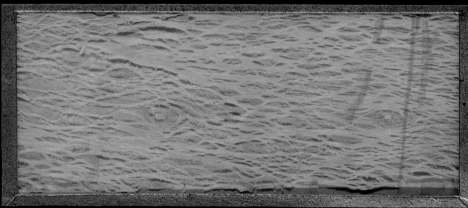

TANGENTIAL SECTION.

Ger. Indische Feige.　　*Fr.* Figue d'Indien.　　*Sp.* Nopal.

C | CACTUS FAMILY
Saguaro

FAMILIE DER KAKTUSGEWÄCHSE
Säulenkaktus

FAMILLE DES CACTUS
Carnégie géante, Cierge géant

CACTACEAE
Carnegiea gigantea (Engelm.) Britton & Rose
Syn.: *Cereus giganteus* Engelm.

Description: 50 to 60 ft. (15 to 18 m) high and 2 ft. (0.6 m) in diameter. Trunk columnar, thicker in the middle, gradually tapering above and below. Branch-free or with few, usually two or three branches in the upper half. Eight to twelve ribs at the base. Ribs carry groups of large spines. Flowers in groups at tip of trunk. Fruit ovoid, bursting open in three to four segments, seeds smooth.

Habitat: Crevices in low, stony hills as well as dry, rocky mesas in the deserts of southern North America from Arizona to the Sonora.

Wood: Very light, soft, strong, silky-sheened and rather coarse-grained, polishable and almost indestructible in the ground. Many distinct medullary rays and bands of open cells along the margins of annual rings.

Use: Roof beams, fences, used by American Indians for spears, bows etc. Sap gathered with long, curved sticks, dried, and either consumed immediately or pressed and kept as a molasses-like drink for the winter.

Beschreibung: 15 bis 18 m hoch bei 0,6 m Durchmesser. Stamm säulenförmig, oberhalb der Mitte am dicksten, nach beiden Richtungen allmählich verschmälert. Astfrei oder in der oberen Hälfte mit wenigen, meist 2 oder 3 Ästen. An der Basis mit 8 bis 12 Rippen. Auf den Rippen Gruppen langer Dornen. Blüten in Gruppen an der Spitze des Stammes. Frucht oval, in 3 bis 4 Teile aufplatzend, Samen glatt.

Vorkommen: Spalten niedriger, steiniger Hügel sowie trockene, steinige Mesas der Wüste im südlichen Nordamerika von Arizona bis in die Sonora-Wüste.

Holz: Sehr leicht, weich, fest, seidig und eher mit grober Textur; polierfähig und im Boden fast unzerstörbar. Zahlreiche deutliche Markstrahlen und Bänder offener Zellen längs der Grenzen der Jahresringe.

Nutzung: Dachsparren, Zäune, von den Indianern für Lanzen, Bögen etc. verwendet. Der Saft wird mit langen Stöcken gesammelt, getrocknet und direkt verzehrt oder gepret und als molasseähnliches Getränk für den Winter aufbewahrt.

Description: 15 à 18 m de hauteur, avec un tronc columnaire, atteignant 0,6 m de diamètre aux trois quarts de sa hauteur, se rétrecissant ensuite vers le haut et le bas. Dépourvu de ramifications ou portant peu de rameaux, 2 ou 3 le plus souvent, dans sa moitié supérieure. Longues épines réunies en faisceaux sur les côtes. Groupes de fleurs à l'extrémité du tronc. Fruits ovales, éclatant en 3 ou 4 parties, avec des graines lisses.

Habitat: fissures de collines basses et caillouteuses, mesas sèches et caillouteuses du désert du sud de l'Amérique du Nord depuis l'Arizona jusque dans le désert de Sonora

Bois: très léger, tendre, résistant, satiné et à grain plutôt grossier; susceptible d'être poli et quasi indestructible sur pied. Nombreux rayons médullaires distincts et bandes de cellules de type ouvert le long des cernes annuels.

Intérêts: chevrons, clôtures; utilisé par les Indiens pour les arcs, les lances etc. Sève récoltée au moyen de perches courbées, séchée pour être consommée telle quelle ou conservée comme boisson siru-peuse pour l'hiver.

Giant Cactus, Saguaro, Suwarro.

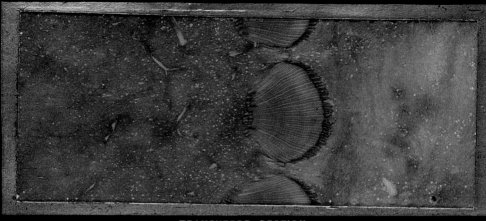

TRANSVERSE SECTION.

RADIAL SECTION.

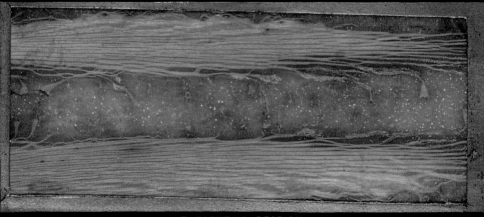

TANGENTIAL SECTION.

Ger. Riesencactus.   *Fr.* Cactus gigantesque.

Sp. Saguaro.

C | CASHEW FAMILY

American Smoke Tree, Chittam wood

FAMILIE DER SUMACHGEWÄCHSE
Amerikanische Pistazie

FAMILLE DE LA MANGUE
Fustet d'Amérique, Arbre à perruque

ANACARDIACEAE
Cotinus obovatus Raf.
Syn.: *C. americanus* Nutt.

Description: 25 to 30 ft. (7.5 to 9 m) high and 12 to 15 in. (0.3 to 0.35 m) in diameter. Branching only occurs from 12 to 14 ft. (3.7 to 4 m) above the ground, when the main stem divides into numerous, erect, partial stems with wide-spreading, branches. Bark thin, light gray, and furrowed. Inner bark white, becoming orange in the air, exuding an unpleasant-smelling juice when damaged. Leaves more or less ovate, thin, dark green. Inflorescences in the form of narrow panicles, with long-stalked, few-flowered, partial inflorescences. Sterile branches of the inflorescences lightly covered with pale purple to brown hairs.

Habitat: On limestone in mountain regions in southern North America and in forests as thickets or small groups. Not common anywhere.

Wood: Light, soft, and rather coarse-grained; very durable in the ground. Heartwood a bright rich orange; sapwood almost white. Annual rings marked by rows of open pores with medullary rays numerous and very distinct.

Use: Fence-posts. An orange dye can be obtained from the wood. Also as a decorative plant.

Beschreibung: 7,5 bis 9,0 m hoch bei 0,3 bis 0,35 m Durchmesser. Verzweigungen erst in 3,7 bis 4,0 m Höhe über dem Boden in zahlreiche aufrechte Teilstämme mit weit ausladenden Ästen. Borke dünn, hellgrau und rissig. Innere Rinde weiß, an der Luft aber orange werdend, bei Verletzung einen überriechenden Saft absondernd. Blätter eiförmig bis oval, dünn, dunkelgrün. Blütenstände schlanke Rispen mit lang gestielten, wenigblütigen Teilblütenständen. Sterile Zweige der Blütenstände mit locker stehenden, blassvioletten bis braunen Haaren.

Vorkommen: Über Kalkgestein der Bergländer im südlichen Nordamerika, im Inneren dichter Wälder als Dickichte oder in kleinen Gruppen. Nirgendwo häufig.

Holz: Leicht, weich und eher mit grober Textur; im Boden sehr haltbar. Kern hell, tieforange; Splint fast weiß. Jahresringe durch einige Reihen offener Gefäße markiert, mit vielen, sehr deutlichen Markstrahlen.

Nutzung: Zaunpfähle. Aus dem Holz lässt sich eine orange Farbe gewinnen. Auch als Zierpflanze.

Description: arbuste de 7,5 m à 9,0 m de hauteur sur un diamètre de 0,3 à 0,35 m. Le tronc se ramifie très haut, entre 3,7 et 4,0 m au-dessus du sol, en de nombreuses tiges secondaires dressées, portant des branches très étalées. Ecorce mince, gris clair et fissurée. Ecorce interne de couleur blanche, devenant orange à l'air et excrétant un suc nauséabond à la suite d'un traumatisme. Feuilles ovoïdes à ovales, fines, vert sombre. Fleurs en fines panicules de grappes élémentaires à long pédicule et peu fournies. Les rameaux stériles des inflorescences sont garnis de poils épars, violet pâle à brun.

Habitat: sur les calcaires dans les régions élevées du sud de l'Amérique du Nord, en fourrés ou en petites formations au sein de forêts denses. Nulle part très répandu.

Bois: léger, tendre, à grain grossier très durable sur pied. Bois de cœur orange clair; aubier presque blanc. Cernes annuels délimités par quelques faisceaux vasculaires de type ouvert avec de et nombreux rayons médullaires distincts.

Intérêts: pieux. Extraction d'une teinture orange. Ornement.

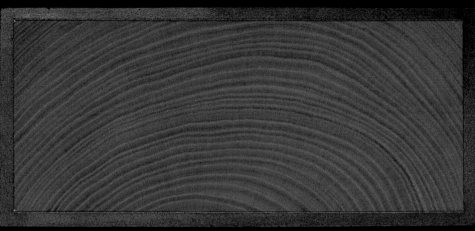

TRANSVERSE SECTION.

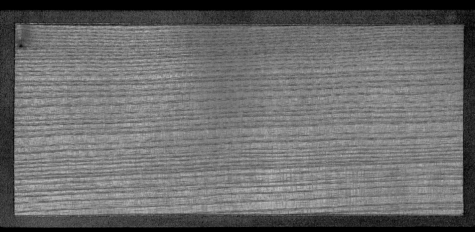

RADIAL SECTION.

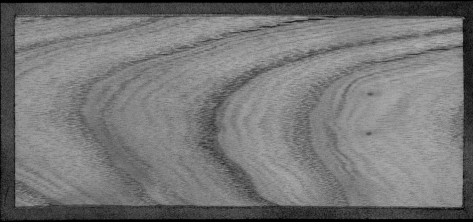

TANGENTIAL SECTION.

C | CASHEW FAMILY
Dwarf Sumac, Common Sumac

FAMILIE DER SUMACHGEWÄCHSE
Sumach, Kopall-Sumach

FAMILLE DE LA MANGUE
Sumac brillant, Copal, Sumac ailé

ANACARDIACEAE
Rhus copallinum L.

Description: 25 to 30 ft. (7 to 9 m) high and 8 to 10 in. (0.2 to 0.25 m) in diameter. Trunk short, stout. Shrubby in the north. Contains a colorless, watery juice. Bark moderately thick, light brown with dark reddish brown, rounded excrescences. Branches and leaf stalks hairy, leaves divided into 9 to 21 leaflets. Flowers dioecious in pyramidal panicles. Fruit slightly flattened, with a thin red skin covered with colorless hairs.

Habitat: Dry hills and mountain ridges in eastern North America, developing into a tree only in southern Arkansas and eastern Texas. Forming thickets on poor, stony soils.

Wood: Light, soft, with a silky sheen, and close-grained. Heartwood light brown, streaked with green, and often tinged red; sapwood of four or five annual rings, paler.

Use: The leaves are rich in tannin, which is used for tanning leather and as a dye. Also its fruits are useful. Its foliage is more attractive than that of the other Sumacs, and it is well suited to be a decorative tree or shrub, especially as it quickly spreads, even on rocky slopes.

Beschreibung: 7 bis 9 m hoch bei 0,2 bis 0,25 m Durchmesser. Stamm kurz, kräftig. Im Norden als Strauch. Enthält farblosen, wässrigen Milchsaft. Borke mäßig dick, hellbraun mit dunkel rotbraunen, runden Auswüchsen. Äste und Blattstiele behaart, Blätter in 9 bis 21 Fiedern geteilt. Blüten diözisch in pyramidalen Rispen. Frucht leicht abgeflacht mit dünner roter Hülle, die von farblosen Haaren bedeckt ist.

Vorkommen: Trockene Hügel und Bergkämme im östlichen Nordamerika, baumförmig nur Süd-Arkansas und Ost-Texas. Auf steinigen, armen Böden Dickichte bildend.

Holz: Leicht, weich, seidig und mit grober Textur. Kern hellbraun, mit grünen Streifen und oft rot getönt; Splint 4 bis 5 Jahresringe, heller.

Nutzung: Blätter sind reich an Tannin und werden zur Ledergerbung und Färberei genutzt. Auch die Früchte sind nutzbar. Das Laub ist noch attraktiver als jenes der anderen Sumachs, gut als Ziergehölz geeignet, zumal sie sich auch auf steinigen Hängen rasch ausbreitet.

Description : arbuste de 7 à 9 m de hauteur, avec un tronc court et vigoureux de 0,2 à 0,25 m de diamètre. Sous forme de buisson dans le nord. Secrète un suc laiteux incolore et aqueux. Ecorce moyennement épaisse, brun clair, présentant des excroissances liégeuses brun rouge foncé. Rameaux et pédoncules pubescents, feuilles composées de 9 à 21 folioles. Fleurs dioïques, groupées en panicules. Fruits en drupes légèrement aplaties, rouges et tomenteuses.

Habitat : collines et crêtes sèches dans l'est de l'Amérique du Nord ; forme arborescente dans le sud de l'Arkansas et l'est du Texas uniquement ; constitue des fourrés sur les sols pauvres et caillouteux.

Bois : léger, tendre, satiné et à grain grossier. Bois de cœur brun clair, à rubans verts et souvent teinté de rouge ; aubier de 4 à 5 cernes annuels, plus clair.

Intérêts : feuilles riches en tannin, pour le tannage et la teinture, fruits comestibles. Le feuillage est encore plus attrayant que celui des autres sumacs. Beau sujet pour l'ornement, il se propage très vite sur les pentes caillouteuses.

Dwarf Sumach.

TRANSVERSE SECTION.

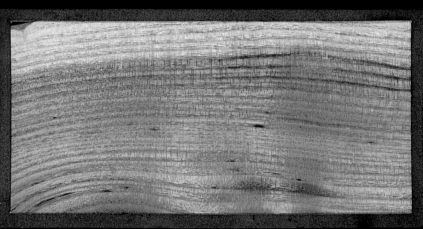

RADIAL SECTION

TANGENTIAL SECTION.

Ger. Zwerg-Sumach.　　*Fr.* Sumac nain.

Sp. Zumaque enano.

C | CASHEW FAMILY
Florida Poison Tree, Bum Wood, Coral Sumac

FAMILIE DER SUMACHGEWÄCHSE
–

FAMILLE DE LA MANGUE
Metopium toxique, Bois mulâtre, Machandeuse

ANACARDIACEAE
Metopium toxiferum (L.) Krug & Urban

Description: The tree reaches a height of up to 45 ft. (13 m) and the trunk is up to 2 ft. (0.6 m) in diameter. The widely spreading branches form a low crown. The light red-brown outer layers of the bark separate into large scales, exposing the bright orange-colored inner parts of the bark. The dark spots that are often visible on the bark are caused by the exudation of a gum. The leaves consist of five to seven leaflets. The base of the leaf stalks is swollen and enlarged. The yellow flowers are arranged in clusters at the ends of the branches. The inside of the petals is marked by dark lines. The orange-red fruit contains a stone-like seed, surrounded by a resinous covering.

Habitat: On the coast of Florida and its islands.

Wood: The dark brown, red-streaked heartwood is hard, heavy, strong, and durable, and is surrounded by distinctly lighter sapwood. In general it is difficult to work, but takes polish well.

Use: The wood is very suitable for the manufacture of cabinets. The exuded gum has medicinal uses.

Beschreibung: Der bis zu 13 m hohe Baum erreicht einen Stammdurchmesser von bis zu 0,6 m. Die weit ausgebreiteten Äste formen eine flache Krone. Die hell rotbraunen äußeren Schichten der Borke lösen sich in großen Schuppen ab, wobei die kräftig orange gefärbten inneren Teile der Borke freigelegt werden. Auf der Borke sind dunkle Flecken von Gummiausscheidungen zu sehen. Die Blätter bestehen aus fünf bis sieben Blattfiedern. Die Basis der Blattstiele ist blasig vergrößert. Die gelben Blüten gruppieren sich am Ende der Äste. Die Innenseite der Blütenblätter trägt eine dunkle Linienzeichnung. Die Frucht ist orangerot. In ihr liegt der von einer harzigen Hülle umgebene Steinsamen.

Vorkommen: An der Küste und auf den Inseln Floridas.

Holz: Das dunkelbraune, rotstreifige und von deutlich hellerem Saftholz umgebene Kernholz ist hart, fest, schwer und haltbar. Es lässt sich gut polieren.

Nutzung: Möbeltischlerei. Das ausgeschiedene Gummi findet arzneiliche Verwendung.

Description: arbre atteignant 13 m de hauteur, avec un tronc pouvant mesurer jusqu'à 0,6 m de diamètre et armé de branches très étalées, constituant un houppier bas. L'écorce brun rouge clair s'exfolie en grandes écailles, mettant à nu une écorce interne orange vif. Elle présente souvent des taches sombres correspondant à une exsudation de gomme. Feuilles composées de 5 à 7 folioles, avec un pétiole enflé à la base. Groupes de fleurs jaunes à l'extrémité des branches. Face inférieure des feuilles marquée de lignes sombre Fruit rouge orangé, avec un noyau entouré de résine.

Habitat: littoral et îles de la Floride

Bois: dur, lourd, résistant; durable difficile à travailler mais susceptible d'être poli. Bois de cœur brun sombre avec des rubans de couleu rouille; aubier nettement plus clai

Intérêts: excellent bois d'ébénisterie fine. La gomme est utilisée en médecine.

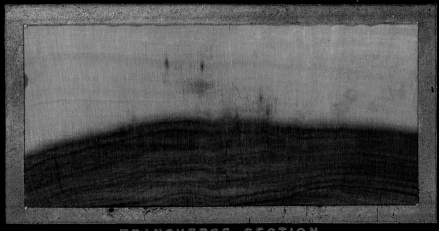

TRANSVERSE SECTION.

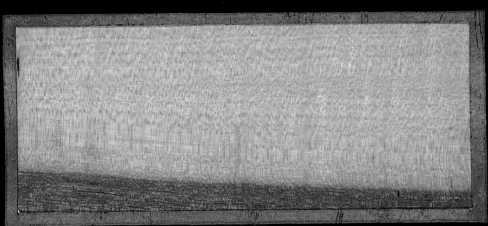

RADIAL SECTION.

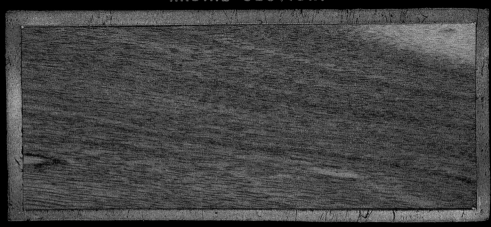

TANGENTIAL SECTION.

Gr. Korallen. Fr. Mancenillier.
Sp. Cedro prieto (Porto Rico); Guao de Costa (Cuba)

C | CASHEW FAMILY
Laurel Sumac

FAMILIE DER SUMACHGEWÄCHSE
Lorbeerblättriger Sumach

FAMILLE DE LA MANGUE
Malosma faux-laurier

ANACARDIACEAE
Malosoma laurina (Nutt.) Nutt. ex Abrams
Syn.: *Rhus laurina* Nutt.

Description: Usually a shrub; only in sheltered places on the island of Santa Catalina forming a tree up to 23 to 26 ft. (7 to 8 m) high, with a trunk 12 in. (0.3 m) in diameter that is covered with smooth, thin, dark gray bark. The leathery, evergreen leaves are ovate to lanceolate. The flowers are small, yellowish in color, and crowded into dense, compact inflorescences. The whitish fruit has a stone surrounded by thin flesh, and is ovoid to oblong in shape.

Habitat: On mesas (tablelands) and hills of southern California and its offshore islands.

Wood: Very soft and not very strong, pale, and polishes well, but apparently frequently consists only of sapwood.

Use: No commercial use known.

Beschreibung: Normalerweise ein Busch, nur in geschützten Lagen der Insel Santa Catalina auch baumförmig mit bis zu 7 oder 8 m Höhe und einem 0,3 m dicken Stamm, der von einer glatten, dünnen und dunkelgrauen Borke umhüllt ist. Die ledrigen, immergrünen Blätter sind oval bis lanzettlich. Die Blüten sind klein, von gelblicher Farbe und stehen in dichten kompakten Blütenständen beieinander. Die weißliche, von dünnem Fleisch umgebene Steinfrucht besitzt eine eiförmige bis rundliche Gestalt.

Vorkommen: In Mesas und auf den Hügeln Süd-Kaliforniens und der vorgelagerten Inseln.

Holz: Sehr weich und wenig belastbar, hell und gut polierbar, aber offenbar häufig nur aus Saftholz bestehend.

Nutzung: Keine gewerbliche Nutzung bekannt.

Description : buisson le plus souvent, croissant dans des lieux abrités de l'île Santa Catalina, ou sous forme arborescente, avec un tronc pouvant atteindre 7 à 8 m de hauteur et un diamètre de 0,3 m. Ecorce lisse, mince et gris sombre. Feuilles coriaces, persistantes, ovales à lancéolées. Petites fleurs jaunâtres, en cymes ramifiées, compactes. Fruits en drupes, blanchâtres et peu charnues, ovoïdes à sphériques.

Habitat : mesas et collines du sud de la Californie et des îles côtières

Bois : très tendre et peu résistant mécaniquement ; clair et pouvant se polir mais réduit visiblement souvent à un aubier.

Intérêts : utilisation commericale non déterminée.

Laurel Sumach, Sumach.

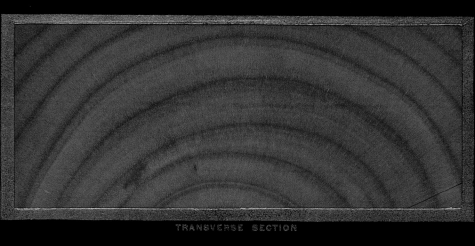

TRANSVERSE SECTION

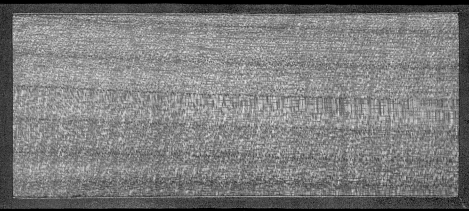

RADIAL SECTION.

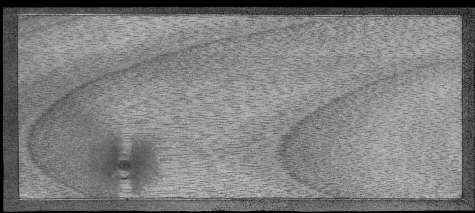

TANGENTIAL SECTION.

Ger. Lorberblättriger Sumach. *Fr.* Sumac de laurier.
Sp. Zumaque de laurel.

C | CASHEW FAMILY
Mango

FAMILIE DER SUMACHGEWÄCHSE
Mangobaum

FAMILLE DE LA MANGUE
Manguier

ANACARDIACEAE
Mangifera indica L.

Description: A tree up to 85 ft. (25 m) high with a roundish crown and enlarged base of the trunk. It has a gray, rough bark, which is smoother and red-tinged only on the branches. The evergreen, leathery leaves are entire and dark green. The many-flowered panicles can become very large. The large, oblong fruits are pale green or yellowish green with a juicy, aromatic flesh and a fibrous stone.

Habitat: It is native to tropical Asia, from where it was introduced into the Old and New Worlds. In North America it has escaped from cultivation and now grows wild in some regions of Florida.

Wood: The light brown heartwood is surrounded by grayish brown sapwood. It is heavy, moderately hard, and easy to work.

Use: The wood is very robust in salt water but quickly rots in fresh water. Nails hold tight in mango wood. Only some of the varieties have palatable fruits. The bark is used as a febrifuge.

Beschreibung: Der bis zu 25 m hohe Baum mit rundlicher Krone und verdickter Stammbasis besitzt eine graue, raue Borke, die nur auf den Ästen glatter und rötlich gefärbt ist. Die immergrünen, ledrigen Blätter sind ganzrandig und dunkelgrün. Die vielblütige Blütenrispe kann sehr groß werden. Die großen länglichen Früchte sind hellgrün oder gelbgrün mit saftigem, aromatischem Fleisch und einem faserigen Kern.

Vorkommen: Seine Heimat ist das tropische Asien, von wo aus er in weite Teile der Alten und Neuen Welt eingeführt wurde. In Nordamerika verwilderte er in einigen Gebieten Floridas.

Holz: Das hellbraune Kernholz ist von graubraunem Saftholz umgeben. Es ist schwer, mäßig hart und leicht zu bearbeiten.

Nutzung: Das Holz ist in Salzwasser sehr widerstandsfähig, fault aber rasch in Süßwasser. Nägel halten in Mango-Holz sehr gut. Die Früchte sind nur bei einem Teil der Varietäten genießbar. Rinde zur Fieberbekämpfung verwendbar.

Description: arbre pouvant atteind 25 m de hauteur, avec un tronc élargi à la base et une couronne ronde. Ecorce grise et rugueuse su le tronc mais lisse et rougeâtre su les branches. Fleurs groupées en une panicule florifère, souvent trè grande. Fruits volumineux et allor gés, vert clair ou vert jaune, à pul juteuse et savoureuse, renfermant un noyau fibreux.

Habitat: originaire de l'Asie tropicale; introduit dans le Vieux Monde et le Nouveau. En Amérique du Nord, il est retourné à l'état sauvage dans quelques endroits de la Floride.

Bois: lourd, moyennement dur; facile à travailler. Bois de cœur brun clair; aubier brun gris.

Intérêts: le bois est très résistant dans l'eau salée mais pourrit rapidement dans l'eau douce. Les clou tiennent extrêmement bien dans l bois de manguier. Seuls les fruits de quelques espèces sont comestibles. L'écorce a des propriétés fébrifuges.

TRANSVERSE SECTION.

RADIAL SECTION.

TANGENTIAL SECTION.

Gr. Mangobaum. Fr. Manguier. Sp. Mango.

C | CASHEW FAMILY
Pepper Tree

FAMILIE DER SUMACHGEWÄCHSE
Pfefferbaum

FAMILLE DE LA MANGUE
Poivrier d'Amérique

ANACARDIACEAE
Schinus molle L.

Description: Trunk up to 3 ft. (1 m) in diameter and grayish brown bark that becomes fissured longitudinally as it ages. The pinnate leaves consist of 12 to 15 leaflets or more that, like the young shoots, are covered with a sticky substance. The small yellowish green flowers are arranged in large inflorescences.

Habitat: In North America it has been introduced primarily into California, where it thrives, in places escaping cultivation and growing wild.

Wood: Rather soft and light, mottled brown in color, with clearly visible annual rings, and fine medullary rays.

Use: A very popular ornamental tree. The fruits have a laxative effect and have been successfully used to treat gonorrhea. The leaves, bark, and the sticky substance on the young branches and leaves have also been used for medicinal purposes.

Beschreibung: Baumförmig mit bis zu einem Meter dickem Stamm und graubrauner, im Alter längsrissiger Borke. Die Fiederblätter mit 12 bis 15 oder noch mehr Blättchen sind wie auch die jungen Triebe von einer gummiartigen Substanz bedeckt. Die kleinen gelblich grünen Blüten werden in großen Blütenständen hervorgebracht. Die Früchte sind kleine, sehr scharf schmeckende Schoten.

Vorkommen: In Nordamerika vor allem nach Kalifornien eingeführt, wo er sehr gut gedeiht und stellenweise verwildert.

Holz: Eher weich und leicht, mit deutlich erkennbaren Jahresringen, feinen Markstrahlen und fleckig brauner Farbe.

Nutzung: Ein sehr beliebtes Ziergehölz. Die abführend wirkenden Früchte werden erfolgreich gegen Gonorrhö eingesetzt, und auch die Blätter, die Rinde und die gummiartige Substanz auf jungen Ästen und Blättern werden für medizinische Zwecke genutzt.

Description: se présente sous form arborescente, avec un tronc pouvant atteindre 1 m de diamètre et une écorce brun gris qui se creuse de fissures longitudinales avec l'âge. Feuilles composées de 12 à 15 folioles, recouvertes comme les jeunes rameaux d'une substance gommeuse. Petites fleurs vert jaunâtre groupées en grandes inflorescences. Fruits en petites gousse d'une saveur très épicée.

Habitat: introduit en Amérique du Nord, en particulier en Californie, où il prospère et est retourné çà et là à l'état sauvage.

Bois: plutôt tendre et léger, avec des cernes annuels bien visibles et des rayons médullaires fins et d'un brun taché.

Intérêts: espèce ornementale très recherchée. Les fruits ont des propriétés purgatives employées avec succès contre la gonorrhée; les feuilles, l'écorce et la substance gommeuse recouvrant les jeunes rameaux et les feuilles sont utilisées également en médecine.

Pepper-tree, Chili Pepper, False Pepper.

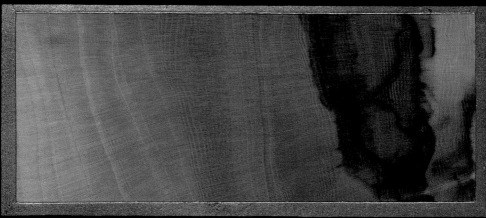

TRANSVERSE SECTION

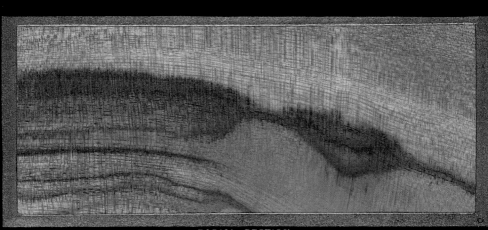

RADIAL SECTION.

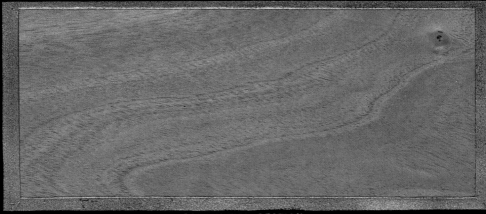

TANGENTIAL SECTION.

Ger. Pfefferbaum. *Fr.* Poivrier faux. *Sp.* Pimiento falso.

C | CASHEW FAMILY

Poison Sumac, Poison Dogwood, Poison Elder

FAMILIE DER SUMACHGEWÄCHSE
Sumpf-Sumach, Gift-Sumach

FAMILLE DE LA MANGUE
Toxicodendron à vernis, Bois-chandelle,
Sumac vénéneux

ANACARDIACEAE
Toxicodendron vernix (L.) Kuntze
Syn.: *Rhus vernix* L.

Description: 20 to 25 ft. (6 to 7.5 m) high and 6 to 7 in. (0.12 to 0.15 m) in diameter, mostly shrubby and springing directly from the ground in groups of thin stems. Branches slender, rather drooping. Bark thin, smooth, and pale gray. Milky sap bitter, poisonous, and turning black in the air. Leaves pinnate, at first covered with orange hairs. In October they are a brilliant scarlet or orange. Flowers arranged in panicles. Fruit globular and white.

Habitat: Wet, often flooded, swampy places in eastern North America.

Wood: Light, soft, and close-grained. Heartwood light yellow streaked with brown; sapwood paler. Three or four rows of large open pores mark cleanly the edges of the annual rings, but the medullary rays are thin and indistinct.

Use: A popular garden plant. The leaves and young branches are used in homeopathy. The juice can be made into a durable black varnish.

One of the most poisonous plants in North America. The juice contains toxicodendrin which is used in medicine.

Beschreibung: 6,0 bis 7,5 m hoch bei 0,12 bis 0,15 m Durchmesser, meist als Busch, der direkt aus der Erde Gruppen schmaler Stämme hervorbringt. Äste schlank, fast herabhängend. Borke dünn, glatt und blass hellgrau. Bitterer, giftiger Milchsaft, der an der Luft schwarz wird. Blätter gefiedert. Anfangs orange behaart. Im Oktober leuchtend scharlachrot oder orange. Blüten in Rispen angeordnet. Frucht kugelig und weiß.

Vorkommen: Nasse, oft überschwemmte Sümpfe im östlichen Nordamerika.

Holz: Leicht, weich und mit feiner Textur. Kern hellgelb mit braunen Streifen; Splint heller. 3 bis 4 Reihen großer offener Gefäße, sauber die Grenzen der Jahresringe markierend, sowie dünne, sehr undeutliche Markstrahlen.

Nutzung: Sehr beliebte Gartenpflanze. Blätter und junge Äste werden in der Homöopathie verwendet. Saft zu schwarzem, haltbarem Lack verarbeitbar.

Eine der giftigsten Pflanzen Nordamerikas. Im Saft enthalten ist Toxicodendrinsäure, die großen medizinischen Nutzen verspricht.

Description: 6,0 à 7,5 m de haute sur un diamètre de 0,12 à 0,15 m. Souvent arbrisseau buissonnant, groupes de tiges fines issues de racines traçantes et des branches grêles, presque retombantes. Ecorce mince, lisse, et gris pâle. S laiteux amer et toxique qui noirci à l'air. Feuilles composées pennée orange et veloutées à l'état juvénile. Fleurs en panicules orange o d'un pourpre lumineux en octobr Fruits globuleux et blancs.

Habitat: marais souvent inondés e l'est de l'Amérique du Nord.

Bois: léger, tendre et à grain fin. Bois de cœur jaune clair, rubanné de brun; aubier plus clair. 3 ou 4 faisceaux vasculaires de type ouvert, délimitant parfaitement les cernes annuels, et rayons médullaires fins, très indistincts.

Intérêts: très décoratif dans les jardins. Les feuilles et les jeunes rameaux sont employés en homéo pathie. Le suc est transformé en une laque noire durable.

Une des plantes les plus toxiques d'Amérique du Nord. Le suc renferme de la toxicodendrine trè utile en médecine.

TRANSVERSE SECTION.

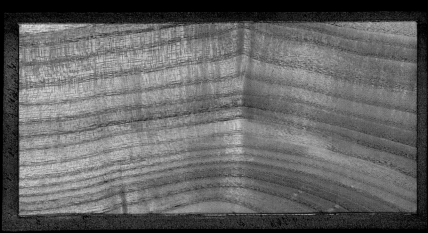

RADIAL SECTION.

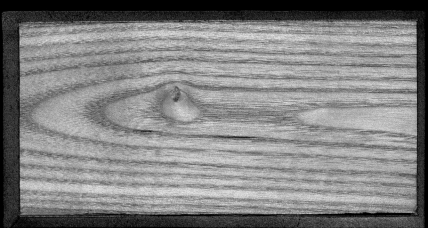

C | CASHEW FAMILY
Staghorn Sumac

FAMILIE DER SUMACHGEWÄCHSE
Staghorn-Sumach, Essigbaum

FAMILLE DE LA MANGUE
Sumac vinaigrier, Sumac de Virginie,
Sumac amarante (Canada)

ANACARDIACEAE
Rhus typhina L.
Syn.: *R. hirta* (L.) Sudw.

Description: Only occasionally up to 35 or 40 ft. (10 or 12 m) in height, usually much lower, and 10 to 16 in. (0.3 to 0.4 m) in diameter at the base. Several stems curving upwards with few large branches. Produces suckers and so forms thickets. Foliage ornamental, in autumn orange-red to scarlet. Fruit clusters erect, with a dense velvety covering of crimson hairs, highly decorative. The plant contains a white milky sap which turns black on exposure to air.

Habitat: In addition to the better soils, also found on dry sandy and stony soils at higher altitudes, and on slopes that it quickly covers. Eastern North America.

Wood: Light and soft. Heartwood golden yellow with brown and green streaks; sapwood white.

Use: Bark, like the foliage, rich in tannin. An ancient treatment for fever.

Grown in the parks and gardens of Europe since the first third of the 17th century.

Beschreibung: Nur gelegentlich bis zu 10, 12 m hoch, meist weit darunter, bei 0,3 bis 0,4 m Durchmesser an der Basis. Mehrstämmig bogig aufwachsend, mit wenigen großen Auszweigungen. Ausläufer treibend und dadurch Dickichte bildend. Dekoratives Laub, im Herbst orangerot bis scharlachrot aufflammend. Aufrechte, plüschartig dicht karminrot behaarte Fruchtstände von hohem Zierwert. Die Pflanze führt weißen Milchsaft, der an der Luft schwarz wird.

Vorkommen: Neben besseren Böden auch höher gelegene, trockene Sand- und Kiesböden, auch Hänge, die er schnell überzieht. Östliches Nordamerika.

Holz: Leicht und weich. Kern goldgelb („Goldholz") mit braunen und grünen Einschlägen; Splint weiß.

Nutzung: Rinde (wie das Laub) stark gerbsäurehaltig. Altes Fiebermittel.

Seit dem ersten Drittel des 17. Jh. in Parks und Gärten Europas.

Description: arbre atteignant 10, 12 m de hauteur, généralement beaucoup moins, avec un tronc de 0,3 à 0,4 m de diamètre à la base. Développe plusieurs tiges arquées portant quelques grandes ramifications. Espèce stolonifère, constitua des fourrés. Le feuillage flamboie en automne, avec des teintes allan du rouge orangé au rouge vif. Infrutescences pourpres en forme de torche, composées de fruits tomenteux à l'aspect velouté, d'ur grand effet décoratif. Sécrète un s laiteux blanc qui noircit à l'air.

Habitat: croît aussi bien sur les meilleurs sols que sur les terres sèches sableuses ou graveleuses, même sur les pentes, qu'il recouv rapidement. On le trouve dans l'e de l'Amérique du Nord.

Bois: léger et tendre. Bois de cœu jaune doré («bois d'or») avec un maillure brune et verte; aubier blanc.

Intérêts: écorce (comme le feuillage) très riche en tannin. Ancien fébrifuge.

Planté dans les parcs et les jardin européens depuis le premier tiers du 17ᵉ siècle.

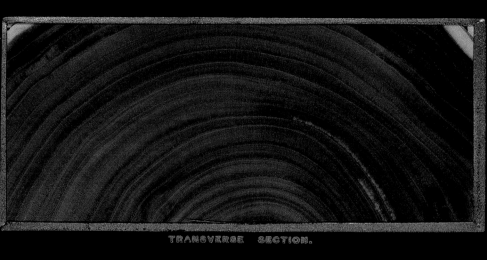

TRANSVERSE SECTION.

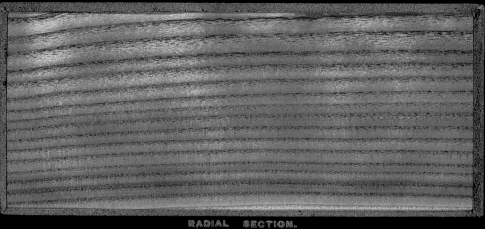

RADIAL SECTION.

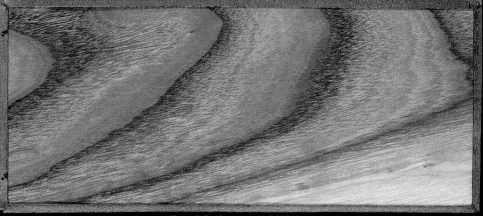

TANGENTIAL SECTION.

C | CASHEW FAMILY
Sugar Sumac

FAMILIE DER SUMACHGEWÄCHSE
Limonadenbaum

FAMILLE DE LA MANGUE
Sumac ovaté

ANACARDIACEAE
Rhus ovata S. Watson

Description: Normally growing as a shrub, and only rarely occurring as a tree up to 20 ft. (6 m) high, having a trunk with a maximum diameter of 10 in. (0.25 m), and reddish brown bark. The leathery, evergreen leaves are ovate and usually entire, the upper surface a dark yellowish green. The pale yellow flowers are clustered in dense inflorescences. The fruits are ovoid, and laterally compressed.

Habitat: Hills and mountains in south-western North America.

Wood: The rich red or yellowish brown heartwood is very hard, heavy, strong, with numerous pores, and polishable. It is surrounded by paler sapwood.

Use: The wood is used mainly for firewood. A refreshing drink is made from the fruits. In addition, the Sugar Sumac is suitable as a decorative plant.

Beschreibung: Normalerweise strauchig wachsend und nur selten als bis zu 6 m hoher Baum mit maximal 0,25 m dickem Stamm und rötlichbrauner Borke vorkommend. Die ledrigen, immergrünen Blätter sind oval und meist ganzrandig. Die Oberseite ist dunkel gelbgrün gefärbt. Die hellgelben Blüten sind in dichten Blütenständen vereinigt. Die Früchte sind oval und seitlich zusammengedrückt.

Vorkommen: Hügel und Berge im Südwesten Nordamerikas.

Holz: Das kräftig rot- oder gelbbraune Kernholz ist sehr hart, schwer, fest, dichtporig und gut polierbar. Es wird von hellerem Saftholz umgeben.

Nutzung: Das Holz wird hauptsächlich als Brennholz verwendet. Aus den Früchten wird ein Erfrischungsgetränk hergestellt. Daneben ist der Limonadenbaum als Ziergehölz geeignet.

Description : ordinairement arbuste à port buissonnant, atteignant rarement 6 m de hauteur, avec un tronc de 0,25 m maximum de diamètre et une écorce brun rougeâtre. Feuilles coriaces, persistantes, ovales et entières le plus souvent, vert jaune foncé à la face supérieure. Fleurs jaune clair groupées en inflorescences denses. Fruits ovales et aplatis sur les côté.

Habitat : étage collinéen et montagnard dans le sud-ouest de l'Amérique du Nord.

Bois : lourd, très dur, résistant et dense ; susceptible d'être poli. Bois de cœur d'un brun rouge ou brun jaune intense ; aubier plus clair.

Intérêts : utilisé essentiellement comme bois de chauffage. Les fruits servent à préparer une boisson rafraîchissante. Espèce ornementale également.

Lemonade Tree.

TRANSVERSE SECTION.

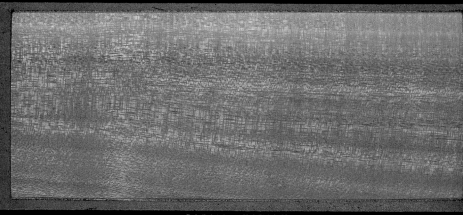

RADIAL SECTION.

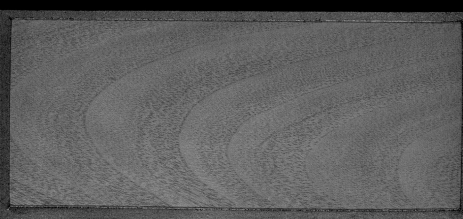

TANGENTIAL SECTION

Ger. Limonadenbaum.   *Fr.* Arbre de limonade.
Sp. Arbol de limonada.

C | CITRUS FAMILY, RUE FAMILY
Lemon

FAMILIE DER ZITRUSGEWÄCHSE
Limone

FAMILLE DE L'ORANGE
Citronnier

RUTACEAE
Citrus x limonia Osbeck

Description: "Between 20 and 40 ft. (5 to 12 m) high and 8 and 20 in. (0.2 to 0.5 m) in diameter" (Begemann, 1981). The genus provides, under the collective name of citrus fruits, fruit that is cultivated in all suitable climates.

Habitat: Cultivated ground in the warmer parts of North America. Grows wild in places.

Wood: Heavy, hard, tough, very strong, the planed surfaces very smooth and slightly shiny; polishable, resistant to changes in weather, and durable. Yellowish to reddishyellow, partly with a light greenish luster.

Use: For turnery, small luxury articles, and marquetry. Only of local importance, in small quantities.

Under the name of citrus wood, it is sometimes confused with the similar sounding cedar wood, and also with the wood of *Callitris quadrivalvis,* syn. *Tetraclinis articulata,* a member of the cypress family from north-west Africa, which provides sandarac wood and resin.

Beschreibung: „Zwischen 5 und 12 m hoch bei 0,2 bis 0,5 m Durchmesser." (Begemann 1981). Die Gattung liefert unter dem Sammelbegriff Agrumen die in allen geeigneten Klimaten kultivierten Obstarten.

Vorkommen: Kulturböden in den wärmeren Teilen Nordamerikas. Bisweilen auch verwildert.

Holz: Schwer, hart, zäh, sehr fest, gehobelte Flächen sehr glatt und mattglänzend; polierfähig, witterungsbeständig und dauerhaft. Gelblich bis rötlich gelb, teilweise mit leichtem grünlichem Schimmer.

Nutzung: Für Drechslerarbeiten, kleine Luxusartikel, Intarsien. Immer nur in kleinen Mengen von lokaler Bedeutung.

Unter dem Namen Citrus-Holz aufgrund einer Lautähnlichkeit mit Zedernholz sowie dem Holz von *Callitris quadrivalvis,* syn.: *Tetraclinis articulata,* einem Zypressengewächs aus Nordwestafrika (Sandarakholz und -harz), verwechselt.

Description: « De 5 à 12 m de hauteur sur un diamètre de 0,2 à 0,5 m. » (Begemann 1981). Le genre *Citrus* regroupe les arbustes cultivés pour leurs fruits sous tous les climats favorables.

Habitat: sol cultivé des régions chaudes de l'Amérique du Nord. Dans certaines régions à l'état sauvage.

Bois: lourd, dur, tenace, très résistant; surfaces rabotées très lisses et mates; susceptible d'un beau poli, durable en extérieur et à l'état sec. Jaunâtre à jaune rougeâtre, avec, çà et là, un léger reflet verdâtre.

Intérêts: tournerie, marqueterie, petits objets de luxe. Exploitation locale et en petites quantités uniquement.

Sous le nom de « bois de citronnier », on comprenait aussi en raison de confusions et d'une ressemblance phonémique le cèdre, le bois de *Callistris quadrivalvis,* syn.: *Tetraclinis articulata,* un thuya originaire du nord-ouest de l'Afrique, dont on extrait la résine ou sandaraque.

Lemon.

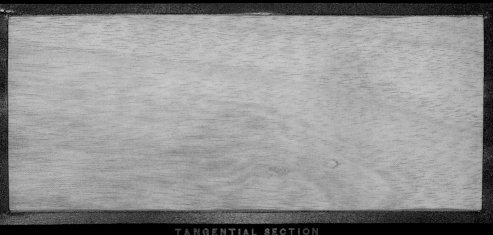

TRANSVERSE SECTION.

RADIAL SECTION.

TANGENTIAL SECTION

C | CITRUS FAMILY, RUE FAMILY

Prickly Ash, Toothache Tree, Hercules' Club

FAMILIE DER ZITRUSGEWÄCHSE
Gelbholz-Herkuleskeule

FAMILLE DE L'ORANGE
Clavalier massue-d'Hercule, Bois épineux jaune

RUTACEAE
Zanthoxylum clava-herculis L.

Description: 25 to 40 (50) ft. (8 to 10 (15) m) high and 10 to 18 in. (0.25 to 0.35 m) in diameter. Commonly smaller, often shrubby. The numerous branches are almost horizontal. Bark a characteristic blue-gray, smooth, but covered with rounded, corky protuberances tipped by a sharp point which later breaks off.

Habitat: Swamp margins and light, sandy soils with pines, Evergreen Oak, Water Oak, Red Bay, Dwarf Palmetto, etc., in south-eastern North America. In the Gulf region it grows further inland west of the Mississippi River on better sandy soils with Florida Anise, the so-called Poison Laurel, Styrax, Symplocos, holly, and tupelo.

Wood: Light, soft, and close-grained.

Use: Used by the black Americans as a remedy for toothache, and in places almost wiped out.

Bark "hot," stimulating the production of saliva.

Beschreibung: 8 bis 10 (15) m hoch bei 0,25 bis 0,35 m Durchmesser. Gewöhnlich kleiner, oft strauchig. Die zahlreichen Äste fast horizontal. Die eigentümlich blaugraue, glatte Borke mit zitzenförmigen, korkigen Buckeln besetzt, die in eine scharfe, zuletzt abfallende Spitze ausmünden.

Vorkommen: Sumpfränder und leichte Sandböden mit Kiefern, Immergrüner Eiche, Wasser-Eiche, Roter Avocado, Zwerg-Palmetto u. a. im südöstlichen Nordamerika. Am Golf auch weiter landeinwärts, westlich des Mississippi auf besseren Sandböden zusammen mit Florida-Sternanis, dem sog. „Giftlorbeer", *Styrax, Symplocos,* Stechhülse und Tupelo.

Holz: Leicht, weich und mit feiner Textur.

Nutzung: Von Teilen der afrikanischstämmigen Bevölkerung gegen Zahnschmerzen benutzt, stellenweise fast ausgerottet.

Rinde stechend scharf („hot"), speichelfördernd.

Description : 8 à 10 m, voire 15 m de hauteur sur un diamètre de 0,25 à 0,35 m. En général plus petit, souvent buissonnant. Les nombreuses branches sont presque horizontales. Ecorce très particulière : gris bleu, lisse, elle est garnie de bosses liégeuses en forme de mamelon dont les pointes acérées tombent avec l'âge.

Habitat : bords des marais et sols sablonneux légers dans le sud-est de l'Amérique du Nord, en association avec les pins, le chêne toujours vert, le chêne noir, l'avocatier rouge, le palmetto nain etc. Dans de meilleurs sols sablonneux sur le pourtour du Golfe ainsi qu'à l'ouest du Mississippi, en remontant le fleuve, en compagnie du sternani de Floride, appelé aussi « Giftlorbeer », des *Styrax,* des *Symplocos,* du houx et du nyssa.

Bois : léger, tendre et à grain fin.

Intérêts : utilisé par les Noirs américains pour calmer les maux de dents ; a presque disparu à certains endroits.

Ecorce très pimentée (« hot »), augmentant la sécrétion salivaire.

Prickly Ash, Sea Ash, Toothache Tree, Pepper-wood.

TRANSVERSE SECTION.

RADIAL SECTION.

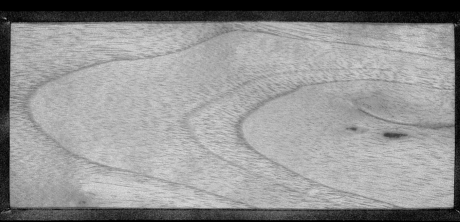

TANGENTIAL SECTION

Ger. Eschenblättriger Gelbholz.   *Fr.* Frêne des épines.

Sp. Jantoxaro.

C | CITRUS FAMILY, RUE FAMILY

Seville Orange, Sour Orange

FAMILIE DER ZITRUSGEWÄCHSE
Orange, Apfelsinenholz

FAMILLE DE L'ORANGE
Bigaradier, Oranger amer

RUTACEAE
Citrus aurantium L.

Description: "Between 20 and 40 ft. (5 to 12 m) high and 8 and 20 in. (0.2 to 0.5 m) in diameter" (Begemann, 1981). The same is true of *Citrus x limonia* Osbeck, loc. cit. The genus provides, under the collective name of citrus fruits, fruit that is cultivated in all suitable climates.

Habitat: Cultivated ground in the warmer parts of North America. Grows wild in places.

Wood: Heavy, hard, tough, very strong, the planed surfaces very smooth and slightly shiny; polishable, resistant to changes in weather, and durable. Yellowish to reddish yellow, partly with a light greenish luster.

Use: For turnery, small luxury articles, and marquetry. Only of local importance, in small quantities.

Under the name of citrus wood, it is sometimes confused with the similar sounding cedar wood, and also with the wood of *Callitris quadrivalvis,* syn. *Tetraclinis articulata,* a member of the cypress family from north-west Africa, which provides sandarac wood and resin.

Beschreibung: „Zwischen 5 und 12 m hoch bei 0,2 bis 0,5 m Durchmesser." (Begemann 1981). Das Gleiche gilt für *Citrus x limonia* Osbeck, l. c. Die Gattung liefert unter dem Sammelbegriff Agrumen die in allen geeigneten Klimaten kultivierten Obstarten.

Vorkommen: Kulturböden in den wärmeren Teilen Nordamerikas. Bisweilen auch verwildert.

Holz: Schwer, hart, zäh, sehr fest, gehobelte Flächen sehr glatt und mattglänzend; polierfähig, witterungsbeständig und dauerhaft. Gelblich bis rötlich gelb, teilweise mit leichtem grünlichem Schimmer.

Nutzung: Für Drechslerarbeiten, kleine Luxusartikel, Intarsien. Immer nur in kleinen Mengen von lokaler Bedeutung.

Unter dem Namen Citrus-Holz aufgrund einer Lautähnlichkeit mit Zedernholz sowie dem Holz von *Callitris quadrivalvis,* syn.: *Tetraclinis articulata,* einem Zypressengewächs aus Nordwestafrika (Sandarakholz und -harz), verwechselt.

Description: «De 5 à 12 m de hauteur sur un diamètre de 0,2 à 0,5 m.» (Begemann 1981). Mêmes dimensions pour *Citrus x limonia* Osbeck, l. c. Le genre *Citru* regroupe les arbustes cultivés pou leurs fruits sous tous les climats favorables.

Habitat: sol cultivé des régions chaudes de l'Amérique du Nord. Dans certaines régions à l'état sauvage.

Bois: lourd, dur, tenace, très résistant; surfaces rabotées très lisses e mates; susceptible d'un beau poli, durable en extérieur et à l'état sec Jaunâtre à jaune rougeâtre, avec, ç et là, un léger reflet verdâtre.

Intérêts: tournerie, marqueterie, petits objets de luxe. Exploitation locale et en petites quantités uniquement.

Sous le nom de «bois de citronnier», on comprenait aussi en raison de confusions et d'une ressemblance phonémique le cèdre, le bois de *Callistris quadrivalvis,* syn.: *Tetraclinis articulata,* u thuya originaire du nord-ouest de l'Afrique, dont on extrait la résine ou sandaraque.

TRANSVERSE SECTION.

RADIAL SECTION.

TANGENTIAL SECTION

C | CITRUS FAMILY, RUE FAMILY

Wafer Ash, Hop Tree, Stinking Ash

FAMILIE DER ZITRUSGEWÄCHSE
Klee-Ulme

FAMILLE DE L'ORANGE
Ptéléa trifolié, Arbre houblon, Orme de Samarie (Canada)

RUTACEAE
Ptelea trifoliata L.

Description: Only occasionally 20 to 27 ft. (6 to 8 m) high and 8 to 10 in. (0.2 to 0.25 m) in diameter. More often a shrub. Foliage an attractive dark green. Fruits in America are waffle-like in shape, appearing in pale green clusters in late summer and after leaf-fall in winter. Leaves and bark, when rubbed, smell like hops.

Habitat: Rocky slopes on forest margins, often in the shade of trees. Mostly only in the interior of eastern North America; along the Gulf as far as Mexico.

Wood: Fairly heavy and hard, close-grained. Light brown, the sapwood barely distinguishable.

Use: In medicine, an extract from the bark is used as a tonic. In brewing, the leaves and bark are used as a substitute for hops.

Beschreibung: Nur gelegentlich 6 bis 8 m hoch bei 0,2 bis 0,25 m Durchmesser. Eher ein Strauch. Zierend durch dunkelgrünes Laub, Früchte in Amerika in waffelähnlicher Form, in hellgrünen Büscheln im Spätsommer und nach dem Laubfall im Winter. Blätter und Borke, wenn gerieben, mit Hopfengeruch.

Vorkommen: Felsige Hänge an Waldrändern, oft im Baumschatten. Meist nur im Landesinneren des östlichen Nordamerika; entlang des Golfes bis nach Mexiko.

Holz: Ziemlich schwer und hart, mit feiner Textur. Hellbraun, Splint kaum zu unterscheiden.

Nutzung: Ein Extrakt aus der Rinde als Tonikum in der Medizin. Blätter und Borke werden als Hopfenersatz beim Bierbrauen benutzt.

Description : atteint parfois 6 à 8 m de hauteur sur un diamètre de 0,2 à 0,25 m. Plus un gros arbrseau qu'un arbre. Le feuillage ver foncé est ornemental. Les fruits vert clair, comparés en Amérique à des motifs de gaufres, sont réunis en touffes ; ils sont tardifs et persistent sur l'arbre une partie de l'hiver. Les feuilles et l'écorce dégagent une odeur de houblon quand on les froisse.

Habitat : pentes rocheuses en lisière forestière, souvent à l'ombr d'autres arbres. On le trouve le plus souvent dans les terres à l'est de l'Amérique du Nord ; croît éga lement le long du Golfe jusqu'au Mexique.

Bois : assez lourd et dur, à grain fin. Bois de cœur brun clair ; aubie à peine distinct.

Intérêts : on extrait de l'écorce des substances toniques utilisées en médecine. Les feuilles et l'écorce constituent un succédané du houblon dans la fabrication de la bière

TRANSVERSE SECTION.

RADIAL SECTION.

TANCENTIAL SECTION.

Ger. Dreyblättrige Lederblume.   Fr. Arbre de houblon.

Sp. Arbol de lupulo.

C | COCOPLUM FAMILY
Coco Plum, Fat Pork, Gopher Plum

FAMILIE DER GOLDPFLAUMENGEWÄCHSE
Kokos-Pflaume

FAMILLE DE L'ICAQUE
Icaquier d'Amérique, Gros icaque

CHRYSOBALANACEAE
Chrysobalanus icaco L.

Description: Rarely 18 to 20 ft. (7.5 to 9.0 m) high and 12 in. (0.3 m) in diameter, more frequently a shrub, in exposed places also prostrate and creeping. Bark thin, peeling off in long strips, with a light gray surface tinged red. Leaves broadly elliptic to roundish ovate, leathery, dark green above, pale yellow green beneath. Fruit bright pink, yellow, purple, creamy white or sometimes almost black, with a sweet and juicy, white flesh. Only occasionally does a single fruit develop to maturity from each inflorescence. Fruits oily.

Habitat: Shores and coasts, sometimes forming extensive thickets beyond the influence of the tides. Widespread across the Caribbean, Central America, South America, and West Africa. In North America confined to Florida.

Wood: Heavy, hard, strong, and close-grained.

Use: At the time of the landing of Christopher Columbus, the fruits were one of the favorite foods of the Caribs, and they were also used as torches. The bark, leaves, and roots have an astringent effect, and help in the treatment of diarrhea, leucorrhea, and hemorrhage.

Beschreibung: Selten 7,5 bis 9,0 m hoch bei 0,3 m Durchmesser, häufiger als Strauch, in exponierten Lagen auch niederliegend-kriechend. Borke dünn, in langen Streifen abschuppend, mit hellgrauer Oberfläche und roter Tönung. Blätter breit elliptisch bis rundlich oval, ledrig, dunkelgrün auf der Ober-, hell gelbgrün auf der Unterseite. Frucht hellrosa, gelb, violett, cremeweiß oder manchmal fast schwarz mit weißem, süßem und saftigem Fleisch. Nur jeweils eine Frucht pro Blütenstand entwickelt sich zur Reife. Früchte ölig.

Vorkommen: Strände und Küsten, außerhalb des Gezeiteneinflusses z.T. ausgedehnte Dickichte bildend. Weit verbreitet über die Karibik, Mittelamerika, Südamerika und Westafrika. Nordamerika nur in Florida erreichend.

Holz: Schwer, hart, fest und mit feiner Textur.

Nutzung: Früchte eine der Lieblingsspeisen der Kariben zur Zeit der Landung von Christopher Kolumbus, auch als Fackeln verwendet. Borke, Blätter und Wurzeln adstringierend, helfen gegen Diarrhö, Leukorrhö und Hämorrhagie.

Description: arbuste atteignant rarement 7,5 à 9,0 m de hauteur et 0,3 m de diamètre; souvent so forme de buisson. Ecorce mince, gris clair teinté de rouge, avec de longues lanières se desquamant. Feuilles larges-elliptiques à arrondies-ovales, coriaces, vert sombre dessus, vert jaune clair dessous. Fruits de couleur variable, rose clair, jaunes, violets, blanc crème ou parfois presque noirs, à chair juteuse, blanche et douce. Un se fruit par inflorescence arrive à maturité. Fruits oléagineux.

Habitat: plages et côtes, formant en dehors des zones soumises aux influences des marées de vastes fourrés. Très répandu dans les Antilles, l'Amérique centrale, l'Amérique du Sud et l'Afrique d l'Ouest. En Amérique du Nord, uniquement en Floride.

Bois: lourd, dur, résistant et à gra fin.

Intérêts: les fruits étaient consommés dans les Caraïbes à l'arrivée de Christophe Colomb et servaie aussi pour les torches. L'écorce, le feuilles et les racines sont astringentes et soignent la diarrhée, la leucorrhée et les hémorragies.

COCO PLUM. COCOA PLUM. GROPHER PLUM.

TRANSVERSE SECTION.

RADIAL SECTION.

TANGENTIAL SECTION.

Ger. COCOS-PFLAUME.   Fr. ICAQUIER.

Sp. ICACO.

C | COFFEE FAMILY, GARDE- NIA FAMILY, MADDER FAMILY, QUNINE FAMILY
Princewood

FAMILIE DER RÖTEGEWÄCHSE
Prinzholz

FAMILLE DU CAFÉIER
Exostema des Caraïbes

RUBIACEAE
Exostema caribaeum (Jacq.) Schult.

Description: 20 to 25 ft. (6 to 7.5 m) high and 10 to 12 in. (0.25 to 0.3 m) in diameter, with slender, erect branches that form a narrow crown, often also a shrub. Bark thin, deeply furrowed, and divided into almost white, square scales. Leaves oblong-ovate to lanceolate, thick and leathery, dark green. Flowers solitary on unbranched stalks arising from the axils of the opposite leaves, delicate and wonderfully scented. Lower half of petals united into a narrow tube, the upper half of the narrowly linear petals spreading apart from each other and displaying the stamens with their large anthers. Fruit cylindrical, dark brown, becoming black after drying.

Habitat: In North America confined to small islands off southern Florida. Chiefly found in the northern Caribbean.

Wood: Very heavy, extremely hard, strong, with a silky sheen, and close-grained; polishable.

Use: A rare plant, so hardly used.

Beschreibung: 6 bis 7,5 m hoch bei 0,25 bis 0,3 m Durchmesser. Mit schlanken, aufrechten Ästen, die eine schmale Krone bilden, oft auch als Busch. Borke dünn, tief rissig und in quadratische, fast weiße Schuppen zergliedert. Blätter länglich oval bis lanzettlich, dick und ledrig, dunkelgrün. Blüten einzeln auf unverzweigten Blütenstielen aus den Achseln der gegenständigen Blätter wachsend. Außerordentlich wohlriechend und zart. Blütenkronblätter zur Hälfte zu einem schmalen Tubus verwachsen, in der oberen Hälfte die schmal linealischen Kronblätter auseinanderweichend und die mit großen Staubbeuteln besetzten Staubblätter präsentierend. Frucht zylindrisch, dunkelbraun, nach dem Trocknen schwarz.

Vorkommen: In Nordamerika nur auf kleinen Inseln vor Süd-Florida. Hauptverbreitung in der Nord-Karibik.

Holz: Sehr schwer, außerordentlich hart, fest, seidig und von feiner Textur; polierfähig.

Nutzung: Sehr selten, daher kaum genutzt.

Description : arbuste de 6 à 7,5 m de hauteur, avec un tronc de 0,25 0,3 m de diamètre et des branche fines, dressées, constituant un houppier étroit ; souvent aussi sou forme de buisson. Ecorce mince, profondément fissurée et craque- lée en écailles carrées, presque blanches. Feuilles allongées-ovales à lancéolées, épaisses et coriaces, vert foncé. Les fleurs solitaires sor portées par des pédoncules non ramifiés, à l'aisselle des feuilles opposées ; elles sont délicates et e halent une odeur suave. Les pétal linéaires, à moitié soudés, formen un tube étroit puis s'évasent pour présenter les étamines coiffées de grandes anthères. Fruit cylindriqu brun sombre, noir quand il est sec

Habitat : en Amérique du Nord uniquement sur les petites îles en face du sud de la Floride. Surtout répandu dans le nord des Caraïbe

Bois : très lourd, extrêmement dur, résistant, satiné et à grain fin susceptible d'être poli.

Intérêts : essence très rare, et peu utilisée par conséquent.

PRINCE-WOOD.

TRANSVERSE SECTION.

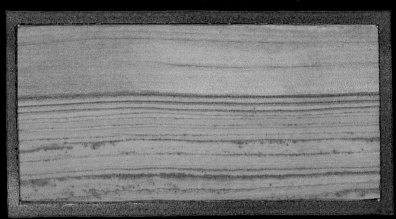

RADIAL SECTION.

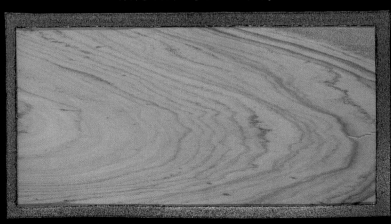

TANGENTIAL SECTION.

Ger. Prinz-holz.　　　　　Fr. Quinquina Caraïbe.

Sp. Cuero de sapo, Macagua de costa, Falsa quina (Mex.)

C | CUSTARD-APPLE FAMILY
Pawpaw

FAMILIE DER FLASCHENBAUMGEWÄCHSE
Papau

FAMILLE DE L'YLANG-YLANG
Asiminier trilobé, Asiminier à trois lobes,
Corossol (Canada)

ANNONACEAE
Asimina triloba (L.) Dunal

Description: In favorable conditions 33 to 40 ft. (10 to 12 m) high and 6 to 10 in. (0.15 to 0.25 m) in diameter. Often only a large shrub. Bark dark brown, thin, and very smooth, on older trunks with scattered, longitudinal fissures. Recommended as an ornamental tree for its beautiful flowers in early spring, and its strangely shaped fruits.

Habitat: Forms a dense undergrowth in forest thickets and sometimes grows here and there in the undergrowth on rich lowland soils. Eastern North America, and especially in the south more abundant in the interior.

Wood: Light, and when freshly cut tinged a beautiful greenish or yellow color.

Use: Wood devoid of commercial value. Fibers from the inner bark were used in earlier times for making fishing nets. The fruits, when ripe, are pleasant-tasting and nourishing, and are found at local markets.

Beschreibung: Wenn begünstigt 10 bis 12 m hoch bei 0,15 bis 0,25 m Durchmesser. Oft nur ein Großstrauch. Borke dunkelbraun, dünn und recht glatt, an alten Stämmen mit spärlichen Längsrissen. Mit schönen Blüten im zeitigen Frühjahr und seltsam geformten Früchten als Zierbaum empfohlen.

Vorkommen: Bewohnt als geschlossener Unterwuchs die Dickichte der Wälder, manchmal zerstreut im Unterholz auf reichen Flachlandböden. Östliches Nordamerika, besonders im Süden mehr im Landesinneren.

Holz: Leicht und wenn frisch geschnitten, mit schönen grünlichen und gelben Tönungen.

Nutzung: Holz ohne Handelswert. Fasern der inneren Rinde früher für Fäden von Fischnetzen verwendet. Früchte, wenn reif, wohlschmeckend und nahrhaft. Auf Märkten der Region.

Description: 10 à 12 m de hauteur dans des conditions favorables, su un diamètre de 0,15 à 0,25 m. Ne forme souvent qu'un gros buisson Ecorce brun sombre, mince et très lisse, se creusant de quelques fissures longitudinales sur les vieux sujets. Arbuste conseillé en ornement pour sa floraison préco très attrayante et ses fruits oblong en forme d'outre.

Habitat: forme le sous-étage ferm des fourrés forestiers; parfois disséminé dans le sous-bois des forê installées sur des sols de plaine riches. On le trouve dans l'est de l'Amérique du Nord, surtout dan la partie sud plus dans les terres.

Bois: léger et joliment teinté de vert et de jaune lorsqu'il est fraîchement coupé.

Intérêts: bois sans valeur économique. Les fibres de l'écorce interne étaient autrefois utilisées pour confectionner les filets de pêche. Les fruits, savoureux et nourrissants à maturité, sont ven dus sur les marchés locaux.

Papaw, Custard Apple.

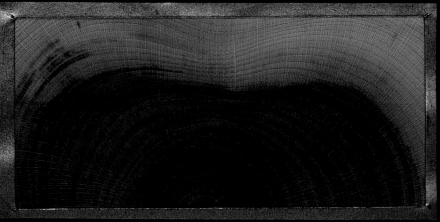

TRANSVERSE SECTION.

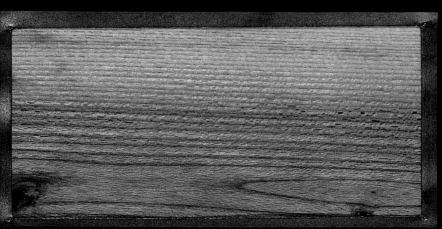

RADIAL SECTION.

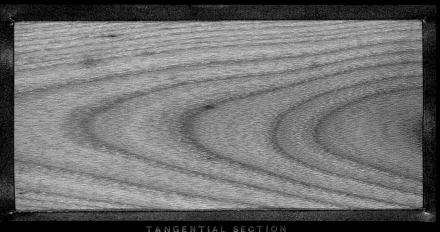

TANGENTIAL SECTION

C | CUSTARD-APPLE FAMILY

Pond Apple, Anone, Cimarron, Custard Apple

FAMILIE DER FLASCHENBAUMGEWÄCHSE
Flaschenbaum

FAMILLE DE L'YLANG-YLANG
Annone glabre, Annone des marais, Bois flot

ANNONACEAE
Annona glabra L.

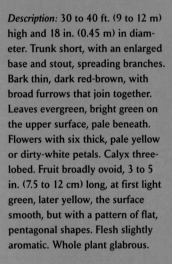

Description: 30 to 40 ft. (9 to 12 m) high and 18 in. (0.45 m) in diameter. Trunk short, with an enlarged base and stout, spreading branches. Bark thin, dark red-brown, with broad furrows that join together. Leaves evergreen, bright green on the upper surface, pale beneath. Flowers with six thick, pale yellow or dirty-white petals. Calyx three-lobed. Fruit broadly ovoid, 3 to 5 in. (7.5 to 12 cm) long, at first light green, later yellow, the surface smooth, but with a pattern of flat, pentagonal shapes. Flesh slightly aromatic. Whole plant glabrous.

Habitat: Shallow, fresh water ponds, swampy hummocks, banks of small, fresh-water streams in the everglades of Florida as well as the Bahamas and the Lesser Antilles.

Wood: Light, soft, and not strong. Light brown streaked with yellow. Numerous, scattered, large, and open pores.

Use: Fruits edible, but of little value.

Beschreibung: 9 bis 12 m hoch bei 0,45 m Durchmesser. Stamm kurz, mit verdickter Basis und kräftigen, ausladenden Ästen. Borke dünn, dunkel rotbraun, mit breiten, zusammenfließenden Rissen. Blätter immergrün, hellgrün auf der Ober- und blass auf der Unterseite. Blüte mit 6 dicken, blassgelben oder schmutzig weißen Kronblättern. Kelch dreilappig. Frucht breit oval, 7,5 bis 12 cm lang, zunächst hellgrün, später gelb, Oberfläche glatt, flach pentagonal gefeldert. Fleisch schwach aromatisch. Ganze Pflanze unbehaart.

Vorkommen: Flache Süßwasserteiche, sumpfige Hügel, Ufer kleiner Süßwasserbäche in den Everglades in Florida, den Bahamas und Inseln der Kleinen Antillen.

Holz: Leicht, weich und nicht fest. Hellbraun mit gelben Streifen. Zahlreiche verstreute, große und offene Gefäße.

Nutzung: Früchte essbar, aber von geringem Wert.

Description: 9 à 12 m de hauteur sur un diamètre de 0,45 m. Tronc court, élargi à la base, portant des branches vigoureuses et étalées. Ecorce mince, brun rouge foncé, parcourue de larges fissures confluentes. Feuilles persistantes, vert clair dessus, pâles dessous. Fleur composée de 6 gros pétales jaune pâle ou blanc sale et d'un calice trilobé. Fruits oblongs de 7,5 à 12 cm, vert clair puis jaunes avec une surface lisse, pentagonale plate. Pulpe légèrement aromatique. Toutes les parties de la plante sont pubescentes.

Habitat: étangs d'eau douce peu profonds, collines marécageuses, rives des ruisseaux d'eau douce dans les Everglades en Floride, au Bahamas et dans les îles des Petites Antilles.

Bois: léger, tendre et non résistant. Brun clair avec des rubans jaunes. Nombreux vaisseaux de type ouvert disséminés dans le bois.

Intérêts: fruits comestibles mais de peu d'importance sur le plan économique.

POND APPLE.   WILD CUSTARD APPLE.

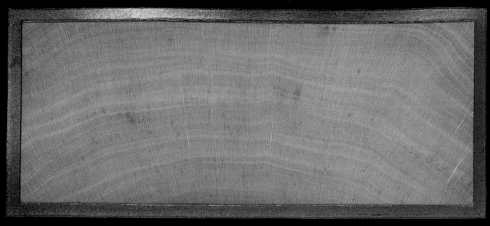

TRANSVERSE SECTION.

RADIAL SECTION.

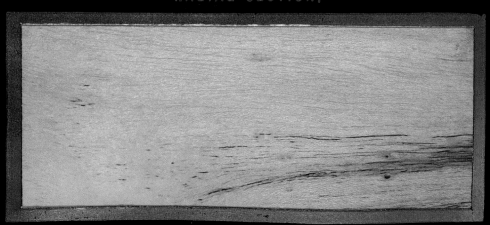

TANGENTIAL SECTION.

Ger. Flaschenbaum.   Fr. Anone. Corossol.

Sp. Corazon. Cimarron. Baga (Cuba). Anona (Mexico).

C | CYPRESS FAMILY
Alligator Juniper, Checkered-bark Juniper

FAMILIE DER ZYPRESSENGEWÄCHSE
Dickrindiger Wacholder

FAMILLE DU CYPRÈS
Genévrier alligator, Genévrier de Deppe, Genévrier gercé

CUPRESSACEAE
Juniperus deppeana var. pachyphloea (Torr.) Martinez
Syn.: *J. pachyphloea* Torr.

Description: 50 to 60 ft. (15 to 18 m) high and 5 ft. (1.5 m) in diameter. Trunk short and stout. Crown broadly pyramidal, open, but in old trees sometimes also rounded and compact. Bark thick, dark brown in color with a red tinge, separated into angular plates by deep fissures. Leaves scale-like, closely appressed, pale blue-green. Flowers insignificant at the ends of short side branches. Fruits globular, maturing only in the second year, reddish brown, at first glaucous whitish blue, often with an irregularly dented surface.

Habitat: Dry, desert-like mountain slopes in southern North America, with pines and evergreen oaks. Usually at altitudes of at least 3,800 to 5,500 ft. (1200 to 1800 m), at its best in canyons with a good supply of water, but also occurs on dry slopes and rocky ridges.

Wood: Soft and close-grained. Heartwood a clear light red, often streaked with yellow; sapwood thin, almost white.

Use: The fruits serve as food for the Indians, and the tree is found in gardens and parks, especially in Europe. One of the most beautiful trees in western North America.

Beschreibung: 15 bis 18 m hoch bei 1,5 m Durchmesser. Stamm kurz und kräftig. Krone breit pyramidal, offen, im Alter auch rundlich geschlossen. Borke dick, von dunkelbrauner Farbe mit roter Tönung. Tiefrissig in eckige Platten zergliedert. Blätter schuppenartig eng anliegend, blass blaugrün. Blüten unauffällig an den Enden kurzer Seitenzweige. Früchte kugelig, erst im zweiten Jahr reifend, rotbraun, zunächst weißlichblau bereift, oft mit unregelmäßig eingedrückter Oberfläche.

Vorkommen: Trockene, wüstenartige Gebirgshänge im südlichen Nordamerika mit Kiefern und immergrünen Eichen. Normalerweise in Höhen über 1200 bis 1800 m, optimal in Cañons mit guter Wasserversorgung, kommt aber auch auf trockenen Hängen und felsigen Graten vor.

Holz: Weich und von feiner Textur. Kern rein hellrot, oft mit gelben Streifen; Splint dünn, fast weiß.

Nutzung: Früchte dienten Indianern als Nahrung; der Baum ist als Garten- und Parkbaum besonders in Europa beliebt. Schönster Baum im westlichen Nordamerika.

Description: 15 à 18 m de hauteur tronc court et vigoureux de 1,5 m de diamètre et un houppier ampl pyramidal et ouvert, souvent ferm et rond sur les vieux sujets. Ecorc épaisse, brun sombre teinté de rouge, profondément fissurée et craquelée en plaques anguleuses. Feuilles squamiformes plaquées sur les rameaux, vert bleu pâle. Fleurs peu marquantes, groupées l'extrémité de petits rameaux laté raux. Fruits sphériques brun roug à maturité bisannuelle, couverts de pruine bleue jeunes; surface aux creux irréguliers.

Habitat: versants de montagnes, désertiques du sud de l'Amériqu du Nord, avec les pins et les chên toujours verts, habituellement en 1200 et 1800 m. Dans les canyon bien alimentés en eau, mais égale ment sur les pentes sèches et les arêtes rocheuses.

Bois: tendre et à grain fin. Bois de cœur d'un rouge clair pur, présen tant souvent des rubans jaunes; aubier mince, presque blanc.

Intérêts: Fruits consommés par le Indiens; arbre de parcs et de jardins en Europe. Un des plus beau arbres de l'ouest nord-américain.

198

Alligator Juniper, Thick-bark Juniper.

TRANSVERSE SECTION.

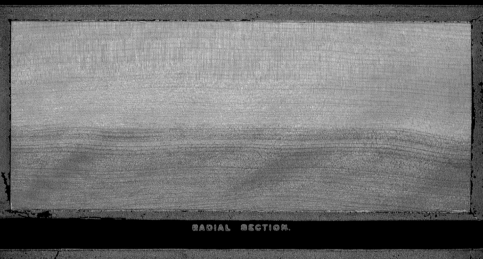

RADIAL SECTION.

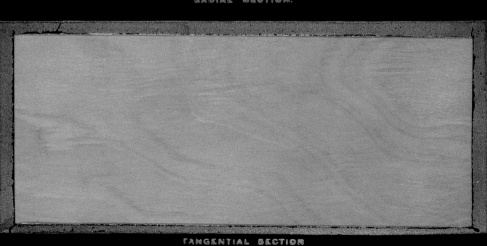

TANGENTIAL SECTION

Ger. Dickborke Wachholder. *Fr.* Genevrier a ècorce épais.
Sp. Enebro de corteza espesa.

C | CYPRESS FAMILY
Arizona Cypress

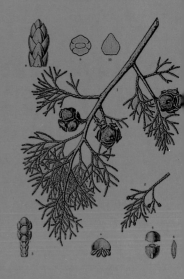

FAMILIE DER ZYPRESSENGEWÄCHSE
Arizona-Zypresse

FAMILLE DU CYPRÈS
Cyprès de l'Arizona

CUPRESSACEAE
Cupressus arizonica Greene

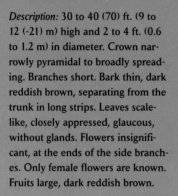

Description: 30 to 40 (70) ft. (9 to 12 (-21) m) high and 2 to 4 ft. (0.6 to 1.2 m) in diameter. Crown narrowly pyramidal to broadly spreading. Branches short. Bark thin, dark reddish brown, separating from the trunk in long strips. Leaves scale-like, closely appressed, glaucous, without glands. Flowers insignificant, at the ends of the side branches. Only female flowers are known. Fruits large, dark reddish brown.

Habitat: Extensive stands on the northern slopes of mountains in south-western North America from Arizona to northern Mexico at altitudes between 5,000 and 7,500 ft. (1,500 to 2,400 m).

Wood: Light, soft, and close-grained; easy to work.

Use: Introduced successfully into English gardens in 1882. Since then grown as a park tree.

Beschreibung: 9 bis 12 (bis 21) m hoch bei 0,6 bis 1,2 m Durchmesser. Krone schmal pyramidal bis breit ausladend. Äste kurz. Borke dünn, dunkel rotbraun, sich in langen Streifen vom Stamm lösend. Blätter schuppenartig eng anliegend, bereift, drüsenlos. Blüten unauffällig, an den Enden von Seitenästen. Es sind nur weibliche Blüten bekannt. Früchte groß, dunkel rotbraun.

Vorkommen: Ausgedehnte Bestände auf Nordhängen der Gebirge im südwestlichen Nordamerika von Arizona bis Nordmexiko in Höhen zwischen 1500 bis 2400 m.

Holz: Leicht, weich und mit feiner Textur; leicht zu bearbeiten.

Nutzung: 1882 erfolgreich in englische Gärten eingeführt. Seitdem als Parkbaum.

Description : petit arbre de 9 m à 12 m de hauteur (21 m max.) sur un diamètre de 0,6 à 1,2 m. Houppier étroit et pyramidal ou ample et étalé, à branches courtes. Ecorce mince, brun rouge sombre, se détachant en longues lanières. Feuilles squamiformes, plaquées sur les rameaux, avec des bandes blanches pruineuses et dépourvue de glandes dorsales. Fleurs peu marquantes, terminant les rameaux latéraux. Cônes de grande taille et d'un brun rouge sombre.

Habitat : en vastes peuplements sur les versants d'ubac, à une altitude de 1500 à 2400 m dans le sud-ouest de l'Amérique du Nord depuis l'Arizona jusqu'au nord du Mexique.

Bois : léger, tendre et à grain fin ; facile à travailler.

Intérêts : acclimatation réussie en 1882 dans des jardins anglais ; arbre de parcs depuis lors.

Arizona Cypress.

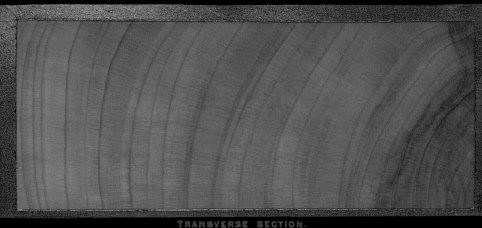

TRANSVERSE SECTION.

RADIAL SECTION.

TANGENTIAL SECTION.

Ger. Arizonische Cypresse. *Fr.* Cypres d'Arizona.
Sp. Cipres de Arizona.

C | CYPRESS FAMILY
Bigtree, Wellingtonia, Giant Sequoia

FAMILIE DER ZYPRESSENGEWÄCHSE
Mammutbaum

FAMILLE DU CYPRÈS
Séquoiadendron géant, Séquoia géant, Mammouth

CUPRESSACEAE
Sequoiadendrum giganteum (Lindl.) J. Buchholz
Syn.: *Sequoia gigantea* (Lindl.) Decne.

Description: 300 to 350 ft. (90 to 106 m) high, 20 to 36 ft. (6 to 11 m) in basal diameter. Age: according to annual rings on average 2,000 to 2,300, to a known maximum of 4,000 years. Base broadened by massive, spreading roots. Branches lost with age to a height of 100 to 165 ft. (30 to 50 m). Bark 1 to 2 ft. (0.3 to 0.6 m) thick, in four to five shallow depressions around the trunk. Pale cinnamon red, tinged purple towards the outside. Attractive when young. Later peeling off in fibrous strips. Fire-resistant.

Habitat: Mountain woodland in California with Sugar Pine, Douglas Fir, and Incense Cedar.

Wood: Very light, soft, not strong, and coarse-grained; durable in the ground. Heartwood an attractive, clear red, darkening later; sapwood almost white.

Use: Fencing, building, roof shingles.

Probably not the tallest, but certainly the most massive, North American tree.

Beschreibung: 90 bis 106 m hoch bei 6 bis 11 m Durchmesser. Alter: Nach Jahresringzählungen durchschnittlich 2000 bis 2300, bisher max. 4000 Jahre alt. Basis durch mächtige Wurzelanläufe verbreitert. Im Laufe der Jahre Abwurf der Äste auf 30 bis 50 m Höhe. Borke 0,3 bis 0,6 m dick, in vier bis fünf Buchten, flach abgerundet um das Holz. Hell zimtrot, außen purpurn überflogen. Jung ansehnlich. Zerfällt später in lose streifenförmige Faserschuppen. Feuerhemmend.

Vorkommen: Bergwälder Kaliforniens mit der Zucker-Kiefer, der Douglasie und Flusszeder.

Holz: Sehr leicht, weich, nicht fest und von grober Textur; sehr dauerhaft im Boden. Kern schönes, klares Rot, dunkelt nach; Splint fast weiß.

Nutzung: Für Zäune, Bauwerke, Dachschindeln.

Wahrscheinlich nicht der höchste, wohl aber der massivste Baum Nordamerikas.

Description: 90 à 106 m de hauteu sur un diamètre de 6 à 11 m à la base. Vit 2000 à 2300 ans en moyenne selon le nombre de cern annuels; maximum 4000 ans. La base du tronc est élargie par de gros contreforts. Les branches à ur hauteur de 30 à 50 m tombent av les années. Ecorce de 0,3 à 0,6 m d'épaisseur, creusée d'indentation profondes, formant quatre ou cinq grosses cannelures arrondies et aplaties autour du bois; couleur cannelle, avec une touche pourpre en surface. Bel aspect à l'état juvénile, s'exfoliant en lanières fibreus souples avec l'âge. Protège l'arbre des incendies de forêt.

Habitat: forêts montagnardes de Californie en compagnie du pin à sucre, du douglas de Menzies et du cèdre des rivières.

Bois: très léger, tendre, pas résistant et à grain grossier; très durab sur pied. Bois de cœur d'un beau rouge franc, se rembrunissant; aubier presque blanc.

Intérêts: palissades, bâtiments et bardeaux.

Sans doute pas le plus grand arbr nord-américain mais le plus mass

Big Tree, Giant Redwood, Redwood of the Mountains.

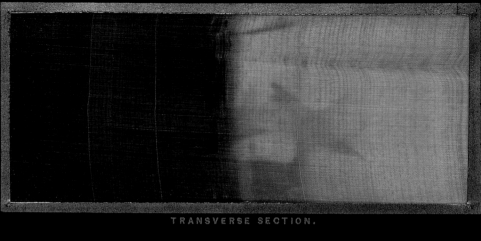

TRANSVERSE SECTION.

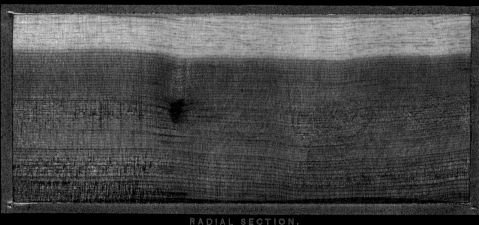

RADIAL SECTION.

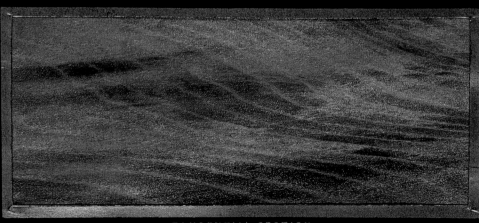

TANGENTIAL SECTION

Ger. Riesenbaum. *Fr.* Arbre gigantesque.

Sp. Arbol giganteo.

C | CYPRESS FAMILY
California Juniper

FAMILIE DER ZYPRESSENGEWÄCHSE
Kalifornischer Sadebaum

FAMILLE DU CYPRÈS
Genévrier de Californie

CUPRESSACEAE
Juniperus californica Carrière

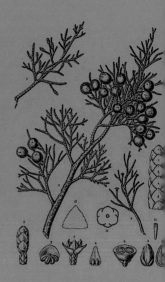

Description: To 45 ft. (13 m) high, 1 to 2 ft. (0.3 to 0.6 m) in diameter. Trunk straight, heavily fluted, asymmetrical in cross section, often shrubby, with many sturdy, twisted, erect branches, forming a broad, open crown. Bark thin, made up of long, loose ash-gray scales like wood shavings, which last many years. Detached scales reveal a red-brown inner side. Leaves scaly, in threes, densely packed and pale yellow-green.

Habitat: Mountain slopes and hills of California between 400 (130) and 4,000 ft. (1,300 m) in altitude. Best developed in desert areas.

Wood: Light, soft, and close-grained; durable in ground. Heartwood pale brown, with slight red tinge; sapwood thin, almost white.

Use: In southern California for fencing and as firewood.

The fruits (berry-like cones) gathered in large quantities by American Indians and eaten raw, or milled and baked into nourishing cakes.

Beschreibung: Gelegentlich 13 m hoch bei 0,3 bis 0,6 m Durchmesser. Gerader, großbuchtiger Stamm mit unsymmetrischem Querschnitt, öfter strauchig mit zahlreichen gedrungenen und gedrehten, aufrechten Ästen, die eine breite, offene Krone bilden. Borke dünn und in langen, losen, hobelspanartigen Schuppen, die, aschgrau, lange Jahre halten. Bei ihrem Abgehen Inneres rötlich braun. Blätter schuppig, zu dritt, dicht anliegend, hell gelbgrün.

Vorkommen: Berghänge und Hügel Kaliforniens zwischen 130 und 1300 m. Beste Entwicklung an Wüstenstandorten.

Holz: Leicht, weich und mit feiner Textur; dauerhaft im Boden. Kern hellbraun, etwas rot getönt; Splint dünn, fast weiß.

Nutzung: In Süd-Kalifornien für Zäune und als Brennholz.

Die Früchte (Beerenzapfen) von Indianern in großen Mengen gesammelt und frisch gegessen oder gemahlen und zu nahrhaften Kuchen gebacken.

Description : arbre atteignant parfois 13 m de hauteur, avec un tronc de 0,3 à 0,6 m de diamètre, profondément indenté et à section transversale asymétrique. Il est plus souvent buissonnant avec de nombreuses branches ascendantes, trapues et contournées, formant un houppier ample et ouvert. Ecorce mince, recouverte de longues écailles gris cendré, persistantes qui, en se détachant, révèlent un endoderme brun rougeâtre. Feuilles vert jaune clair, to en écailles, verticillées par trois et rapprochées des rameaux.

Habitat : versants de montagnes et collines de Californie, entre 130 e 1300 mètres. Développement opti mal dans les zones désertiques.

Bois : léger, tendre et à grain fin ; durable sur pied. Bois de cœur brun clair légèrement teinté de rouge; aubier mince, presque blan

Intérêts : en Californie pour les clô tures et comme bois de chauffage

Les fruits (cônes formés d'écailles charnues, ou «galbules») étaient consommés par les Indiens, soit frais, soit moulus et préparés sous forme de gâteaux nourrissants.

California Juniper, Sweet-fruited Juniper.

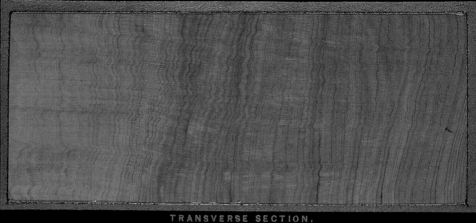

TRANSVERSE SECTION.

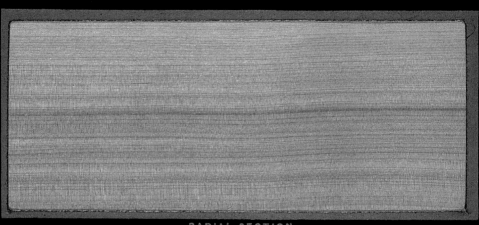

RADIAL SECTION.

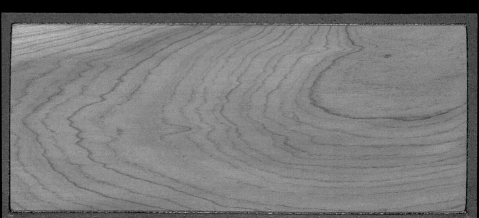

TANGENTIAL SECTION

Ger. Californischer Wachholder. *Fr.* Genievre de Californie.
Sp. Enebro de California.

C | CYPRESS FAMILY

Eastern Red Cedar, Pencil Cedar

FAMILIE DER ZYPRESSENGEWÄCHSE
Bleistiftholz

FAMILLE DU CYPRÈS
Genévrier de Virginie, Cèdre rouge (Canada)

CUPRESSACEAE
Juniperus virginiana L.

Description: Commonly 40 to 50 ft. (12 to 15 m) high but exceptionally twice that height, and 2 to 3 ft. (0.6 to 0.9 m) in diameter. With buttresses at the base of the trunk. Fibrous bark separating into long, narrow strips.

Habitat: Dry, gravely slopes, rocky ridges, and less often good soils in the lowlands. Sand hills on the coast, cliffs on the coast of New England. Withstands the wind, even near the ocean (salt spray). (North-)Eastern North America.

Wood: Light and aromatic; very durable. Purple-red.

Use: Especially for mothproof cupboards, pencils, fence posts, etc. The berry-like fruit is used in cooking and for medicinal purposes.

Beschreibung: Gewöhnlich 12 bis 15 m hoch, ausnahmsweise aber doppelt so hoch bei 0,6 bis 0,9 m Durchmesser. Mit Wurzelanläufen an der Stammbasis. Faserige Borke mit längs abblätternden Streifen.

Vorkommen: Trockene, kiesige Hänge, Felskanten und weniger häufig gute Böden im Flachland. Sandhügel an der Küste, Klippen der Küste Neuenglands. Widersteht dem Wind auch nahe dem Ozean (Salzgischt). (Nord-) Östliches Nordamerika.

Holz: Leicht und duftend; sehr haltbar. Purpurrot.

Nutzung: Vorzugsweise für mottendichte Schränke, Bleistifte, Zaunpfähle usw. Die (Schein-) Beeren eignen sich als Küchengewürz und für medizinische Zwecke.

Description: arbre s'élevant d'ordinaire entre 12 à 15 m mais qui pe atteindre quelquefois le double de cette hauteur. Tronc de 0,6 à 0,9 m de diamètre, armé de contreforts à la base. Ecorce fibreuse pelant en longues lanières.

Habitat: pentes sèches graveleuse arêtes rocheuses et, moins souver sols favorables de plaine. Dunes e écueils sur la côte de la Nouvelle-Angleterre. Résiste au vent même à proximité de l'océan (embruns). On le trouve dans le (nord-)est de l'Amérique du Nord.

Bois: léger et odorant; très durabl Rouge pourpre.

Intérêts: utilisé de préférence pou la fabrication d'armoires à linge (l'odeur du bois éloigne les mites) de crayons, de pieux de clôtures etc. Les galbules, appelées à tort «baies», sont utilisées en médecir et comme condiment.

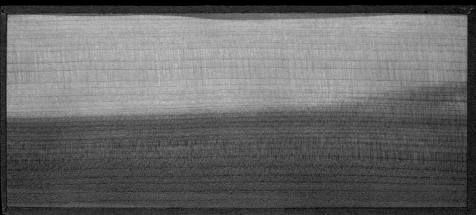

TRANSVERSE   SECTION.

RADIAL   SECTION.

TANGENTIAL   SECTION.

Gen. Virginischer Wachholder. *Fr.* Genevrier. *Sp.* Sabina.

C | CYPRESS FAMILY
Gowen Cypress, Californian Cypress

FAMILIE DER ZYPRESSENGEWÄCHSE
Gowens Zypresse

FAMILLE DU CYPRÈS
Cyprès de Gowen, Cyprès de Californie

CUPRESSACEAE
Cupressus goveniana Gordon

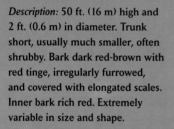

Description: 50 ft. (16 m) high and 2 ft. (0.6 m) in diameter. Trunk short, usually much smaller, often shrubby. Bark dark red-brown with red tinge, irregularly furrowed, and covered with elongated scales. Inner bark rich red. Extremely variable in size and shape.

Habitat: Near mountain streams with Douglas Fir and Ponderosa Pine, sandy wasteland or rocky slopes only a few miles inland in the California coastal region. Shrubby in the north of its range. Nowhere very common.

Wood: Light, soft, not strong, brittle, and close-grained.

Use: No commerical use known.

Hardy in western and southern Europe. English gardens.

Beschreibung: 16 m hoch bei 0,6 m Durchmesser. Kurzer Stamm, gewöhnlich viel kleiner, oft strauchig. Borke dunkel rotbraun, rot getönt und unregelmäßig in schmale, flache Rillen und Riefen geteilt, die in längliche Schuppen übergehen. Inneres kräftig rotbraun. Außerordentlich variabel in Größe und Tracht.

Vorkommen: In der Nachbarschaft von Bergbächen gemeinsam mit Douglasien und Westlicher Gelb-Kiefer, sandiges Ödland oder felsige Hänge nur wenige Kilometer landeinwärts in der kalifornischen Küstenregion. Im Norden des Gebiets strauchig. Nirgendwo sehr häufig.

Holz: Leicht, weich, nicht fest, spröde und mit feiner Textur.

Nutzung: Keine gewerbliche Nutzung bekannt.

Winterhart in West- und Südeuropa. Englische Gärten.

Description : 16 m de hauteur sur un diamètre de 0,6 m. Tronc cou d'ordinaire bien plus petit. Souve à port de buisson. Ecorce d'un brun rouge foncé teinté de rouge texture entrelacée, présentant de sillons et des crêtes étroits et plat qui font place à des écailles allongées. Endoderme d'un brun roug intense. Essence extrêmement variable en taille et en parure.

Habitat : non loin des torrents de montagne, en association avec le douglas et le pin jaune de l'Oues terrains sablonneux ingrats ou pentes rocheuses à quelques kilo mètres seulement vers l'intérieur dans la région côtière de la Califo nie. Port buissonnant au nord de son aire. Peu fréquent.

Bois : léger, tendre, non résistant, cassant et à grain fin.

Intérêts : utilisation commerciale non déterminée.

Résiste bien à l'hiver en Europe occidentale et méridionale. Jardin anglais.

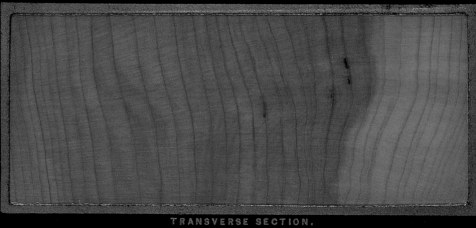

TRANSVERSE SECTION.

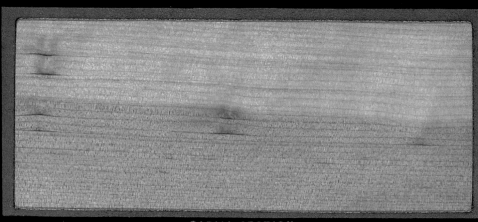

RADIAL SECTION.

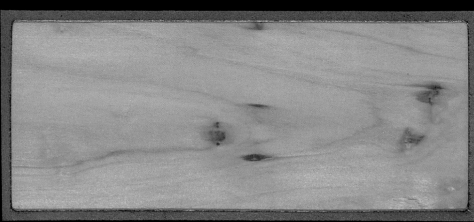

TANGENTIAL SECTION

Ger. Cypresse von Gowen. **Fr.** Cypres de Gowen.

Sp. Ciprés de Gowen.

C | CYPRESS FAMILY

Incense Cedar, White Cedar

FAMILIE DER ZYPRESSENGEWÄCHSE
Kalifornische Flusszeder

FAMILLE DU CYPRÈS
Calocèdre à encens, Arbre à encens,
Cèdre blanc de Californie

CUPRESSACEAE
Calocedrus decurrens (Torr.) Florin
Syn.: *Libocedrus decurrens* Torr.

Description: 165 ft. (50 m) high, 7 to 8 ft. (2.1 to 2.4 m) in diameter. Trunk tall and straight, irregularly indented in section. In first hundred years, branches are slim and upright, bending downwards. In older trees, branches at first horizontal, then growing upwards as secondary trunks. Bark cinnamon red, irregularly furrowed and with a dense covering of scales. Inner chamber of seed coat with red liquid balsamic resin.

Habitat: All types of soil in western North America; warm, dry hills, and high ground as well as water meadows.

Wood: Light, soft, not strong; close-grained; durable in the ground. Heartwood pale red-brown; sapwood thin, almost white.

Use: Fences, slats, roof shingles, water channels, interior construction, furniture. Bark tannin-rich. Ornamental park tree.

Discovered by Frémont in 1846 on the upper Sacramento River.

Beschreibung: 50 m hoch bei 2,1 bis 2,4 m Durchmesser. Hoher, gerader Stamm, im Querschnitt unregelmäßig gebuchtet. Im ersten (Lebens-) Jahrhundert schlanke Äste, aufwärts gebildet, abwärts geneigt. Im hohen Alter Äste zuerst horizontal, dann aufwärts und als Sekundärstämme. Borke zimtrot, in unregelmäßigen Riefen mit dicht anliegenden Schuppen. Innere Kammer der Samenschale mit rotem, flüssig-balsamischem Harz.

Vorkommen: Alle Arten von Böden im westlichen Nordamerika, auf warmen, trockenen Hügeln und Hochflächen ebenso wie in Flussauen.

Holz: Leicht, weich, nicht fest und mit feiner Textur; sehr dauerhaft im Boden, Kern hell rötlichbraun; Splint dünn, fast weiß.

Nutzung: Für Zäune, Latten und Schindeln, Wassergerinne, Innenausbau, Möbel. Rinde gerbstoffreich. Ziergehölz für Parks und Gärten.

Von Frémont 1846 am oberen Sacramento River entdeckt.

Description: 50 m de hauteur sur un diamètre de 2,1 à 2,4 m. Tronc élancé, droit, irrégulièrement indenté dans sa section transversale. Branches fines, développées verticalement et infléchies vers le bas durant le premier siècle d'âge; horizontales puis dressées devenant des tiges secondaires se les sujets très âgés. Ecorce cannel se fissurant en côtes irrégulières, recouvertes d'écailles appliquées. La graine possède une poche de résine balsamique liquide de couleur rouge.

Habitat: tous les sols de l'ouest d l'Amérique du Nord, les collines hauteurs chaudes et sèches, et le berges.

Bois: léger, tendre, peu résistant à grain fin; très durable sur pied. Bois de cœur brun rougeâtre clai aubier mince, presque blanc.

Intérêts: fabrication de clôtures, lattes, de bardeaux, de gouttières menuiserie intérieure et d'ameublement. Ecorce riche en tannin.

Découvert par Frémont en 1846 en amont du Sacramento. Arbre de parcs ornemental.

210

Californian White Cedar, Post Cedar, Incense Cedar.

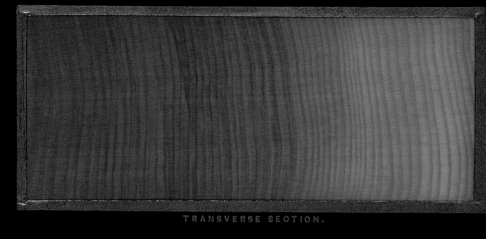

TRANSVERSE SECTION.

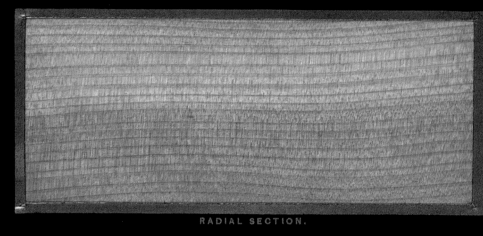

RADIAL SECTION.

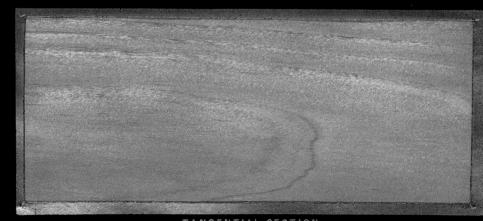

TANGENTIAL SECTION

𝔯. Californian Weisze Zeder.　　　　𝔉𝔯. Thuya blanc de Califo

𝔖p. Tuya blanco de California.

C | CYPRESS FAMILY

Lawson's Cypress, Port Orford Cedar, Oregon cedar

FAMILIE DER ZYPRESSENGEWÄCHSE

Lawsons Scheinzypresse, Lawsons Lebensbaumzypresse

FAMILLE DU CYPRÈS

Faux-cyprès de Lawson, Cyprès de Lawson

CUPRESSACEAE

Chamaecyparis lawsoniana (A. Murray) Parl.

Description: Up to 200 ft. (60 m) high and 12 ft. (3.7 m) in diameter. Often without branches for the lower 150 ft. (45 m). Top inclined to one side. Branches slender, flattened. Bark up to 10 in. (25 cm) thick, deeply furrowed, dark reddish brown, clearly differentiated into two layers; the inner is compact and dark, the outer paler, looser, and separated into scales. Leaves distinctly glandular, scale-like, closely appressed. Flowers insignificant, at the ends of the side branches. Male flowers with bright red connectives.

Habitat: In biodiverse coastal forests in Pacific North America, often with dense undergrowth.

Wood: Light, hard, strong, and very close-grained; polishable, durable in the ground, easy to work. Heartwood pale yellow to almost white; sapwood thin, barely distinguishable. Much aromatic resin.

Use: One of the most important sources of timber. Perhaps the largest and most valuable representative of the Cupressaceae in North America. One of the most beautiful garden trees, cultivated in all parts of Europe since 1854.

Beschreibung: Bis 60 m hoch bei 3,7 m Durchmesser. Häufig bis in 45 m Höhe astfrei. Wipfel zur Seite geneigt. Äste schlank, abgeflacht. Borke bis 25 cm dick, tief rissig, dunkel rotbraun, deutlich in 2 Lagen differenziert; die innere kompakt und dunkel, die äußere heller, locker und Schuppen abgliedernd. Blätter deutlich drüsig, schuppenförmig, eng anliegend. Blüten unauffällig, an den Enden von Seitenästen. Männliche Blüten mit hellroten Konnektiven.

Vorkommen: Artenreiche Küstenwälder im pazifischen Nordamerika, oft mit reichem Unterwuchs.

Holz: Leicht, hart, fest und mit sehr feiner Textur; polierfähig, im Boden sehr fäulnisresistent und leicht zu bearbeiten. Kern hellgelb bis fast weiß; Splint dünn, kaum unterscheidbar. Reich an wohlriechendem Harz.

Nutzung: Einer der wichtigsten Holzlieferanten. Der vielleicht größte und wertvollste Vertreter der Cupressaceen in Nordamerika. Einer der schönsten Gartenbäume, der bereits seit 1854 in weiten Teilen Europas kultiviert wird.

Description: s'élevant jusqu'à 60 tronc de 3,7 m de diamètre, souvent non ramifié jusqu'à 45 m de hauteur, garni de branches mince et aplaties et d'une cime inclinée. Ecorce de 25 cm d'épaisseur, profondément fissurée, brun rou foncé, formée de deux couches bien distinctes, compacte et sombre dessous, plus claire, lâche et craquelée en écailles dessus. Feuilles squamiformes, couvrant les ramules et pourvues de gland très marquées. Fleurs insignifiant terminant les rameaux latéraux. Fleurs mâles formant des cônes rouge clair.

Habitat: forêts côtières polymorphes du Pacifique en Amériq du Nord, souvent avec sous-bois.

Bois: léger, dur, résistant et à gra fin; peut être poli, résiste à la décomposition et se travaille bien Bois de cœur jaune clair à presqu blanc; aubier mince, à peine distinct. Riche en résine odorante.

Intérêts: un des principaux producteurs de bois. Peut-être le plus grand et le plus précieux des Cupressacées en Amérique du Nord. Très bel arbre de jardins. Cultivé en Europe depuis 1854.

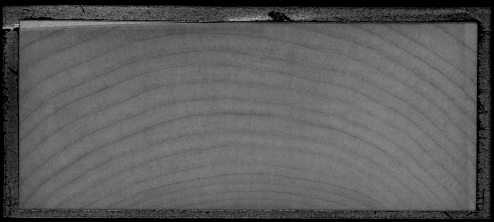

TRANSVERSE SECTION.

RADIAL SECTION.

TANGENTIAL SECTION

Ger. Cypresse von Lawson.   *Fr.* Cypres de Lawson.
Sp. Cipres de Lawson.

C | CYPRESS FAMILY
MacNab Cypress

FAMILIE DER ZYPRESSENGEWÄCHSE
MacNabs Zypresse

FAMILLE DU CYPRÈS
Cyprès de MacNab

CUPRESSACEAE
Cupressus macnabiana A. Murray

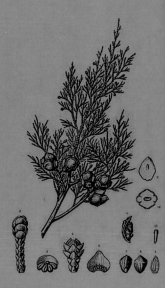

Description: Maximum 30 ft. (9 m) high and 12 to 15 in. (0.3 to 0.38 m) in diameter. Small, rather bushy tree with a short trunk. Most commonly as much-branched bush, only 6 to 12 ft. (1.8 to 3.7 m) high, with broad, open crown. Bark thin, dark red-brown, with the surface marked by many fissures into diamond-shaped areas, peeling into lightly attached strips. Leaves scale-like, spreading only on young shoots, later closely appressed. Branches slender. Flowers inconspicuous, at ends of lateral branches, unisexual.

Habitat: Dry slopes in few locations in California. One of the rarest trees in the state.

Wood: Light, soft, and very close-grained. Heartwood pale brown; sapwood thick, almost white. Conspicuous rings of dark summer wood and thin, indistinct medullary rays.

Use: Occasionally grown in parks in England.

Beschreibung: Maximal 9 m Höhe bei 0,3 bis 0,38 m Durchmesser. Kleiner, eher buschartiger Baum, Stamm kurz. Häufiger als Busch mit zahlreichen Stämmen und von nur 1,8 bis 3,7 m Höhe, mit breiter, offener Krone. Borke dünn, dunkel rotbraun, mit zahlreichen Rissen, welche die Borke mit diamantförmigen Vertiefungen überziehen und aus denen sich nur leicht angeheftete, dünne Borkenstreifen abschälen. Blätter schuppenartig, nur an jungen Ästen abspreizend, später eng anliegend. Äste schlank. Blüten unauffällig, an den Enden von Seitenästen, eingeschlechtlich.

Vorkommen: Trockene Hänge an wenigen Stellen in Kalifornien. Einer der seltensten Bäume Kaliforniens.

Holz: Leicht, weich und mit sehr feiner Textur. Kern blassbraun; Splint dick, fast weiß. Auffällige Ringe dunkler Sommerzellen und dünne, undeutliche Markstrahlen.

Nutzung: In England gelegentlich als Parkbaum kultiviert.

Description : arbuste à port plutôt touffu de 9 m de hauteur maximum, avec un tronc court de 0,3 à 0,38 m de diamètre. Constitue plus souvent une masse arbustive, d'une hauteur ne dépassant pas 1,8 à 3,7 m, à cime ample et ouverte. Ecorce mince, brun rouge foncé, présentant de nombreuses fissures et des plaques en forme de diamant, pelant en lanières fines plus ou moins lâches. Seuls les rameaux juvéniles portent des feuilles en écailles, d'abord écarté puis plaquées. Ramifications fines Fleurs insignifiantes, terminales s les branches latérales, unisexuées.

Habitat : pentes sèches dans de rares points de Californie. Un des arbres les plus rares de Californie

Bois : léger, tendre et à grain très fin. Cœur brun pâle ; aubier épais presque blanc. Anneaux visibles cellules sombres formées en été e fins rayons médullaires indistincts

Intérêts : cultivé de temps à autre Angleterre comme arbre de parcs

Macnab Cypress.

TRANSVERSE SECTION.

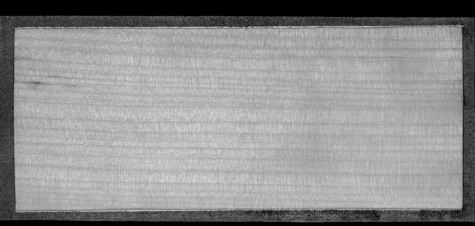

RADIAL SECTION.

TANGENTIAL SECTION.

C | CYPRESS FAMILY
Monterey Cypress

FAMILIE DER ZYPRESSENGEWÄCHSE
Monterey-Zypresse

FAMILLE DU CYPRÈS
Cyprès de Monterey

CUPRESSACEAE
Cupressus macrocarpa Hartw. ex Gordon

Description: 70 to 80 ft. (20 to 23 m) high and 2 to 3 (5 to 6) ft. (0.6 to 0.9 (1.5 to 1.8) m) in diameter. Trunk short. Bark with broad, flat, irregularly connected ridges, with long, narrow, adhering scales. Almost white on old, wind-blown specimens. Crown flat, branches twisted by Pacific storms.

Habitat: High, red granite cliffs, exposed to sea spray; with sentry-like isolated trees in the south of Monterey Bay, California.

Wood: Heavy, hard, strong, rather brittle, silky-sheened, somewhat aromatic, and close-grained; polishable and durable in the ground.

Use: Fast-growing. The most commonly grown conifer in the Pacific states. Major ornamental tree in the parks and gardens of central California. Used as hedge and windbreak. Also grown in Europe, and in temperate South America and Australia. Many cultivated forms.

Beschreibung: 20 bis 23 m hoch bei 0,6 bis 0,9 (1,5 bis 1,8) m Durchmesser. Kurzer Stamm. Borke mit breiten, flachen, unregelmäßig verbundenen Riefen, in schmale, längliche, haftende Schuppen übergehend. An alten, von Stürmen mitgenommenen Exemplaren fast weiß. Die Kronen flach, die Äste verdreht, als Folge der Stürme des Pazifiks.

Vorkommen: Hohe, rote Granitklippen, dauernd vom Meer übersprüht, mit wachtpostenartig einzelnen Bäumen im Süden der Bucht von Monterey/Kalifornien.

Holz: Schwer, hart, fest, ziemlich spröde, seidig, etwas duftend und mit feiner Textur; polierfähig und sehr dauerhaft im Boden.

Nutzung: Schnellwüchsig. Die am meisten kultivierte Konifere in den Pazifikstaaten. Hauptzierbaum in den Parks und Gärten von Mittel-Kalifornien. Auch als Hecke und Windschutz. In Europa so gut wie in Kalifornien. Desgleichen im gemäßigten Südamerika und in Australien. Zahlreiche Kulturformen.

Description : arbre de 20 à 23 m de hauteur, avec un tronc court de 0,6 à 0,9 m (1,5 à 1,8 m) de diamètre. Ecorce à texture entrelacée, formée de fissures et de large crêtes plates, se craquelant dans l sens longitudinal en écailles minc et adhérentes ; presque blanche sur les vieux spécimens malmené par les tempêtes. La couronne est plate, les branches tordues par les tempêtes du Pacifique.

Habitat : hauts écueils en granit rouge, balayés en permanence par la mer et habités par quelque arbres postés en « sentinelle » dan le sud de la baie de Monterey en Californie.

Bois : lourd, dur, résistant, assez cassant, satiné, légèrement odora et à grain fin ; susceptible d'être poli et très durable sur pied.

Intérêts : essence à croissance rapide ; le plus cultivé des conifer de la façade pacifique. Principal arbre des parcs et des jardins de moyenne Californie. Employé po constituer des haies protectrices e des brise-vent. Utilisation similair en Europe et dans les régions tem pérées de l'Amérique du Sud et e Australie. Nombreux cultivars.

Monterey Cypress.

TRANSVERSE SECTION.

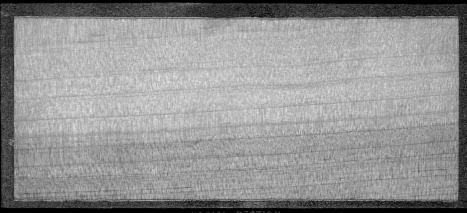

RADIAL SECTION.

TANGENTIAL SECTION.

Ger. Cypresse von Monterey. *Fr.* Cypres de Monterey.

Sp. Cipres de Monterey.

C | CYPRESS FAMILY

Nootka Cypress, Sitka Cypress, Yellow Cypress

FAMILIE DER ZYPRESSENGEWÄCHSE
Nootka-Scheinzypresse

FAMILLE DU CYPRÈS
*Faux-cyprès de Nootka, Cèdre de l'Alaska,
Cèdre jaune (Canada)*

CUPRESSACEAE
Chamaecyparis nootkatensis (D. Don) Spach

Description: Up to 120 ft. (36 m) high and 5 to 6 ft. (1.5 to 1.8 m) in diameter. Crown narrowly pyramidal. Branches short, somewhat flattened. Branchlets stout, two-ranked, with crowded, short shoots that are often deciduous. Leaves usually without glands, scale-like, closely appressed, and dark blue-gray in color. Flowers insignificant on side branches. Fruits broadly globular with erect, pointed bosses, dark reddish brown with a thick glaucous bloom.

Habitat: In forests in north-western North America from sea level up to 5,400 ft. (1,700 m), in the south mainly on rocky slopes and cliffs.

Wood: Light, hard, tends to crumble, with a silky sheen and a pleasant scent, very close-grained; polishable, extremely durable in the ground, and easy to work.

Use: The wood is unsurpassed as a basis for cabinet construction. The wood is used in shipbuilding, for the interiors of houses, and for the manufacture of furniture. Introduced into Europe in 1858 and subsequently cultivated in numerous forms.

Beschreibung: Bis 36 m hoch bei 1,5 bis 1,8 m Durchmesser. Krone schmal pyramidal. Äste kurz, etwas abgeflacht. Kräftige, zweireihig angeordnete Zweige mit gedrängten, nicht ausdauernden Kurztrieben. Blätter normalerweise drüsenlos, schuppenartig eng anliegend und dunkel blaugrau gefärbt. Blüten unauffällig auf Seitenästen. Früchte breit kugelig mit aufgesetzter Spitze, dunkel rotbraun mit dicker blauer Reifschicht.

Vorkommen: In Wäldern des nordwestlichen Nordamerika auf Meereshöhe bis 1700 m, im Süden vor allem auf felsigen Hängen und Kliffen.

Holz: Leicht, hart, eher brüchig, seidenartig, mit angenehmem Geruch und mit sehr feiner Textur; polierfähig, im Boden außerordentlich fäulnisresistent und leicht zu bearbeiten.

Nutzung: Das Holz ist unübertroffen als Grundlage für die Kabinettherstellung. Auch für Schiffbau, die Inneneinrichtung von Häusern und für die Möbelherstellung verwendet. Seit 1858 in Europa eingeführt und in zahlreichen Formen kultiviert.

Description : arbre pouvant atteind 36 m de hauteur, avec un tronc de 1,5 à 1,8 m de diamètre et des branches courtes, légèrement aplaties, constituant un houppier étroit, pyramidal. Rameaux vigoureux, disposés sur deux rangs, ave des rameaux courts, serrés et non persistants. Feuilles squamiformes appliquées, gris bleu foncé, dépou vues de glandes le plus souvent. Fleurs discrètes sur les rameaux latéraux. Cônes larges-sphériques à écailles mucronées, d'un brun rouge sombre couvert de pruine bleue.

Habitat : dans les forêts du nordouest de l'Amérique du Nord, jusqu'à 1700 m d'altitude ; dans le sud, surtout sur les falaises et les pentes rocheuses du littoral.

Bois : léger, dur, plutôt cassant, satiné, aromatique et à grain fin ; susceptible d'être poli, résiste extrê mement bien à la décomposition e est facile à travailler.

Intérêts : bois sans égal en ébénisterie. Utilisé dans la construction navale, en menuiserie intérieure e d'ameublement. Cultivé en Europe depuis son introduction en 1858 ; nombreuses variétés.

TRANSVERSE SECTION.

RADIAL SECTION.

TANGENTIAL SECTION.

Ger. Gelbe Zeder. *Fr.* Cedre jaune.

Sp. Cedro amarillo.

C | CYPRESS FAMILY
Red Cedar

FAMILIE DER ZYPRESSENGEWÄCHSE
Südlicher Wacholder

FAMILLE DU CYPRÈS
Genévrier des Bermudes, Bois scie

CUPRESSACEAE
? Juniperus barbadensis L.

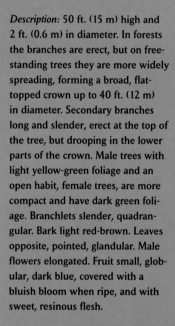

Description: 50 ft. (15 m) high and 2 ft. (0.6 m) in diameter. In forests the branches are erect, but on free-standing trees they are more widely spreading, forming a broad, flat-topped crown up to 40 ft. (12 m) in diameter. Secondary branches long and slender, erect at the top of the tree, but drooping in the lower parts of the crown. Male trees with light yellow-green foliage and an open habit, female trees, are more compact and have dark green foliage. Branchlets slender, quadrangular. Bark light red-brown. Leaves opposite, pointed, glandular. Male flowers elongated. Fruit small, globular, dark blue, covered with a bluish bloom when ripe, and with sweet, resinous flesh.

Habitat: Swamps beside rivers on the Atlantic coast of Georgia to the Gulf coast of Florida. Widespread in the Caribbean.

Wood: Color and smell similar to those of *Juniperus virginiana*.

Use: Used in large measure in the manufacture of pencils, which is exhausting supplies of the tree. One of the most beautiful of all the species of Juniperus.

Beschreibung: 15 m hoch bei 0,6 m Durchmesser. Äste im Waldbestand aufrecht, bei frei stehenden Bäumen dagegen stärker ausgebreitet, eine breite, oben flache Krone von bis zu 12 m Durchmesser bildend. Sekundäräste lang und schlank, an der Spitze des Baumes aufrecht, in den tieferen Teilen der Krone herabhängend. Männliche Bäume mit hellem, gelbgrünem Laub und von offenerem Habitus als die kompakteren weiblichen Bäume mit dunkelgrünem Laub. Zweige schlank, vierkantig. Borke hell rotbraun. Blätter gegenständig, spitz, drüsig. Männliche Blüten verlängert. Frucht klein, kugelig, dunkelblau, reif von bläulichem Reif bedeckt, mit süßem, harzigem Fleisch.

Vorkommen: Überschwemmte Sümpfe an Flüssen der Atlantikküsten von Georgia bis zur Golfküste Floridas. Ferner auf den Westindischen Inseln verbreitet.

Holz: Farbe und Geruch ähnlich *Juniperus virginiana*.

Nutzung: Im großem Maßstab zur Bleistiftherstellung, die Vorkommen des Baumes erschöpfend. Eine der schönsten aller *Juniperus*-Arten.

Description : 15 m de hauteur sur un diamètre de 0,6 m. En peuplement forestier, les branches sont fastigiées mais, en situation isolée elles sont plus étalées et forment un houppier ample, aplati au sommet. Les branches secondaires sont longues et fines, dressées au sommet de l'arbre, retombantes dans la partie inférieure du houppier. Les arbres mâles ont un feuillage clair, vert jaune et un aspect plus ouvert que les arbres femelles plus compacts et garnis d'un feuillage vert foncé. Rameaux fins, anguleux. Ecorce brun rouge clair. Feuilles opposées, pointues à l'apex et glanduleuses. Fleurs mâles prolongées. Petit fruit globuleux, bleu sombre couvert de pruine bleue lorsqu'il est mûr, à chair douce et résineuse.

Habitat : marécages inondés au bord des cours d'eau de la côte atlantique depuis la Géorgie jusqu'à sur le littoral du Golfe en Floride. Répandu aussi sur les îles Caraïbes.

Bois : couleur et odeur similaires à celles de *Juniperus virginiana*.

Intérêts : soumis à une exploitation dévastatrice pour la fabrication de crayons. Un des plus beaux genévriers.

SOUTHERN RED CEDAR. PENCIL-WOOD.

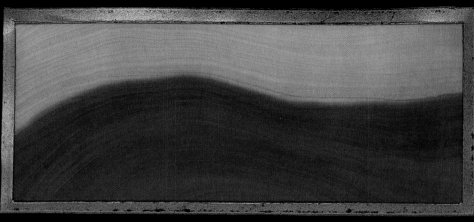

TRANSVERSE SECTION.

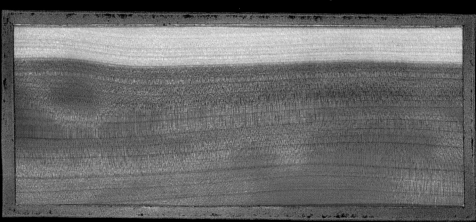

RADIAL SECTION.

TANGENTIAL SECTION.

C | CYPRESS FAMILY
Redwood

FAMILIE DER ZYPRESSENGEWÄCHSE
Küstensequoie

FAMILLE DU CYPRÈS
Séquoia côtier, Séquoia toujours vert, Arbre à amadou

CUPRESSACEAE
Sequoia sempervirens (Lamb. ex D. Don) Endl.

Description: 220 to 330 ft. (65 to 100 m) high, 10 (20 m) to 16 (30 m) ft. (3 (6 m) to 5 (9) m) in diameter; said to reach 400 (430) ft. (120 to 130 m). Tallest North American tree, but less massive than the Bigtree. Spreading considerably at the base (root buttresses). Bark 5 to 10 in. (0.12 to 0.24 m) thick, in furrows separated by ridges 2 to 3 ft. (0.6 to 0.9 m) across. Resprouts vigorously from dormant buds following forest fires, so-called "fire-columns."

Habitat: Riverbanks in the coastal mountain ranges of California and southern Oregon, forms homogeneous woods on damp sandstone.

Wood: Light, soft, not very strong, brittle, and close-grained; polishable, very durable in the ground, and easy to split.

Use: Most valued timber of Pacific states. For roof shingles, fence posts, telegraph poles, railroad sleepers, wine barrels, tanning and watertanks, coffins. Wavy-grained texture suitable for veneers. Bark used as stuffing. Not hardy in central Europe.

Beschreibung: 65 bis 100 m hoch bei 3 (6) bis 5 (9) m Durchmesser, nach anderen Angaben 120 und 130 m. Höchster Baum Nordamerikas, aber weniger massiv als der Riesenmammutbaum. Basis sehr verbreitert (Wurzelanläufe). Borke 0,12 bis 0,24 m dick, in abgerundeten, 0,6 bis 0,9 m breiten, dem Stammquerschnitt und seinen Buchtungen entsprechenden Riefen. Nach Waldbränden am ganzen verbliebenen Stamm kräftiger Austrieb aus „schlafenden Augen", sog. „Feuersäulen".

Vorkommen: Flussufer der Küstengebirge Kaliforniens und Süd-Oregons, bildet auf feuchtem Sandstein reine Wälder.

Holz: Leicht, weich, nicht sehr fest, spröde und mit feiner Textur; polierfähig, sehr dauerhaft im Boden und leicht zu spalten.

Nutzung: Wertvollstes Bauholz der pazifischen Staaten. Für Schindeln, Zaunpfähle, Telegrafenmasten, Eisenbahnschwellen, Weinbottiche, Gerber- und Wassertanks, Särge. Mit gekräuselter Textur Furnierholz für Schränke. Borke als Stopfmaterial. Nicht winterhart in Mitteleuropa.

Description: 65 à 100 m de haute sur un diamètre de 3 m (6 m) à 5 m (9 m); selon d'autres source 120 m ou 130 m. Le plus grand arbre d'Amérique du Nord mais moins massif que le séquoia géan Base élargie (contreforts). Ecorce 0,12 à 0,24 m d'épaisseur, en car nelures arrondies, de 0,6 à 0,9 m de largeur, correspondant à ses profondes indentations et à la se tion transversale du tronc. Rejett vigoureusement de souche après incendie à partir d'yeux dorman ou «colonnes de feu».

Habitat: berges des côtes montagneuses de la Californie et du su de l'Oregon; peuplements purs s le grès humide. Versants de ravin pourvus de nappes d'eau.

Bois: léger, tendre, pas très résistant, cassant et à grain fin; pouva se polir, très durable sur pied et facile à fendre.

Intérêts: bois de construction le plus précieux des Etats de la côte pacifique. Bardeaux, pieux, teaux, traverses de chemin de fer cuves, réservoirs d'eau, cercueils. Bois de placage. Ecorce utilisée pour le bouchage. Ne résiste pas l'hiver en Europe centrale.

Redwood of the Coast.

TRANSVERSE SECTION.

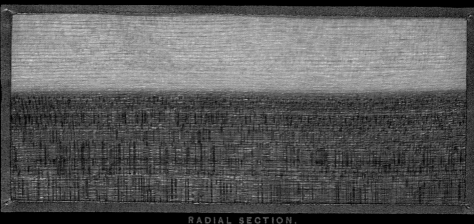

RADIAL SECTION.

TANGENTIAL SECTION

Ger. Californianischer Rothholz.   **Fr.** Bois rouge.

Sp. Madera roja.

C | CYPRESS FAMILY
Sierra Juniper

FAMILIE DER ZYPRESSENGEWÄCHSE
Westamerikanischer Sadebaum

FAMILLE DU CYPRÈS
Genévrier occidental

CUPRESSACEAE
Juniperus occidentalis Hook.

Description: To 70 ft. (20 m) high, 2 to 3 ft. (0.6 to 0.9 m) in diameter. Tall, straight tree, but mostly with only a short trunk of 20 ft. (6 m) and sometimes 10 ft. (3 m) in diameter and with enormous branches set almost at right angles and forming a low, broad crown. Usually smaller. Towards the north, low-growing and shrubby, with many erect or semi-creeping branches. Bark a rich cinnamon-red, divided into broad, low, irregularly connected ridges by broad, shallow furrows; thin, shiny scales on upper surface. Leaves scaly, close together, in threes, gray-green.

Habitat: Mountain slopes and high prairies of western North America, rarely as high as 6,500 ft (2,000 m).

Wood: Light, soft, and very close-grained; very durable in the ground. Heartwood pale red or pale brown; sapwood thick, almost white.

Use: Fencing and firewood. Fruits (cone-like berries) eaten by Californian Indians.

Beschreibung: Gelegentlich 20 m hoch bei 0,6 bis 0,9 m Durchmesser. Hoher, gerader Baum, aber meist mit nur kurzem Stamm von 6 m hoch bei manchmal 3 m Durchmesser und enormen Ästen, die fast rechtwinklig abgehen und eine niedrige, breite Krone bilden. Gewöhnlich kleiner. Gegen Norden des Gebiets strauchig niederliegend mit vielen aufrechten oder halb kriechenden Ästen. Borke kräftig zimtrot, durch breite, flache Rillen in breite, flache, unregelmäßig verbundene Riefen geteilt, die an der Oberfläche dünne, glänzende Schuppen abgeben. Blätter schuppig, zu dritt, eng anliegend, graugrün.

Vorkommen: Berghänge und hoch gelegene Prärien des westlichen Nordamerika, selten auf 2000 m absteigend.

Holz: Leicht, weich und sehr feine Textur; ungemein dauerhaft im Boden. Kern hellrot oder -braun; Splint dick, fast weiß.

Nutzung: Für Zäune und als Brennholz. Die Früchte (Beerenzapfen) dienten den kalifornischen Indianern als Nahrungsmittel.

Description: quelquefois 20 m de hauteur sur un diamètre de 0,6 à 0,9 m. Tronc droit, élancé, mais r mesurant d'ordinaire que 6 m de haut sur un diamètre atteignant quelquefois 3 m, avec d'énormes branches insérées à angle presqu droit, formant un houppier bas et ample. D'ordinaire plus petit. Vers le nord de son aire, à port prostré et buissonnant, avec de nombreuses branches droites ou semi-rampantes. Ecorce à texture entrelacée d'un rouge cannelle vi formée de larges crêtes et sillons plats et s'exfoliant en minces écailles brillantes. Feuilles squamiformes vert gris, verticillées pa trois et appliquées sur les rameau

Habitat: versants de montagnes et prairies alpines de l'ouest de l'Amérique du Nord rarement en dessous de 2000 m d'altitude.

Bois: léger, tendre et à grain très fin; durable sur pied. Bois de cœu rouge clair ou brun clair; aubier épais, presque blanc.

Intérêts: fabrication de clôtures, bois de chauffage. Les fruits («ga bules») étaient consommés par le Indiens de la Californie.

Western Juniper, Yellow Cedar.

TRANSVERSE SECTION.

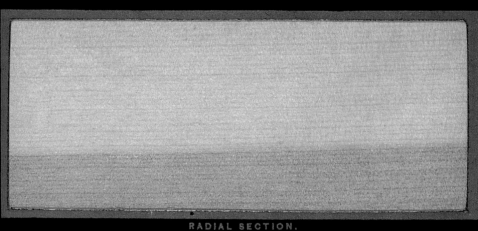

RADIAL SECTION.

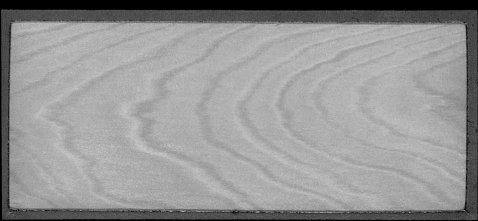

TANGENTIAL SECTION

Ger. Westlicher Wachholder.   Fr. Genievre occidental.
Sp. Enebro occidental.

C | CYPRESS FAMILY

Swamp Cypress, Bald Cypress

FAMILIE DER ZYPRESSENGEWÄCHSE
Zweizeilige Sumpfzypresse

FAMILLE DU CYPRÈS
Taxode chauve, Cyprès à feuilles caduques, Cyprès chauve

CUPRESSACEAE
Taxodium distichum (L.) Rich.

Description: 165 ft. (50 m) high and 8 to 10 ft. (2.4 to 3.0 m) in diameter. Trunk has large root buttresses and often develops knee-like bumps from the roots, especially when growing in water. These are often (wrongly) referred to as "pneumatophores." They perhaps help to stabilize the tree in swampy ground. Short shoots with their needles are shed entire, hence the name "bald" cypress.

Habitat: Forms woods in swamps, in open water or in association with other wetland trees of south-eastern North America.

Wood: Rather light; does not shrink or warp; extremely durable.

Use: Resists insect attack, used for all kinds of weather-resistant construction, also for roof shingles, water containers, cooperage, etc. Useful for barrels, lacking pigments or substances that affect flavor.

Beschreibung: 50 m hoch bei 2,4 bis 3,0 m Durchmesser. Stamm mit mächtigen Wurzelanläufen und besonders wenn im Wasser stehend mit hindernisartigen Vorsprüngen der Wurzeln, bekannt als „Wurzel-knie". Als „Atemwurzeln" gedeutet bzw. missdeutet. Vielleicht zur Erhöhung der Standfestigkeit des gewaltigen Baums im sumpfigen Untergrund. Benadelte Kurztriebe werden als Ganze abgeworfen, daher „bald", d. h. „kahle" Zypresse.

Vorkommen: Waldbildend in Sümpfen im blanken Wasser stehend oder in Gesellschaft von verschiedenen anderen sumpfliebenden Bäumen im südöstlichen Nordamerika.

Holz: Eher leicht; schrumpft nicht, verzieht sich nicht und ist ungemein haltbar.

Nutzung: Frei von Insektenbefall, für alle Arten von wetterbeständigem Holzbau, auch für Dachschindeln, Wasserbehälter, Böttcherei usw. Für Fässer geeignet. Keine Verfärbungen oder geschmacksverändernde Bestandteile.

Description : 50 m de hauteur sur un diamètre de 2,4 à 3,0 m. Le tronc présente de puissants contreforts et des «pneumophores», racines aériennes émergeant du s et dont le rôle fonctionnel est ma connu : considérés comme assura la respiration de l'appareil souterrain sous l'eau, peut-être jouent-i un rôle dans la stabilisation de ce arbre monumental dans les fonds marécageux. Les rameaux courts, garnis de feuilles en aiguilles, tombent en entier à l'automne, d'où le qualificatif de «chauve».

Habitat : forme des peuplements forestiers dans les marécages, en partie immergé ou en compagnie de diverses autres espèces aiman les milieux gorgés d'eau du sud-c de l'Amérique du Nord.

Bois : plutôt léger; ne rétrécit pas ne gauchit pas et est extrêmeme durable.

Intérêts : aucune attaque d'insecte utilisé pour toutes les constructions exposées aux intempéries, les bardeaux, les réservoirs d'eau en tonnellerie etc. Convient bien pour les tonneaux car il ne chang pas de couleur et est exempt de substances altérant le goût.

TRANSVERSE SECTION.

RADIAL SECTION.

TANGENTIAL SECTION

Ger. Zweizeilche Eibencypresse.　　Fr. Cypres afeuille.

C | CYPRESS FAMILY

White Cedar, American Arbor-vitae

FAMILIE DER ZYPRESSENGEWÄCHSE
Abendländischer Lebensbaum

FAMILLE DU CYPRÈS
Thuja occidental, Cèdre blanc (Canada)

CUPRESSACEAE
Thuja occidentalis L.

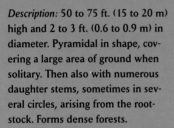

Description: 50 to 75 ft. (15 to 20 m) high and 2 to 3 ft. (0.6 to 0.9 m) in diameter. Pyramidal in shape, covering a large area of ground when solitary. Then also with numerous daughter stems, sometimes in several circles, arising from the rootstock. Forms dense forests.

Habitat: Homogeneous stands in swampy places and on riverbanks. Mostly in association with Black Ash, Swamp Spruce, Paper Birch, Red and Silver Maple, Eastern Larch, Balsam Fir, etc., in northeastern North America. In large numbers in the north; in the south only in uplands.

Wood: Light and aromatic.

Use: For light boats, canoes, etc. Very well suited for roof shingles. The main source of fence posts and telegraph poles for the northeastern states of the USA and in Canada.

Arrived in Europe in the first half of the 16th century. Since then has been used as an ornamental park tree and as a hedge plant (hard to trim) with numerous nursery-bred varieties.

Beschreibung: 15 bis 20 m hoch bei 0,6 bis 0,9 m Durchmesser. Pyramidenform auf weiter Grundfläche, wenn solitär. Dann auch zahlreiche Tochterstämme, u. U. in mehreren Kreisen, Stockausschlag. Dichte Wälder.

Vorkommen: Reinbestände an sumpfigen Orten und Flussufern. Meist in Gesellschaft von Schwarz-Esche, Sumpf-Fichte, Papier-Birke, Rot- und Silber-Ahorn, Östlicher Lärche, Balsam-Tanne u. a. im nordöstlichen Nordamerika. Massenhaft im Norden, im Süden nur in den Hochlagen.

Holz: Leicht und duftend.

Nutzung: Für leichte Boote, Kanus usw. Bestens geeignet für Dachschindeln. Hauptlieferant von Zaunpfählen und Telegrafenmasten der nordöstlichen Staaten Amerikas und in Kanada.

In der ersten Hälfte des 16. Jh. nach Europa eingeführt. Seither als ornamentaler Parkbaum und als Heckenpflanze (schnittfest) mit zahlreichen Spielarten der Baumschulen.

Description: 15 à 20 m de hauteu sur un diamètre de 0,6 à 0,9 m. Arbre à port pyramidal qui occup une surface étendue lorsqu'il est solitaire, avec de nombreuses tige filles disposées en une successior de cercles. Rejette de souche et forme des forêts denses.

Habitat: en peuplements purs da les zones marécageuses et sur les berges de rivières. Le plus souve en compagnie du frêne noir, du p des marais, du bouleau blanc, de érables rouge et argenté, du mélè de l'Est, du sapin baumier etc. da le nord-est de l'Amérique du No Grandes colonies dans le nord; uniquement en altitude au sud.

Bois: léger et odorant.

Intérêts: employé pour la constru tion de bateaux légers, de canots etc. Excellent pour les bardeaux. Principal fournisseur de pieux de clôtures et de poteaux télégraphiques dans le nord-est des Etats-Unis et au Canada.

Introduit en Europe dans la secon moitié du 16e siècle. Bel arbre de parcs et de haies (résiste à la taille ayant donné naissance à de nombreux cultivars ornementaux.

Arbor-Vitae, White Cedar.

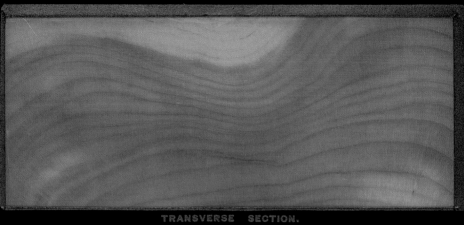

TRANSVERSE SECTION.

RADIAL SECTION.

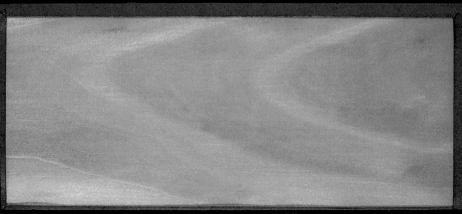

TANGENTIAL SECTION.

Ger. Amerikanische Lebendbaum. *Sp.* Tuya Occidental.
Fr. Thuja d'Occident.

C | CYPRESS FAMILY

White Cedar, White Cypress, Coast White Cedar

FAMILIE DER ZYPRESSENGEWÄCHSE
Weißzeder

FAMILLE DU CYPRÈS
Faux-cyprès blanc, Faux thuya, Cyprès blanc (Canada)

CUPRESSACEAE
Chamaecyparis thyoides (L.) B. S. P.

Description: 75 to 85 ft. (23 to 25 m) high and 2 ft. (0.6 m), occasionally 3 to 4 ft. (0.9 to 1.2 m) in diameter. Trunk straight, with reddish brown bark separating in thin strips of elongated fibers. Foliage of dark bluish-green scales, closely appressed on the outermost twigs. Cones globose, bluish green when ripe.

Habitat: In almost pure stands in the cold swamps of the Atlantic coastal region, especially in New England, where it forms thick forests. Further to the south it appears in association with other trees such as Swamp Cypress, Swamp Bay, Water Tupelo, holly, Sweet Gum, Swamp Oak, Laurel Oak, etc.

Wood: Very light; durable.

Use: For buckets, tubs, in boatbuilding and for railroad sleepers.

Beschreibung: 23 bis 25 m hoch bei 0,6, gelegentlich 0,9 bis 1,2 m Durchmesser. Gerader Stamm mit rotbrauner Borke, in dünnen Streifen von Längsfasern sich ablösend. Belaubung durch dunkel blaugrüne Schuppen, flach an die äußersten Zweige angepresst. Zapfen kugelig, bei der Reife blaugrün.

Vorkommen: In fast reinen Beständen in kalten Sümpfen der Küstenregion am Atlantik, vor allem in Neuengland, wo sie dichte Wälder bildet. Weiter südlich tritt sie in Gesellschaft anderer Bäume auf wie Sumpfzypresse, Sumpf-Avocado, Wasser-Tupelo, Stechhülse, Amberbaum, Sumpf-Eiche, Lorbeer-Eiche u. a.

Holz: Sehr leicht; dauerhaft.

Nutzung: Für Eimer, Kübel, Eisenbahnschwellen und im Bootsbau.

Description: 23 à 25 m de hauteur sur un diamètre de 0,6 m, parfois de 0,9 à 1,2 m. Tronc droit recouvert d'une écorce brun roug se détachant en minces lanières fibreuses longitudinales. Feuilles plaquées sur le rameau par des écailles gris-vert foncé. Cônes glo buleux, vert bleu à maturité.

Habitat: en peuplements presque purs dans les marais froids de la côte atlantique, principalement e Nouvelle-Angleterre où il constitu des forêts denses. Se rencontre aussi plus au sud en compagnie d'autres arbres tels le taxode chauve, l'avocatier palustre, le nyssa aquatique, le houx, le liqui dambar, le chêne des marais, le chêne à feuilles de laurier etc.

Bois: très léger, durable.

Intérêts: utilisé pour la fabrication de seaux, de traverses de chemin de fer et pour la construction de bateaux.

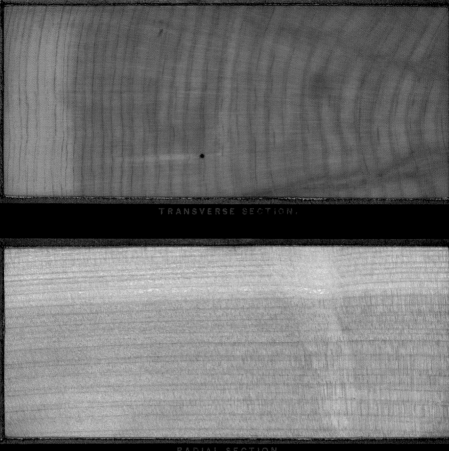

TRANSVERSE SECTION.

RADIAL SECTION.

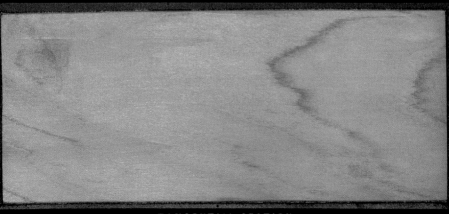

TANGENTIAL SECTION

Ger. Weisze Zeder. *Sp.* Cedro blanco.

Fr. Cedre blanc.

C | CYRILLA FAMILY

Buckwheat Tree, Titi, Ironwood

FAMILIE DER LEDERHOLZGEWÄCHSE
Buchweizenbaum

FAMILLE DU CLIFTONIA
Cliftonie à une feuille

CYRILLACEAE
Cliftonia monophylla (Lam.) Britton ex Sarg.

Description: 40 to 50 ft. (12 to 15 m) high and 12 to 14 in. (0.3 to 0.35 m) in diameter. Trunk stocky, crooked, or twisted, branching at 13 to 16 ft. Bark deeply furrowed, dark red-brown, ridges broken into short, broad scales.

Habitat: Swamps and low-lying areas in south-eastern North America that are flooded for months at a time; with Red Bay, White Cedar, Water Oak, gums (*Nyssa* spp.), and Slash Pine. Forms extensive, dense thickets. In drier open lowland, thickets with Pacific Bayberry, Swamp Bay, and Large Gallberry.

Wood: Heavy, fairly hard, close-grained, brittle. Heartwood brown with red tinge; sapwood thick and paler.

Use: Firewood.

Small trees are ornamental in spring. Sweet-scented flowers. Dark green foliage.

Beschreibung: 12 bis 15 m hoch bei 0,3 bis 0,35 m Durchmesser. Gedrungener, dabei krummer oder schiefer Stamm. Teilung des Stammes in 4 bis 5 m Höhe. Borke tief gefurcht, dunkel rotbraun, Riefen in kurze, breite Schuppen gebrochen.

Vorkommen: Sümpfe und Niederungen im südöstlichen Nordamerika, die monatelang unter Wasser stehen, mit Roter Avocado, Weißzeder, Wasser-Eiche, den Tupelos und Kuba-Kiefer in meilenweit undurchdringlichen Dickichten. Wenn trockenere, offene Niederungen: Dickichte mit dem Gagelstrauch, mit Sumpf-Avocado und Lederblättriger Stechhülse.

Holz: Schwer, mäßig hart, mit feiner Textur, spröde. Von brauner Farbe, rot getönt, dicker, hellerer Splint.

Nutzung: Brennmaterial.

Kleiner Zierbaum im Frühling. Blüten wohlriechend. Dunkelgrünes Laub.

Description: arbre de 12 à 15 m d hauteur, avec un tronc court, tors ou de travers de 0,3 à 0,35 m de diamètre, se divisant à 4 ou 5 m de hauteur. Ecorce brun rouge foncé à texture profondément fis rée et craquelée en écailles court et larges.

Habitat: marécages et secteurs déprimés du sud-est de l'Amériq du Nord, immergés longtemps da l'année, où il forme avec l'avocatier rouge, le faux-cyprès blanc, l● chêne noir, les nyssas et le pin de Cuba de vastes étendues de four impénétrables. Dans les secteurs déprimés plus secs, ouverts, il es● associé au myrica, à l'avocatier palustre et au houx.

Bois: lourd, moyennement dur, à grain fin, cassant. Bois de cœur brun teinté de rouge; aubier épa● plus clair.

Intérêts: utilisé comme combustik

Petit arbre ornemental au printemps. Fleurs odorantes. Feuillag vert foncé.

Titi, Buckwheat Tree.

TRANSVERSE SECTION.

RADIAL SECTION.

TANGENTIAL SECTION

Ger. Buchweizenbaum. Fr. Cliftonie à feuilles de Troene

C | CYRILLA FAMILY
Ironwood, Leatherwood

FAMILIE DER LEDERHOLZGEWÄCHSE
Traubenblättrige Cyrille

FAMILLE DU CLIFTONIA
Cyrilla des marais, Bois de fer

CYRILLACEAE
Cyrilla racemiflora L.

Description: 40 to 45 ft. (10 to 12 m) high and 8 to 12 in. (0.20 to 0.30 m) in diameter. Various growth forms: slender; or with a stouter, eccentric trunk; or 50 to 100 reed-like stems of all sizes, sometimes even from a common rootstock, forming impenetrable thickets.

Habitat: Low, swampy ground, and ponds in pine-woods in south-eastern North America, in the Gulf region with Water Oak, Water Tupelo, Cliftonia, and Yaupon.

Wood: Heavy, hard, not strong, and close-grained.

Use: The spongy bark is used as a styptic.

Beschreibung: 10 bis 12 m hoch bei 0,20 bis 0,30 m Durchmesser. Verschiedene Wuchsformen: schlanker oder gedrungener und exzentrisch wachsender Stamm, oder 50 bis 100 wie Schilf aufschießende Teilstämme aller Größen manchmal sogar aus gemeinsamem Wurzelstock, die undurchdringliche Dickichte bilden.

Vorkommen: Flache Sumpfböden und Teiche der Kiefernwälder im südöstlichen Nordamerika, in der Golfregion zusammen mit Wasser-Eiche, Wasser-Tupelo, Cliftonia und Yaupon-Stechhülse.

Holz: Schwer, hart, nicht stark und mit feiner Textur.

Nutzung: Die schwammige Rinde wird als Tupfer verwendet.

Description : 10 à 12 m de hauteur sur un diamètre de 0,20 à 0,30 m. Forme de croissance variable : tro▮ soit élancé, soit court et très singu▮ lier ; quelquefois aussi, il constitue▮ telle une formation de roseaux, u▮ cépée composée de 50 à 100 reje▮ de souche de toutes les tailles et formant un taillis dense.

Habitat : sols plats des marécages et étangs des forêts de pins dans l▮ sud-est de l'Amérique du Nord, s▮ le pourtour du Golfe en associatio▮ avec le chêne noir, le nyssa aqua-tique, la cliftonie et le houx vomit▮

Bois : lourd, dur, pas fort et à grain fin.

Intérêts : l'écorce spongieuse est utilisée comme tampons.

Red Titi. Leather-wood, Iron-wood.

TRANSVERSE SECTION.

RADIAL SECTION.

TANGENTIAL SECTION

Ger. Traubenblättrige Cyrille. Fr. Cyrille de Caroline.
Sp. Madera de hierro.

D | DOGWOOD FAMILY
Alternate-leaf Dogwood, Blue-fruited Dogwood

FAMILIE DER HARTRIEGELGEWÄCHSE
Wechselblättriger Hartriegel

FAMILLE DU CORNOUILLER
Cornouiller à feuilles alternes

CORNACEAE
Cornus alternifolia L. f.

Description: In favorable conditions occasionally as much as 25 to 35 ft. (8 to 10 m) high, and usually 5 to 6 in (0.12 to 0.15 m), exceptionally 10 in. (0.25 m), in diameter. Easily recognizable by its distinctive appearance: Branches spreading horizontally or curving upward, the straight branches with numerous branchlets growing upwards and only very few growing downwards. Tree, therefore, with a flattened crown, interspersed with umbrella-shaped clusters of white flowers. Also decorative are the small, blue stone fruits, commonly called "berries," at the ends of red stalks.

Habitat: Rich, well-drained soils on forest margins, partially cleared ground, and along stretches of fencing in (north-)eastern North America.

Wood: Heavy, hard, and close-grained

Beschreibung: Wenn begünstigt, gelegentlich bis zu 8 bis 10 m hoch bei 0,12 bis 0,15 m, ausnahmsweise 0,25 m Durchmesser. Durch seine eigentümliche Wuchsform leicht zu erkennen: Äste horizontal ausgebreitet, oder aufwärts gebogene, gerade Äste mit zahlreichen nach oben, aber nur ganz wenigen nach unten abgehenden Zweigen. Die dadurch flach ausgebildete Belaubung von schirmartigen weißen Blütenständen unterbrochen. Zierend auch die blauen, gemeinhin als „Beeren" bezeichneten kleinen Steinfrüchte an roten Stielen.

Vorkommen: Reiche, gut drainierte Böden an Waldrändern, teilweise gerodetes Land und an Zaunreihen im (nord-) östlichen Nordamerika.

Holz: Schwer, hart und mit sehr feiner Textur.

Nutzung: Für Drechslerarbeiten herangezogen

Description : peut atteindre, dans des conditions favorables, 8 à 10 de hauteur sur un diamètre de 0,12 à 0,15 m, exceptionnellemer 0,25 m. Facilement identifiable à son port caractéristique : branche étalées à l'horizontale ou branche droites et arquées vers le haut, ga nies de nombreux rameaux dress ou, pour quelques-uns seulement retombants. Le feuillage, disposé ainsi en plans horizontaux, est émaillé de petites fleurs blanches en ombelles. Les drupes, appelée communément «baies», sont éga lement très décoratives, bleu fonc sur un pédoncule rouge.

Habitat : sols riches, bien drainés lisière forestière, quelquefois auss sur les terrains défrichés et le lon des clôtures dans le (nord-) est de l'Amérique du Nord.

Bois : lourd, dur et à grain très fin

Intérêts : employé en tournerie

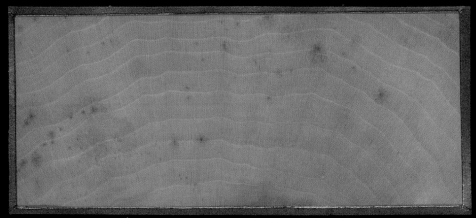

TRANSVERSE SECTION.

RADIAL SECTION.

TANGENTIAL SECTION.

Ger. Wechselblättriger Hartriegel. Fr. Cornuiller alternifeuille

D | DOGWOOD FAMILY

Dogwood

FAMILIE DER HARTRIEGELGEWÄCHSE
Nuttalls-Benthamie

FAMILLE DU CORNOUILLER
Cornouiller de Nuttall (Canada)

CORNACEAE
Cornus nuttallii Audubon ex Torr. & A. Gray

Description: 40 to 70 ft. (13 to 20 m) (exceptionally 100 ft., 30 m) high and 1 to 2 ft. (0.3 to 0.6 m) in diameter. Bark brown, with reddish tinge. Surface with layer of thin scales. Leaves turn a brilliant orange in autumn. The large involucral bracts of the flowers are white or white with pink.

Habitat: Well-drained soils in the shade of conifers, e.g. Redwoods, in northern California. Widespread in the coastal regions of Pacific North America. Found up to 3,300 ft., (1,000 m) and to 4,000 or 5,000 ft. (1,300 to 1,600 m) towards the southern limit.

Wood: Heavy, unusually hard, strong, silky-sheened, and close-grained; polishable.

Use: Cupboards, hammers, and tool handles.

Flowers are the most attractive and impressive of all the woody species of the Pacific region. Not in cultivation.

Beschreibung: 13 bis 20 (ausnahmsweise 30) m hoch bei 0,3 bis 0,6 m Durchmesser. Borke braun, rot getönt. Oberfläche aus dünnen Schuppen geschichtet. Herbstfärbung der Blätter brillantorange. Die großen Involukralblätter der Blüten weiß oder weiß mit Rosa.

Vorkommen: Gut drainierte Böden im Schatten von Koniferen, z. B. den Küstenmammutbäumen in Nordkalifornien. Verbreitet in den küstennahen Regionen des pazifischen Nordamerika. Aufsteigend bis 1000 m, gegen die Südgrenze bis zu 1300 bis 1600 m Seehöhe.

Holz: Schwer, außerordentlich hart, fest, seidig und mit feiner Textur; polierfähig.

Nutzung: Für Schränke, Schlägel, Handgriffe von Werkzeug.

Blütenstände schöner und auffälliger als die irgendeines anderen Gehölzes auf der pazifischen Seite. Geht nicht in Kultur.

Description: 13 à 20 m de hauteur (très rarement 30 m) sur un diamètre de 0,3 à 0,6 m. Ecorce brun teintée de rouge, recouverte de fines écailles stratifiées. Feuillage orange brillant en automne. Fleur entourées de grandes bractées blanches ou blanches mêlées de rose.

Habitat: sols bien drainés, sous le couvert de conifères (séquoias géants par exemple) dans le nord de la Californie. Répandu dans les régions côtières de la façade pacifique de l'Amérique du Nord. Croît jusqu'à 1000 m au-dessus du niveau de la mer, mais entre 1300 et 1600 m vers la frontière sud.

Bois: lourd, extrêmement dur, résistant, satiné et à grain fin; pou vant se polir.

Intérêts: utilisé pour la fabrication de meubles, de masses et de manches d'outils.

Sa floraison est la plus extraordinaire de toutes les espèces ligneuses de la façade pacifique. Ne se prête pas à la culture.

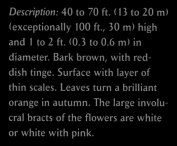

238

Western Dogwood, Flowering Dogwood.

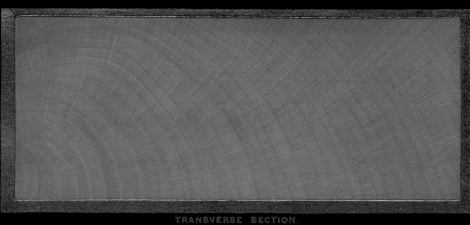

TRANSVERSE SECTION.

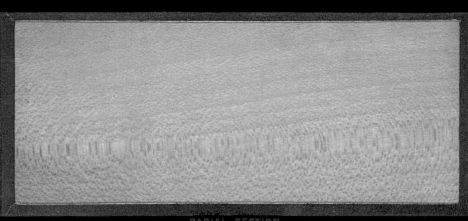

RADIAL SECTION.

TANGENTIAL SECTION.

Ger. Westlicher Hartriegel.   *Fr.* Cornuiller occidental.

D | DOGWOOD FAMILY

Flowering Dogwood

FAMILIE DER HARTRIEGELGEWÄCHSE

Blumen-Hartriegel

FAMILLE DU CORNOUILLER

Cornouiller fleuri, Cornouiller à fleurs,
Bois de buis (Canada)

CORNACEAE

Cornus florida L.

Description: 40 to 50 ft. (12 to 15 m) high and 10 to 15 in. (0.25 to 0.35 m) in diameter. Also with several stems and shrub-like in the north. Beautiful in flower with four large, white bracts round each inflorescence. Leaves colored scarlet in fall. Shiny red fruits.

Habitat: Often in the shade of taller trees on rich, well-drained soils from the Atlantic coast to the peaks of the Allegheny Mountains, especially in the south.

Wood: Heavy, hard, strong, with a silky sheen and close-grained; polishable.

Use: For turnery, also printing blocks. The bark and especially the roots have aromatic and medicinal properties.

Probably cultivated in Europe since 1730, introduced via England.

Beschreibung: 12 bis 15 m hoch bei 0,25 bis 0,35 m Durchmesser. Auch mehrstämmig und nordwärts strauchig. Schönblühend mit vier großen, weißen Hochblättern um jeden Blütenstand. Herbstfärbung der Laubblätter scharlachrot. Glänzend rote Früchte.

Vorkommen: Oft im Schatten höherer Bäume auf reichen, gut drainierten Böden von der Atlantikküste bis in Gipfellagen der Alleghenies, besonders der südlichen.

Holz: Schwer, hart, fest, seidig glänzend und mit sehr feiner Textur, polierfähig.

Nutzung: Für Drechslerarbeiten, auch Druckstöcke. Rinde und besonders die Wurzeln verfügen über aromatische und medizinische Eigenschaften.

Wohl seit 1730 über England in Europa in Kultur.

Description : 12 à 15 m de hauteu● sur un diamètre de 0,25 à 0, 35 r Développe aussi plusieurs tiges et se présente vers le nord sous une forme buissonnante. Magnifique floraison : les petites fleurs en glomérules sont entourées de quatre grandes bractées blanches Le feuillage automnal est écarlate Les fruits sont d'un rouge brillan

Habitat : pousse souvent à l'ombr● des grands arbres sur les sols riches, bien drainés du littoral atlantique jusqu'aux sommets de Alleghanys, en particulier dans le sud.

Bois : lourd, dur, résistant, satiné et à grain très fin, susceptible de prendre un beau poli.

Intérêts : utilisé en tournerie et p● fabriquer des planches d'imprim● rie. L'écorce et surtout les racines ont des propriétés aromatiques e médicinales.

Probablement cultivé depuis 173● en Europe via l'Angleterre.

TRANSVERSE SECTION.

RADIAL SECTION.

D | DOGWOOD FAMILY
Silk Tassel Bush, Quinine Bush

FAMILIE DER HARTRIEGELGEWÄCHSE
Ellipsenblättrige Garrya, Kalifornische Garrya

FAMILLE DU CORNOUILLER
Garrya elliptique

CORNACEAE
Garrya elliptica Douglas ex Lindl.

Description: Mostly growing only as a shrub, more rarely as a tree up to 20 ft. (6 m) high with a trunk 12 in. (0.30 m) in diameter. Bark brownish gray and rough. Leaves elliptical with a flat or slightly wavy margin; the upper side smooth and lustrous dark green; the lower surface, leaf stalks, and young shoots covered with felted hairs. Flowers in pendant catkin-like inflorescences, the sepals and petals with silky hairs. Fruits also with silky hairs, roundish, with juicy red flesh, tipped by the remains of the style, in dense clusters.

Habitat: Hills and slopes by flowing water near the coast. In western North America.

Wood: The reddish brown heartwood is soft, brittle, and shrinks considerably as it dries. It has fine pores, contains numerous distinct medullary rays, and is surrounded by whitish sapwood.

Use: Hardly used, although it is a very decorative shrub.

Beschreibung: Meist nur als Busch wachsend, seltener als maximal 6 m großer Baum mit 0,30 m dickem Stamm. Borke braungrau und rau. Blätter elliptisch mit glattem oder leicht gewelltem Rand. Oberseite glatt und dunkelgrün glänzend. Die Blattunterseite, Blattstiele und die jungen Triebe sind filzig behaart. Blüten in hängenden Rispen mit seidig behaarten Kelch- und Blütenblättern. Früchte rundlich, mit rotem saftigem Fleisch, seidig behaart, an der Spitze mit den Resten der Griffel, in sehr dicht stehenden Trauben.

Vorkommen: Hügel und Abhänge an Fließgewässern in der Nähe der Küste. Westliches Nordamerika.

Holz: Das rotbraune Kernholz ist weich, brüchig und schrumpft beträchtlich beim Trocknen. Es ist feinporig, enthält zahlreiche deutliche Markstrahlen und wird von viel weißlichem Saftholz umgeben.

Nutzung: Kaum genutzt, obwohl es sich um ein sehr dekoratives Gehölz handelt.

Description : arbuste à port touffu le plus souvent, de 6 m de haute maximum, avec un tronc de 0,30 de diamètre. Ecorce gris brun et rugueuse. Les feuilles elliptiques entières ou crénelées sont glabre vert foncé et brillantes dessus, tomenteuses dessous, ainsi que les pétioles et les jeunes rameau: Fleurs en panicules pendantes, à corolle duvetée. Fruits globuleu en grappes très serrées, à chair juteuse et rouge, veloutés et surmontés des restes du style.

Habitat : collines et pentes au bord des cours d'eau situés près de la côte. Dans l'ouest de l'Ame rique du Nord.

Bois : tendre, cassant ; à pores fin fort retrait au séchage. Bois de cœur brun rouge ; aubier épais, blanchâtre: Pores fins et nombre rayons médullaires distincts.

Intérêts : très peu utilisé malgré son caractère très ornemental.

Silk-tassel Tree, Quinine Tree.

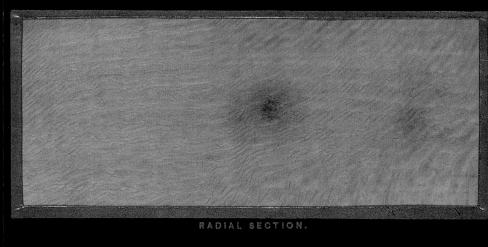

TRANSVERSE SECTION.

RADIAL SECTION.

TANGENTIAL SECTION

Ger. Seidenquastenbaum.   *Fr.* Arbre à signets de so

Sp. Arbol de borlita de seda.

E | EBONY FAMILY

American Persimmon, Possumwood

FAMILIE DER EBENHOLZGEWÄCHSE
Persimone

FAMILLE DE L'ÉBÈNE
Plaqueminier de Virginie, Persimon

EBENACEAE
Diospyros virginiana L.

Description: 33 to 50 ft. (10 to 15 m), only in especially favorable situations (Mississippi basin) up to 100 ft. (30 m) high, and 2 to 3 ft. (0.6 to 0.9 m) in diameter. On roadsides and on exhausted soils it is often shrubby, with suckers growing up from the roots. Fruits very rich in tannin, like the very closely related Japanese Persimmon, good to eat only when fully ripe. An ornamental tree, both when its orange-red bark contrasts with the green foliage, and also during the winter when the leaves have fallen and the branches are bare.

Habitat: Light, well-drained, sandy soils, but also rich, low-lying ground. Eastern North America

Wood: Heavy, hard, strong, and close-grained.

Use: Favored for wickerwork chairs, turnery, small wooden articles. An antiscorbutic of the Indians (De Soto, 1577), and since then also cultivated.

Most of the 450 species in this family are tropical and subtropical, including the Black Ebony.

Beschreibung: 10 bis 15 m, nur bei günstigen Bedingungen (Mississippi-Becken) bis 30 m hoch bei 0,6 bis 0,9 m Durchmesser. Durch wurzelbürtige Sprosse oft strauchigbuschig an Straßenrändern und auf erschöpften Böden. Früchte wie die nächstverwandten Kaki-Pflaumen sehr gerbstoffreich, gut essbar nur wenn voll ausgereift. Ornamental orangerot im Kontrast zum grünen Laub oder nach dessen Fall an den bloßen Ästen („Wintersteher").

Vorkommen: Leichte, gut drainierte Sandböden, aber auch tiefgründige, reiche Niederungen im östlichen Nordamerika.

Holz: Schwer, hart, fest und von feiner Textur.

Nutzung: Bevorzugt für Webstühle, Drechslerarbeiten, kleinere Artikel aus Holz. Antiskorbutmittel der Indianer (De Soto 1577), seither auch kultiviert.

Die meisten der etwa 450 Arten der Familie tropisch und subtropisch. Von daher das schwarze Ebenholz.

Description: arbre de 10 à 15 m d hauteur, pouvant atteindre, lorsc les conditions sont particulièrem favorables (bassin du Mississipp 30 m de haut sur un diamètre d 0,6 à 0,9 m. Constitue souvent d fourrés le long des routes et sur sols épuisés car il est doué d'une grande capacité de drageonneme Fruits très riches en tannins com ceux de son proche parent, le ka mais comestibles seulement qua ils sont blets; leur couleur orang offre un beau contraste avec les feuilles d'un vert intense ou, apr leur chute, sur les rameaux nus.

Habitat: sols sablonneux légers, bien drainés, mais aussi sols rich et profonds des secteurs déprim dans l'est de l'Amérique du Nor

Bois: lourd, dur, résistant et à grain fin.

Intérêts: employé de préférence pour le tournage et la fabrication de métiers à tisser et de petits objets en bois. Remède antiscorb tique des Indiens (De Soto 1577) en culture depuis sa découverte.

Cette famille a env. 450 espèces dont la plupart tropicales et sub picales d'où la couleur d'ébène.

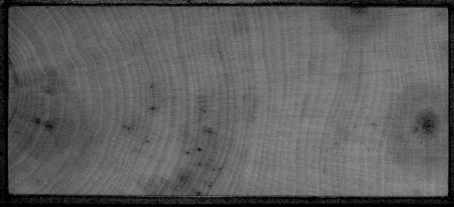

TRANSVERSE SECTION.

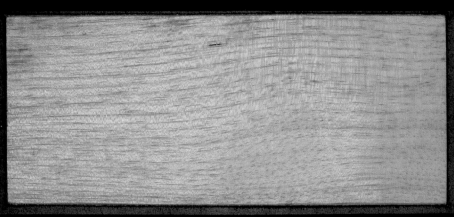

RADIAL SECTION.

TANGENTIAL SECTION

Ger. Virginische Dattelpflaume. Sp. Persimon.
Fr. Plaqueminier de Virginie.

E | ELM FAMILY

American Elm, White Elm, Water Elm

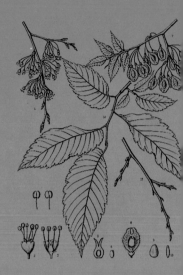

FAMILIE DER ULMENGEWÄCHSE
Weiß-Ulme

FAMILLE DE L'ORME
Orme d'Amérique, Orme blanc (Canada)

ULMACEAE
Ulmus americana L.

Description: 100 to 135 ft. (30 to 40 m) high and 6 to 10 ft. (1.8 to 3 m) in diameter. Trunk columnar, undivided until 35 to 70 ft. (10 to 20 m) above the ground, or dividing between 20 and 35 ft. (6 to 10 m) up with the several branches growing upwards to form a cup-shaped head. Enlarged at the base by great buttresses. One of the commonest forest trees in eastern North America. Regarded by the early settlers from England as "an old friend" and a "greeting from the home country" and so individual trees were spared in the "great clearing" for agricultural land.

Habitat: On riverbanks and low-lying land near rivers together with Box Elder, Green Ash, Cottonwood, etc., in eastern North America.

Wood: Heavy, hard, and strong; difficult to split. Heartwood light brown; sapwood somewhat paler.

Use: Hubs of wheels, floorboards, barrels, and in boat and shipbuilding. Bark used by the Indians for canoes when birch bark was not available. Fibers from the inner bark made into ropes. One of the main ornamental trees of the northern states of the USA.

Beschreibung: 30 bis 40 m hoch bei 1,8 bis 3 m Durchmesser. Säulenförmiger Stamm ungeteilt bis zwischen 10 und 20 m, oder schon zwischen 6 und 10 m über dem Boden mit mehreren Teilstämmen kelchförmig aufstrebend. An der Basis weitreichende Wurzelanläufe. Einer der gewöhnlichsten Waldbäume im östlichen Nordamerika. Von den ersten Siedlern aus England als „alter Freund" und „Gruß aus der Heimat" betrachtet und bei dem „great clearing" für Ackerland in Einzelexemplaren verschont.

Vorkommen: Uferbegleiter, im Flachland in Flussnähe gemeinsam mit Eschen-Ahorn, Grün-Esche, Kanadischer Schwarz-Pappel u. a. im östlichen Nordamerika.

Holz: Schwer, hart und fest; schwer zu spalten. Kern hellbraun; Splint etwas heller.

Nutzung: Radnaben, Dielen, Fässer, Boots- und Schiffbau. Rinde von den Indianern alternativ zu Birkenrinde für Kanus benutzt. Fasern der inneren Rinde lassen sich zu Seilen drillen. Hauptzierbaum der nördlichen Staaten der USA.

Description: 30 à 40 m de hauteu sur un diamètre de 1,8 à 3 m. Tronc columnaire, non divisé jusqu'à une hauteur de 10 à 20 m ou se divisant dès 6 à 10 m en plusieurs tiges secondaires calici-formes. Longs drageons sur la pa tie basale du tronc. Un des arbre les plus courants de l'est du con nent nord-américain. Appelé par les premiers colons britanniques « vieil ami» et «bonjour du pays» Individus épargnés lors du «grea clearing» (grand défrichement).

Habitat: lié aux berges des cours d'eau; en plaine, à proximité des rivières, en compagnie de l'érabl négundo, du frêne rouge, du peuplier deltoïde etc. On le trou dans l'est de l'Amérique du Nor

Bois: lourd, dur et résistant. Bois de cœur brun clair; aubier un pe plus clair.

Intérêts: utilisé en tonnellerie, da la construction de bateaux et de navires, pour fabriquer des moye et des lames de parquet. Ecorce autrefois employée par les Indie pour la construction de canots et écorce interne pour fabriquer les cordes. Principal arbre d'orneme du nord des Etats-Unis.

White Elm, Water Elm, American Elm.

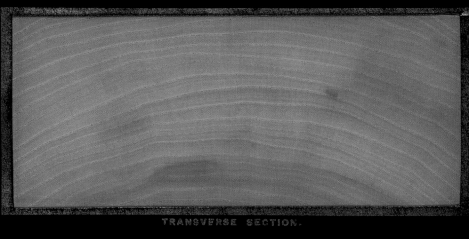

TRANSVERSE SECTION.

RADIAL SECTION.

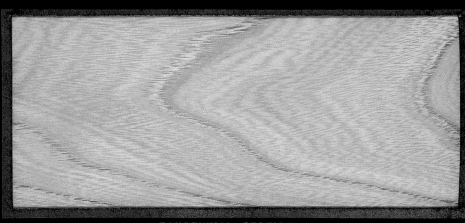

TANGENTIAL SECTION.

Ger. Weisse Ulme, Rüster. Fr. Orme parasol.

Sp. Olmo blanco.

E | ELM FAMILY

Cork Elm, Rock Elm, Hickory Elm

FAMILIE DER ULMENGEWÄCHSE
Felsen-Ulme

FAMILLE DE L'ORME
Orme liège, Orme de Thomas (Canada)

ULMACEAE
Ulmus thomasii Sarg.
Syn.: *U. racemosa* Thomas

Description: 85 to 100 ft. (25 to 30 m) high and 3 to 4 ft. (0.9 to 1.2 m) in diameter. Trunk straight, sometimes free of branches for the first 70 ft. (20 m). Branchlets with corky wings. Branches drooping when the tree is free standing.

Habitat: Dry, gravely uplands, heavy clay soils in lowlands, rocky slopes and steep riverbanks in association with Sugar Maple, Hop Hornbeam, Butternut, lime, White Ash, Beech, etc. Not so abundant and widespread as the White Elm. Interior of (north-)eastern North America.

Wood: Heavy, hard, very tough and strong, close-grained; polishable. Heartwood light clear brown, often tinged with red; sapwood paler.

Use: Widely used for heavy agricultural equipment (plows, mowing and threshing machines), poles for grubbing up tree stumps, thresholds of doorways.

Beschreibung: 25 bis 30 m hoch bei 0,9 bis 1,2 m Durchmesser. Gerader Stamm, manchmal bis 20 m astfrei. Korkflügel an den Zweigen. Im Freistand Äste hängend.

Vorkommen: Trockene, kiesige Hochlagen, schwere Tonböden im Flachland, felsige Hänge und steile Flussufer in Gesellschaft von Zucker-Ahorn, Hopfenbuche, Butternuss, Linde, Weiß-Esche, Buche u. a. Weniger häufig und verbreitet als die Weiß-Ulme. Im Innern des (nord-) östlichen Nordamerika.

Holz: Schwer, hart, sehr zäh und fest, mit feiner Textur; polierfähig. Kern helles reines Braun, oft rot gefärbt; Splint heller.

Nutzung: In großem Umfang für schweres Ackergerät (Pflüge, Mäh- und Dreschmaschinen), Balken beim Roden von Baumstümpfen, Schwellen verwendet.

Description: 25 à 30 m de hauteur sur un diamètre de 0,9 à 1,2 m. Tronc droit, parfois dénudé jusqu' 20 m de haut. Rameaux garnis d'ailes liégeuses. Branches retombantes en situation isolée.

Habitat: sols secs graveleux des hautes terres, sols argileux et lourds des plaines, pentes rocheuses et berges escarpées en association avec l'érable à sucre, l'ostryer, le noyer cendré, le tilleul le frêne blanc, le bouleau etc. Moins fréquent et répandu que l'orme blanc. On le trouve dans les régions intérieures (nord-)est de l'Amérique du Nord.

Bois: lourd, dur, très tenace et résistant, à grain fin; pouvant se polir. Bois de cœur d'un brun pur et clair, souvent teinté de rouge; aubier plus clair.

Intérêts: utilisé principalement pour fabriquer des machines agricoles (charrues, moissonneuse batteuses), des traverses et des poutres, après essouchements.

Cork Elm,   Rock Elm,   Cliff Elm,   White Elm.

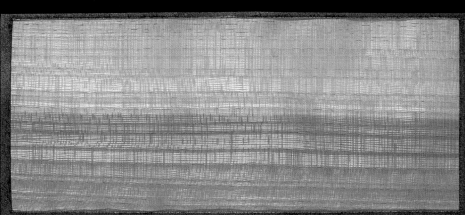

TRANSVERSE SECTION.

RADIAL SECTION.

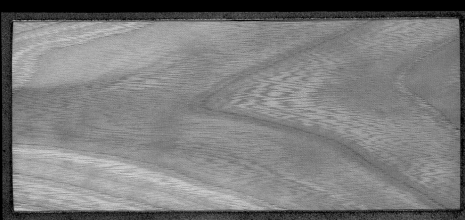

TANGENTIAL SECTION.

Ger. Trauben Ulme.   Fr. Orme à grappa.

E | ELM FAMILY
Florida Trema

FAMILIE DER ULMENGEWÄCHSE
Trema

FAMILLE DE L'ORME
Capulin

ULMACEAE
Trema micrantha (L.) Blume
Syn.: *T. mollis* Lour.

Description: This small tree reaches a height of up to 35 ft. (10 m) in North America, and has a trunk 2 ft. (0.6 m) in diameter. It is quick-growing, but only short-lived. The brown bark is thin and rough. The ovate leaves have a pointed tip and a heart-shaped base. The upper surface is dark green and very rough, and the lower surface is paler and densely hairy. The small, insignificant flowers are clustered in inflorescences. The roundish, light-brown fruit remains small, and is crowned by the remains of the style.

Habitat: The tree occurs commonly in southern Florida, where it acts as a pioneer plant and rapidly colonizes areas of land that have been cleared.

Wood: The pale brown wood is light, soft, and brittle.

Use: The branchlets of the tree can be woven into baskets, and the fibers of the bark can be used in the manufacture of ships' ropes.

Beschreibung: Der kleine Baum wird in Nordamerika bis zu 10 Meter hoch mit einem 0,6 m dicken Stamm. Er ist raschwüchsig und kurzlebig. Die braune Borke ist dünn und rau. Die ovalen Blätter sind vorne zugespitzt und besitzen eine herzförmige Basis. Die Oberseite ist dunkelgrün und sehr rau, die Unterseite dagegen blasser und dicht behaart. Die unauffälligen kleinen Blüten sind in Blütenständen vereinigt. Die hellbraune rundliche Frucht bleibt klein und wird von den Resten des Griffels gekrönt.

Vorkommen: Der Baum kommt häufig in Süd-Florida vor, wo er als Pioniergehölz gerodete Flächen rasch besiedelt.

Holz: Das hellbraune Holz ist leicht, weich und brüchig.

Nutzung: Die Zweige des Baumes können zu Körben geflochten werden. Die Fasern der Borke eignen sich zur Herstellung von Schiffstauen.

Description: arbuste atteignant en Amérique du Nord jusqu'à 10 m de hauteur, avec un tronc de 0,6 de diamètre. Essence à croissance rapide et de brève durée de vie. Ecorce brune, mince et rugueuse. Feuilles ovales, pointues à l'apex cordiformes à la base, vert foncé très rugueuses à la face supérieu plus pâles et très pubescentes à la face inférieure. Petites fleurs insignifiantes, groupées en inflorescences. Le fruit globuleux bru clair demeure petit et est surmon des restes du style.

Habitat: très répandu dans le sud de la Floride et espèce pionnière, colonisant rapidement les surface défrichées.

Bois: léger, tendre et cassant; brun clair.

Intérêts: les rameaux peuvent servir en vannerie, les fibres de l'écorce, pour fabriquer des câble de marine.

TREMA.

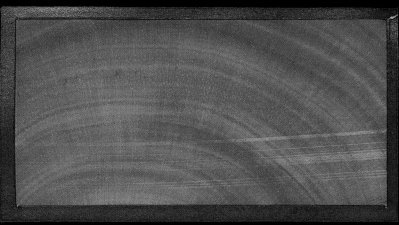

TRANSVERSE SECTION.

RADIAL SECTION.

TANGENTIAL SECTION.

E | ELM FAMILY
Hackberry, Sugarberry

FAMILIE DER ULMENGEWÄCHSE
Zürgelbaum

FAMILLE DE L'ORME
Micocoulier occidental, Bois inconnu (Canada)

ULMACEAE
Celtis occidentalis L.

Description: Sometimes over 100 ft. (30 m) or even 130 ft. (40 m) high, and 2 to 3 or even 4 to 5 ft. (0.7 to 0.9 or even 1.2 to 1.5 m) in diameter. Trunk straight, slender, often without branches for 70 ft. (20 m). Fruits dark purple with yellow sweetish pulp.

Habitat: Rich lowland in association with oaks, hickories, and walnut trees. One of the most common forest trees, mainly in lower Ohio and the Mississippi basin. Extends also on to dry, gravely soils and rocky hills, but then much smaller. Eastern North America.

Wood: Heavy, fairly soft, not very strong, and coarse-grained.

Use: For fencing, cheap furniture, tools, and firewood.

Beschreibung: Manchmal über 30 m hoch, bis 40 m und mehr bei 0,7 bis 0,9 m, aber auch 1,2 bis 1,5 m Durchmesser. Gerader, schlanker, oft bis 20 m astloser Stamm. Dunkelpurpurne Früchte mit gelbem, süßlichem Fleisch.

Vorkommen: Reiches Flachland in Gesellschaft von Eichen, Hickories und Nussbäumen als einer der gewöhnlichsten Waldbäume, vor allem im unteren Ohio- und im Mississippi-Becken. Gedeiht auch auf trockenen, kiesigen Böden und Felshügeln, dann aber viel kleiner. Östliches Nordamerika.

Holz: Schwer, ziemlich weich, nicht sehr fest und von grober Textur.

Nutzung: Für Zäune, billige Möbel, Gerätschaften, Brennmaterial.

Description: s'élève quelquefois à plus de 30 m, voire 40 m ou plu avec un tronc de 0,7 à 0,9 m de diamètre, mais pouvant atteindre entre 1,2 et 1,5 m. Tronc droit, élancé, souvent dénudé jusqu'à 20 m de haut. Drupes violet sombre à maturité, à chair jaune et sucrée.

Habitat: plaines riches en compagnie de chênes, de caryers et de noyers. Un des arbres forestiers les plus communs, surtout dans le bassin inférieur de l'Ohio et du Mississippi. S'accommode aussi de sols secs cailouteux et des colline rocheuses, mais sa taille y demeur bien plus modeste. On le trouve dans l'est de l'Amérique du Nord.

Bois: lourd, assez tendre, pas très résistant et à grain grossier.

Intérêts: utilisé pour fabriquer des clôtures, des meubles bon marché des outils, et comme combustible.

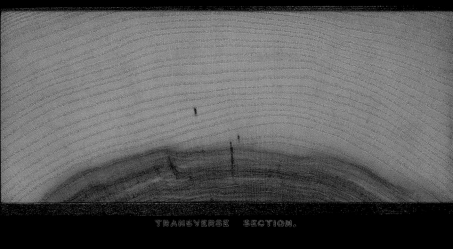

TRANSVERSE SECTION.

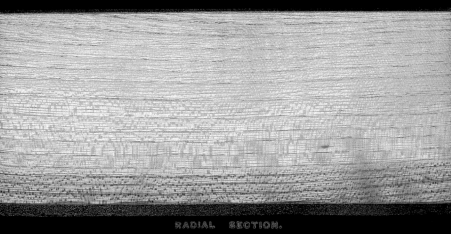

RADIAL SECTION.

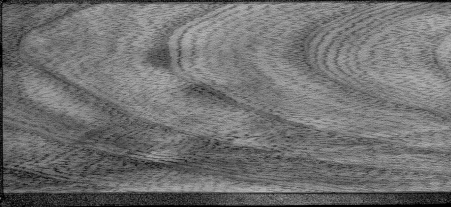

TANGENTIAL SECTION.

Fr. Abendländischer Zürgelbaum. Sp. Almez American.
It. Micocoulier occidental.

E | ELM FAMILY
Mississippi Hackberry, Sugarberry

FAMILIE DER ULMENGEWÄCHSE
Mississippi-Zürgelstrauch

FAMILLE DE L'ORME
Micocoulier lisse, Micocoulier du Mississippi

ULMACEAE
Celtis tenuifolia Nutt.
Syn.: *C. mississippiensis* Bosc.

Description: 65 to 85 ft. (18 to 24 m) high and 2 to 3 ft. (0.6 to 0.9 m) in diameter. Often also much smaller and only a shrub. Trunk short, with sometimes drooping branches that form a broad crown. Bark moderately thick, light blue-gray and covered with prominent excrescences. Leaves ovate-lanceolate, ovate, or oblong-lanceolate, somewhat sickle-shaped, entire or serrate, dark green above and pale beneath. Flowers in small clusters with few flowers. Fruit ovoid, small, with a thin, dry flesh, pale orange-red in color.

Habitat: Rich, low-lying soils and riverbanks, occasionally also on dry limestone hills. Widespread in the interior of eastern North America and in the states on the Gulf coast.

Wood: Rather soft, not strong, and close-grained. Heartwood light, clear yellow; sapwood thick, paler. Bands of large, open pores along the annual rings, and between them groups of smaller pores in concentric rings; thin medullary rays.

Use: Not distinguished from *Celtis orientalis* in commerce, used for fences and cheap furniture.

Beschreibung: 18 bis 24 m hoch bei 0,6 bis 0,9 m Durchmesser. Oft auch viel kleiner und nur als Strauch. Stamm kurz mit manchmal herabhängenden Ästen, die eine breite Krone bilden. Borke mäßig dick, hell blaugrau und mit Auswüchsen bedeckt. Blätter oval-lanzettlich, oval oder länglich-lanzettlich, etwas sichelförmig, ganzrandig oder gesägt, dunkelgrün auf der Ober- und blass auf der Unterseite. Blüten in kleinen, wenigblütigen Büscheln. Frucht oval, klein, mit dünnem, trockenem Fleisch und hell orangeroter Farbe.

Vorkommen: Reiche Tieflandböden und Flussufer, gelegentlich auch trockene Kalkhügel. Im Inneren des östlichen Nordamerika sowie in den Golfstaaten verbreitet.

Holz: Eher weich, nicht fest und mit feiner Textur. Kern hell, klargelb; Splint dick, heller. Bänder großer offener Gefäße längs der Jahresringe, dazwischen Gruppen kleinerer Gefäße in konzentrischen Ringen, dünne Markstrahlen.

Nutzung: Wird kommerziell nicht von *Celtis orientalis* unterschieden, wie diese für Zäune und billige Möbel verwendet.

Description : 18 à 24 m de hauteu sur un diamètre de 0,6 à 0,9 m. Souvent aussi seulement sous forme de buisson. Tronc court, garni de branches parfois retombantes, constituant un houppier ample. Ecorce gris bleu clair, moyennement épaisse et recouverte d'excroissances liégeuses. Feuilles ovales-lancéolées, ovales lancéolées allongées, légèrement lunulaires, entières ou dentées, v sombre dessus et pâles dessous. Inflorescences en petites grappes pauciflores. Drupes ovales, petite à chair sèche et mince, orangé cla

Habitat : sols riches de basses-terr et rives de cours d'eau, parfois aussi collines calcaires sèches ; da l'intérieur à l'est de l'Amérique d Nord et dans les Etats du Golfe.

Bois : plutôt tendre, non résistant et à grain fin. Bois de cœur jaune et clair ; aubier épais, plus pâle. Bandes de grands vaisseaux ouve le long des cernes annuels, sépar par des faisceaux de vaisseaux pl petits, disposés en anneaux conc triques ; rayons médullaires fins.

Intérêts : aucune différence dans le commerce avec *Celtis orientalis*. clôtures et meubles bon marché.

Mississippi Hackberry or Sugarberry.

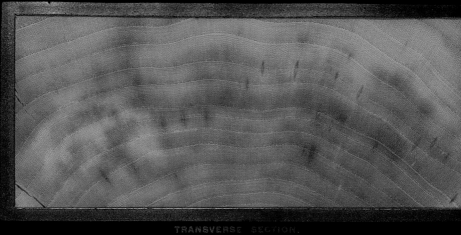

TRANSVERSE SECTION.

RADIAL SECTION.

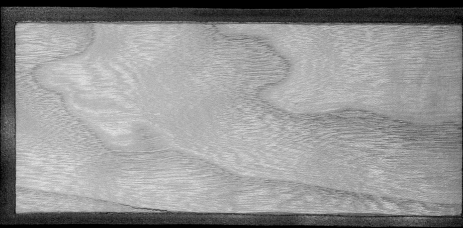

TANGENTIAL SECTION.

Ger. Mississippi Zürgelbaum. Fr. Micocoulier de Mississippi. Sp. Almez del Mississippi.

E | ELM FAMILY

Planer Tree, Water Elm

FAMILIE DER ULMENGEWÄCHSE
Wasser-Ulme

FAMILLE DE L'ORME
Planéra des marécages, Orme des marécages

ULMACEAE
Planera aquatica J. F. Gmel.

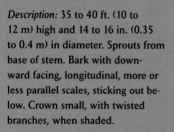

Description: 35 to 40 ft. (10 to 12 m) high and 14 to 16 in. (0.35 to 0.4 m) in diameter. Sprouts from base of stem. Bark with downward facing, longitudinal, more or less parallel scales, sticking out below. Crown small, with twisted branches, when shaded.

Habitat: Regularly flooded lowlands and deep marshes in (south-)eastern North America on sites that are too wet for almost all other trees.

Wood: Light and soft.

Use: Little ornamental value.

Beschreibung: 10 bis 12 m hoch bei 0,35 bis 0,4 m Durchmesser. Stockausschlag an der lebenden Stammbasis. Borke mit abwärts gerichteten, länglichen, ziemlich parallelen, unten abstehenden Streifen-Schuppen. Kleinkronig mit gedrehten Ästen, wenn überschattet.

Vorkommen: Häufig überschwemmte Niederungen und tiefe Sümpfe, im (süd-) östlichen Nordamerika an Orten, die für nahezu alle anderen Bäume zu nass sind.

Holz: Leicht und weich.

Nutzung: Von geringem Zierwert.

Description : 10 à 12 m de hauteu sur un diamètre de 0,35 à 0,4 m. Rejette de souche à la base vivant du tronc. Ecorce à lanières longit dinales assez parallèles, allongée et dressées, se desquamant. Petit houppier, branches entrelacées si à l'ombre.

Habitat : secteurs déprimés souve inondés et marais profonds du (sud-) est de l'Amérique du Nor des lieux trop humides pour presque toutes les autres espèces

Bois : léger et tendre.

Intérêts : peu de valeur ornementale.

Planer Tree.

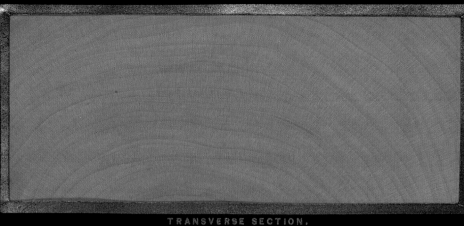

TRANSVERSE SECTION.

RADIAL SECTION.

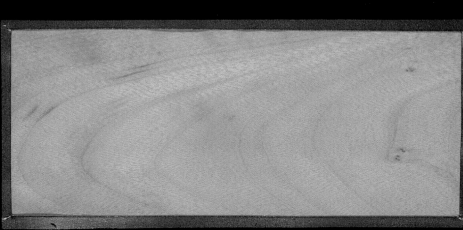

TANGENTIAL SECTION

Ger. Ulmenblättrige Planera.   **Fr.** Planera aquatique.

Sp. Planera acuatica.

E | ELM FAMILY

Red Elm, Slippery Elm

FAMILIE DER ULMENGEWÄCHSE
Rot-Ulme

FAMILLE DE L'ORME
Orme rouge, Orme gras (Canada)

ULMACEAE
Ulmus rubra Muhl.
Syn.: *U. fulva* Michx.

Description: 65 to 75 ft. (20 to 23 m) high and 2 to 3 ft. (0.6 to 0.9 m) in diameter. When standing alone, the trunk soon divides. Inner bark pleasantly scented, containing mucilage (hence "slippery"). Leaves larger and rougher in comparison with the otherwise similar White Elm.

Habitat: Mainly on rich lowland soils, riverbanks. Found with the Burr Oak and Swamp White Oak, the Black Maple, Silver Maple and Red Maple, Mississippi Hackberry, etc. In addition, but smaller growing, on rocky ridges and slopes in eastern North America.

Wood: Heavy, tough, strong, and very close-grained; durable in the ground, easy to split when fresh. Heartwood dark brown or red; sapwood thin and paler.

Use: For fence posts, railway sleepers, thresholds of doorways, cart wheels, agricultural implements. Inner bark used medicinally, but also nutritious.

Beschreibung: 20 bis 23 m hoch bei 0,6 bis 0,9 m Durchmesser. Wenn freistehend, mit früh geteilten Stämmen. Innere, wohlriechende Rinde schleimig („slippery"). Laub größer und rauer als bei der sonst ähnlichen Weiß-Ulme.

Vorkommen: Vorzugsweise nährstoffreiche Flachlandböden, Flussufer. Zusammen mit Großfrüchtiger Eiche und Sumpf-Weiß-Eiche, dem Schwarz-Ahorn, Silber-Ahorn und Rot-Ahorn, Mississippi-Zürgelstrauch u. a. Außerdem, aber kleiner, auf Felsgraten und Hängen im östlichen Nordamerika.

Holz: Schwer, zäh, fest und mit sehr feiner Textur; dauerhaft im Boden, leicht zu spalten, wenn frisch. Kern dunkelbraun oder rot; Splint dünn und heller.

Nutzung: Für Zaunpfähle, Bahn- und Türschwellen, Wagenräder, Ackergerät. Innere Rinde medizinisch genutzt, aber auch nahrhaft.

Description: de 20 à 23 m de hauteur sur un diamètre de 0,6 0,9 m. Développe très tôt des tig distinctes s'il pousse isolément. Ecorce interne mucilagineuse («slippery»), dégageant une ode agréable. Semblable à l'orme d'Amérique, hormis par son feu lage qui est plus grand et plus rugueux.

Habitat: préfère les sols riches de plaine et les berges de cours d'eau, en compagnie du chêne à gros fruits, du chêne bicolore, d érables noir, argenté et rouge, d micocoulier du Mississipi etc. Se rencontre aussi sur les pentes et les arêtes rocheuses de l'est de l'Amérique du Nord, mais dans ce cas plus petit.

Bois: lourd, tenace et résistant, à grain très fin; durable sur pied Frais, facile à fendre. Bois de cœ brun foncé ou rouge; aubier fin clair.

Intérêts: utilisé pour fabriquer d pieux de clôtures, des traverses chemin de fer, des seuils de por des roues de charrettes et des in truments aratoires. Ecorce inter employée en médecine, mais au comestible.

Slippery Elm,   Ped Flm,   Moose Elm.

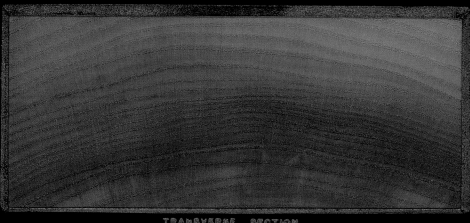

TRANSVERSE SECTION.

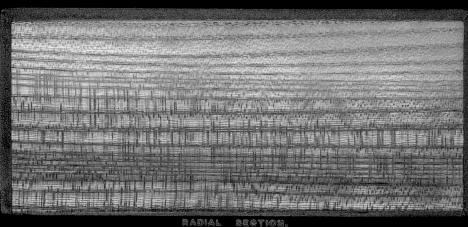

RADIAL SECTION.

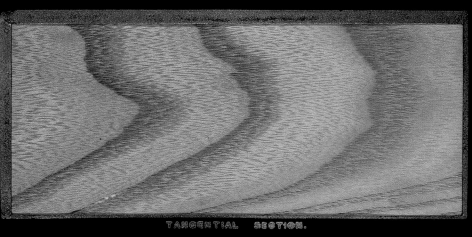

TANGENTIAL SECTION.

Ger. Rothe Ulme.   Sp. Olmo colorado.

Fr. Orme gras.

E | ELM FAMILY
Winged Elm, Wahoo

FAMILIE DER ULMENGEWÄCHSE
Wahoo-Ulme

FAMILLE DE L'ORME
Orme ailé, Orme à petites feuilles, Orme liège ailé

ULMACEAE
Ulmus alata Michx.

Description: 40 to 50 ft. (12 to 15 m) high and 2 ft. (0.6 m) in diameter, frequently much smaller. Branches stout, forming a narrowly oblong or rather open, rounded crown. Bark thin, light brown with a red tinge, and shallow fissures. Branches with conspicuous corky wings. Leaves ovate-oblong to lanceolate, dark green above, and covered with soft downy hairs beneath. Leaf often asymmetric. Flowers appearing before the leaves, in groups on slender stalks. Fruits ripening shortly before or at the same time as the leaves begin to unfold. Fruit oblong, with two pincer-like horns, densely hairy.

Habitat: Dry, stony, upland soils in south-eastern North America, rarely in rich soils subject to flooding, on swamp margins and riverbanks.

Wood: Heavy, hard, not strong, and close-grained; difficult to split.

Use: Occasionally for tool-handles and the hubs of small wheels. Ropes for fastening cotton bales used to be made from the inner bark. Frequently planted as a shade tree in the southern USA.

Beschreibung: 12 bis 15 m hoch bei 0,6 m Durchmesser, häufig viel kleiner. Äste kräftig, eine schmale, längliche oder eher offene, kugelige Krone bildend. Borke dünn, hellbraun mit roter Tönung und flachen Rissen. Äste mit kräftigen Korkleisten. Blätter oval-länglich bis -lanzettlich, oberseits dunkelgrün und kahl, unterseits mit weichem Haarflaum. Blattgrund oft asymmetrisch. Blüten vor dem Laubaustrieb, in Gruppen auf schlanken Stielchen. Früchte reifen kurz vor oder zu Beginn des Laubaustriebes. Frucht länglich, mit 2 zangenähnlichen Hörnern, dicht behaart.

Vorkommen: Trockene, steinige Hochlandböden im südöstlichen Nordamerika, selten in reichen Schwemmböden an den Rändern von Sümpfen und Flussufern.

Holz: Schwer, hart, nicht fest und mit feiner Textur; schwer spaltbar.

Nutzung: Gelegentlich für Werkzeugstiele und als Radnabe kleiner Räder. Aus der inneren Rinde wurden Stricke zum Umwickeln von Baumwollballen gewonnen. In den südlichen USA häufig als Schattenspender.

Description: 12 à 15 m de hauteur sur un diamètre de 0,6 m, mais souvent beaucoup plus petit. Branches vigoureuses, constituant un houppier soit étroit et allongé soit globuleux et plutôt ouvert. Ecorce mince, brun clair avec une teinte de rouge et creusée de fissures plates. Branches garnies grosses crêtes liégeuses. Feuilles ovales-allongées à ovales-lancéolées, vert foncé et glabres dessus, duvetées dessous. Limbe souvent asymétrique à la base. Fleurs apparaissant avant la feuillaison, réunies en glomérules et portées par des pédicelles. Samares mûres peu avant ou au début de la feuillaison allongées, composant deux ailes en forme de pince, très velues.

Habitat: sols caillouteux et secs des hautes terres du sud-est de l'Amérique du Nord, rarement sols alluviaux en bordure de marais et de berges.

Bois: lourd, dur, non résistant et à grain fin, difficile à fendre.

Intérêts: manches d'outils et petits moyeux. L'écorce fournissait des cordes pour lier les balles de coton. Souvent arbre d'ombrage dans le sud des Etats-Unis.

Winged Elm.

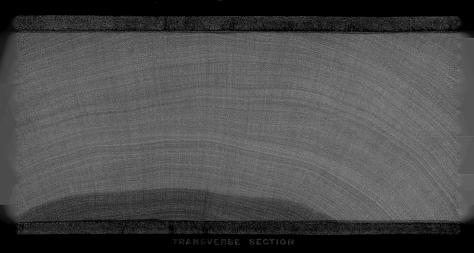

TRANSVERSE SECTION

RADIAL SECTION.

F | FOGWORT FAMILY

Foxglove Tree

FAMILIE DER RACHENBLÜTLER
Paulownie

FAMILLE DE LA SCROFULAIRE
Paulownia impérial

SCROPHULARIACEAE
Paulownia tomentosa (Thumb.) Siebold & Zucc. Steud

Description: The tree reaches a height of 40 ft. (12 m), and has a trunk up to 2 ft. 6 in. (7.5 m) in diameter. The soft, brownish gray bark is smooth at first, but becomes deeply furrowed with age. Its open crown is formed by a few stout main branches. The large, broadly ovate leaves are covered with felted hairs directly after they have sprouted. The large, blue-violet, pleasantly scented flowers appear before the leaves, in loose panicles.

Habitat: It is native to eastern Asia, from where it was introduced into North America, and it has now spread and grows wild in the central and southern parts of the continent.

Wood: Light, soft, not strong, and easy to work. The mottled reddish brown heartwood has a beautiful satin-like sheen on its surface after being polished.

Use: The tree is a favorite ornamental plant.

Beschreibung: Der Baum erreicht eine Höhe von 12 m bei einem Stammdurchmesser von 7,5 m. Die weiche, braungraue Borke ist anfangs glatt, wird aber im Alter tief rissig. Seine lichte Krone wird von nur wenigen kräftigen Hauptästen gebildet. Die großen breit ovalen Blätter sind direkt nach dem Austrieb pelzig behaart. Die großen blauvioletten und wohlriechenden Blüten blühen vor dem Laubaustrieb in lockeren Rispen.

Vorkommen: Seine Heimat ist Ostasien, von wo aus er nach Nordamerika eingeführt wurde und in den mittleren und südlichen Teilen des Kontinents verwilderte.

Holz: Leicht, weich, nicht fest und leicht zu bearbeiten. Das fleckige, rotbraune Kernholz zeigt nach dem Polieren eine schöne satinartig glänzende Oberfläche.

Nutzung: Der Baum ist ein beliebtes Ziergehölz.

Description : arbre de 12 m de hauteur, avec un tronc de 0,75 m de diamètre. Ecorce gris brun, à texture molle et lisse, se creusant de fissures profondes avec l'âge. Couronne claire, constituée seulement de quelques grosses branches maîtresses. Grandes feuilles larges-ovales, pubescentes dès leur apparition. Grandes fleurs d'un beau bleu violacé, au parfum délicat, groupées en panicules légères avant la feuillaison.

Habitat : originaire d'Asie orientale, introduit en Amérique du Nord et retourné à l'état sauvage dans les régions centrales et méridionales du sous-continent.

Bois : léger, tendre, non résistant et facile à travailler. Bois de cœur brun rouge, taché, prenant un bel aspect satiné après polissage.

Intérêts : belle espèce ornementale appréciée pour sa floraison printanière et son feuillage.

Paulownia. Princess-tree.

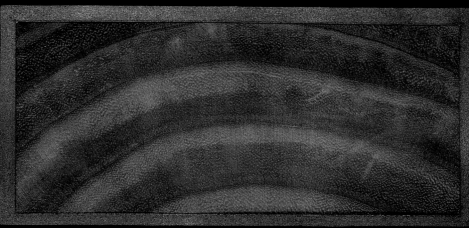

TRANSVERSE SECTION

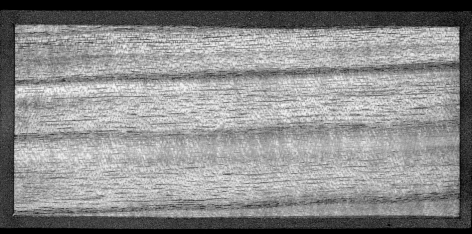

RADIAL SECTION.

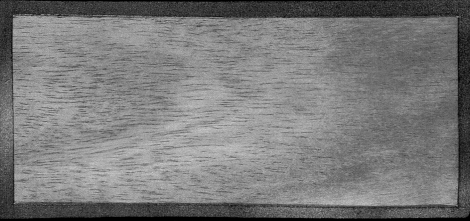

TANGENTIAL SECTION.

Ger. Princessinn-Baum.   *Fr.* Arbre de princesse.

G | GRAPEVINE FAMILY
Summer Grape

FAMILIE DER WEINREBENGEWÄCHSE
Sommer-Weinrebe

FAMILLE DE LA VIGNE
Vigne estivale, Vigne cultivée

VITACEAE
Vitis aestivalis Michx.

Description: These woody lianes vary in size according to the structures supporting them, but are usually 2 to 3 in. (0.08 to 0.1 m) in diameter, and exceptionally may be as much as 12 in. (0.3 m) at the base.

Habitat: Open forests in water meadows with oaks acting as their supports. Occur naturally in eastern North America.

Wood: Fairly light, very hard, very tough, elastic, and difficult to split. Dries very slowly and has a tendency to warp, without cracking. Texture extremely fine and even. Surface very smooth and polishable.

Use: Rarely used, except for walking sticks and turnery. Of no commercial value (Begemann, 1994). American rootstocks were used to re-establish European vineyards.

Beschreibung: Länge der Holzliane („Kletterstrauch") variiert entsprechend dem Trägerbaum von 0,08 bis 0,1 m Durchmesser, ausnahmsweise wohl auch 0,3 m an der Basis.

Vorkommen: Lichte Wälder der Flussauen mit Eichen als Trägerbaum. Natürliche Vorkommen im östlichen Nordamerika.

Holz: Ziemlich leicht, sehr hart, sehr zäh, elastisch und schwer zu spalten. Trocknung sehr langsam mit einer Tendenz zu Deformationen, ohne Rissbildung. Textur äußerst fein und gleichmäßig. Flächen sehr glatt und polierfähig.

Nutzung: Selten verarbeitet, bisweilen zu Spazierstöcken oder Drechslerwaren. Kein Handelsgut (Begemann 1994). Amerikanische Weinreben dienten als Pfropfunterlagen in europäischen Weinkulturen.

Description : variable selon l'hôte pour cette liane ligneuse dont la tige mesure 0,08 à 0,1 m de diamètre, voire 0,3 m à la base.

Habitat : forêts claires des vallées fluviales avec le chêne pour hôte. On la trouve dans l'est de l'Amérique du Nord.

Bois : assez léger, très dur, très tenace, élastique et difficile à fendre. Séchage très lent, avec une tendance aux déformations, sans éclatement. Grain fin et régulier. Surface très lisse, pouvant se polir.

Intérêts : bois rarement employé, sauf pour fabriquer des cannes et des objets de tournage. Non commercialisé (Begemann 1994). Les cépages américains ont été largement utilisés comme porte-greffe dans la viticulture européenne.

Summer Grape.

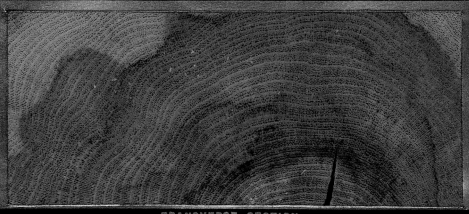

TRANSVERSE SECTION.

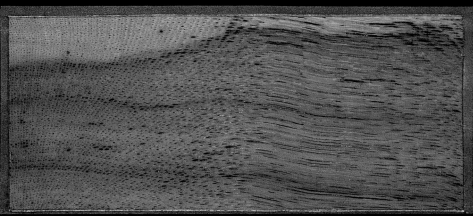

RADIAL SECTION.

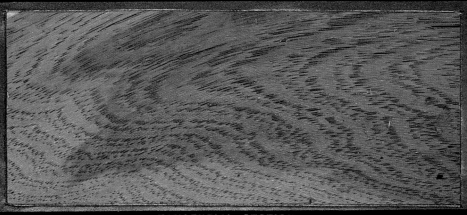

TANGENTIAL SECTION.

Ger. Weinstock.　　Fr. Vigne sauvage de l'Amerique.

FAMILIE DER HEIDEKRAUTGEWÄCHSE
Kalifornische Alprose

FAMILLE DE LA BRUYÈRE OU DE LA MYRTILLE
Rhododendron à grandes feuilles

ERICACEAE
Rhododendron macrophyllum D. Don ex G. Don
Syn.: *R. californicum* Hook.

Description: Normally a many-stemmed shrub, but in the forests of Mendocino County, California, also a tree 33 ft. (10 m) high with a trunk 8 in. (0.20 m) in diameter that is covered by a very thin, reddish brown bark. The evergreen, dark-green leaves are oblong. The flowers are arranged in attractive inflorescences. The corolla is pink with yellowish dots on the upper lobe. The fruit capsules are oblong.

Habitat: Coastal regions of Pacific North America.

Wood: The uniformly brownish-cream wood is moderately heavy, strong, and very close-grained. It is easy to work, but is not used.

Use: A beautiful ornamental shrub that deserves to be distributed more widely.

Beschreibung: Normalerweise ein vielstämmiger Strauch, in den Wäldern des Mendocino-County (Kalifornien) aber auch als Baum mit 10 m Höhe und einem 0,20 m dicken Stamm, der von einer sehr dünnen rotbraunen Borke bedeckt ist. Die immergrünen dunkelgrünen Blätter sind länglich. Die Blüten werden in dekorativen Blütenständen gebildet. Die Blütenkrone ist rosa mit gelblichen, gepunkteten oberen Blütenlappen. Die Fruchtkapseln sind länglich.

Vorkommen: Küstenregionen des pazifischen Nordamerika.

Holz: Das einheitlich cremefarben-bräunliche Holz ist mäßig schwer, fest und sehr feinkörnig. Es ließe sich leicht bearbeiten, wird aber nicht genutzt.

Nutzung: Ein schöner Zierstrauch, der eine weitere Verbreitung verdiente.

Description : se présente généralement sous forme de buisson à tronc multiple dans les forêts du Mendocino County (Californie), mais également sous une forme arborescente de 10 m de hauteur, avec un tronc de 0,20 m de diamètre et une écorce très fine brun rouge. Feuilles allongées, ver sombre, persistantes. Fleurs groupées en corymbes décoratifs. Fruit en capsules allongées.

Habitat : façade pacifique de l'Ame rique du Nord.

Bois : brunâtre teinté de blanc crème, moyennement lourd, résistant et à grain très fin ; se travaille bien mais il est peu employé.

Intérêts : belle espèce ornementale qui mériterait d'être plus cultivée.

California Rose Bay, Rhododendron.

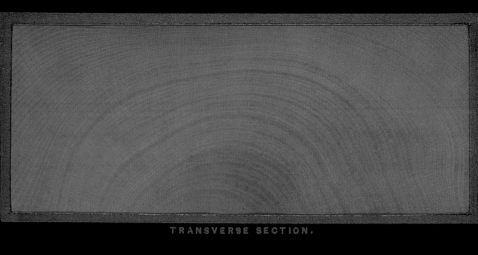

TRANSVERSE SECTION.

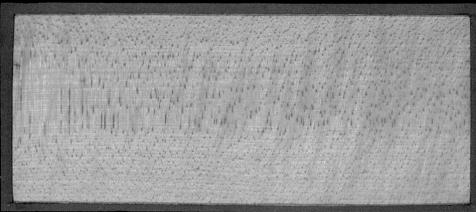

RADIAL SECTION.

TANGENTIAL SECTION

Ger. Californische Alprose.　*Fr.* Rhododendron de Californie.

Sp. Rhododendron de California.

H | HEATHER FAMILY
Madrona, Madrono

FAMILIE DER HEIDEKRAUTGEWÄCHSE
Arbutus, Westamerikanischer Erdbeerbaum

FAMILLE DE LA BRUYÈRE OU DE LA MYRTILLE
Arbousier de Menzies, Arbousier d'Amérique,
Madroño (Canada)

ERICACEAE
Arbutus menziesii Pursh

Description: 85 to 120 ft. (25 to 35 m) high and 4 to 7 ft. (1.2 to 2.1 m) in diameter. Trunk tall and straight. Old bark dark reddish brown, broken up into small, thick scales. Young bark shiny red, flaking into long, thin scales. Evergreen. Numerous white flowers, ornamental.

Habitat: Coastal regions of Pacific North America. Misty coastal forests in northern California, together with the Redwood, the evergreen Sequoia. Much smaller further north and south. South of San Francisco only shrubby.

Wood: Heavy, hard, strong, and close-grained. Heartwood pale brown with reddish tints; sapwood thin, paler.

Use: For furniture and charcoal. Tannin-rich bark for tanning leather.

The most decorative of its genus in the forests of Pacific North America, comparable with Evergreen Magnolia in the southern Atlantic states and Kalmia in Pennsylvania. Occasionally in parks and gardens in southern and western Europe.

Beschreibung: 25 bis 35 m hoch bei 1,2 bis 2,1 m Durchmesser. Hoher, gerader Stamm. Alte Borke dunkel rötlich braun, in kleine, dicke Schuppen gebrochen. Junge Borke glänzend rot, in langen dünnen Schuppen abblätternd. Immergrün. Weiße Blüten in Fülle, ornamental.

Vorkommen: Küstenregionen des pazifischen Nordamerika. Nebelfeuchte Küstenwälder in Nord-Kalifornien, gemeinsam mit der immergrünen Sequoia, dem Redwood. Weiter nach Norden als nach Süden und dort viel kleiner. Südlich von San Francisco nur noch strauchig.

Holz: Schwer, hart, fest und mit feiner Textur. Kern hellbraun mit roter Tönung; Splint dünn, heller.

Nutzung: Für Möbel und für Holzkohle. Gerbstoffreiche Rinde zum Gerben von Leder.

Gilt als besondere Zierde ihrer Gattung in den Wäldern des pazifischen Nordamerika, vergleichbar mit der Großen Magnolie in den südlichen Atlantikstaaten und der *Kalmia* in Pennsylvanien. In Süd- und Westeuropa gelegentlich in Parks oder Gärten.

Description : 25 à 35 m de hauteur ; grand tronc droit de 1,2 à 2,1 m de diamètre. L'écorce brun rougeâtre foncé se craquelle en petites écailles épaisses sur les vieux sujets ; elle est rouge brillan et s'exfolie sur les jeunes arbres. Feuilles persistantes. Nombreuses fleurs blanches ornementales.

Habitat : façade pacifique de l'Am rique du Nord. Forêts côtières de brouillard dans le nord de la Californie, avec le séquoia sempe virent. Plus au nord et plus au su beaucoup plus petit. Au sud de San Francisco, forme buissonnant seulement.

Bois : lourd, dur, résistant et à gra fin. Bois de cœur brun clair teinté de rouge ; aubier mince, plus clai

Intérêts : fabrication de meubles e le charbon de bois. L'écorce riche en tannins est utilisée en tannage

Espèce très ornementale dans les forêts de la façade pacifique de l'Amérique du Nord, comparable au magnolia à grandes feuilles de Etats du Sud-Est et à la kalmie de Pennsylvanie. Quelquefois arbre de parcs ou de jardins en Europe occidentale et méridionale.

Madroña, Madroña Laurel, Strawberry Tree.

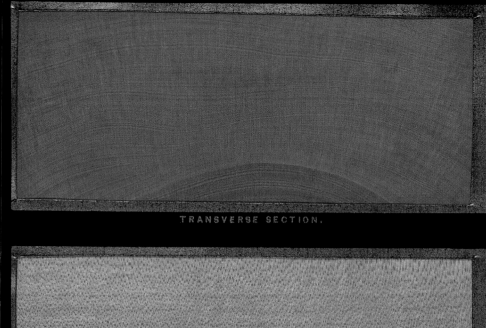

TRANSVERSE SECTION.

RADIAL SECTION.

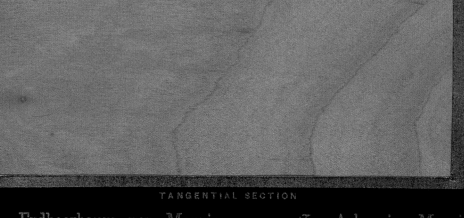

TANGENTIAL SECTION

Ger. Erdbeerbaum von Menzies. Fr. Arbousier Menzi

Sp. Madroña.

H | HEATHER FAMILY
Madroña, Madrono

FAMILIE DER HEIDEKRAUTGEWÄCHSE
Erdbeerbaum

FAMILLE DE LA BRUYÈRE OU DE LA MYRTILLE
Arbousier de l'Arizona

ERICACEAE
Arbutus arizonica (A. Gray) Sarg.

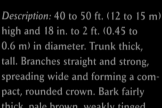

Description: 40 to 50 ft. (12 to 15 m) high and 18 in. to 2 ft. (0.45 to 0.6 m) in diameter. Trunk thick, tall. Branches straight and strong, spreading wide and forming a compact, rounded crown. Bark fairly thick, pale brown, weakly tinged red on the surface. Bark of young trees and branches thin and darker. Leaves lanceolate or more rarely narrowly ovate, pale green above, paler below. Petals of globular, white flowers fused almost to the tips. Flowers short-stalked, in panicles. Fruits rounded to elongate, orange, with rough surface.

Habitat: Dry stony sites, between 5,500 and 7,500 ft. (1,800 to 2,400 m), in southern North America from Arizona to Chihuahua.

Wood: Heavy, soft, brittle, close-grained. Heartwood pale brown, tinged red; sapwood paler.

Use: Very decorative plant, with the strongly contrasting bark color of trunk and branches, flowers and foliage, but as yet not cultivated.

Beschreibung: 12 bis 15 m hoch bei 0,45 bis 0,6 m Durchmesser. Stamm dick, hoch. Äste gerade und kräftig, weit ausladend, eine kompakte, runde Krone bildend. Borke mäßig dick, hellgrau mit schwach rötlicher Tönung an der Oberfläche. Borke junger Bäume und der Äste dünn und von dunkelroter Farbe. Blätter lanzettlich oder seltener schmal oval mit hellgrüner Ober- und blasser Unterseite. Kronblätter der kugeligen, weißen Blüten verschmolzen, nur die oberen Enden frei. Blüten in Rispen auf kleinen Stielchen angeordnet. Früchte rund bis länglich, von orangener Farbe, mit rauer Oberfläche.

Vorkommen: Trockene, steinige Bänke, zwischen 1800 bis 2400 m, im südlichen Nordamerika von Arizona bis Chihuahua.

Holz: Schwer, weich, brüchig und mit feiner Textur. Kern hellbraun mit rötlicher Tönung; Splint heller.

Nutzung: Mit den deutlichen Farbkontrasten der Stammborke und der Äste, der Blüten und des Laubes eine sehr dekorative Pflanze, bislang aber nicht kultiviert.

Description: 12 à 15 m de hauteur sur un diamètre de 0,45 à 0,6 m. Tronc haut, épais avec des branch droites, vigoureuses, très étalées, formant une couronne ronde et compacte. Ecorce moyennement épaisse, gris clair avec une touche de rouge en surface; mince et rouge foncé sur les jeunes arbres et les branches. Feuilles lancéolée ou ovales, vert clair dessus, plus pâle dessous. Fleurs blanches à corolle en grelot (pétales soudés à la base, libres à leur extrémité), groupées en panicules et courtement pédicellées. Fruits sphériqu ou allongés, orange et rugueux.

Habitat: bancs caillouteux secs; à une altitude de 1800 à 2400 m dans le sud de l'Amérique du Noi depuis l'Arizona jusqu'à Chihuahua.

Bois: lourd, tendre, cassant et à grain fin. Bois de cœur brun clair teinté de rouge; aubier plus clair.

Intérêts: espèce ornementale, très attrayante par les contrastes de couleurs entre l'écorce du tronc e des branches, les fleurs et le feuillage, mais non cultivée jusqu'ici.

Arizona Madroña.

TRANSVERSE SECTION.

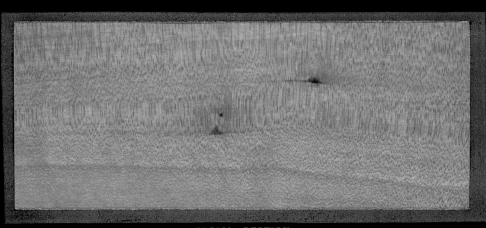

RADIAL SECTION.

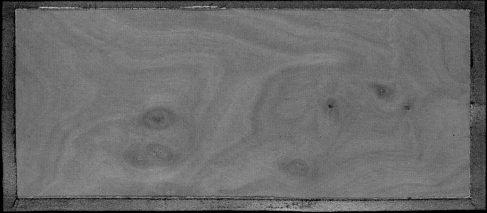

TANGENTIAL SECTION.

Ger. Arizonischer Erdbeerbaum. *Fr.* Arbousier d'Arizona.
Sp. Madroña de Arizona.

H | HEATHER FAMILY
Mexican Manzanita

FAMILIE DER HEIDEKRAUTGEWÄCHSE
Kalifornische Bärentraube

FAMILLE DE LA BRUYÈRE OU DE LA MYRTILLE
Arctostaphyle à feuilles pointues

ERICACEAE
Arctostaphylos pungens Kunth.

Description: Usually a shrub no more than 3 ft. (1 m) high, only rarely a tree, with a maximum height of 17 ft. (4.5 m) Many-stemmed, with twisted secondary stems. Bark very thin, brownish red or mahogany-colored, separating into small, papery scales. Leaves oblong-lanceolate, very stiff, pale, and light bluish-green, entire or (on young branches) toothed. Flowers in short clusters, white or pink. Fruits roundish, laterally flattened, at first yellowish, later red.

Habitat: Dry ridges and mountainsides in Pacific North America.

Wood: The brownish red heartwood is heavy, hard, and easy to split. It is surrounded by thin, whitish sapwood.

Use: Occasionally used for small objects such as buttons, boxes, etc. The fruits are eaten by the indigenous tribes and are an important food for bears and certain birds.

Beschreibung: Meist als Strauch, nicht höher als einen Meter, nur selten als maximal 4,5 m hoher Baum. Vielstämmig mit krummen Teilstämmen. Borke sehr dünn braunrot oder mahagonifarben, sich in papierartigen Schüppchen abschälend. Blätter länglich-lanzettlich, sehr steif, blass und leicht bläulich grün, ganzrandig oder (an jungen Ästen) gezähnt. Blüten in kurzen Trauben, weiß oder rosa. Früchte rundlich und seitlich abgeflacht, anfangs gelblich, später rot.

Vorkommen: Trockene Grate und Berghänge im pazifischen Nordamerika.

Holz: Das braunrote Kernholz ist schwer, hart und leicht zu spalten. Es wird von einem dünnen weißlichen Saftholz umgeben.

Nutzung: Wenig genutzt für kleine Gegenstände wie Knöpfe, Schachteln etc. Die Früchte wurden von den indigenen Völkern gegessen und stellen eine wichtige Nahrung für Bären und bestimmte Vögel dar.

Description : ordinairement sous forme de buisson d'un mètre de hauteur, atteignant rarement 4,5 de haut. Tronc multiple avec des tiges secondaires tortueuses. Ecor très fine, rouge brun ou acajou, s'exfoliant en petites pellicules écailleuses. Feuilles allongées-lancéolées, très rigides, pâles et légèrement vert bleuté, entières ou dentées sur les jeunes rameau Fleurs en courtes grappes blanche ou roses. Fruits globuleux, aplatis sur les côtés, jaunâtres puis rouge à maturité.

Habitat : crêtes et versants de mor tagnes secs sur la façade pacifique de l'Amérique du Nord.

Bois : lourd, dur et facile à fendre. Bois de cœur rouge brun ; aubier mince, blanchâtre.

Intérêts : peu utilisé, hormis pour la fabrication de petits objets (boutons, boîtes etc.). Les fruits sont consommés par les Indiens et constituent aussi une nourritur importante pour les ours et certai

Common Manzanita.

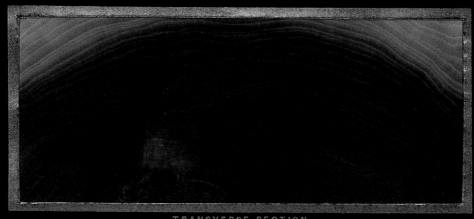

TRANSVERSE SECTION.

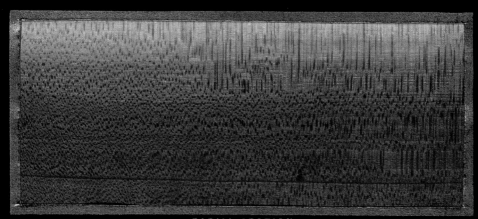

RADIAL SECTION.

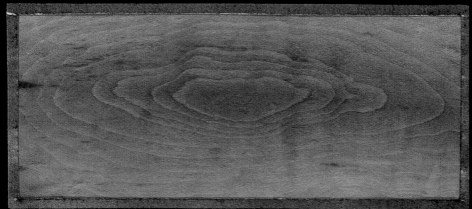

TANGENTIAL SECTION

r. Californianische Bärentraube.   *Fr.* Busserole de Califorï

Sp. Manzanita comun.

H | HEATHER FAMILY
Mountain Laurel, Calico Bush

FAMILIE DER HEIDEKRAUTGEWÄCHSE
Berg-Lorbeer, Breitblättrige Kalmie

FAMILLE DE LA BRUYÈRE OU DE LA MYRTILLE
Laurier des montagnes, Kalmia à feuilles larges,
Laurier américain, Laurier des montagnes

ERICACEAE
Kalmia latifolia L.

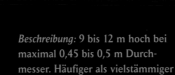

Description: 30 to 40 ft. (9 to 12 m) high and a maximum of 15 to 16 in. (0.45 to 0.5 m) in diameter. More frequently a dense, broad, many-stemmed shrub, 6 to 10 ft. (1.8 to 3 m) in height. Trunk short and crooked, branches stout and forked, crown round and compact. Bark thin, dark brown with a red tinge, and mainly longitudinal furrows. Leaves oblong or elliptic-lanceolate, pointed or rounded at the apex, thick and stiff, dark green above, paler beneath. Flowers in clusters; petals white or pink with wavy red lines and fine purple dots on the inner surface. Fruit capsule subglobose, bearing the remains of the style.

Habitat: Swamp margins or shady, drier slopes in eastern North America. In the southern mountains it grows on rich soils on rocky slopes of hills and mountains up to 3,000 to 4,000 ft. (900 to 1,200 m) above sea level, and forms extensive thickets.

Wood: Heavy, hard, strong, rather brittle, and close-grained.

Use: In flower it is one of the most beautiful ornamental woody plants in the North American flora.

Beschreibung: 9 bis 12 m hoch bei maximal 0,45 bis 0,5 m Durchmesser. Häufiger als vielstämmiger dichter und breiter Strauch von 1,8 bis 3 m Höhe. Stamm kurz, gekrümmt, Äste kräftig verzweigt, Krone rund, kompakt. Borke dünn, dunkelbraun mit roter Tönung und vorwiegend längs orientierten Rissen. Blätter länglich oder elliptisch-lanzettlich, spitz oder abgerundet, dick und steif, dunkelgrün, heller auf der Unterseite. Blüten in Büscheln, mit weißen oder rosa Kronblättern. Auf der Innenseite mit wellenförmigen roten Linien und feinen violetten Punkten. Fruchtkapsel zusammengedrückt rundlich, an der Spitze mit Rest des Griffels.

Vorkommen: Ränder von Sümpfen oder schattige, trockenere Hänge im östlichen Nordamerika. In den südlichen Gebirgen auf nährstoffreichen Böden steiniger Hügel- und Berghänge bis auf 900 bis 1200 m, ausgedehnte Dickichte bildend.

Holz: Schwer, hart, fest, eher brüchig und mit feiner Textur.

Nutzung: Blühend eines der schönsten Ziergehölze der nordamerikanischen Flora.

Description: arbuste de 9 à 12 m hauteur sur un diamètre de 0,45 à 0,5 m. Souvent gros buisson, à tronc multiple, de 1,8 à 3 m de haut. Tronc court, tordu, avec des branches très ramifiées, formant houppier rond et compact. Ecorce mince, brun sombre teinté de rouge et parcourue de fissures, longitudinales pour la plupart. Feuill allongées ou elliptiques-lancéolée pointues ou arrondies, épaisses et rigides, vert foncé dessus, plus claires dessous. Fleurs blanches o roses en corymbes, avec des tach violettes et des lignes sinueuses rouges à l'intérieur. Fruits en capsules rondes comprimées, sur montées des restes du style.

Habitat: bords de marécages ou pentes sèches abritées de l'est de l'Amérique du Nord. Dans les chaines méridionales, en vastes fourrés sur les sols riches des versants collinéens et montagneu entre 900 et 1200 m.

Bois: lourd, dur, plutôt cassant et grain fin.

Intérêts: une des plus belles espè ornementales nord-américaine a moment de la floraison.

Mountain Laurel.

TRANSVERSE SECTION.

RADIAL SECTION.

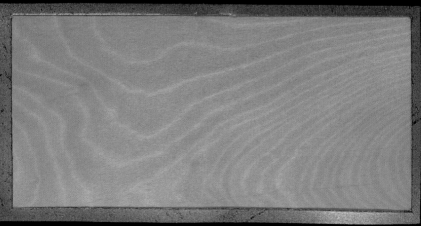

TANGENTIAL SECTION.

Ger. Berg-Lorbeer.　　*Fr.* Kalmie de Montagne.

Sp. Laurel de la Montana.

H | HEATHER FAMILY

Rose Bay, Big Laurel, Bigleaf Laurel

FAMILIE DER HEIDEKRAUTGEWÄCHSE
Große amerikanische Alpenrose

FAMILLE DE LA BRUYÈRE OU DE LA MYRTILLE
Rhododendron géant, Laurier des marais

ERICACEAE
Rhododendron maximum L.

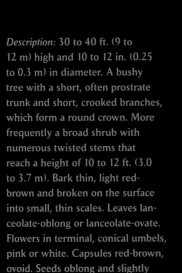

Description: 30 to 40 ft. (9 to 12 m) high and 10 to 12 in. (0.25 to 0.3 m) in diameter. A bushy tree with a short, often prostrate trunk and short, crooked branches, which form a round crown. More frequently a broad shrub with numerous twisted stems that reach a height of 10 to 12 ft. (3.0 to 3.7 m). Bark thin, light red-brown and broken on the surface into small, thin scales. Leaves lanceolate-oblong or lanceolate-ovate. Flowers in terminal, conical umbels, pink or white. Capsules red-brown, ovoid. Seeds oblong and slightly flattened.

Habitat: (North-)eastern North America. Rare in the north, in deep, cold swamps; more frequent towards the south on rocky riverbanks and mountain slopes, sometimes forming extensive thickets.

Wood: Heavy, hard, strong, rather brittle, and close-grained.

Use: Tool handles, and as a substitute in engraving. A decoction of the leaves is occasionally used in the treatment of rheumatism.

Beschreibung: 9 bis 12 m hoch bei 0,25 bis 0,3 m Durchmesser. Buschartiger Baum mit kurzem, oft niederliegenden Stamm und kurzen, krummen Ästen, die eine runde Krone bilden, häufiger noch als breiter Strauch mit zahlreichen verdrehten Stämmen, die eine Höhe von 3 bis 3,7 m erreichen. Borke dünn, hell rotbraun und an der Oberfläche in dünne Schüppchen zerfallend. Blätter lanzettlich-länglich oder lanzettlich-oval. Blüten in endständigen, konischen Dolden, rosa oder weiß. Kapsel rotbraun, oval. Samen länglich und leicht abgeflacht.

Vorkommen: (Nord-) Östliches Nordamerika. Im Norden selten in tiefen und kalten Sümpfen, nach Süden hin häufiger und auf steinigen Flussufern sowie Berghängen z. T. ausgedehnte Dickichte bildend.

Holz: Schwer, hart, fest, eher brüchig und mit feiner Textur.

Nutzung: Werkzeuggriffe und als Ersatz für Gravurarbeiten. Der abgekochte Sud der Blätter gelegentlich als Rheumamittel.

Description: arbre à port touffu d 9 à 12 m de hauteur, doté d'un tronc court, souvent prostré de 0,25 à 0,3 m de diamètre, avec d branches courtes, tortes, forman un houppier rond. Plus souvent encore sous forme de buisson d 3,0 à 3,7 m de hauteur, avec de nombreuses tiges tortueuses et contournées. Ecorce mince, brur rouge clair et s'exfoliant en peti écailles minces. Feuilles lancéole allongées ou lancéolées-ovales. Fleurs en ombelles terminales, coniques, roses ou blanches. Ca sules brun rouge, ovales. Graine allongées et légèrement aplaties.

Habitat: dans le (nord-) est de l'Amérique du Nord. Au nord, r ment dans les marécages profor et froids; vers le sud, plus fréqu sur les rives caillouteuses des c d'eau ainsi que sur les versants de montagnes où il constitue de fourrés souvent étendus.

Bois: lourd, dur, résistant, plutô cassant et à grain fin.

Intérêts: utilisé pour fabriquer d manches d'outils et comme bois de substitution pour la gravure. décoction de feuilles sert quelqu fois à soigner les rhumatismes.

TRANSVERSE SECTION

RADIAL SECTION.

TANGENTIAL SECTION.

H | HEATHER FAMILY

Sorrel Tree, Sour Wood

FAMILIE DER HEIDEKRAUTGEWÄCHSE
Gemeiner Sauerbaum, Sauerampferbaum

FAMILLE DE LA BRUYÈRE OU DE LA MYRTILLE
Oxydendron arborescent, Andromède en arbre,
Arbre à oseille

ERICACEAE
Oxydendrum arboreum (L.) DC.

Description: 50 to 60 ft. (15 to 18 m) high and 12 to 20 in. (0.3 to 0.5 m) in diameter. Branches slender, spreading, forming a narrowly oblong crown, rounded at the top. Bark moderately thick, gray with a red tinge, and deeply furrowed. Leaves at first bronze-green, then bright scarlet. Flowers white, in long panicles, with a dilated, ovoid corolla tube. Flowering in June and July. Fruit capsule, like the flowers, oblong-ovoid to conical in form.

Habitat: Well-drained, stony soils on ridges above the riverbanks in deciduous, broad-leaved forests in south-eastern North America.

Wood: Heavy, hard, strong, and fairly flexible; excellent for taking polish. Heartwood light to reddish brown; sapwood pale brownish white.

Use: For turnery and as a decorative plant. The color of the foliage in fall can scarcely be surpassed, and the flowers appear at a time of the year when only very few other woody plants are in flower. Cultivated in England since 1752.

Seldom subject to insect or fungal attack to any great extent.

Beschreibung: 15 bis 18 m hoch bei 0,3 bis 0,5 m Durchmesser. Äste schlank, ausladend, eine schmale längliche, an der Spitze rundliche Krone bildend. Borke mäßig dick, grau mit roter Tönung, mit tiefen Längsrissen. Blätter anfangs bronzegrün, später hell scharlachrot. Blüten weiß, in langen Doppeltrauben, mit bauchiger, ovaler Kronblattröhre. Blüte Juni/Juli. Fruchtkapsel ähnlich den Blüten länglich oval bis konisch geformt.

Vorkommen: Gut drainierte, steinige Böden auf Anhöhen über Flussufern in Wäldern sommergrüner Laubbäume im südöstlichen Nordamerika.

Holz: Schwer, hart, fest und ziemlich elastisch; vorzüglich polierfähig. Kern hell- bis rötlich braun; Splint hell bräunlich weiß.

Nutzung: Zur Drechslerei und als Zierstrauch. Die Farben des Herbstlaubes werden kaum übertroffen und die Blüte liegt in einer Jahreszeit, in der nur sehr wenige Gehölze noch blühen. Bereits seit 1752 in England kultiviert.

Das Laub wird nicht nennenswert von Insekten oder Pilzen befallen.

Description : 15 à 18 m de hauteur sur un diamètre de 0,3 à 0,5 m. Branches fines, étalées, constitua un houppier allongé et étroit, à cime arrondie. Ecorce moyennement épaisse, grise teintée de rouge et creusée de fissures longitudinales. Feuilles vert bronz puis pourpres. Fleurs blanches en longues panicules, avec des pétal ovales soudés en un tube renflé. Floraison en juin-juillet. Fruits secs en capsules, allongés-ovales coniques comme les fleurs.

Habitat : sols caillouteux, bien drainés sur les surplombs de berg dans les forêts d'arbres caducs du sud-est de l'Amérique du Nord.

Bois : lourd, dur, résistant et asse élastique, offrant un beau poli. Bois de cœur brun clair à brun ro geâtre ; aubier d'un blanc brunâtr clair.

Intérêts : utilisé en tournerie et en ornement. Ses coloris en automne sont d'une beauté excep tionnelle et la floraison se produi à une période où très peu d'arbre fleurissent encore. Cultivé en Angleterre depuis 1752.

Pas de parasite dangereux signalé

Sorrel-Tree.   Sour-Wood.

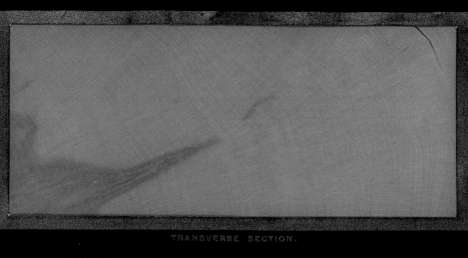

TRANSVERSE SECTION.

RADIAL SECTION.

TANGENTIAL SECTION.

H | HEATHER FAMILY
Sparkleberry

FAMILIE DER HEIDEKRAUTGEWÄCHSE
Baumartige Heidelbeere

FAMILLE DE LA BRUYÈRE OU DE LA MYRTILLE
Airelle arborée

ERICACEAE
Vaccinium arboreum Marshall

Description: 20 to 30 ft. (6 to 9 m) high, with a short, often crooked trunk occasionally 8 to 10 in. (0.2 to 0.25 m) in diameter, and slender branches forming an irregular rounded crown. Bark thin, light reddish brown and covered with small scales. Leaves ovate to oblong, leathery, dark green above, pale beneath. Corolla of united petals, five-lobed only at the apex, white. Berries black, round, dry, and astringent.

Habitat: Damp, sandy soils by ponds and streams, in the south along the Gulf coast, and in the interior of North America.

Wood: Heavy, hard, with a silky sheen, and very close-grained; polishable. Heartwood light brown, tinged red; sapwood thick.

Use: Occasionally for tool handles or other small objects. A decotion of the root bark and leaves is used in the treatment of diarrhea. The bark is a source of tannin. Introduced into England in 1805, but later disappeared.

Beschreibung: 6 bis 9 m hoch, mit kurzem, oft krummem Stamm von gelegentlich 0,2 bis 0,25 m Dicke und schlanken Ästen, die eine unregelmäßige runde Krone bilden. Borke dünn, hell rotbraun und mit kleinen Schüppchen bedeckt. Blätter oval bis länglich, ledrig und dunkelgrün auf der Oberseite, blass auf der Unterseite. Blütenkrone verwachsen, nur an der Spitze fünflappig, weiß. Beeren schwarz, rund, trocken und adstringierend.

Vorkommen: Feuchte, sandige Böden an Teichen und Flüssen im Süden längs der Golfküste sowie im Inneren Nordamerikas.

Holz: Schwer, hart, seidig und mit sehr feiner Textur; polierfähig. Kern hellbraun, rot getönt; Splint dick.

Nutzung: Gelegentlich für Werkzeuggriffe oder andere kleine Gegenstände. Der abekochte Sud der Wurzelrinde und Blätter werden für die Behandlung der Diarrhö verwendet. Aus der Borke wird Tannin gewonnen. 1805 nach England eingeführt, später wieder verschwunden.

Description : arbuste de 6 à 9 m de hauteur avec un tronc court, souvent tordu, atteignant parfois de 0,2 à 0,25 m de diamètre et des branches fines, constituant un houppier rond et irrégulier. Ecorce mince, rouge brun clair, recouvert de petites écailles. Feuilles ovales à allongées, coriaces, vert sombre dessus, pâles dessous. Corolle formée de pétales soudés, à marge découpée en 5 lobes. Baies noires globuleuses, sèches et astringente

Habitat : sols sablonneux humides près des étangs et des rivières dar le sud le long du Golfe sur la côte ainsi que dans l'intérieur de l'Ame rique du Nord.

Bois : lourd, dur, satiné et à grain très fin ; susceptible d'être poli. Bc de cœur brun clair teinté de rouge aubier épais.

Intérêts : pour manches d'outils ou autres petits objets. L'écorce et les feuilles sont utilisées en décoction pour soigner la diarrhée. Extractio du tannin de l'écorce. Introdüit en Angleterre en 1805 puis disparaît de la culture.

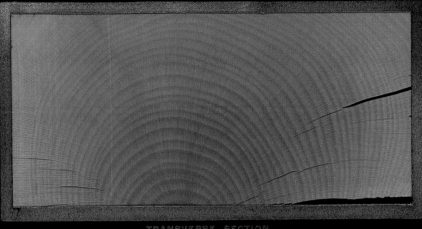

TRANSVERSE SECTION

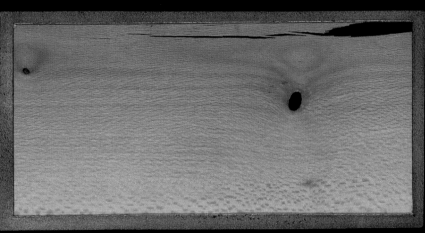

RADIAL SECTION.

TANGENTIAL SECTION

H | HOLLY FAMILY

American Holly

FAMILIE DER STECHPALMENGEWÄCHSE
Amerikanische Hülse

FAMILLE DU HOUX
Houx d'Amérique

AQUIFOLIACEAE
Ilex opaca Aiton

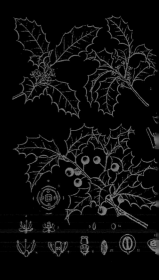

Description: In favorable conditions 40 to 50 ft. (12 to 15 m) high and a maximum of 2 to 3 ft. (0.6 to 0.9 m) in diameter. Numerous pendent or horizontal lateral branches. Bark smooth. Leaves dark green, with undulate margins and a few stout, spinose teeth. Evergreen. Decorative red, globose fruits, usually borne singly.

Habitat: Well-drained slopes and lowland in association with various oaks and hickories, Eastern Red Cedar, Tulip Tree, magnolias, Water Beech, etc., in eastern North America.

Wood: Light, tough, and close-grained. Almost white.

Use: For turnery, cupboards. Twigs used in America at Christmas for decorative purposes.

Radical decline owing to overuse for Christmas decorations.

Beschreibung: Unter günstigen Bedingungen 12 bis 15 m hoch bei maximal 0,6 bis 0,9 m Durchmesser. Zahlreiche hängende oder waagerechte Seitenäste. Borke glatt. Ledrige, dunkelgrüne, grob buchtig gezähnte Blätter mit stechender Spitze. Immergrün. Zierende rote, kugelige Früchte meist einzeln.

Vorkommen: Gut drainierte Hänge und Flachland in Gesellschaft verschiedener Eichen und Hickories, Virginischer Rotzeder, Tulpenbaum, Magnolien, Wasser-Buche u. a. im östlichen Nordamerika.

Holz: Leicht, zäh und von feiner Textur. Fast weiß.

Nutzung: Für Drechslerarbeiten, Schränke. Die dekorativen Zweige in Amerika zu Weihnachten.

Radikaler Rückgang durch Übernutzung als Weihnachtsdekoration.

Description : peut atteindre, si les conditions sont favorables, 12 à 15 m sur un diamètre de 0,6 à 0,9 m maximum. Nombreuses branches latérales retombantes o insérées à angle droit sur le tron Ecorce lisse. Feuilles persistantes, coriaces, vert foncé et piquantes. Fruits globuleux rouge brillant, attrayants, généralement non groupés.

Habitat : pentes bien drainées et plaines de l'est de l'Amérique du Nord, en association avec divers chênes et cayers, le genévrier de Virginie, le tulipier de Virginie, le magnolias, le hêtre aquatique etc

Bois : léger, tenace et à grain fin. Presque blanc.

Intérêts : utilisé en tournerie et pour la fabrication d'armoires. Le rameaux servent de décoration c Noël.

Raréfaction depuis son utilisation excessive au moment de Noël.

American Holly.

TRANSVERSE SECTION.

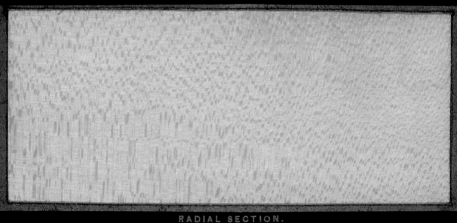

RADIAL SECTION.

TANGENTIAL SECTION

H | HOLLY FAMILY
Dahoon

FAMILIE DER STECHPALMENGEWÄCHSE
Lorbeerblättrige Hülse

FAMILLE DU HOUX
Houx d'Ahon, Houx de Dahoon

AQUIFOLIACEAE
Ilex cassine L.

Description: 25 to 30 ft. (7.5 to 9.0 m) high and 12 to 18 in. (0.3 to 0.45 m) in diameter or a low shrub. Bark thin, dark gray in color, and with numerous lenticels. Young branches until their second or third year covered with whitish felt. Leaves very variable, ovate-oblong to oblong-lanceolate, entire or sharply serrate towards the tip, dark green above and pale beneath. Flower-parts in fours, sepals pointed with ciliate margins. Fruit round, bright red (rarely pale or almost yellow).

Habitat: Cold swamps, especially their margins, in rich moist soil in the south-eastern coastal states of North America, occasionally on the Gulf coast on high sandbanks along rivers.

Wood: Light, soft, not strong, and close-grained. Heartwood light brown; sapwood thick, almost white. Numerous medullary rays.

Use: For Christmas decorations (Harrar, 1971).

Beschreibung: 7,5 bis 9,0 m hoch bei 0,3 bis 0,45 m Durchmesser oder als niedriger Busch. Borke dünn, von dunkelgrauer Farbe und mit zahlreichen Atemporen. Junge Äste bis ins zweite oder dritte Jahr mit weißlichem Filz bedeckt. Blätter sehr unterschiedlich, oval-länglich bis länglich-lanzettlich, ganzrandig oder an der Spitze scharf gesägt, mit dunkelgrüner Ober- und blasser Unterseite. Blütenorgane vierzählig, Kelchblätter spitz mit gewimperten Rändern. Frucht rund, hellrot (seltener blass oder fast gelb).

Vorkommen: Kalte Sümpfe, besonders deren Ränder, in reicher, feuchter Erde in den südöstlichen Küstenstaaten Nordamerikas, an der Golfküste gelegentlich an hohen Sandbänken längs der Flüsse.

Holz: Leicht, weich, nicht stabil und von feiner Textur. Kern blassbraun; Splint dick, fast weiß. Viele Markstrahlen.

Nutzung: Weihnachtsdekoration (Harrar 1971).

Description : arbuste de 7,5 à 9,0 r de hauteur sur un diamètre de 0, à 0,45 m ou arbrisseau bas. Ecorc mince, gris sombre et couverte de lenticelles. Jeunes rameaux tomenteux jusqu'à la seconde ou troisième année. Feuilles très différentes, ovales-allongées à allongées-lancéolées, entières ou apex acuminé, vert sombre dessu pâles dessous. Fleurs à tétramères sépales terminées en pointe et à marge ciliée. Fruits globuleux, rouge clair (rarement pâles ou presque jaunes).

Habitat : marais froids, en particulier sur leurs bords, sur substrat riche et humide dans les Etats côtiers du sud-est de l'Amérique du Nord, parfois aussi hauts banc de sable le long des fleuves sur le pourtour du golfe du Mexique.

Bois : léger, tendre, non stable et à grain fin. Bois de cœur brun pâle ; aubier épais, presque blanc Nombreux rayons médullaires.

Intérêts : rameaux utilisés comme décoration de Noël (Harrar 1971)

TRANSVERSE SECTION.

RADIAL SECTION.

TANGENTIAL SECTION.

Ger. Cassena Stechpalme.   *Fr.* Houx de Cassena.

H | HOLLY FAMILY
Mountain Holly, Large-leaf Holly

FAMILIE DER STECHPALMENGEWÄCHSE
Berg-Hülsen

FAMILLE DU HOUX
Houx de montagne

AQUIFOLIACEAE
Ilex montana Torr & A. Gray ex A. Gray
Syn.: *I. monticola* Gray

Description: 30 to 40 ft. (9 to 12 m) high and 12 to 15 in. (0.25 to 0.3 m) in diameter. Trunk short. A tree with a narrow-pyramidal crown or, more frequently, a many-stemmed shrub. Bark thin, light brown in color, with many lenticels. Leaves ovate to oblong-lanceolate with a serrate margin, glabrous or finely hairy along the veins, light green above and paler beneath, deciduous. Floral parts in fours or fives, sepals acute and ciliate. The monoecious flowers open when the leaves are half grown. Fruits globular, bright red.

Habitat: On rich soils, e.g. on riverbanks, in eastern North America.

Wood: Heavy, hard, and close-grained. Creamy white with numerous medullary rays.

Use: The large fruits and the attractive foliage make the species one of the most popular garden plants among the deciduous hollies.

Beschreibung: 9 bis 12 m hoch bei 0,25 bis 0,3 m Durchmesser. Stamm kurz. Als Baum mit schmalpyramidaler Krone oder (häufiger) als vielstämmiger Busch. Borke dünn, von hellbrauner Farbe mit zahlreichen Atemporen. Blätter oval bis länglich-lanzettlich, am Rand gesägt, nackt oder fein behaart entlang der Blattrippen, mit hellgrüner Ober- und blasser Unterseite. Sommergrün. Blütenorgane vier- bis fünfzählig, Kelchblätter spitz und bewimpert. Die einhäusigen Blüten öffnen sich, wenn die Blätter halb ausgewachsen sind. Früchte kugelig, hellrot.

Vorkommen: Auf reichen Böden z. B. an Flussufern im östlichen Nordamerika.

Holz: Schwer, hart und mit feiner Textur. Cremefarben-weißlich, mit zahlreichen Markstrahlen.

Nutzung: Die großen Früchte und das attraktive Laub machen die Art zu einer der beliebtesten Gartenpflanzen unter den sommergrünen *Ilex*-Gewächsen.

Description: 9 à 12 m de hauteur, avec un tronc court de 0,25 à 0,3 de diamètre. Port arborescent, avec un houppier étroit, pyramidal ou (plus souvent) arbrisseau buissonnant. Ecorce mince, brun clair, ponctuée de nombreuses lenticelles. Feuilles ovales à lancéolées-allongées, dentées, glabre ou duvetées sur les nervures, vert clair dessus, pâles au revers, et caduques. Fleurs tétra- ou pentamères, sépales pointus et ciliés. Fleurs monoïques s'épanouissant avant que la pousse des feuilles soit achevée. Fruits globuleux, rouge clair.

Habitat : sols riches, par exemple de berges dans l'est de l'Amériqu du Nord.

Bois : lourd, dur et à grain fin. Bla crème, avec de nombreux rayons médullaires.

Intérêts : un des houx à feuilles caduques les plus recherchés pou les jardins en raison de ses gros fruits et de son feuillage attrayant

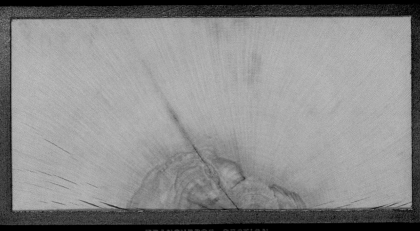

TRANSVERSE SECTION.

RADIAL SECTION.

TANGENTIAL SECTION.

Ger. Berg-Stechpalme. *Fr.* Houx de Montagne.

Sp. Acebo de la Montaña.

H | HOLLY FAMILY

Swamp Holly, Deciduous Holly, Bear Berry

FAMILIE DER STECHPALMENGEWÄCHSE
Walddistel, Sommergrüner Hülsen

FAMILLE DU HOUX
Houx décidu, Apalanche d'Amérique

AQUIFOLIACEAE
Ilex decidua Walter

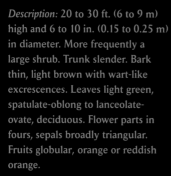

Description: 20 to 30 ft. (6 to 9 m) high and 6 to 10 in. (0.15 to 0.25 m) in diameter. More frequently a large shrub. Trunk slender. Bark thin, light brown with wart-like excrescences. Leaves light green, spatulate-oblong to lanceolate-ovate, deciduous. Flower parts in fours, sepals broadly triangular. Fruits globular, orange or reddish orange.

Habitat: Margins of rivers and swamps in slightly moist soil in south-eastern North America.

Wood: Heavy, hard, and close-grained. Heartwood creamy white; sapwood somewhat paler. Numerous thin medullary rays.

Use: Introduced into England by 1760, but nowadays only very rarely grown as a park tree.

Beschreibung: 6 bis 9 m hoch bei 0,15 bis 0,25 m Durchmesser. Häufiger als großer Strauch. Stamm schlank. Borke dünn, hellbraun mit warzenähnlichen Auswüchsen. Blätter hellgrün, spatelförmig-länglich bis lanzettlich-oval. Sommergrün. Blütenorgane vierzählig, Kelchblätter breit dreieckig. Früchte kugelig, orange oder orange-rötlich.

Vorkommen: Ufer von Flüssen und Sümpfen in gering feuchtem Boden, im südöstlichen Nordamerika.

Holz: Schwer, hart und von feiner Textur. Kern cremeweiß; Splint etwas heller. Zahlreiche dünne Markstrahlen.

Nutzung: Bereits 1760 nach England eingeführt, heute jedoch als Parkgehölz nur noch sehr selten.

Description : arbuste de 6 à 9 m d hauteur, avec un tronc fin de 0,1 à 0,25 m de diamètre. Plus souv sous forme de buisson. Ecorce mince, brun clair et verruqueuse Feuilles vert clair, spatulées-allor gées à lancéolées-ovales, persis-tantes. Fleur tétramère, à sépale larges et triangulaires. Fruits glol leux, orange ou rouge orangé.

Habitat : sol peu humide en bor-dure de cours d'eau et de marai dans le sud-est de l'Amérique d Nord.

Bois : lourd, dur et à grain fin. Bc de cœur blanc crème ; aubier un peu plus clair. Nombreux rayons médullaires fins.

Intérêts : introduit en Angleterre 1760 mais devenu très rare de n jours comme arbre de parcs.

SWAMP HOLLY. DECIDUOUS HOLLY.

TRANSVERSE SECTION.

RADIAL SECTION.

TANGENTIAL SECTION.

H | HOLLY FAMILY
Yaupon, Cassena, Apalachen Tree

FAMILIE DER STECHPALMENGEWÄCHSE
Brech-Hülse

FAMILLE DU HOUX
Houx vomitif

AQUIFOLIACEAE
Ilex vomitoria Aiton

Description: 20 to 25 ft. (6.0 to 7.5 m) high and rarely more than 6 in. (0.15 m) in diameter. Trunk slender, often inclined. More frequently a many-stemmed shrub. Leaves elliptic, slightly toothed to serrate, evergreen. Flower parts in fours. Sepals obtuse. Fruits very numerous, bright red.

Habitat: Mostly near salt water along the Atlantic coast (as far north as southern Virginia) and the North American Gulf coast.

Wood: Heavy, hard, and close-grained; turning yellow on exposure to light. Sapwood thick, pale. Numerous distinct medullary rays.

Use: Branches bearing fruit are used as winter decoration in the towns of the northern USA. The leaves were much sought after by the Indians, who used to make an emetic and laxative infusion from them.

Beschreibung: 6,0 bis 7,5 m hoch, bei selten mehr als 0,15 m Durchmesser. Stamm schlank, oft schräg. Häufiger als vielstämmiger Busch. Blätter elliptisch, gezähnelt bis gesägt. Immergrün. Blütenorgane vierzählig. Kelchblätter stumpf. Früchte sehr zahlreich, hell rötlich.

Vorkommen: Meist in der Nähe von Salzwasser entlang der Atlantikküste (bis ins südliche Virginia) und der nordamerikanischen Golfküste.

Holz: Schwer, hart und von feiner Textur; unter Lichteinwirkung gelb werdend. Kern fast weiß; Splint dick, hell. Zahlreiche deutliche Markstrahlen.

Nutzung: Früchte tragende Äste werden in den Städten der nördlichen USA als winterliche Dekoration verwendet. Bei den Indianern war ein Tee aus den Blättern äußerst begehrt, der das Erbrechen fördert und abführend wirkt.

Description: arbuste de 6,0 à 7,5 de hauteur, avec un tronc fin, souvent oblique, ne dépassant p 0,15 m de diamètre. Se présente plus souvent sous forme d'arbri seau touffu. Feuilles elliptiques, à marge dentée ou acérée, persi tantes. Fleur tétramère, à sépale obtus. Fruits très nombreux, rougeâtre clair.

Habitat: le plus souvent près de eaux salines le long du littoral atlantique (jusque dans le sud c la Virginie) et dans les Etats du Golfe en Amérique du Nord.

Bois: lourd, dur et à grain fin; ja nissant à la lumière. Bois de cœ presque blanc; aubier épais et c Nombreux rayons médullaires distincts.

Intérêts: les rameaux enjolivés de fruits sont utilisés l'hiver à des fins décoratives dans les vill septentrionales des Etats-Unis. Indiens employaient les feuilles pour préparer une infusion éme tique et purgative.

TRANSVERSE SECTION.

RADIAL SECTION.

TANGENTIAL SECTION.

H | HONEYSUCKLE FAMILY
Blueberry Elder

FAMILIE DER GEISSBLATTGEWÄCHSE
Blaubeeriger Holunder

FAMILLE DU CHÈVREFEUILLE
Sureau noir bleu, Sureau bleu, Sureau velu

CAPRIFOLIACEAE
Sambucus cerulea Raf. var. *cerulea*
Syn.: *S. glauca* Nutt.

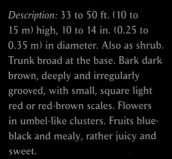

Description: 33 to 50 ft. (10 to 15 m) high, 10 to 14 in. (0.25 to 0.35 m) in diameter. Also as shrub. Trunk broad at the base. Bark dark brown, deeply and irregularly grooved, with small, square light red or red-brown scales. Flowers in umbel-like clusters. Fruits blueblack and mealy, rather juicy and sweet.

Habitat: Valleys, on rather dry, gravely soils in western North America. Best developed in western Oregon. Rare as a tree further north, in the eastern Cascade Mountains, and in Sierra Nevada.

Wood: Light, soft, weak, and coarsegrained. Heartwood yellow and brown; sapwood thin and paler.

Use: Occasionally as ornamental tree. The fruits (the best in the genus) used in pies and preserves.

Beschreibung: 10 bis 15 m hoch bei 0,25 bis 0,35 m Durchmesser. Auch als Strauch. Stammbasis verbreitert. Borke dunkelbraun, tief geteilt in unregelmäßige Rillen und in kleine, quadratische, anliegende, leicht rot oder rotbraun getönte Schuppen. Blüten in Scheindolden. Früchte blauschwarz mit mehligem Anflug, ziemlich saftig und süß.

Vorkommen: Täler auf ziemlich trockenem, kiesigem Boden im westlichen Nordamerika. Erreicht die Bestform in West-Oregon. Weiter nördlich und im Osten des Kaskadengebirges sowie in der Sierra Nevada selten als Baum.

Holz: Leicht, weich, schwach und mit grober Textur. Kern gelb und braun getönt; Splint dünn und heller.

Nutzung: Gelegentlich als Zierbaum. Die Früchte, die besten der Gattung, für Pasteten und zum Einkochen.

Description: 10 à 15 m de hauteur sur un diamètre de 0,25 à 0,35 m. Sous forme de buisson également. Tronc élargi à la base Ecorce brun sombre, creusée de sillons irréguliers et recouverte de petites écailles adhérentes, à section carrée, d'un rouge léger brun rouge. Fleurs groupées en c rymbes. Fruits noir bleu légèrem farineux, plutôt juteux et doux.

Habitat: sols de plaine graveleux, plutôt secs dans l'ouest de l'Amérique du Nord. Optimum de développement dans l'ouest d l'Oregon. Rarement à port arborescent plus au nord et à l'est de chaîne des Cascades ainsi que d la Sierra Nevada.

Bois: léger, tendre, faible et à gr grossier. Bois de cœur teinté de jaune et de brun; aubier mince et plus clair.

Intérêts: parfois planté pour l'orr ment. Les fruits, les meilleurs du genre, sont utilisés en pâtisserie mis en conserve.

Elder, Pale Elder, Elderberry.

TRANSVERSE SECTION.

RADIAL SECTION.

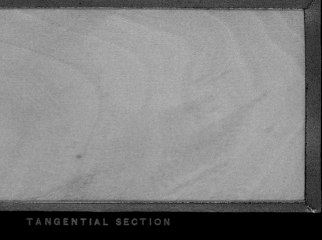

TANGENTIAL SECTION

Ger. Holunder.   Fr. Sureau.   Sp. Sauco.

H | HONEYSUCKLE FAMILY
Elder, Elderberry

FAMILIE DER GEISSBLATTGEWÄCHSE
Mexikanischer Holunder

FAMILLE DU CHÈVREFEUILLE
Sureau noir d'Amérique, Sureau du Canada

CAPRIFOLIACEAE
Sambucus mexicana C. Presl ex DC.

Description: 25 to 30 ft. (7.5 to 9.0 m) high and 12 in. (0.3 m) in diameter. Trunk short, with strong, spreading branches, forming a compact, rounded crown. Bark thin, pale brown, tinged red and peeling into long strips. Young shoots and leaves finely hairy. Leaves with five leaflets, the latter ovate-lanceolate, with toothed margin, dark yellow-green. Flowers creamy white, in flat, long-stalked inflorescences. Fruit almost black, juicy.

Habitat: Lowland and riverbanks on moist, stony loams. From Central America to the south of North America.

Wood: Light, soft, and coarse-grained. Heartwood pale brown; sapwood thin (two to three annual rings), paler. Many distinct medullary rays.

Use: Attractive decorative tree, with its thick crown foliage and large inflorescences. Often planted near houses in north Mexico and Lower California as a shade tree and for its fruit.

Beschreibung: 7,5 bis 9 m hoch bei 0,3 Durchmesser. Stamm kurz, mit kräftigen, ausladenden Ästen, die eine kompakte, runde Krone bilden. Borke dünn, hellbraun mit roter Tönung und sich in langen Streifen abschälend. Junge Sprosse und Blätter fein behaart. Blätter in 5 Fiedern geteilt, Fiedern oval-lanzettlich, mit gesägtem Rand, dunkel gelbgrün. Blüten cremeweiß, in flachen, lang gestielten Blütenständen. Frucht fast schwarz, saftig.

Vorkommen: Tiefländer und Flussufer auf feuchtem, von Steinen durchsetztem Lehm. Von Mittelamerika bis in den Süden Nordamerikas verbreitet.

Holz: Leicht, weich und mit grober Textur. Kern hellbraun; Splint dünn (zwei bis drei Jahresringe), heller. Zahlreiche deutliche Markstrahlen.

Nutzung: Die dicht belaubte Krone und die großen Blütenstände machen die Pflanze zu einem reizvollen Ziergehölz, das in Nord-Mexiko und Niederkalifornien oft in der Nähe von Häusern als Schattenspender und zur Obstgewinnung angepflanzt wird.

Description : 7,5 à 9,0 m de haute sur un diamètre de 0,3 m. Tronc court, avec des branches vigoureuses et étalées, formant un houppier rond et compact. Ecorce mince, d'un brun clair teinté de rouge, s'exfoliant en longues lanières. Jeunes pousses et feuilles finement pubescentes. Feuilles ovales-lancéolées, à marge dentée, ve jaune sombre. Fleurs blanc crème en corymbes étalés, longuement pédonculés. Fruits presque noirs, juteux.

Habitat : basses-terres et berges dans des sols argileux-caillouteux humides. Répandu depuis l'Amérique centrale jusque dans le sud de l'Amérique du Nord.

Bois : léger, tendre et à grain grossier. Bois de cœur brun clair ; aubier mince (2 à 3 cernes annuels), plus clair. Rayons médu laires nombreux et distincts.

Intérêts : bel arbre ornemental à cause de son houppier dense et s grands corymbes ; planté près des maisons dans le nord du Mexiqu et en Basse-Californie pour son ombrage et ses fruits.

TRANSVERSE SECTION.

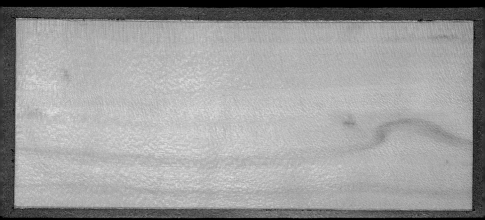

RADIAL SECTION.

TANGENTIAL SECTION.

Ger. Mexicanischer Holunder. *Fr.* Sureau de Mexico.

H | HORSE CHESTNUT FAMILY

Californian Buckeye

FAMILIE DER ROSSKASTANIENGEWÄCHSE
Kalifornische Rosskastanie

FAMILLE DU MARRONNIER
Marronnier de Californie, Pavier de Californie

HIPPOCASTANACEAE
Aesculus californica (Spach.) Nutt.

Description: 33 to 40 ft. (10 to 12 m) high and 2 to 3 ft. (0.6 to 0.9 m) in diameter. Also as thicket-forming shrub, 10 to 16 ft. (3 to 5 m) high. Trunk short, stocky, often twice the width at the base.

Habitat: Riverbanks in California.

Wood: Soft, light, and very close-grained. Heartwood white to pale yellow; sapwood thinner, scarcely distinguishable.

Use: Grown in the Pacific states and occasionally in southern and western Europe, for its decorative inflorescences.

Beschreibung: 10 bis 12 m hoch bei 0,6 bis 0,9 m Durchmesser. Auch als 3 bis 5 m hoher Strauch, Dickichte bildend. Kurzer, gedrungener Stamm, an der Basis oft um das Doppelte verbreitert.

Vorkommen: Ufer von Wasserläufen in Kalifornien.

Holz: Weich, leicht und mit sehr feiner Textur. Kern weiß bis schwach gelb; Splint dünner, kaum unterscheidbar.

Nutzung: Aufgrund der zierenden Blütenstände in den pazifischen Staaten, in Süd- und Westeuropa (nicht oft) in Kultur.

Description: 10 à 12 m de hauteur sur un diamètre de 0,6 à 0,9 m. Existe aussi sous une forme buissonnante de 3 à 5 m de hau constituant des fourrés. Tronc court, trapu, souvent deux fois plus large à la base.

Habitat: berges des cours d'eau de la Californie.

Bois: tendre, léger et à grain très fin. Bois de cœur blanc à jaune faible; aubier plus mince, à pein distinct.

Intérêts: cultivé dans les Etats de façade pacifique, quelquefois au en Europe méridionale et occide tale en raison de sa belle floraiso en panicules.

California Buckeye.

TRANSVERSE SECTION.

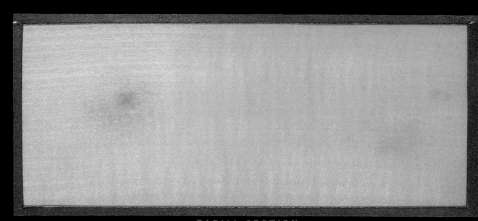

RADIAL SECTION.

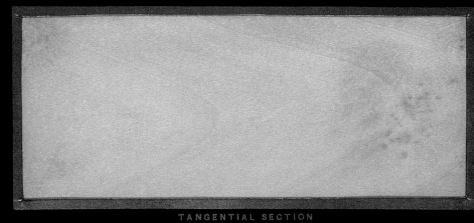

TANGENTIAL SECTION

Gr. Californianische Roszkastanie.　　*Fr.* Marronier de Calif

Sp. Esculo Californiano.

H | HORSE CHESTNUT FAMILY
Horse Chestnut

FAMILIE DER ROSSKASTANIENGEWÄCHSE
Rosskastanie

FAMILLE DU MARRONNIER
Marronnier

HIPPOCASTANACEAE
Aesculus hippocastanum L.

Description: 80 ft. (25 m) and 2 to 3 ft. (0.6 to 0.9 m) in diameter. Decorative white flowers flecked with red, in pyramidal clusters ("candles"). Abundant crop of green spiny fruits in autumn, bursting open when ripe. The large seeds (known as "conkers" in Britain), colored Titian-red to brown, have a large paler "eye."

Habitat: Forms forests in northern Greece, Bulgaria, and Albania. As early as 1567 it had reached the Imperial Court in Vienna from the court of the Sultan in Constantinople through the envoy Busbeck. Introduced into North America in the middle of the 18th century and since then naturalized in many places.

Wood: Light and soft.

Use: For artificial limbs and concealed wood in furniture. In Europe, a winter food for game. Rich in water-soluble bitter substances. After their removal, used as flour and food in emergencies.

Related to the Ohio Buckeye (*Aesculus glabra* Willd.), native to North America. Many varieties. A widespread park and garden tree.

Beschreibung: 25 m hoch bei 0,6 bis 0,9 m Durchmesser. Dekorative weiße, rot gefleckte Blüten in Pyramiden („Kerzen"). Reiche Bildung von grünen, bestachelten Fallfrüchten, aufspringend. Der große Samen ist tizianrot bis braun gemasert mit einem großen, helleren „Auge".

Vorkommen: Waldbildend in Nord-Griechenland, Bulgarien, Albanien. Schon 1567 vom Hofe des Sultans in Konstantinopel durch den Gesandten Busbeck an den Kaiserhof in Wien gelangt. In Nordamerika seit Mitte des 18. Jh. eingeführt und seither vielerorts eingebürgert.

Holz: Leicht und weich.

Nutzung: Für Prothesen, Blindholz im Möbelbau. In Europa zur Wildfütterung im Winter. Reich an wasserlöslichen Bitterstoffen. Nach deren Entfernung: Notmehl, Notnahrung.

Mit *Aesculus glabra* Willd., dem in Nordamerika heimischen Ohio-Buckeye, verwandt. Viele Varietäten. Weitverbreiteter Park- und Gartenbaum.

Description : 25 m de hauteur sur un diamètre de 0,6 à 0,9 m. Belles fleurs blanches, tachées de rouge et groupées en grandes panicules dressées. Fruits verts, couverts d'aiguillons, qui libèrent une ou deux graines, dont la plus grosse est brun cuivré ou marron avec une grande marque pâle.

Habitat : endémique du nord de la Grèce, de la Bulgarie et de l'Albanie, où il constitue des peuplements forestiers. Découvert Constantinople et rapporté à Vienn par Busbeck dès 1567. Introduit en Amérique du Nord au milieu du 18 siècle et naturalisé aujourd'hui dan nombre de régions.

Bois : léger et tendre.

Intérêts : utilisé en menuiserie d'ameublement (bois de bâtis) et pour fabriquer des prothèses. Rich en substances amères solubles dan l'eau. Comestible (farine et nourri ture en cas de disette) lorsqu'il est débarrassé de ces substances.

Proche parent d'*Aesculus glabra* Willd., originaire d'Amérique du Nord. Nombreuses espèces voi sines. Arbre très répandu dans les parcs et les jardins.

Horse Chestnut.

TRANSVERSE SECTION.

RADIAL SECTION.

TANGENTIAL SECTION.

Ger. Roszkastanie.   **Sp.** Castaño de caballo.

Fr. Marronier d'Inde.

H | HORSE CHESTNUT FAMILY
Ohio Buckeye, Fetid Buckeye

FAMILIE DER ROSSKASTANIENGEWÄCHSE
Ohio-Rosskastanie

FAMILLE DU MARRONNIER
Marronnier glabre, Marronnier de l'Ohio

HIPPOCASTANACEAE
Aesculus glabra Willd.

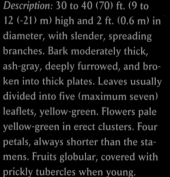

Description: 30 to 40 (70) ft. (9 to 12 (-21) m) high and 2 ft. (0.6 m) in diameter, with slender, spreading branches. Bark moderately thick, ash-gray, deeply furrowed, and broken into thick plates. Leaves usually divided into five (maximum seven) leaflets, yellow-green. Flowers pale yellow-green in erect clusters. Four petals, always shorter than the stamens. Fruits globular, covered with prickly tubercles when young.

Habitat: Rich, moist plains formed by the lower reaches of rivers or on riverbanks in Eastern North America.

Wood: Light, soft, not strong, and close-grained. Heartwood almost white, often veined; sapwood (10 to 12 annual rings) darker!

Use: Is not distinguished in commerce from *Aesculus octandra*, and like this species is mainly used in the manufacture of artificial limbs, for which purpose it remains unsurpassed among all the North American timbers. It is also used for household objects, hats, and in paper-making. It was introduced into England in 1821. Not a very attractive chestnut and so only rarely cultivated.

Beschreibung: 9 bis 12 (bis 21) m hoch bei 0,6 m Durchmesser und schlanken ausladenden Ästen. Borke mäßig dick, aschgrau, tiefrissig und in dicke Schollen zerfallend. Blätter meist in 5 (maximal 7) Blättchen geteilt, gelbgrün. Blüten blass gelbgrün in aufrechten Kerzen. 4 Kronblätter, die kürzer als die Staubblätter bleiben. Früchte kugelig, jung mit stacheligen Warzen.

Vorkommen: Reiche, feuchte Erde in den Ebenen der Flussunterläufe oder an Flussufern im östlichen Nordamerika.

Holz: Leicht, weich, nicht fest und mit feiner Textur. Kern fast weiß, oft von dunklen, fauligen Linien durchzogen; Splint (zehn bis zwölf Jahresringe) dunkler!

Nutzung: Wird im Handel nicht von *Aesculus octandra* unterschieden und wie dieses vor allem zur Herstellung von Prothesen verwendet, für das es unter allen nordamerikanischen Hölzern unübertroffen bleibt. Ferner für hölzerne Haushaltsgegenstände, Hüte und zur Papierherstellung verwendet. 1821 nach England eingeführt. Wenig reizvolle Kastanie und daher nur selten kultiviert.

Description : arbre de 9 à 12 m de hauteur (21 m max.) avec un tro de 0,6 m de diamètre, portant de branches fines et étalées. Ecorce moyennement épaisse, gris cend à texture profondément fissurée en plaques épaisses. Feuilles à 5 folioles (7 au maximum), vert jaune. Fleurs vert jaune en panicules dressées, à étamines dépassant de la corolle composée de 4 pétales. Fruits globuleux, à verrues épineuses.

Habitat : sols riches, humides des plaines alluviales ou des berges de l'est de l'Amérique du Nord.

Bois : léger, tendre, non résistant à grain fin. Bois de cœur presque blanc, souvent parcouru de ligne sombres, putrides ; aubier (10 à cernes annuels) plus sombre !

Intérêts : équivalent à *Aesculus octandra* dans le commerce et util de la même manière, en particul pour fabriquer des prothèses, ce en quoi aucun autre arbre nordaméricain ne peut le concurrenc Egalement ustensiles de cuisine, papier et chapeaux. Introduit en Angleterre en 1821. Marronnier peu attrayant, et peu cultivé par conséquent.

TRANSVERSE SECTION

RADIAL SECTION.

TANGENTIAL SECTION.

Ger. Ranzige-Kastanie. *Fr.* Marronnier fetide.

Sp. Castaño fetido.

H | HORSE CHESTNUT FAMILY
Sweet Buckeye, Yellow Buckeye

FAMILIE DER ROSSKASTANIENGEWÄCHSE
Gelbe Rosskastanie

FAMILLE DU MARRONNIER
Marronnier jaune, Marronnier à fleurs jaunes, Pavier jaune

HIPPOCASTANACEAE
Aesculus flava Aiton
Syn.: *A. octandra* R. S. Marsh

Description: 90 ft. (27 m) high and 2 ft. 6 in. to 3 ft. (0.75 to 0.9 m) in diameter. Small, almost pendant branches. Bark thick, dark brown, divided up by fissures, the surface separating into thin scales. Leaves composed of five to seven leaflets. The pale or deep yellow flowers (in one variety also purple or red) open when the leaves are half grown, and are arranged in erect, candle-like inflorescences 1 to 2 in. (12.5 to 17.5 cm) high. Fruits 2 to 3 in. (5 to 7.5 cm) long, with two thin, smooth, or somewhat rough valves.

Habitat: On rich soils by the lower reaches of rivers and on damp slopes in the interior of Eastern North America.

Wood: Light, soft, and close-grained; difficult to split. Heartwood creamy white, sapwood barely distinguishable. Numerous, medullary rays.

Use: A favorite garden tree, particularly the variety with reddish flowers, and less susceptible to fungal attack than the European species. Its inflorescences are less attractive and the tree usually smaller than its European relative.

Beschreibung: 27 m hoch bei 0,75 bis 0,9 m Durchmesser. Kleine, fast herabhängende Äste. Borke dick, dunkelbraun, durch Risse gefeldert. Oberfläche sich in dünnen Schuppen abschälend. Blätter aus 5 bis 7 Blättchen zusammengesetzt. Die blass- oder dunkelgelben (in einer Varietät auch violetten oder roten) Blüten öffnen sich, wenn die Blätter halb ausgewachsen sind, in aufrechten, kerzenartigen Blütenständen von 12,5 bis 17,5 cm Höhe. Früchte 5–7,5 cm lang und von 2 dünnen, glatten oder etwas rauen Valven umgeben.

Vorkommen: Auf reichen Böden an Flussunterläufen und an feuchten Hängen im Inneren des östlichen Nordamerika.

Holz: Leicht, weich und mit feiner Textur; schwer spaltbar. Kern cremeweiß, Splint kaum unterscheidbar. Zahlreiche Markstrahlen.

Nutzung: Besonders die Varietät mit rötlichen Blüten ist ein beliebter Gartenbaum, zumal die Art weniger anfällig gegen Pilzbefall ist als die europäische. Ihre Blütenstände sind weniger attraktiv und der Baum bleibt meistens kleiner als sein europäischer Verwandter.

Description : arbre de 27 m de hauteur, avec un tronc de 0,75 à 0,9 m de diamètre. Petites branch presque retombantes. Ecorce épaisse, brun sombre, à texture fissurée et craquelée, s'exfoliant fines écailles. Feuilles composées de 5 à 7 folioles. Fleurs jaune pâl ou foncé (il existe une variété à fleurs violettes ou rouges) éclosa lorsque les feuilles sont à moitié déployées, groupées en panicules dressées de 12,5 à 17,5 cm de ha Bogue de 5 à 7,5 cm de long, liss ou légèrement rugueuse, avec 2 valves renfermant les graines.

Habitat : sols riches dans le cours inférieur des fleuves et pentes humides à l'intérieur des régions est de l'Amérique du Nord.

Bois : léger, tendre et à grain fin ; difficile à fendre. Bois de cœur blanc crème, aubier à peine distinct. Rayons médullaires nombreux.

Intérêts : la variété à fleurs rouges est très appréciée dans les jardins (moins sensible aux maladies que le marronnier européen). Ses infl rescences sont moins attrayantes sa taille est inférieure à celle de s parent européen.

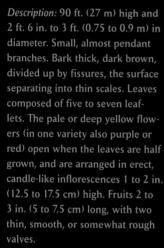

Yellow Buckeye. Sweet Buckeye.

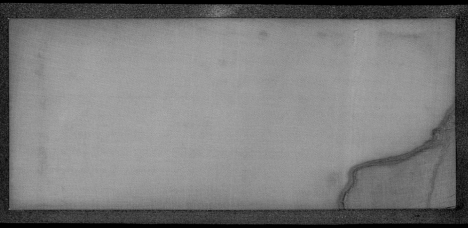

TRANSVERSE SECTION

RADIAL SECTION.

TANGENTIAL SECTION.

Ger. Gelbe Rosskastanie.   *Fr.* Marronnier jaune.

Sp. Castano de caballo amarillo.

I | INDIAN ALMOND FAMILY
Black Mangrove, Blackwood, Green Turtle Bough

FAMILIE DER FLÜGELSAMENGEWÄCHSE
Parawa

FAMILLE DU BADAMIER
Avicennia noir

COMBRETACEAE
Avicennia germinans (L.) L.
Syn.: *A. nitida* Jacq.

Description: 60 to 70 ft. (18 to 21 m) high and rarely 2 to 3 ft. (0.6 to 0.9 m) in diameter. Trunk short, with spreading branches, which form a broad, round crown, but mostly smaller and not more than 20 to 30 ft. (6 to 9 m) high, in the north also a low shrub. Bark moderately thick, roughened by numerous dark brown scales tinged red. When these fall, the orange-red inner bark becomes visible. Leaves oblong or lanceolate-elliptic, thick and leathery, dark green. Flowers in short inflorescences with a pointed outline, appearing throughout the year. Fruits formed only in Florida, and then rarely.

Habitat: Shores and riverbanks of Florida and the Mississippi delta, forming large thickets together with Rhizophora mangle. Also widespread across the Antilles as far as Brazil.

Wood: Very heavy, hard, and rather coarse-grained. Numerous medullary rays and eccentric annual rings, marked by a few rows of large, open pores.

Use: Railroad sleepers (Bärner, 1962).

Beschreibung: 18 bis 21 m hoch bei selten 0,6 bis 0,9 m Durchmesser. Stamm kurz, mit ausladenden Ästen, die eine breite runde Krone bilden, meist aber kleiner und nicht höher als 6 bis 9 m, im Norden auch als niedriger Busch. Borke mäßig dick, durch dunkelbraune Schüppchen mit roter Tönung aufgeraut. Nach deren Verlust wird die orangerote innere Rinde sichtbar. Blätter länglich oder lanzettlich-elliptisch, dick und ledrig, dunkelgrün. Blüte ganzjährig, in kurzen Blütenständen von zugespitztem Umriss. Früchte werden in Florida nur selten gebildet.

Vorkommen: Strände und Flussufer Floridas und des Mississippi-Deltas, zusammen mit *Rhizophora mangle* große Dickichte bildend. Weitverbreitet über die Antillen bis nach Brasilien.

Holz: Sehr schwer, hart, und von eher grober Textur. Zahlreiche Markstrahlen und exzentrische Jahresringe, die durch einige Reihen großer, offener Gefäße markiert werden.

Nutzung: Eisenbahnschwellen (Bärner 1962).

Description : arbre de 18 à 21 m de hauteur, avec un tronc court atteignant rarement 0,6 à 0,9 m diamètre et des branches étalées formant un houppier ample et rond, mais d'ordinaire plus petit, pas plus de 6 à 9 m de haut, et dans le nord sous forme de buisson bas. Ecorce moyennement épaisse rendue rugueuse par des petites écailles brun sombre teinté de rouge, qui laissent apparaître dessous une écorce rouge orange. Feuilles allongées à lancéolées-elliptiques, épaisses et coriaces, vert foncé. Floraison tout au long de l'année, en courtes inflorescences terminées en pointe. Fructification rare en Floride.

Habitat : plages et rives des cours d'eau de la Floride et du delta du Mississippi ; constitue avec *Rhizophora mangle* de vastes fourrés. Très répandu des Antilles jusqu'au Brésil.

Bois : très lourd, dur et à grain plutôt grossier. Rayons médullaires et cernes annuels excentriques, marqués de quelques alignements de grands vaisseaux ouverts.

Intérêts : traverses de chemin de fer (Bärner 1962).

BLACK MANGROVE. NATIVE OAK (JAMAICA).

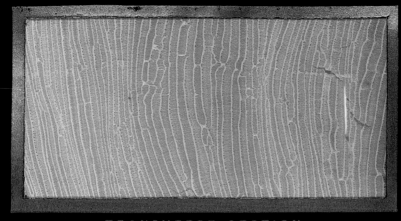

TRANSVERSE SECTION.

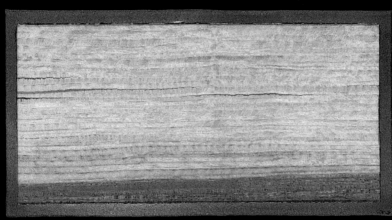

RADIAL SECTION.

TANGENTIAL SECTION.

Ger. Schwarzer Mangelbaum. Fr. Palétuvier blanc (Fr.W.I.).

Sp. Mangle bobo (Sp.W.I.). Palo de sal. Culumata (Cent.Amer.).

I | INDIAN ALMOND FAMILY

Buttonwood, Florida Buttonwood, Button Mangrove

FAMILIE DER FLÜGELSAMENGEWÄCHSE
Zaragoza-Mangrove

FAMILLE DU BADAMIER
Conocarpe dressé, Arbre bouton, Palétuvier, Manglier gris

COMBRETACEAE
Conocarpus erectus L.

Description: 40 to 60 ft. (12 to 18 m) high and 2 ft. 6 in. to 3 ft. 6 in. (0.5 to 0.75 m) in diameter. Branches slender, forming a narrow, regular crown, occasionally also a shrub with almost creeping stems. Bark dark brown, with irregular furrows. Leaves oblong, dark green, which leave behind large, circular scars after they fall. Flowers in globular heads, produced throughout the year.

Habitat: Grows together with *Rhizophora mangle* on low-lying, muddy, tidal soils in lagoons and bays on tropical coasts, e.g. the Antilles, Central and South America. In North America confined to southern Florida, where it is common.

Wood: Heavy, hard, dense, and strong; polishable. Olive-brown (Begemann, 1981 to 1994).

Use: Firewood and charcoal (Bärner, 1962). Cultivated in England in 1755.

Beschreibung: 12 bis 18 m hoch bei 0,5 bis 0,75 m Durchmesser. Schmale Äste, die eine schmale, gleichmäßige Krone bilden, gelegentlich auch als Strauch mit fast kriechenden Stämmen. Borke dunkelbraun, von unregelmäßigen Rissen durchzogen. Blätter länglich, dunkelgrün, hinterlassen nach dem Laubfall große kreisrunde Narben. Blüten ganzjährig in kugeligen Köpfchen.

Vorkommen: Zusammen mit *Rhizophora mangle* auf tief liegenden, schlammigen Tidenböden in Lagunen und Buchten an tropischen Küsten z. B. der Antillen, Mittel- und Südamerikas. Nordamerika nur in Süd-Florida erreichend, dort aber häufig.

Holz: Schwer, hart, dicht und fest; gut polierfähig. Olivbraun (Begemann 1981 bis 1994).

Nutzung: Brennholz, Holzkohle (Bärner 1962). Wurde 1755 in England kultiviert.

Description: 12 à 18 m de haute sur un diamètre de 0,5 à 0,75 m Branches étroites, constituant u houppier régulier et étroit; que quefois aussi sous forme de bui à port presque rampant. Ecorce brun sombre, parcourue de fiss irrégulières. Feuilles allongées, foncé, laissant de grandes cicatr rondes sur les rameaux après le chute. Floraison toute l'année, e petits fleurs globuleuses.

Habitat: vases côtières profonde dans les lagunes et les baies des côtes tropicales comme celles des Antilles, de l'Amérique cen trale et du Sud, en association a *Rhizophora mangle*. En Amériqu du Nord, seulement dans le sue la Floride où il est très répandu

Bois: lourd, dur, dense et résist susceptible de prendre un beau poli; brun olivâtre (Begemann à 1994).

Intérêts: combustible et charbon de bois (Bärner 1962). Cultivé en Angleterre en 1755.

FLORIDA BUTTONWOOD. BUTTON MANGROVE.

TRANSVERSE SECTION.

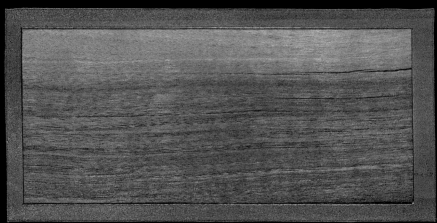

RADIAL SECTION.

TANGENTIAL SECTION.

Ger. Knopfbaum.　　　　Fr. Conocarpe droit. Palétuvier.

Sp. Mangle boton (Sp.W.I.). Mangle prieto (Mex.)

I | INDIAN ALMOND FAMILY

White Buttonwood, White Mangrove

FAMILIE DER FLÜGELSAMENGEWÄCHSE
Weiße Mangrove

FAMILLE DU BADAMIER
Manglier blanc

COMBRETACEAE
Laguncularia racemosa (L.) Gaertn. f.

Description: 30 to 60 ft. (9 to 18 m) high and 12 to 18 in. (0.3 to 0.5 m) in diameter. Branches stout, spreading, forming a narrow, round crown. Also grows as a low shrub on the northern limit of its range. Bark thin, brown tinged red, and broken into long, thick pieces. Leaves ovate to broadly ovate, at first slightly tinged red, but soon becoming dark green, leaf stalks red. Flowers in loose spikes, with short petals, densely covered with hairs, half as long as the ovoid, keeled fruits. Flowering throughout the whole year.

Habitat: Grows together with *Rhizophora mangle* and *Conocarpus erectus* on the muddy, tidal shores of tropical bays and lagoons. In North America confined to southern Florida.

Wood: Moderately heavy, hard, strong, somewhat brittle, and liable to break. Heartwood grayish yellow to grayish pink; the sapwood is somewhat paler.

Use: Many and diverse, such as for building timber, construction work, coffins, posts, and small items of everyday use.

Beschreibung: 9 bis 18 m hoch bei 0,3 bis 0,5 m Durchmesser. Kräftige, ausladende Äste, die eine schmale runde Krone bilden. An der Nordgrenze der Verbreitung auch als niedriger Busch. Borke dünn, braun mit roter Tönung und in lange, wulstige Stücke zerreißend. Blätter eiförmig bis breit oval, anfangs schwach rot getönt, bald aber dunkelgrün mit roten Blattstielen. Blüten in lockeren Ähren, mit kurzen Kronblättern, dicht mit Haaren bedeckt. Halb so lang wie die ovalen, gekielten Früchte. Während des ganzen Jahres blühend.

Vorkommen: Zusammen mit *Rhizophora mangle* und *Conocarpus erecta* auf schlammigen Gezeitenufern tropischer Buchten und Lagunen. Nordamerika nur in Süd-Florida erreichend.

Holz: Mittelschwer, hart, fest, etwas spröde und nicht schlagfest. Kern graugelb bis graurosa; Splint etwas heller.

Nutzung: Vielfältig, z. B. als Bauholz, für Konstruktionen, Särge, Pfosten und kleine Gebrauchsgegenstände.

Description: arbre de 9 à 12 m de hauteur, avec un tronc de 0,3 à 0,5 m de diamètre, armé de branches vigoureuses, étalées, constituant un houppier étroit et rond. Se rencontre aussi dans le nord de son aire sous forme de buisson. Ecorce mince, brune teintée de rouge et se craquelan▪ en longs secteurs protubérants. Feuilles ovoïdes à larges-ovales, d'une teinte rougeâtre au début mais devenant bientôt vert fonc▪ et portées par des pétioles rouge▪ Fleurs disposées en épis légers, avec des pétales courts, très velo▪ tés, et deux fois plus petits que ▪ fruits ovales, carénés. Floraison ▪ au long de l'année.

Habitat: associé à *Rhizophora mangle* et à *Conocarpus erecta* da▪ les vasières des baies et des lagu▪ tropicales. En Amérique du Nor▪ dans le sud de la Floride.

Bois: moyennement lourd, dur, résistant, un peu cassant; ne ri▪ pas au choc. Bois de cœur jaune gris à rose gris; aubier un peu plus clair.

Intérêts: bois d'œuvre, constructions, cercueils et petits objets d'usage courant.

White Mangrove.   White Buttonwood.

TRANSVERSE SECTION.

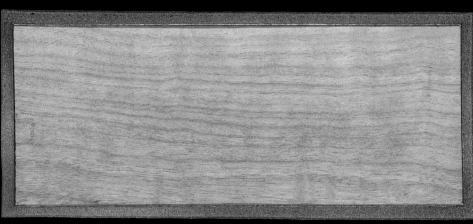

RADIAL SECTION.

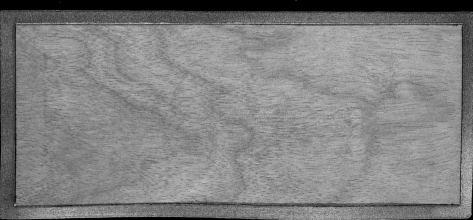

TANGENTIAL SECTION.

I | IVY FAMILY

Hercules' Club, Angelica Tree, Devil's Walking Stick

FAMILIE DER ARALIENGEWÄCHSE, EFEUGEWÄCHSE
Herkuleskeule

FAMILLE DU GINSENG
Gourdin d'Hercule, Bâton du diable

ARALIACEAE
Aralia spinosa L.

Description: Rarely more than 30 to 40 ft. (10 or 12 m) high, if that, and 4 to 5 in. (0.1 to 0.15 m) in diameter. A prickly, small tree or shrub, with stiff stems, some arising from the roots, with no or few branches. Leaves doubly pinnate, long-stalked, the largest in the tree's range. Flowers numerous, in many-branched clusters of umbels. Fruits take the form of small, blue berries.

Habitat: Deep, moist soils near rivers in eastern North America.

Wood: Light, soft, brittle, and hollow in the center. Heartwood brown, streaked with yellow; sapwood paler.

Use: Root bark and berries used as a diuretic.

A garden curiosity, with close relatives in Manchuria and Japan. Reached England from Virginia towards the end of the 17th century.

Beschreibung: Selten mehr als 10 bis 12 m hoch, wenn überhaupt, bei vielleicht 0,1 bis 0,15 m Durchmesser. Stacheliger Kleinbaum oder Strauch mit unverzweigten oder wenig verzweigten, z. T. wurzelbürtigen Stämmchen, starr. Laub doppelt gefiedert, lang gestielt, das größte im Verbreitungsgebiet. Zahllose Blüten in mehrfach verzweigten Dolden („Sträußen"). Früchte in Form kleiner blauer Beeren.

Vorkommen: Tiefer, feuchter Boden in Flussnähe im östlichen Nordamerika.

Holz: Leicht, weich und brüchig. Mit zentraler Markhöhle. Stichweise braun mit Gelb, hellerer Splint.

Nutzung: Rinde der Wurzel und Beeren als Diuretika.

Gartenschaustück. Nahe Verwandte in der Mandschurei und in Japan. Gegen Ende des 17. Jh. von Virginia aus nach England eingeführt.

Description : rarement plus de 10 12 m de hauteur sur un diamètre d'environ 0,1 à 0,15 m. Arbuste épineux ou buisson présentant de petites tiges - certaines sont des rejets de souche - peu ou non ramifiées, à port rigide. Feuilles t découpées, à long pétiole, la plus grande dans la zone de propagation. Fleurs nombreuses et groupées en ombelles («bouquets»). Petites baies bleues.

Habitat : sols profonds et humide à proximité des cours d'eau dans l'est de l'Amérique du Nord.

Bois : léger, tendre et cassant. Ca médullaire central. Bois de cœur tacheté de brun avec du jaune ; aubier plus clair.

Intérêts : l'écorce des racines et les baies sont employées comme diurétique.

Plante d'expositions. Proches parents en Mandchourie et au Japon. Découverte en Virginie et connue en Angleterre dès la fin du 17ᵉ siècle.

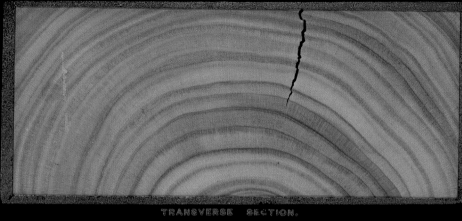

TRANSVERSE SECTION.

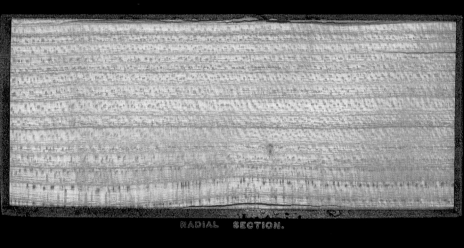

RADIAL SECTION.

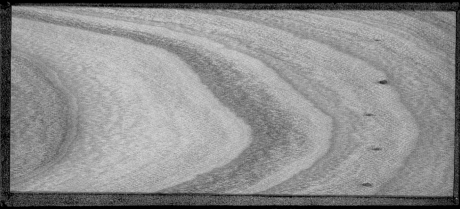

TANGENTIAL SECTION.

Ger. Dornige Bergangelike.   Sp. Aralia espinosa.

Fr. Aralia épineuse.

L | LAUREL FAMILY
Avocado, Avocado Pear, Butter Pear

FAMILIE DER LORBEERGEWÄCHSE
Avocadobaum, Avocado-Birne

FAMILLE DU LAURIER
Avocatier d'Amérique, Perséa d'Amérique,
Poire d'alligator

LAURACEAE
Persea americana Mill.

Description: A tree up to 50 ft. (15 m) high with wide-spreading branches, and a trunk up to 2 ft. 6 in. (0.5 m) in diameter. The bark of young trees is yellowish green and covered with silky hairs, later it becomes pale gray and increasingly fissured. The evergreen leaves are ovate to elliptic. Both sides are dark green. The flowers are arranged in panicles. The fruits are oblong, more or less pear-shaped, with a very oily, edible flesh.

Habitat: Its native area is in the Caribbean. Since its introduction into North America, the Avocado has escaped from cultivation and now grows wild in many parts of southern North America.

Wood: The pale reddish brown wood is easy to work.

Use: The tree is cultivated for its economically important fruits, but its wood is used only to a very

Beschreibung: Bis zu 15 m hoher Baum mit weit ausgebreiteten Ästen und einem bis zu 0,5 m dicken Stamm. Die Borke junger Bäume ist gelbgrün und seidig behaart, später wird sie hellgrau und zunehmend rissiger. Die immergrünen Blätter sind oval bis elliptisch. Beide Seiten sind dunkelgrün. Die Blüten sind in Rispen angeordnet. Die Früchte sind mehr oder weniger birnenförmig-länglich mit einem sehr ölreichen, essbaren Fleisch.

Vorkommen: Seine Heimat liegt in der Karibik. Nach seiner Einführung in Nordamerika verwilderte die Avocado in weiten Teilen des südlichen Nordamerika.

Holz: Das hell rötlich braune Holz ist leicht zu bearbeiten.

Nutzung: Der Baum wird wegen seiner ökonomisch sehr bedeutsamen Früchte kultiviert. Das Holz wird nur in sehr geringem Umfang

Description : arbre s'élevant jusqu 15 m, avec des branches très éta lées et un tronc pouvant atteind 0,5 m de diamètre. Ecorce vert jaune et veloutée à l'état juvénil devenant gris clair et se fissuran de plus en plus avec l'âge. Feuill persistantes, ovales à elliptiques, vert foncé sur les deux faces. Fle groupées en panicules. Le fruit est une volumineuse baie plus o moins piriforme, allongée, conte nant une chair comestible et trè oléagineuse.

Habitat : originaire des Caraïbes. Introduit en Amérique du Nord puis retourné à l'état sauvage ur peu partout dans le sud des Etat Unis.

Bois : brun rougeâtre clair et faci à travailler.

Intérêts : l'arbre est cultivé pour ses fruits d'une grande valeur éc nomique. Le bois présente moin

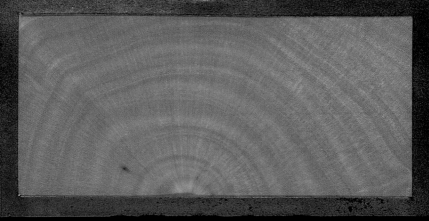

TRANSVERSE SECTION.

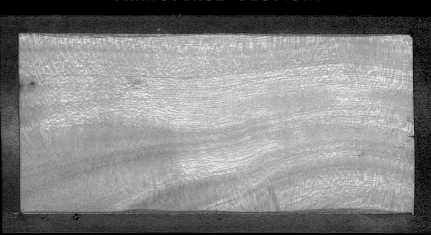

RADIAL SECTION.

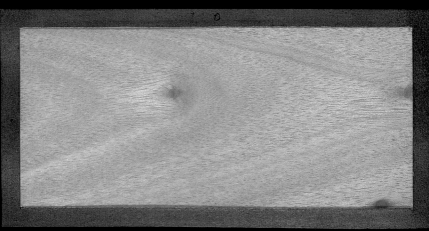

TANGENTIAL SECTION.

Gr. Advocaten. Fr. Avocatier. Sp. Aguacate.

L | LAUREL FAMILY

California Laurel, California Olive, California Bay,
Spice Tree

FAMILIE DER LORBEERGEWÄCHSE
Kalifornischer Lorbeer

FAMILLE DU LAURIER
Umbellaria de Californie, Laurier de Californie,
Myrte de l'Oregon

LAURACEAE
Umbellularia californica (Hook. & Arn.) Nutt.

Description: 85 to 100 ft. (25 to 30 m) high and 4 to 5 ft. (1.2 to 1.5 m) in diameter. Trunk sometimes tall and straight, sometimes dividing into secondary trunks. Shrubby at high altitude and in southern California. Bark dark brown, with red tinge, with layered, appressed scales. Dark, dense crown with shiny foliage. Ornamental. Attractive yellow or orange colors in fall.

Habitat: Banks of watercourses and on low hills with vertical rock strata (allows better root penetration to the deep water table). Widespread in Pacific North America from south-west Oregon, to California. Best in south-west Oregon where it forms woods with Bigleaf Maple.

Wood: Heavy, hard, strong, close-grained.

Use: Very valuable wood for interior construction and fine furniture. The leaves are distilled to produce a pungent oil.

Introduced by Douglas to European parks in 1826.

Beschreibung: Höhe 25 bis 30 m hoch bei 1,2 bis 1,5 m Durchmesser. Stamm teilweise hoch und gerade, bei anderen dagegen früh in Sekundärstämme geteilt. In hohen Gebirgslagen und in Süd-Kalifornien als Strauch. Borke dunkelbraun, rot getönt, in geschichteten, anliegenden Schuppen. Dunkle, dichte Krone mit glänzendem Laub. Ornamental. Laubverfärbung, gelb oder orangefarben.

Vorkommen: Ufer von Wasserläufen und niedrige Hügel mit vertikaler Gesteinsschichtung. Dies ermöglicht den Wurzeln ein besseres Vordringen an tiefe Wasserhorizonte. Im pazifischen Nordamerika von Südwest-Oregon bis Kalifornien. Im Südwesten von Oregon mit dem Breitblättrigen Ahorn waldbildend und in Bestform.

Holz: Schwer, hart, fest, mit feiner Textur.

Nutzung: Sehr wertvolles Holz für Innenausbau und feine Möbel. Die Blätter werden durch Destillation zu einem stechend riechenden Öl verarbeitet.

Durch Douglas 1826 in europäischen Parks angesiedelt.

Description: 25 à 30 m de hauteur sur un diamètre de 1,2 à 1,5 m. Tronc soit élancé et droit, soit très vite divisé en tiges secondaires. Port de buisson en altitude et dans le sud de la Californie. Ecorce d'un brun sombre teinté de rouge recouverte d'écailles stratifiées, adhérentes. Houppier sombre et dense. Feuillage brillant, ornemental, jaune ou orangé à l'automne.

Habitat: berges et basses collines à stratification verticale permettant aux racines d'accéder aux réserves hydriques du sol. Répandu sur la façade pacifique du sud-ouest de l'Oregon à la Californie. Dans le sud-ouest de l'Oregon, il constitue des forêts avec l'érable à grandes feuilles.

Bois: lourd, dur, résistant et à grain fin.

Intérêts: bois très précieux utilisé en menuiserie intérieure et pour fabriquer de beaux meubles. De ses feuilles on extrait par distillation une essence à l'odeur pénétrante.

Planté dans les parcs européens depuis son introduction en 1826 par Douglas.

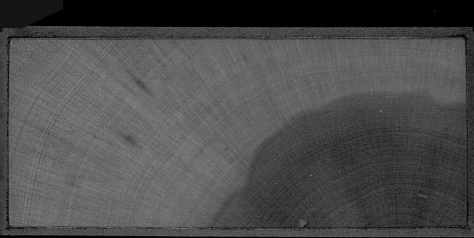

TRANSVERSE SECTION.

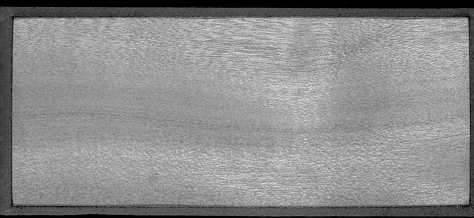

RADIAL SECTION.

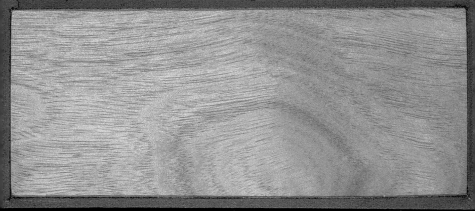

TANGENTIAL SECTION

L | LAUREL FAMILY

Lancewood, Sweetwood

FAMILIE DER LORBEERGEWÄCHSE
Florida-Lanzenholz

FAMILLE DU LAURIER
Nectandra de Floride

LAURACEAE
Ocotea coriaceae (Sw.) Britton
Syn.: *O. catesbyana* (Michx.) Sarg.

Description: 20 to 30 ft. (6 to 9 m) high and rarely more than 8 in. (0.2 m) in diameter. Branches slender, forming a narrow, round crown. Bark dark red-brown, with numerous excrescences. Leaves oblong-lanceolate, entire, at first delicate and light green, later thick, leathery, and dark green above, pale on the underside. Flowers in panicles, petals creamy white. Fruits ovoid or subovoid, lustrous dark blue to almost black. Calyx and fruit stalks red.

Habitat: On mainland shores and islands of southern Florida. Abundant. Also occurs on the Bahamas and probably on the Antilles.

Wood: Heavy, hard, and close-grained. Heartwood rich dark brown; sapwood thick (20 to 30 annual rings), pale yellow. With numerous thin medullary rays and many small, evenly distributed, open pores.

Use: The dense foliage, numerous panicles of flowers, and lustrous fruits of this small tree recommend it for cultivation in tropical gardens.

Beschreibung: 6 bis 9 m hoch bei selten mehr als 0,2 m Durchmesser. Äste schlank, eine schmale runde Krone bildend. Borke dunkel rotbraun, mit zahlreichen Auswüchsen. Blätter länglich-lanzettlich, ganzrandig, anfangs zart und hellgrün, später dick, ledrig und dunkelgrün, Unterseite blass. Blüten in Rispen, Kronblätter cremeweiß. Früchte oval oder kurz-oval, glänzend dunkelblau bis fast schwarz. Kelch und Fruchtstiele rot.

Vorkommen: An Stränden des Festlandes und auf Inseln Süd-Floridas. Häufig. Ferner auch auf den Bahamas und vermutlich auch auf den Antillen.

Holz: Schwer, hart und von feiner Textur. Kern tief dunkelbraun; Splint dick (20 bis 30 Jahresringe), hellgelb. Mit zahlreichen dünnen Markstrahlen und vielen kleinen, gleichmäßig verteilten offenen Gefäßen.

Nutzung: Das kräftige Laub des kleinen Baumes, die zahlreichen Blütenrispen und die glänzenden Früchte empfehlen ihn für die Kultivierung in tropischen Gärten.

Description : arbuste de 6 à 9 m d hauteur, avec un tronc mesuran rement plus de 0,2 m de diamèt et des ramifications fines, consti tuant un houppier étroit et rond Ecorce brun rouge foncé, couver d'excroissances. Feuilles allongée lancéolées, entières, délicates et vert jaune jeunes, puis épaisses, coriaces et vert sombre, pâles à l face inférieure. Fleurs en panicu avec des pétales blanc crème. Fruits ovales ou courts-ovales, bl sombre à presque noir et brillan Calice et pédoncules rouges.

Habitat : plages sur la terre ferme et îles du sud de la Floride. Très répandu. On le trouve aussi aux Bahamas et probablement aux Antilles.

Bois : lourd, dur et à grain fin. Bc de cœur brun profond ; aubier épais (20 à 30 cernes annuels), jaune clair. Nombreux rayons médullaires fins et grand nombre de petits vaisseaux ouverts, répa tis régulièrement.

Intérêts : conseillé pour les jardin tropicaux à cause de son feuillag vigoureux, de ses nombreuses pa cules et ses fruits brillants.

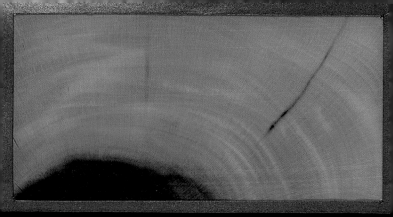

TRANSVERSE SECTION.

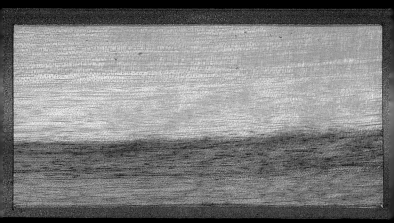

RADIAL SECTION.

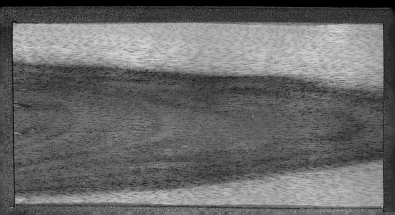

L | LAUREL FAMILY
Red Bay

FAMILIE DER LORBEERGEWÄCHSE
Roter Lorbeerbaum

FAMILLE DU LAURIER
Laurier rouge, Avocatier bourbon, Avocatier de Caroline

LAURACEAE
Persea borbonia (L.) Sprengel

Description: Up to 60 to 70 ft. (18 to 21 m) high and 2 to 3 ft. (0.75 to 0.9 m) in diameter, but normally smaller. Branches stout, erect, forming a dense crown that is a source of shade. Bark moderately thick, dark red, deeply furrowed. Leaves oblong, thick, reddish at first, later bright green, remaining on the branches throughout the year. Flowers two or three in cymes, gathered into small panicles. Fruits globular to subovoid, dark blue to almost black.

Habitat: Frequent on the margins of streams and swamps, together with other broad-leaved trees in rich, moist ground but also on dry, sandy soils shaded by forests of long-needled pines. Its range includes the coastal region of (south-) eastern North America, along the Gulf coast, and the southwestern Mississippi basin.

Wood: Heavy, hard, very strong, rather brittle, and close-grained; polishable.

Use: Occasionally for cabinets and interior construction, formerly used also in building ships and boats. Introduced into England as early as 1739, later hardly cultivated at all.

Beschreibung: 18 bis 21 m hoch bei 0,75 bis 0,9 m Durchmesser. Normalerweise allerdings kleiner. Kräftige, aufrechte Äste, die eine dichte, schattende Krone bilden. Borke mäßig dick, dunkelrot, tiefrissig. Blätter länglich, dick, anfangs rötlich, später hellgrün, das ganze Jahr über am Ast bleibend. Blüten in Zymen zu 2 bis 3 in kleinen Blütenständen. Früchte kugelig bis kurz oval, dunkelblau bis fast schwarz.

Vorkommen: Häufig an den Rändern von Flüssen und Sümpfen, zusammen mit anderen Laubbäumen in reicher, feuchter Erde, aber auch auf trockenen, sandigen Böden im Schatten von Wäldern langnadeliger Kiefern. In der Küstenregion des (süd-) östlichen Nordamerika, entlang der Golfküste und im südwestlichen Mississippi-Becken verbreitet.

Holz: Schwer, hart, sehr fest, eher brüchig und mit feiner Textur; polierfähig.

Nutzung: Gelegentlich für Kabinette und im Innenausbau, früher auch im Schiff- und Bootsbau. Bereits 1739 in England eingeführt, später kaum noch kultiviert.

Description: arbre de 18 à 21 m de hauteur sur un diamètre de 0,75 à 0,9 m, mais généralemen plus petit. Branches vigoureuses et dressées, houppier dense et ample. Ecorce rouge sombre, moyennement épaisse et profond dément fissurée. Feuilles allongé épaisses, d'abord rougeâtres pui vert clair, demeurant sur l'arbre toute l'année. Petites inflorescer composées de 2 à 3 cymes. Frui globuleux à ovales-courts, bleu sombre à presque noirs.

Habitat: souvent sur sols riches humides au bord des cours d'ea des marais, en compagnie d'autr feuillus, mais aussi sur sols sable neux secs, sous le couvert de fo de conifères à longues aiguilles. Répandu sur le littoral dans le (sud-) est de l'Amérique du Nor le long du Golfe et dans le sud-ouest du bassin du Mississippi.

Bois: lourd, dur, très résistant, p tôt cassant et à grain fin; pouvar se polir.

Intérêts: ébénisterie et menuiser intérieure, autrefois aussi constr tion navale. Introduit dès 1739 e Angleterre mais plus guère culti par la suite.

TRANSVERSE SECTION

RADIAL SECTION,

L | LAUREL FAMILY
Sassafras

FAMILIE DER LORBEERGEWÄCHSE
Sassafras

FAMILLE DU LAURIER
Sassafras officinal, Laurier-sassafras (Canada)

LAURACEAE
Sassafras albidum (Nutt.) Nees
Syn.: *S. sassafras* (L.) Karst.

Description: Sometimes 85 to 100 ft. (25 to 30 m) high and 4 to 6 ft. (1.2 to 1.8 m) in diameter. Tends to develop several stems, also found in the form of a shrub. Produces root suckers, which quickly occupy abandoned agricultural land in the southern and Gulf states. Leaves ornamental, variable in shape, beautifully colored in the fall.

Habitat: Dry locations in eastern North America, undemanding.

Wood: Soft and brittle; very durable.

Use: Includes buckets and pumps, fence posts, railings.

Outstandingly beautiful at all seasons of the year and perhaps the first North American tree to come to Europe, although rare in European parks.

Beschreibung: Manchmal 25 bis 30 m hoch bei 1,2 bis 1,8 m Durchmesser. Neigt zur Mehrstämmigkeit, auch Strauchform. Wurzelbürtige Sprosse. Diese besetzen aufgegebenes Ackerland in Kürze (Süd- und Golfstaaten). Ornamentales, vielgestaltiges Laub. Farbenschönheit.

Vorkommen: Trockene Lagen im östlichen Nordamerika, stellt geringe Ansprüche.

Holz: Weich und spröde; sehr dauerhaft.

Nutzung: U. a. für Eimer und Pumpen, Zaunpfähle, Geländer.

Zu allen Jahreszeiten eine auffallende Schönheit und vielleicht der erste nordamerikanische Baum in Europa, allerdings selten in europäischen Parkanlagen.

Description : arbre atteignant parf jusqu'à 25, 30 m de hauteur, ave un tronc de 1,2 à 1,8 m de diamètre, ayant tendance à se divise à la base. Il existe aussi sous une forme buissonnante et rejette de souche. Les drageons colonisent rapidement les terres arables lais sées à l'abandon (Etats du Sud et du pourtour du golfe du Mexiqu Feuilles polymorphes, prenant une belle teinte jaune et pourpre à l'automne.

Habitat : essence peu exigeante q se contente de milieux secs dans l'est de l'Amérique du Nord.

Bois : tendre et cassant ; très durable.

Intérêts : utilisé pour fabriquer de seaux et des pompes à eau, des pieux de clôture et des balustrad etc.

Beauté incomparable en toutes saisons ; peut-être la première essence forestière nord-américain à avoir été introduite en Europe. Arbre de parcs encore peu répan cependant en Europe.

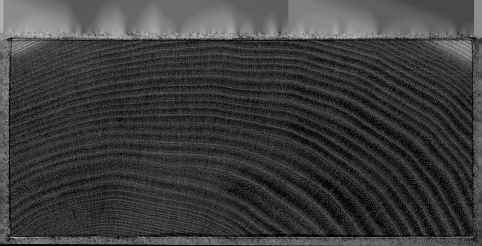

TRANSVERSE SECTION.

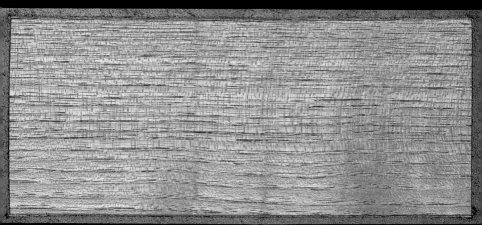

RADIAL SECTION.

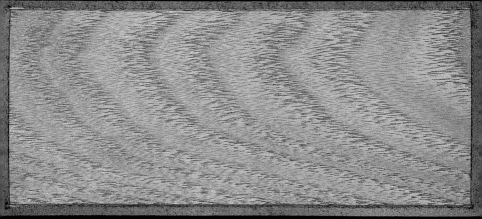

TANGENTIAL SECTION.

Ger. Fenchelholz. *Fr.* Laurier des Iroquois. *Sp.* Sasafras

L | LAUREL FAMILY
Swamp Bay

FAMILIE DER LORBEERGEWÄCHSE
Sumpf-Avocado

FAMILLE DU LAURIER
Avocatier palustre

LAURACEAE
Persea palustris (Raf.) Sarg.
Syn.: *P. pubescens* (Pursh) Sarg.

Description: 35 to 40 ft. (10 to 12 m) high and 18 in. (0.45 m) in diameter. Branches straight. Foliage dense.

Habitat: Marshland, with Swamp Cypress, Red Maple, gums (*Nyssa* spp.), Water Ash, Overcup Oak and Laurel Oak, among others along the Gulf and Atlantic coasts in southeastern North America. Commoner to the south in the marshes of the "pine-barrens," where it forms almost homogeneous stands.

Wood: Rather heavy, soft, and strong.

Use: Similar to Red Bay, for interior construction, furniture, and formerly for boatbuilding.

Beschreibung: 10 bis 12 m hoch bei 0,45 m Durchmesser. Äste gerade. Dichtlaubig.

Vorkommen: Sümpfe in Gesellschaft mit Sumpfzypresse, Rot-Ahorn, den Tupelos, Wasser-Esche, Leier-Eiche und Lorbeer-Eiche u. a. entlang der Golf-und Atlantikküsten im südöstlichen Nordamerika. Südwärts häufiger in den Sümpfen der sog. „pine-barrens", dort auch nahezu allein bestandbildend.

Holz: Ziemlich schwer, weich und stark.

Nutzung: Anwendung bei entsprechenden Abmessungen wie bei der Roten Avocado für Innenausbau, Möbel, früher auch Bootsbau.

Description : petit arbre de 10 à 12 m de hauteur sur un diamètr[e] de 0,45 m. Branches droites. Feu[il]lage dense.

Habitat : marécages le long du G[olfe] et de la côte atlantique dans le sud-est de l'Amérique du Nord, [en] association avec le taxode chauv[e,] l'érable rouge, les nyssas, le frên[e] de la Caroline, le chêne à feuille[s] lyrées, le chêne à feuilles de laur[ier] etc. Plus fréquent vers le sud da[ns] les marécages des «pine barrens[»] où il constitue aussi des peuplements presque purs.

Bois : assez lourd, tendre et fort.

Intérêts : même utilisation, si les mesures sont adéquates, que po[ur] l'avocatier rouge : menuiserie intérieure, meubles et constructi[on]

Swamp Red Bay.

TRANSVERSE SECTION.

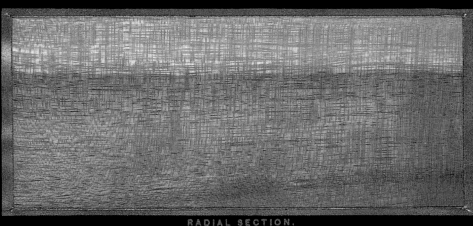

RADIAL SECTION.

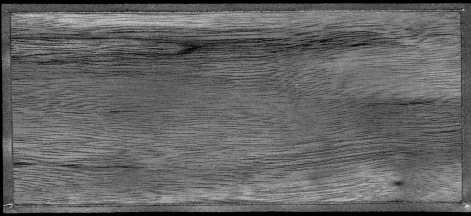

TANGENTIAL SECTION

Ger. Rother Lorberbaum.   *Fr.* Persea de Carolina.

Sp. Laurel colorado.

L | LEGUME FAMILY
Blackwood

FAMILIE DER SCHMETTERLINGSBLÜTLER, LEGUMINOSEN
Schwarzes Holz

FAMILLE DU LÉGUME OU DES LÉGUMINEUSES
Mimosa à bois noir

LEGUMINOSAE
Acacia melanoxylon R. Br. ex W. T. Aiton

Description: A tall, quick-growing tree with reddish brown bark. The evergreen leaves are made up of numerous leaflets, and have two spiny stipules at their base for some time after the leaves have sprouted. The leaf-stalks broaden and spread out as they become older. The flowers are pale yellow and arranged in clusters. The pods are oblong and twisted, often bending into a circle.

Habitat: Its native country is Australia. The tree was first introduced into North America through California, where it escaped into the wild and soon became completely naturalized.

Wood: The rich, dark chocolate-brown heartwood is moderately heavy, hard, strong, very durable, and polishable. It is surrounded by sapwood that is whitish with an olive tinge.

Use: An extremely useful wood that is suitable for many purposes, including construction, joinery, and turnery. The tree was introduced into California as an ornamental

Beschreibung: Großer, schnell-wüchsiger Baum mit rotbrauner Borke. Die immergrünen Blätter bestehen aus zahlreichen Blattfie-dern und tragen an ihrer Basis bis einige Zeit nach dem Laubaustrieb zwei Dornen. Die Blattstiele ver-breitern sich im Alter spreitenartig. Die Blüten sind hellgelb und in Trauben angeordnet. Die Schoten sind länglich und in sich gewun-den, oft sogar kreisförmig gebogen.

Vorkommen: Seine Heimat ist Aus-tralien. Der Baum wurde in Nord-amerika zuerst nach Kalifornien eingeführt, wo er verwilderte und bald vollständig eingebürgert war.

Holz: Das kräftig dunkel schoko-ladenbraune Kernholz ist mäßig schwer, hart, fest, sehr dauerhaft und gut polierbar. Es wird von oliv-weißlichem Saftholz umgeben.

Nutzung: Äußerst vielseitig ein-setzbares Nutzholz, sowohl für Bauzwecke als auch Tischlerei und Drechslerei geeignet. Der Baum wurde als Ziergehölz nach Kalifor-nien eingeführt.

Description : grand arbre à crois-sance rapide et écorce brun roug Les feuilles sont persistantes, composées de nombreuses folio et garnies à l'état juvénile de de stipules épineuses. Les pétioles s'élargissent et s'aplatissent avec l'âge. Les fleurs sont groupées e grappes jaune clair. Les fruits so des gousses allongées, enroulée sur elles-mêmes et prenant mên une forme circulaire.

Habitat : son pays d'origine est l'Australie. Introduit aux Etats-Unis, en Californie d'abord, il s' échappé des jardins et est deven bientôt subspontané.

Bois : moyennement lourd, dur, sistant ; très durable et susceptib d'être poli. Bois de cœur couleu chocolat ; aubier blanc olivâtre.

Intérêts : bois utile se prêtant à d très nombreux usages, que ce so la construction, l'ébénisterie ou tournage. Introduit en Californie comme espèce d'ornement.

Black-wood.

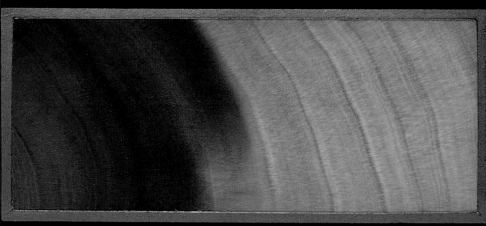

TRANSVERSE SECTION.

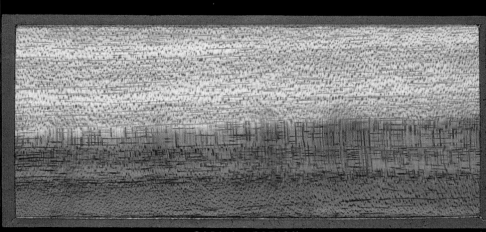

RADIAL SECTION.

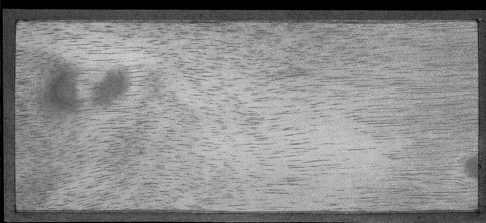

TANGENTIAL SECTION

Ger. Schwarzes Holz. *Fr.* Bois noir. *Sp.* Madera negra.

L | LEGUME FAMILY

False Acacia, Black Locust, Yellow Locust

FAMILIE DER SCHMETTERLINGSBLÜTLER, LEGUMINOSEN
Falsche Akazie, Robinie

FAMILLE DU LÉGUME OU DES LÉGUMINEUSES
Robinier faux-acacia, Acacia blanc, Faux-acacia

LEGUMINOSAE
Robinia pseudacacia L.

Description: 105 to 115 ft. (32 to 35 m) high and 3 to 4 ft. (0.9 to 1.2 m) in diameter. Found especially on poorer, e. g. gravely, soils, with abundant root suckers and daughter stems. Bark deeply furrowed, forming conspicuous, raised scales. Flowers white, in pendent clusters, with a penetratingly sweet scent.

Habitat: Native to the Allegheny Mountains up to 4,000 ft. (1,300 m), but later extensively planted and naturalized.

Wood: Heavy, hard, and extremely strong; very durable.

Use: In shipbuilding, for fence posts, and in turnery. Tree nails. Formerly exported to England in large numbers. Nowadays the wood is used for weatherproof housefronts. It is also a food plant for bees.

Since 1791 it has been recommended in Germany as a valuable timber for forestry (Medikus 1792). Its ability to propagate itself is useful in covering slag-heaps with greenery. However it is problematic from the conservation point of view, as it displaces low-growing plants on poor soils.

Beschreibung: 32 bis 35 m hoch bei 0,9 bis 1,2 m Durchmesser. Besonders auf ärmeren, z. B. kiesigen Böden sehr reichlich sprossende Wurzelbrut und Tochterstämme. Borke tiefrinnig, grubig mit starken, hohen Riefen. Weiße, eindringlich süß duftende Blüten in hängenden Trauben.

Vorkommen: Ursprünglich in den Alleghenies bis 1300 m, später aber weithin gepflanzt und eingebürgert.

Holz: Schwer, hart und außerordentlich fest; sehr dauerhaft.

Nutzung: Im Schiffbau, für Zaunpfähle, Drechslerarbeiten, sehr haltbar. Holznägel. Früher Export der Nägel in großen Mengen nach England, Holz heute für wetterfeste Hausfassaden. Bienenfutterpflanze.

In Deutschland seit 1791 als wertvoller Nutzholzlieferant zum Waldbau empfohlen (Medikus 1792). Hohes Ausschlagsvermögen nützlich für Haldenbegrünung, wegen der Verdrängung niedrigwüchsiger Vegetation auf armen Böden aus Naturschutzsicht mitunter problematisch.

Description: 32 à 35 m de haute sur un diamètre de 0,9 à 1,2 m. Forte capacité à drageonner et à donner des tiges filles, en partic lier dans les sols les plus pauvre Ecorce à texture crevassée, form un réseau de grosses crêtes à sommet arrondi. Grappes de fle blanches très parfumées.

Habitat : spontané à l'origine dar les Alleghanys jusqu'à 1300 m, fut ensuite planté et s'est natura dans d'autres endroits.

Bois : lourd, dur et extrêmement résistant ; très durable.

Intérêts : utilisé dans la construction navale, en tournerie, pour l fabrication de pieux de clôture e de clous en bois. Les clous étaie jadis exportés en grandes quant vers l'Angleterre. Bois employé à jourd'hui pour les façades expos aux intempéries. Plante mellifèr

En Allemagne, essence conseillé en forêt de production depuis 1 (Medikus 1792). Son drageonnement vigoureux est utile pour boiser les crassiers désaffectés ; sols pauvres supplante la végéta tion basse ce qui est problématic pour la protection de la nature.

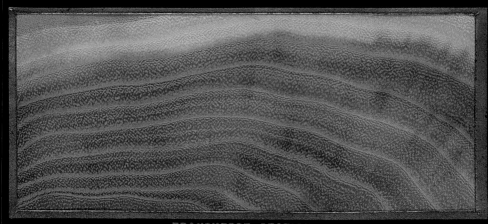

TRANSVERSE SECTION.

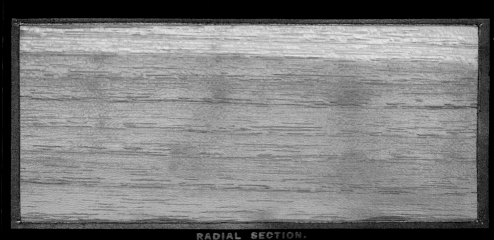

RADIAL SECTION.

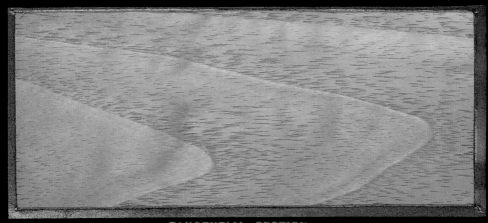

TANGENTIAL SECTION.

Gemine Acacie.   Fr. Robinier faux-acacia.   Sp. Acacia.

ticularly on very poor soils.

Wood: The heartwood is only moderately hard, heavy and strong.

Use: Very good as firewood. The bark contains numerous tannins, which were used in the tanning process. The fibers of the bark are used in the manufacture of paper. The flowers are sold at markets as decoration.

Main Use: the afforestation of poor soil, by reason of its undemanding nature and rapid growth.

eingeführt, wo er sich vor allem auf sehr armen Böden rasch einbürgerte.

Holz: Das Kernholz ist nur mäßig hart, schwer und fest.

Nutzung: Sehr gutes Feuerholz. Das in der Borke enthaltene Tannin fand in der Gerberei Verwendung. Die Borkenfasern eignen sich zur Papierherstellung. Die Blüten werden auf Märkten als Schmuck verkauft. Sehr gut geeignet zur Aufforstung armer Böden.

Bois : moyennement dur, lourd et résistant.

Intérêts : excellent bois de feu. L'écorce, riche en tannins, était employée en tannage et les fibres entrent dans la fabrication de la pâte à papier. Les fleurs sont vendues comme décoration sur le marchés. Mais surtout c'est une essence de reboisement des sols pauvres car il croît rapidement et est peu exigeant.

Green Wattle, Black Wattle.

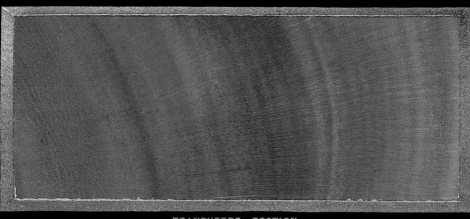

TRANSVERSE SECTION.

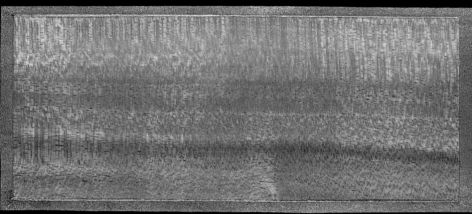

RADIAL SECTION.

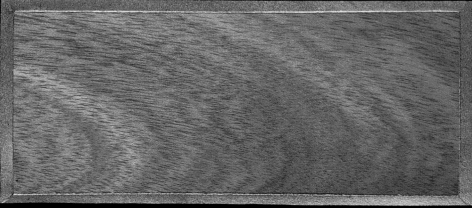

TANGENTIAL SECTION.

L | LEGUME FAMILY
Honey Locust

FAMILIE DER SCHMETTERLINGSBLÜTLER,
LEGUMINOSEN
Christus-Dorn, Amerikanische Gleditschie

FAMILLE DU LÉGUME OU DES LÉGUMINEUSES
Févier à trois épines, Arbre aux escargots

LEGUMINOSAE
Gleditsia triacanthos L.

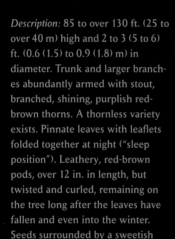

Description: 85 to over 130 ft. (25 to over 40 m) high and 2 to 3 (5 to 6) ft. (0.6 (1.5) to 0.9 (1.8) m) in diameter. Trunk and larger branches abundantly armed with stout, branched, shining, purplish red-brown thorns. A thornless variety exists. Pinnate leaves with leaflets folded together at night ("sleep position"). Leathery, red-brown pods, over 12 in. in length, but twisted and curled, remaining on the tree long after the leaves have fallen and even into the winter. Seeds surrounded by a sweetish pulp.

Habitat: Moist lowland in association with oaks, hickories, Black Walnut, Nettle Tree, Yellow and Ohio Buckeye. Originally only west of the Allegheny Mountains and in the Mississippi Valley, but has since become naturalized over a much wider area.

Wood: Heavy and strong; very durable.

Use: For railroad sleepers, posts, and agricultural implements. The flowers make it a food plant for bees.

A park tree in the USA, where it makes hedges, and in Europe.

Beschreibung: 25 bis über 40 m hoch bei 0,6 (1,5) bis 0,9 (1,8) m Durchmesser. Stamm und größere Äste gespickt mit gewaltigen, verzweigten, glänzend purpurrotbraunen Dornen. Unbedornte Abart bekannt. Fiederlaub mit Blättchen zur Nacht zusammengefaltet („Schlafstellung") und tagsüber ausgebreitet. Über 30 cm lange, ledrige, rotbraune Hülsen, jedoch verwunden und gedreht und auch am entlaubten Baum noch lange bleibend („Wintersteher"). Samen in einem süßlichen Fruchtfleisch.

Vorkommen: Feuchtes Flachland in Gesellschaft von Eichen, Hickories, Schwarznuss, Zürgelbaum, Gelber und Ohio-Rosskastanie. Ursprünglich nur westlich der Alleghenies und im Mississippi-Tal, inzwischen weit darüber hinaus eingebürgert.

Holz: Schwer und fest; sehr dauerhaft.

Nutzung: Für Eisenbahnschwellen, Pfähle und Ackerbaugerät. Bienenfutterpflanze (Blüten).

Parkbaum in den USA und in Europa. In Amerika für undurchdringliche Hecken.

Description: 25 à 40 m de hauteu sur un diamètre de 0,6 (1,5) à 0, (1,8 m). Le tronc et les branches maîtresses sont armés d'épines r; mifiées vigoureuses, brun rougeâ et lustrées. Variété inerme égale-ment. Feuilles pennées, composé de petites folioles qui se replient position de «sommeil» au crépus cule et s'ouvrent le jour. Fruits br rouge en longues gousses dures et pendantes, spiralées à maturit et qui demeurent sur les rameau. tout l'hiver. Graines enchâssées dans une chair sucrée et comes-tible.

Habitat: plaines humides en asso ciation avec les chênes, les carye le noyer noir, le micocoulier, le marronnier glabre et le marronn jaune. S'est implanté entre-temps bien au-delà de son aire d'origine qui se situait à l'ouest des Allegh nys et dans la vallée du Mississip

Bois: lourd et ferme; très durable

Intérêts: utilisé pour la fabricatior de traverses de chemin de fer, de pieux et d'outils agricoles. Espèce mellifère (fleurs).

Arbre de parcs aux Etats-Unis et en Europe. Aussi pour les haies.

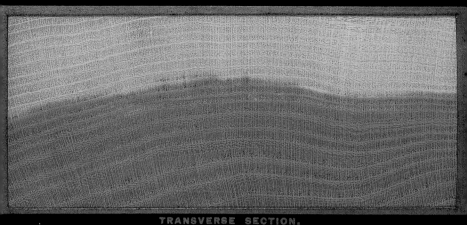

TRANSVERSE SECTION.

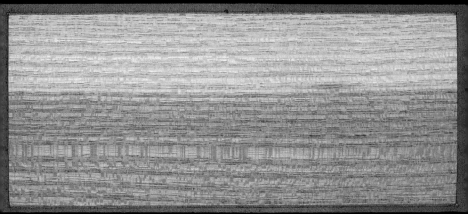

RADIAL SECTION.

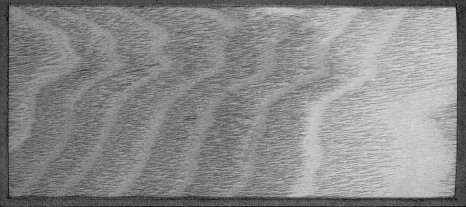

TANGENTIAL SECTION.

L | LEGUME FAMILY
Jamaica Dogwood, Fish Poison

FAMILIE DER SCHMETTERLINGSBLÜTLER,
LEGUMINOSEN
Jamaika-Hundsholz

FAMILLE DU LÉGUME OU DES LÉGUMINEUSES
Piscidia de la Jamaïque, Bois à enivrer,
Mort-à-poissons de Floride

LEGUMINOSAE
Piscidia piscipula (L.) Sarg.
Syn.: *Ichthyomethia piscipula* (L.) Hitchc. ex Sarg.

Description: 40 to 50 ft. (12 to 15 m) high and 2 to 3 ft. (0.6 to 0.9 m) in diameter, with stout, erect branches, which form an irregular crown. Bark thin, with a light brown surface divided into small square scales. Leaves pinnate. Flowering before the leaves. Flowers in loose panicles as much as 10 to 12 in. (25 to 30 cm) long, or more frequently in compact inflorescences only 2 to 3 in. (5 to 7.5 cm) long. Fruits light brown, 3 to 4 in. (7.5 to 10 cm) long, with conspicuous, thin, papery wings on the sides.

Habitat: Shores and coastal strips, and islands of the northern Caribbean. In North America, confined to southern Florida.

Wood: Dark gray when freshly cut, becoming light chocolate-brown in the air. Heartwood and sapwood not distinguishable.

Use: The roots were used as a poison to catch fish. Introduced into England as early as 1690.

Beschreibung: 12 bis 15 m hoch bei 0,6 bis 0,9 m Durchmesser und kräftigen, aufrechten Ästen, die eine unregelmäßige Krone bilden. Borke dünn, mit hellbrauner Oberfläche, die in kleine quadratische Schuppen gefeldert ist. Blätter gefiedert. Blüte vor dem Laubaustrieb. Blüten in maximal 25 bis 30 cm langen, lockeren Rispen oder öfter in nur 5 bis 7,5 cm langen, dichten Blütenständen. Früchte hellbraun, 7,5 bis 10 cm lang, seitlich mit auffallenden, papierartig dünnen Flügeln.

Vorkommen: Strände, Küstenstreifen und Inseln der Nordkaribik. Nordamerika nur in Süd-Florida erreichend.

Holz: An der Luft hell schokoladenbraun werdend. Frisch dunkelgrau, Kern und Splint nicht unterscheidbar.

Nutzung: Wurzeln als Gift zum Fischfang. Bereits 1690 nach England eingeführt.

Description: arbre de 12 à 15 m de hauteur, avec un tronc de 0,6 0,9 m de diamètre et des branche vigoureuses, dressées, constituant un houppier irrégulier. Ecorce mince, brun clair, à texture craquelée en petites écailles carrées. Feuilles composées pennées. Fleu apparaissant avant la feuillaison, en panicules légères de 25 à 30 cm maximum de longueur ou plu souvent en inflorescences denses de 5 à 7,5 cm de long. Fruits brur clair, de 7,5 à 10 cm de long, avec pour particularité des ailes latéral papyracées.

Habitat: plages, franges côtières e îles dans le nord des Caraïbes ; en Amérique du Nord seulement da le sud de la Floride.

Bois: devenant couleur chocolat clair à l'air ; frais, il est gris sombr sans contraste entre le bois de cœur et l'aubier.

Intérêts: les racines toxiques sont utilisées pour la pêche. Introduit en Angleterre dès 1690.

JAMAICA DOGWOOD.

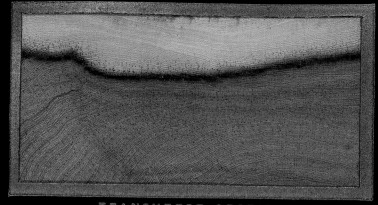

TRANSVERSE SECTION.

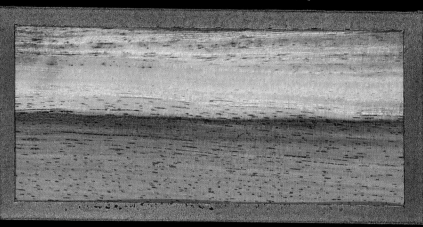

RADIAL SECTION.

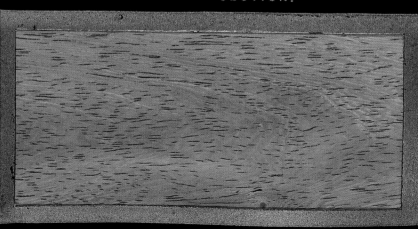

TANGENTIAL SECTION.

Ger. Jamaikischer Hundsholz. Fr. Boisivrant de la Jamaique.

Sp. Borracho (Venezuela). Guama hediondo (Cuba).

L | LEGUME FAMILY
Jerusalem Thorn, Retama, Horse Bean

FAMILIE DER SCHMETTERLINGSBLÜTLER,
LEGUMINOSEN
Jerusalemsdorn

FAMILLE DU LÉGUME OU DES LÉGUMINEUSES
Parkinsonia épineux

LEGUMINOSAE
Parkinsonia aculeata L.

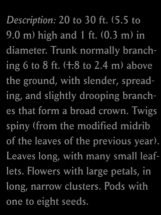

Description: 20 to 30 ft. (5.5 to 9.0 m) high and 1 ft. (0.3 m) in diameter. Trunk normally branching 6 to 8 ft. (1.8 to 2.4 m) above the ground, with slender, spreading, and slightly drooping branches that form a broad crown. Twigs spiny (from the modified midrib of the leaves of the previous year). Leaves long, with many small leaflets. Flowers with large petals, in long, narrow clusters. Pods with one to eight seeds.

Habitat: South-western North America. In Texas only in open situations, on soils with little moisture on the edges, and ends of bays, but common and less specialized in the rest of its range. Introduced into Florida and numerous tropical countries, where it now grows wild.

Wood: Heavy, hard, and very close-grained. Heartwood light brown with a yellow tinge; sapwood very thick.

Use: In southern Europe and western Texas it is a favorite ornamental tree thanks to its rapid growth and large flowers. It is also used as

Beschreibung: 5,5 bis 9,0 m hoch bei 0,3 m Durchmesser. Stamm sich normalerweise 1,8 bis 2,4 m über dem Boden verzweigend. Mit schlanken, ausladenden und leicht herabhängenden Ästen, die eine breite Krone bilden. Zweige dornig (umgewandelte Mittelrippe der vorjährigen Blätter). Blätter lang, mit vielen kleinen Blattfiedern. Blüten in langen schlanken Trauben, mit großen Kronblättern. Hülsen ein- bis achtsamig.

Vorkommen: Südwestliches Nordamerika. In Texas nur in offenen Lagen, auf Böden geringer Feuchte an den Rändern und am Ende von Buchten, im übrigen Verbreitungsgebiet häufig und weniger spezialisiert. In Florida sowie vielen tropischen Ländern eingeführt und verwildert.

Holz: Schwer, hart und von sehr feiner Textur. Kern hellbraun mit gelber Tönung; Splint sehr dick.

Nutzung: In Südeuropa und West-Texas dank des raschen Wuchses und der großen Blüten ein beliebtes Ziergehölz. Wegen der Dornen auch als „lebender Zaun"

Description : 5,5 à 9,0 m de haute sur 0,3 m de diamètre, se ramifia d'ordinaire entre 1,8 et 2,4 m de haut. Branches fines, étalées et légèrement retombantes, consti-tuant un houppier ample. Ramea épineux du fait de la transforma-tion de la nervure médiane des feuilles de l'année précédente. Longues feuilles aux nombreuse petites folioles. Belles fleurs en panicules longues et fines, avec d grands pétales. Gousses renferma de une à huit graines.

Habitat : sud-ouest de l'Amérique du Nord ; au Texas, seulement en situation ouverte et dans des sols peu humides au bord ou au fond des baies ; fréquent et moin spécialisé dans le reste de son air Introduit en Floride et dans de nombreux pays tropicaux, se ren contre aussi à l'état sauvage.

Bois : lourd, dur et à grain très fir Bois de cœur brun clair teinté de jaune ; aubier très épais.

Intérêts : espèce ornementale très appréciée dans le sud de l'Europ et l'ouest du Texas pour sa crois-sance rapide et ses grandes fleur.

RETAMA. HORSE BEAN.

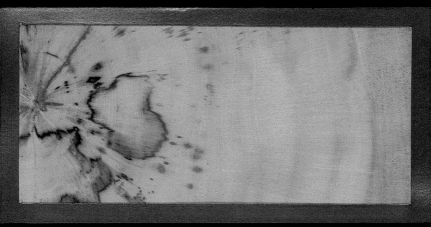

TRANSVERSE SECTION.

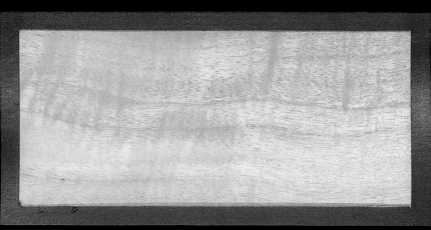

RADIAL SECTION.

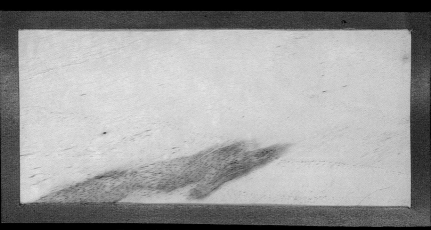

TANGENTIAL SECTION.

Gr. Stachlichter ginsterbaum. Fr. Cateping.
Sp. Flor de rayo (Porto Rico).

L | LEGUME FAMILY
Kentucky Coffee Tree

FAMILIE DER SCHMETTERLINGSBLÜTLER, LEGUMINOSEN
Amerikanischer Geweihbaum

FAMILLE DU LÉGUME OU DES LÉGUMINEUSES
Chicot du Canada, Arbre à café, Gros févier (Canada)

LEGUMINOSAE
Gymnocladus dioicus (L.) K. Koch

Description: Up to 100 ft. (30 m) high and 2 to 3 ft. (0.6 to 0.9 m) in diameter. Straight, columnar trunk. Leaves large, pinnate or bipinnate. Brown fruit, in pods, up to 12 in. (0.5 m) long.

Habitat: Rich, low-lying land in association with Black Walnut, Yellow and Ohio Buckeye, Nettle Tree, Red Elm, gleditsia, oaks, and hickories. Rarely numerous. In the interior of eastern North America.

Wood: Heavy and strong; very durable.

Use: For posts, railroad sleepers, furniture. Seeds in earlier times a substitute for coffee.

A park tree in the USA and Europe, ornamental even in winter.

Beschreibung: Bis 30 m hoch bei 0,6 bis 0,9 m Durchmesser. Gerader, säulenförmiger Stamm. Großes, einfach und zweifach gefiedertes Laub. Braune, bis zu 30 cm lange Früchte (Hülsen).

Vorkommen: Tief gelegenes, nährstoffreiches Flachland in Gesellschaft von Schwarznuss, Gelber und Ohio-Rosskastanie, Zürgelbaum, Rot-Ulme, Gleditschie, Eichen und Hickories. Selten zahlreich. Im Innern des östlichen Nordamerika.

Holz: Schwer und fest; sehr haltbar.

Nutzung: Für Pfosten, Eisenbahnschwellen, Möbel. Samen einst als Kaffee-Ersatz.

Auch im Winter dekorativer Parkbaum in den USA und in Europa.

Description : arbre atteignant jusqu'à 30 m de hauteur, avec u tronc columnaire de 0,6 à 0,9 m de diamètre. Très grandes feuille plusieurs fois pennées, à folioles simples puis composées. Fruits buns (cosses) jusqu'à 0,5 m.

Habitat : basses plaines riches er association avec le noyer noir, le marronnier glabre et le marronn jaune, le micocoulier, l'orme rou le févier, les chênes et les caryer Rarement en grandes colonies. (le trouve dans les régions de l'ir rieur est de l'Amérique du Nord

Bois : lourd et résistant ; très durable.

Intérêts : utilisé pour fabriquer d poteaux, des traverses de chemi de fer, des meubles. Les graines grillées étaient autrefois un succ dané du café.

Arbre de parcs aux Etats-Unis e en Europe, ornemental même e hiver.

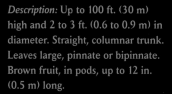

Coffee-Tree.

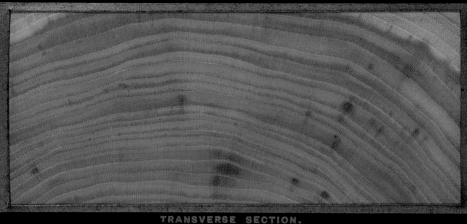

TRANSVERSE SECTION.

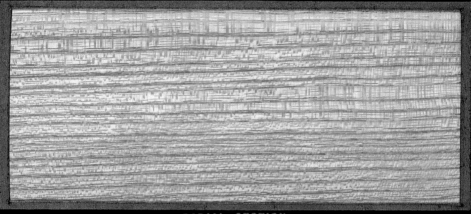

RADIAL SECTION.

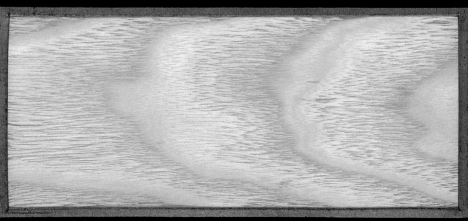

TANGENTIAL SECTION.

Ger. Amerikanischer Schusserbaum. *Fr.* Chicot, Gros Fevier.
Sp. Arbol de café falso.

L | LEGUME FAMILY

Little-leaf Palo Verde, Little-leaf Horse Bean, Foothill Palo Verde

FAMILIE DER SCHMETTERLINGSBLÜTLER, LEGUMINOSEN

–

FAMILLE DU LÉGUME OU DES LÉGUMINEUSES
Cercidium à petites feuilles, Paloverdi jaune

LEGUMINOSAE
Parkinsonia microphylla Torr.

Description: Maximum 20 to 25 ft. (6.0 to 7.5 m) high tree; 12 in. (0.3 m) in diameter, but usually only as shrub. Twigs short, ending in stout thorns. Bark thin, dark orange, mainly smooth and only occasionally marked by fissures. Leaves short, divided into eight to twelve leaflets, at first hairy, later smooth, falling after just a few weeks. Flowers pale yellow, in short clusters. Pod one to three-seeded, narrowing markedly between the seeds. As tree only around Wickenburg, Arizona.

Habitat: Deserts in south-western North America.

Wood: Heavy, hard, and close-grained. Heartwood dark orange-brown with red streaks; sapwood thick (25 to 30 annual rings), pale brown or yellow. Numerous thin medullary rays and many large, open, scattered pores.

Use: Of no commercial value, as large trees are so rare (Begemann, 1985).

Beschreibung: Maximal 6 bis 7,5 m hoher Baum bei 0,3 m Durchmesser, meist aber nur als Strauch. Zweige kurz, in kräftigen Stacheln endend. Borke dünn, dunkelorange, überwiegend glatt und nur stellenweise von Rissen durchzogen. Blätter kurz, in 8 bis 12 Blattfiedern geteilt, anfangs dicht behaart, später kahl, Laubfall bereits nach wenigen Wochen. Blüten blassgelb, in kurzen Trauben. Schoten ein- bis dreisamig, zwischen den Samen stark verschmälert. Als Baum nur in der Nähe von Wickenburg (Arizona).

Vorkommen: Wüsten im südwestlichen Nordamerika.

Holz: Schwer, hart und mit feiner Textur. Kern dunkel orangebraun mit roten Streifen; Splint dick (25 bis 30 Jahresringe), hellbraun oder gelb. Zahlreiche dünne Markstrahlen und viele offene, große, zerstreut liegende Gefäße.

Nutzung: Da die Bäume in nutzbarer Größe selten sind, besitzen sie keinen Handelswert (Begemann 1985).

Description : peut atteindre 6 à 7,5 m de hauteur sur un diamètre de 0,3 m, mais buisson le plus souvent. Rameaux courts, garnis d'épines terminales. Ecor fine, orange foncé, lisse, hormis quelques fissures çà et là. Court feuilles composées de 8 à 12 folioles, très duveteuses puis de nant glabres. Feuillaison très brè (quelques semaines seulement). Fleurs en courtes grappes jaune pâle. Capsules renfermant 1 à 3 graines et présentant un étrangl ment entre les loges. Port arbore cent uniquement dans les envir de Wickenburg (Arizona).

Habitat : déserts du sud-ouest de l'Amérique du Nord.

Bois : lourd, dur et à grain fin. B de cœur brun orangé avec des rubans rouges ; aubier épais (25 30 cernes annuels), brun clair o jaune. Beaucoup de rayons méd laires fins et de grands vaisseau de type ouvert, épars.

Intérêts : aucun intérêt économic étant donné que l'arbre atteint n ment des dimensions adéquates (Begemann 1985).

TRANSVERSE SECTION.

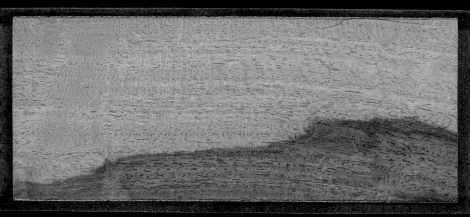

RADIAL SECTION.

TANGENTIAL SECTION.

Ger. Gebirgspaloverde.   *Fr.* Palo Verde de montagne.

Sp. Palo Verde de montaña.

L | LEGUME FAMILY
Mesquito, Honey Locust

FAMILIE DER SCHMETTERLINGSBLÜTLER, LEGUMINOSEN
Mesquitebaum

FAMILLE DU LÉGUME OU DES LÉGUMINEUSES
Prosopis commun, Baya hon

LEGUMINOSAE
Prosopis juliflora (Sw.) DC.

Description: Occasionally 40 to 50 ft. (12 to 15 m) high; 2 ft. (0.6 m) in diameter. Trunk usually only 8 to 10 ft. (2.5 to 3 m) high; 5 to 9 in. (0.12 to 0.25 m) in diameter. A most attractive tree under good conditions, usually a shrub. Irregular, crooked branches arise from near the base. Roots may go down 40 to 50 ft (12 to 15 m).

Habitat: Sheltered river valleys with high water table. Rich prairie; survives annual burning. Dry, rocky hills. Sandy and saline lowland and drifting dunes. Hardy. Unsurpassed resilience. Range extends from South to Central America and into southern North America.

Wood: Heavy, hard, not very strong, coarse-grained; almost indestructible in the ground. Heartwood deep dark brown, sometimes red; sapwood thin, pure yellow.

Use: Main timber of the first settlers. Fence posts, railroad sleepers. Furniture, lumber machinery, heavy wheels, road surfacing, firewood, and charcoal. Not suitable for steam generation because of the high tannin content.

Beschreibung: Gelegentlich 12 bis 15 m hoch bei 0,6 m Durchmesser. Stamm meist nur 2,5 bis 3 m hoch bei 0,12 bis 0,25 m Durchmesser. Bei günstigen Bedingungen ein recht ansehnlicher Baum, meist ein Strauch. Nahe dem Boden in unregelmäßig hakenförmige Äste geteilt. Tiefwurzler bis 12 oder 15 m im Boden.

Vorkommen: Windgeschützte Flussniederungen mit hohem Grundwasserstand. Reiche Prärien; durch jährliche Brände nicht zu vertreiben. Trockene, felsige Hügel. Sandige und salzhaltige Niederungen und Wanderdünen. Frostfest. Unübertroffene Überlebensfähigkeit. Von Süd- und Mittelamerika bis ins südliche Nordamerika verbreitet.

Holz: Schwer, hart, nicht sehr fest und von grober Textur; fast unzerstörbar im Boden. Kern kräftig dunkelbraun, manchmal rot; Splint dünn, reingelb.

Nutzung: Hauptbauholz der ersten Siedler. Zaunpfähle, Eisenbahnschwellen. Möbel, Holzfällmaschinen, schwere Räder, Straßenpflaster, Brennholz und Holzkohle. Wegen des hohen Tanningehalts nicht zur Dampferzeugung.

Description: parfois 12 à 15 m d hauteur sur un diamètre de 0,6 Tronc de 2,5 à 3 m de haut en général sur un diamètre de 0,12 0,25 m. Bel arbre dans des conc tions favorables, buisson la plup du temps. Ramifié dès la base, à des branches en forme de croch irrégulier. Racine pivotante s'en çant jusqu'à 12 ou 15 m dans le

Habitat: lits de basses terres abr du vent et à haut niveau phréatique. Riches prairies; résiste au feux annuels. Collines rocheuse sèches. Zones basses et dunes mouvantes sablonneuses et salé Supporte le gel. Capacité à surv inégalable. Croît de l'Amérique Sud au sud de l'Amérique du N

Bois: lourd, dur, peu résistant e grain grossier; presque imputre cible sur pied. Bois de cœur d'u brun foncé intense, parfois roug aubier mince d'un jaune pur.

Intérêts: bois de construction de premiers colons. Pieux, traverse chemin de fer, meubles, machir pour l'abattage du bois, lourdes roues, pavages, combustible et charbon de bois. Impropre à la duction de vapeur de par sa hau teneur en tannin.

Mesquit, Mesquite, Honey Pod.

TRANSVERSE SECTION.

RADIAL SECTION.

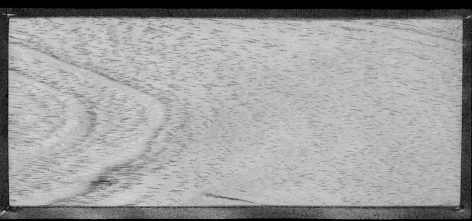

TANGENTIAL SECTION

Ger. Honighülse.   Fr. Cosse de Miel.

Sp. Algaroba.

L | LEGUME FAMILY
Palo Verde

FAMILIE DER SCHMETTERLINGSBLÜTLER, LEGUMINOSEN
Grünrinde Acacie

FAMILLE DU LÉGUME OU DES LÉGUMINEUSES
Cercidium bleu typique

LEGUMINOSAE
Cercidium floridum Benth.
Syn.: *C. Torreyanum* (S. Wats.) Sarg.

Description: 27 to 35 ft. (8 to 10 m) high and 12 to 16 in. (0.35 to 0.4 m) in diameter. Trunk short, often deformed, leafless for most of the year. Bark of older trunks redbrown, branching near the base and shedding thick, plate-like scales. On younger trunks thinner, smooth, pale olive green.

Habitat: Scattered on the sides of low canyons, and in depressions between desert sand dunes in south-western North America.

Wood: Heavy, soft, not strong, silky-sheened, and coarse-grained; polishable. Heartwood pale brown; sapwood pure pale yellow.

Use: Unknown.

Impressive to behold, with its lively trunk and branch color. Attractive in flower.

Beschreibung: 8 bis 10 m hoch bei 0,35 bis 0,4 m Durchmesser. Kurzer, oft schiefwüchsiger Stamm, die meiste Zeit des Jahres blattlos. Borke alter Stämme rotbraun, nahe der Basis gefurcht und dicke, plattige Schuppen abgebend. Jung: dünner, glatt, blass olivgrün.

Vorkommen: Zerstreut an den Seitenwänden niedriger Cañons und in den Mulden zwischen Sandhügeln der Wüsten im südwestlichen Nordamerika.

Holz: Schwer, weich, nicht fest, seidig und von grober Textur; polierfähig. Kern hellbraun; Splint rein hellgelb.

Nutzung: Unbekannt.

Ruft das Entzücken des Betrachters durch die aufmunternde Stamm- und Astfarbe hervor. Schöne Blüte.

Description: 8 à 10 m de hauteur sur un diamètre de 0,35 à 0,4 m. Tronc court, souvent oblique et pourvu de feuilles la majeure partie de l'année. Ecorce brun rouge, creusée de sillons à la base du tronc et se craquelant en plaques écailleuses épaisses sur les vieux sujets; plus fine, lisse et olive pâle à l'état juvénile.

Habitat: disséminé sur les versants de canyons peu élevés et dans les creux entre les dunes dans le désert du sud-ouest de l'Amérique du Nord.

Bois: lourd, tendre, non résistant, satiné et à grain grossier; pouvant se polir. Bois de cœur brun clair, aubier d'un jaune clair pur.

Intérêts: inconnus.

Fait l'admiration du spectateur par la couleur vive de son tronc et d

Green-barked Acacia, Palo Verde.

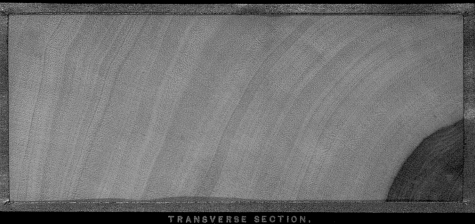

TRANSVERSE SECTION.

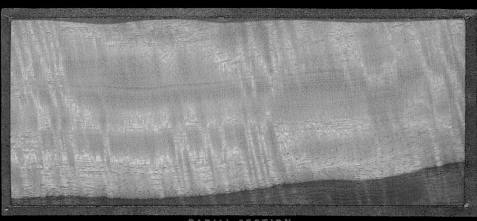

RADIAL SECTION.

TANGENTIAL SECTION

Ger. Grünrinde Acacie.   Fr. Acacia à écorce vert.

Sp. Palo Verde.

L | LEGUME FAMILY

Screwbean Mesquite

FAMILIE DER SCHMETTERLINGSBLÜTLER, LEGUMINOSEN

–

FAMILLE DU LÉGUME OU DES LÉGUMINEUSES

–

LEGUMINOSAE
Prosopis glandulosa var. torreyana (L. D. Benson) M. C. Johnst.
Syn.: *P. odorata* Torr. & Frem.

Description: Mostly occurs only as a shrub, but occasionally also as a small tree reaching a height of up to 30 ft. (9 m) and a trunk diameter of 12 in. (0.3 m) The thin cinnamon-colored bark peels off in papery scales and strips. The char-. acteristic feature of the Screwbean Mesquite, from which it derives its name, is the pod, which has 12 to 20 screw-like twists. The small seeds are enveloped in a sweet juice, which in fact is the liquefied mesocarp.

Habitat: On damp low-lying soils in the arid regions of south-western North America.

Wood: Heavy, hard, somewhat coarse-grained, and not very strong, with fine medullary rays.

Use: The Screwbean Mesquite produces excellent firewood and is also used to a limited extent for fence posts. The sweet mesocarp juice was consumed by the indigenous population but the pods are less productive than those of the related *Prosopis juliflora.*

Beschreibung: Meist nur strauchförmig, gelegentlich aber auch als kleiner Baum, der eine Höhe von bis zu 9 m und einen Stammdurchmesser von 0,3 m erreichen kann. Die zimtfarbene dünne Borke schält sich in papierartigen Schuppen und Streifen. Das charakteristischste und namengebende Merkmal der Schraubenhülse sind die schraubig 12- bis 20-mal gewundenen Schoten. Die kleinen Samen sind von einem süßen Saft (verflüssigtes Mesocarp) umgeben.

Vorkommen: Auf feuchten Tieflandböden in den Trockengebieten des südwestlichen Nordamerika.

Holz: Schwer, hart, eher dichtkörnig und nicht sehr fest, mit feinen Markstrahlen.

Nutzung: Die Schraubenhülse ergibt ein exzellentes Brennholz, daneben wird sie in begrenztem Umfang für Zaunpfähle verwendet. Der süße Mesocarp-Saft wurde von der indigenen Bevölkerung konsumiert, die Schoten sind aber nicht so ergiebig wie jene der verwandten *Prosopis juliflora.*

Description : le plus souvent arbu tif, mais parfois aussi sous la forr d'un arbrisseau pouvant atteindr 9 m de haut avec un tronc de 0,3 m de diamètre. L'écorce min couleur de cannelle se détache e écailles et bandes semblables à d papier. Les gousses sont caractér tiques, souvent tournées 12 à 20 fois. Les petites graines baignent dans une sève sucrée (mésocarp fluidifié).

Habitat : basses terres humides des zones sèches du sud-ouest de l'Amérique du Nord.

Bois : lourd, dur, au grain plutôt épais et pas très dur avec de fins rayons de moelle.

Intérêts : le *Prosopis* donne un excellent combustible, il est auss utilisé de manière limitée pour faire des pieux d'enceinte. La sè sucrée du mésocarpe était conso mée par la population indigène, gousses ne sont pas si productive que celles de son parent *Prosopis juliflora.*

Screwbean, Screw-pod Mesquite.

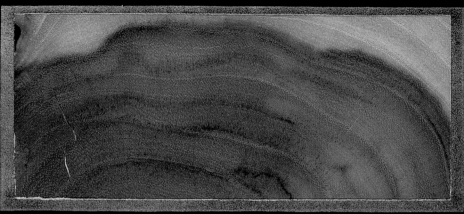

TRANSVERSE SECTION.

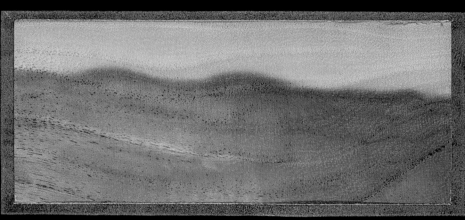

RADIAL SECTION.

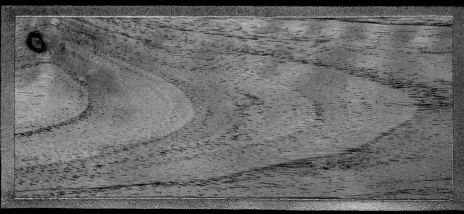

TANGENTIAL SECTION.

L | LEGUME FAMILY
Silver Wattle

FAMILIE DER SCHMETTERLINGSBLÜTLER, LEGUMINOSEN
Gerberakazie

FAMILLE DU LÉGUME OU DES LÉGUMINEUSES
–

LEGUMINOSAE
?Acacia mearnsii De Wild.
A. mollissima Willd.

Description: A tree of small to moderate size, with a trunk up to 15 in. (0.45 m) in diameter covered with smooth, gray-brown bark. The leaves consist of 8 to 15 pairs of leaflets. The flowers, pleasantly scented, are arranged in globular inflorescences. The fruits are flat pods, more or less strongly constricted between the seeds.

Habitat: Its native country is Tasmania and eastern Australia, from where it was introduced into California. There it grows on almost every type of soil and in places has escaped from cultivation into the wild.

Wood: The pale red-brown heartwood is strong, fairly heavy and hard. It is surrounded by a narrow zone of paler sapwood.

Use: Very good for firewood. The bark rich in tannins, which are used in the tanning process. The tree is, however, best suited for afforestation on poor soils, since it is not demanding in its requirements and grows rapidly.

Beschreibung: Der kleine bis mittelgroße Baum kann einen Durchmesser von bis zu 0,45 m erreichen. Er wird von einer glatten, graubraunen Borke bedeckt. Die Blätter bestehen aus 8–15 Fiederblattpaaren. Die angenehm duftenden Blüten werden in kugeligen Blütenständen gebildet. Die Fruchtschoten sind flach und zwischen den Samenfächern mehr oder weniger stark eingeschnürt.

Vorkommen: Seine Heimat ist Tasmanien und Ostaustralien, von wo aus er nach Kalifornien eingeführt wurde. Dort wächst er auf fast allen Bodentypen und verwilderte stellenweise.

Holz: Das blass rotbraune Kernholz ist fest, ziemlich schwer und hart. Es wird von einer schmalen Zone mit hellerem Saftholz umgeben.

Nutzung: Sehr gutes Feuerholz. Die Borke ist reich an Tanninen, die in der Gerberei Verwendung fanden. Vor allem aber eignet sich der Baum aufgrund seiner Anspruchslosigkeit und seines raschen Wuchses zur Aufforstung nährstoffarmer Böden.

Description: arbuste ou arbre de taille moyenne dont le tronc peut atteindre 0,45 m de diamètre. L'écorce est lisse et brun gris. Les feuilles sont composées de 8 à 15 paires de folioles et les fleurs odorantes, groupées en glomérules compacts. Les fruits sont de gousses plates, présentant un étranglement plus ou moins marqué entre les graines.

Habitat: originaire de Tasmanie de l'est de l'Australie; introduit California, où il croît sur presque tous les sols et est devenu subsp tané à certains endroits.

Bois: résistant, assez lourd et dur Bois de cœur brun rouge; aubier mince et plus clair.

Intérêts: très bon bois de chauffa L'écorce, riche en tannins, est em ployée en tannage. Mais il convi surtout au reboisement des sols pauvres car il croît rapidement e est peu exigeant.

Silver Wattle, Black Wattle.

TRANSVERSE SECTION.

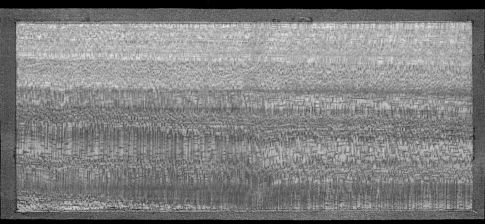

RADIAL SECTION.

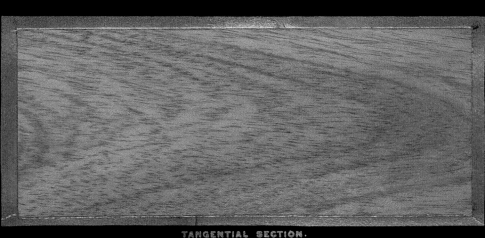

TANGENTIAL SECTION.

L | LEGUME FAMILY

Smoke Tree, Indigo Bush

FAMILIE DER SCHMETTERLINGSBLÜTLER, LEGUMINOSEN

Dornige Dalea

FAMILLE DU LÉGUME OU DES LÉGUMINEUSES

Psorothamnus épineux, Daléa épineux

LEGUMINOSAE

Psorothamnus spinosus (A. Gray) Barneby
Syn.: *Dalea spinosa* Gray

Description: To 20 to 24 ft. (6 to 7 m) high and about 15 in. (0.4 m) in diameter. Small, thorny tree. Trunk short, stocky, and twisted, branching at base, or low shrub. Bark dark gray-brown, deeply furrowed and rough, with small, long-lasting scales. Leaves fine, irregularly distributed on twigs. Flowers dark blue-violet.

Habitat: As undergrowth in the deserts of south-western North America.

Wood: Light, soft, and rather coarse-grained. Heartwood walnut-brown; sapwood thin, almost white.

Use: No commercial use known.

First discovered in 1849 by Frémont.

Beschreibung: Gelegentlich 6 bis 7 m hoch bei etwa 0,4 m Durchmesser. Kleiner, dornenbewehrter Baum. Stamm kurz, gedrungen und gedreht, gleich über dem Boden geteilt oder niedriger Strauch. Borke dunkel graubraun, tief gefurcht und rau mit kleinen, festen Schuppen. Blätter fein und unregelmäßig an den Zweigen verteilt. Blüten dunkel blauviolett.

Vorkommen: Als Unterwuchs in den Wüsten des südwestlichen Nordamerika.

Holz: Leicht, weich und ziemlich grobe Textur. Kern walnussbraun; Splint dünn, fast weiß.

Nutzung: Keine gewerbliche Nutzung bekannt.

1849 von Frémont entdeckt.

Description: atteint de façon occasionnelle 6 à 7 m de hauteur sur un diamètre de 0,4 m enviror Arbuste armé d'épines, avec un tronc court, trapu et tors, divisé d. la base, ou formant un buisson ba Ecorce brun gris sombre profonde ment fissurée et rendue rugueuse par des petites écailles persistante Feuilles fines et dispersées sur les rameaux. Fleurs violet foncé.

Habitat: forme des sous-bois dans les déserts du sud-ouest de l'Ame rique du Nord.

Bois: léger, tendre et à grain assez grossier. Bois de cœur marron; aubier mince, presque blanc.

Intérêts: utilisation commerciale non déterminée.

Découvert par Frémont en 1849 seulement.

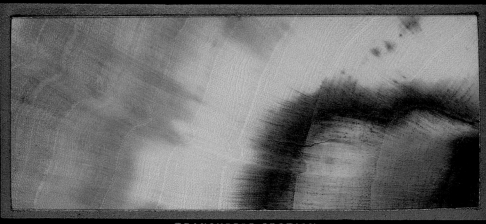

TRANSVERSE SECTION.

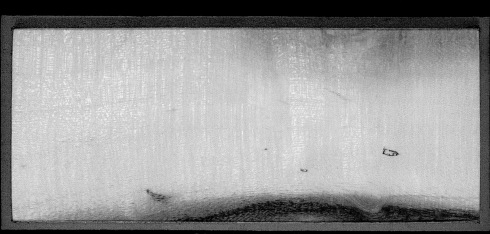

RADIAL SECTION.

TANGENTIAL SECTION

Ger. Dorniger Dalea. *Fr.* Dalea à epines. *Sp.* Dalea espino

L | LEGUME FAMILY

Water Locust, Swamp Locust

FAMILIE DER SCHMETTERLINGSBLÜTLER,
LEGUMINOSEN
Wasser-Gleditschie

FAMILLE DU LÉGUME OU DES LÉGUMINEUSES
Févier aquatique

LEGUMINOSAE
Gleditsia aquatica Marshall

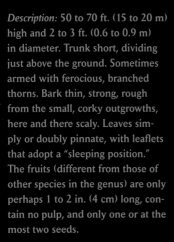

Description: 50 to 70 ft. (15 to 20 m) high and 2 to 3 ft. (0.6 to 0.9 m) in diameter. Trunk short, dividing just above the ground. Sometimes armed with ferocious, branched thorns. Bark thin, strong, rough from the small, corky outgrowths, here and there scaly. Leaves simply or doubly pinnate, with leaflets that adopt a "sleeping position." The fruits (different from those of other species in the genus) are only perhaps 1 to 2 in. (4 cm) long, contain no pulp, and only one or at the most two seeds.

Habitat: Wet locations, and low-lying riverbanks under water for a considerable time, in eastern North America, together with the Western Plane, Swamp Privet, Water Elm, Swamp Cypress, tupelos, and various willows. It reaches its greatest height in the Mississippi Valley.

Wood: Heavy, hard, and strong. Heartwood reddish brown; sapwood thick, pale yellow.

Use: Unknown.

Beschreibung: 15 bis 20 m hoch bei 0,6 bis 0,9 m Durchmesser. Kurzer, dicht über dem Boden geteilter Stamm. Manchmal mit furchterregenden, verzweigten Dornen besetzt. Borke dünn, fest, rau von kleinen korkigen Auswüchsen, hier und da schuppig. Einfach oder zweifach gefiedertes Laub mit „Schlafstellung" der Blättchen. Die Früchte (anders als bei den übrigen Arten der Gattung) sind nur ca. 4 cm lange und meist ein-, höchstens zweisamige Hülsen ohne Fruchtfleisch.

Vorkommen: Nasse Stellen und flache, lange Zeit überschwemmte Flussufer im östlichen Nordamerika, gemeinsam mit der Amerikanischen Platane, der Forestiere, der Wasser-Ulme, der Sumpfzypresse, den Tupelos und verschiedenen Weiden, erreicht ihre größte Höhe im unteren Mississippi-Tal.

Holz: Schwer, hart und stark. Kern rötlich braun; Splint dick, blassgelb.

Nutzung: Unbekannt.

Description: 15 à 20 m de hauteu sur un diamètre de 0,6 à 0,9 m. Tronc court, se divisant dès la base, armé quelquefois d'épines ramifiées impressionnantes. Ecor mince, dure, rendue rugueuse pa des petites excroissances liégeuse écailleuses çà et là. Feuilles simplement ou doublement pennées dont les folioles se replient en position de «sommeil» au crépus cule. Les fruits (à la différence de espèces voisines) n'ont que 4 cm de long et se présentent d'ordina sous la forme de gousses plates, contenant une ou deux graines.

Habitat: lieux humides et berges plates, longtemps inondées de l'e de l'Amérique du Nord, en assoc tion avec le platane d'occident, le forestiéra, le planéra des marécages, le taxode chauve, les nyss et divers saules ; atteint sa taille maximum dans la vallée inférieu du Mississippi.

Bois : lourd, dur et fort. Bois de cœur brun rougeâtre ; aubier épa jaune pâle.

Intérêts : inconnus.

Water Locust.

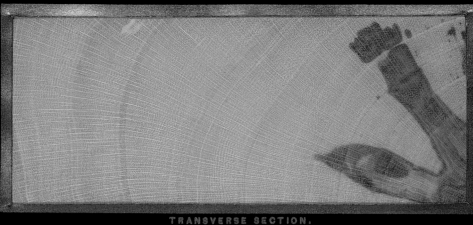

TRANSVERSE SECTION.

RADIAL SECTION.

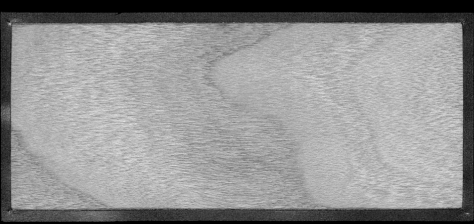

TANGENTIAL SECTION.

Ger. Einsamiger Honigdorn.　　*Fr.* Fevier monosperme.

Sp. Algorrobo acuatico.

or on young trees and branches smooth and light gray with a red tinge. Leaves with two to four pairs of leaflets, these again divided into 10 to 20 ovate to oblong secondary leaflets, light green on the upper surface and pale beneath. Inflorescences globular, flowers with a short corolla tube and very long stamens and carpels. Pods long, thin, and papery, at first bronze-green when fully grown and later red-brown.

Habitat: Islands in the Caribbean, also the Bahamas; in North America only off the coast of Florida, where it is rare and has already disppeared from some places.

Wood: Heavy, hard, not very strong, tough, smooth, and close-grained; polishable. Heartwood rich dark brown, tinged red; sapwood four or five annual rings (1 to 2 in. (2.5 to 4 cm) thick), almost white.

Use: Ship and boat building.

oder an jungen Bäumen und Ästen glatt und hellgrau mit rosa Tönung. Blätter mit 2 bis 4 Fiederpaaren, diese nochmals in 10 bis 20 ovale bis längliche Blättchen zerteilt, hellgrün auf der Ober- und blass auf der Unterseite. Blütenstände kugelig, Blüten mit kurzem Kronblatttubus und sehr langen Staub- und Fruchtblättern. Schote lang, dünn und papierartig, reif zuerst bronzegrün und dann rotbraun.

Vorkommen: Inseln der Karibischen Region und der Bahamas, in Nordamerika nur vor Florida, dort selten und teilweise bereits verschwunden.

Holz: Schwer, hart, nicht sehr fest, stark, glatt und mit feiner Textur; polierfähig. Kern tief dunkelbraun, rot getönt; Splint 4 bis 5 Jahresringe (2,5 bis 4 cm dick), fast weiß.

Nutzung: Gelegentlich zum Schiff- und Bootsbau verwendet.

coupe ou, sur les jeunes arbres e les branches, lisse et gris clair tei de rose. Feuilles composées de 2 à 4 paires de folioles qui portent chacune 10 à 20 foliolules ovales allongées ; limbe vert clair dessus pâle dessous. Inflorescences glob leuses. Les fleurs ont une corolle à tube court, des étamines et des çarpelles très longues. Gousse longue et papyracée, vert bronze maturité puis brun rouge.

Habitat : Caraïbes et Bahamas ; en Amérique du Nord uniquement en face de la Floride où on le rer contre rarement car pratiquemer en voie d'extinction.

Bois : lourd, dur, pas très résistan fort, lisse et à grain fin ; susceptib d'être poli. Bois de cœur brun pr fond teinté de rouge ; aubier forr de 4 à 5 cernes annuels (2,5 à 4 d'épaisseur), presque blanc.

Intérêts : parfois dans la construction navale.

TRANSVERSE SECTION.

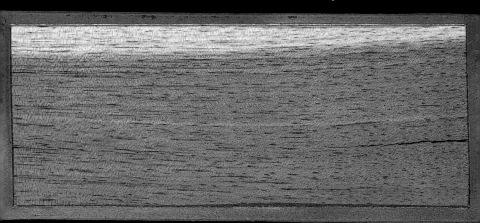

RADIAL SECTION.

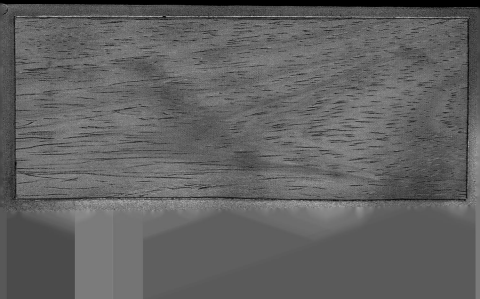

L | LEGUME FAMILY
Yellow Wood, Virgilia

FAMILIE DER SCHMETTERLINGSBLÜTLER, LEGUMINOSEN
Amerikanisches Gelbholz, Gelbholzbaum

FAMILLE DU LÉGUME OU DES LÉGUMINEUSES
Virgilier jaune, Bois jaune

LEGUMINOSAE
Cladrastis kentukea (Dum. Cours) Rudd.
Syn.: *C. lutea* (F. Michx.) K. Koch

Description: 50 to 60 ft. (15 to 18 m) high and 12 to 15 in. (4 ft.) (0.3 to 0.4 (1.2) m) in diameter. Generally 6 to 7 ft. (1.8 to 2.1 m) in height, divided into two or three partial trunks and with rather slender, far-spreading, slightly drooping, brittle branches, which form a broad crown. Bark thin, with a silvery gray or light brown surface. Leaves pinnate, leaflets ovate. Flowers large, in loose, pendant panicles. The fruits are oblong, flat pods. One of the most beautiful flowering trees of the American woods.

Habitat: Limestone cliffs and mountain slopes on rich soils, often by fast-flowing streams in the interior of eastern North America. Here, one of the rarest trees.

Wood: Moderately heavy, fairly hard, and flexible; similar to the acacia. Clear bright yellow.

Use: Decorative interior structures, veneers, plywood, furniture, marquetry, etc. Has been in cultivation for a long time in the USA and the warmer parts of Europe, and is a valuable ornamental tree.

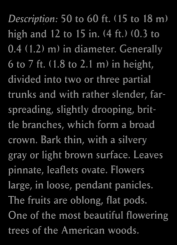

Beschreibung: 15 bis 18 m hoch bei 0,3 bis 0,4 (1,2) m Durchmesser. Meist in 1,8 bis 2,1 m Höhe in zwei bis drei Stämme gespalten und mit eher schlanken, weit ausladenden, leicht herabhängenden, brüchigen Ästen, die eine breite Krone bilden. Borke dünn, mit silbrig grauer oder hellbrauner Oberfläche. Blätter gefiedert, Fiedern oval. Große Blüten in herabhängenden lockeren Doppeltrauben. Früchte längliche, flache Schoten. Einer der am schönsten blühenden Bäume der amerikanischen Wälder.

Vorkommen: Kalkfelsen und Berghänge auf reichen Böden, oft an rasch fließenden Gewässern im Inneren des östlichen Nordamerika. Hier einer der seltensten Bäume.

Holz: Mittelschwer, ziemlich hart und elastisch; ähnlich dem der Akazie. Leuchtend hellgelb.

Nutzung: Dekorative Innenarbeiten, Furniere, Sperrholz, Möbel, Intarsien etc. Seit Langem in den USA und den wärmeren Teilen Europas in Kultur ein wertvolles Ziergehölz.

Description: arbre de 15 à 18 m de hauteur, avec un tronc de 0,3 à 0,4 m (1,2 m) de diamètre, se divisant d'ordinaire en 2 ou 3 tiges. Branches plutôt fines, très étalées, légèrement retombantes, cassantes, constituant un houppe ample. Ecorce mince, gris argent ou brun clair en surface. Feuilles composées de folioles ovales. Grandes fleurs blanches en panicules lâches et tombantes. Gousses longues et plates. Sa floraison est une des plus belles des essences forestières américaines.

Habitat: falaises calcaires et versants de montagnes sur sols riches souvent au bord des rivières à courant rapide. Croit dans l'intérieur l'est de l'Amérique du Nord. Ici des arbres les plus rares.

Bois: moyennement lourd, assez dur et élastique; similaire à celui de l'acacia et d'un jaune clair lumineux.

Intérêts: utilisé pour la décoration intérieure, le placage, le contre-plaqué, l'ameublement, la marqueterie etc. Arbre d'ornement de grande valeur, cultivé depuis longtemps aux Etats-Unis et dans les régions plus chaudes de l'Europe.

Yellow-wood. Gopher-wood. Virgilia.

TRANSVERSE SECTION.

RADIAL SECTION.

TANGENTIAL SECTION.

L | LILY FAMILY
Joshua Tree

FAMILIE DER LILIENGEWÄCHSE
Kurzblättrige Yucca

FAMILLE DU LYSET OU DE LA TULIPE
Yucca à feuilles courtes

LILIACEAE
Yucca brevifolia Engelm.
Syn.: *Y. arborescens* Trel.

Description: 35, 40 ft. (10 to 12 m) Not a true tree botanically. The roots arise together from a basin-shaped primary shoot. These penetrate deep into the soil. The trunk is surrounded by leaves. Leaves sword-shaped, stiff, with sharply serrated edges. After flowering, set at right angles. Later bending backwards, lasting several years. Flowers are large, bell-shaped, greenish-white, drooping in terminal panicles, dull or shiny, with unpleasant smell.

Habitat: High, gravely slopes and lower slopes of dry hills in specific zones and belts, often forming forest-like stands of considerable extent. Between 2,500 and 5,000 ft. (800 and 1,800 m), in Nevada up to 7,500 ft. (2,300 m) Interior of south-western North America.

Wood: Light, soft, and spongy; hard to work. Very pale brown or almost white.

Use: For paper pulp. Seeds gathered and milled by American Indians.

Beschreibung: 10 bis 12 m hoch. Kein regulärer Baum nach botanischer Definition. Die Wurzeln entspringen allesamt einem schüsselförmigen Primärspross. Sie dringen tief in den Boden ein. Der Stamm ist von Blättern umschlossen. Blätter schwertartig, steif, am Rande scharf gesägt. Nach Erscheinen der Blüten rechtwinklig aufgerichtet. Später zurückgebogen, mehrere Jahre andauernd. Große, breit glockige, grünlich weiße Blüten, an endständigen Rispen hängend, stumpf oder glänzend wachsig, mit unangenehmem Duft.

Vorkommen: Hohe, kiesige Hänge, die die Ebene begrenzen und niedrige Hänge trockener Berge in umschriebenen Zonen und Gürteln, oft offene, reine, waldartige Bestände von beträchtlichem Ausmaß bildend. Zwischen 800 und 1800, in Nevada 2300 m. Im Inneren des südwestlichen Nordamerika.

Holz: Leicht, weich und schwammig; schwer zu bearbeiten. Sehr hellbraun oder fast weiß.

Nutzung: Zu Papier-Pulpe. Samen von Indianern gesammelt und zu Mehl vermahlen.

Description: 10 ou 12 m de haute Ne constitue pas un arbre au ser botanique du terme. Les racines toutes issues d'une pousse prima en forme de coupe, s'enfoncent profondément dans le sol. Feuill lancéolées, rigides, à marges coupantes et enveloppant le tror Elles se dressent à angle droit ap l'apparition des fleurs et retrouve leur courbure habituelle après la floraison; elles sont persistantes. Hampe florale portant une panic de grosses fleurs campanulées bl verdâtre, enduites d'une substan cireuse mate ou brillante, et dég geant une odeur désagréable.

Habitat: hautes pentes caillouteuses enfermant les plaines et bas versants de montagnes sèch dans des zones délimitées; form souvent d'immenses peuplemer purs ouverts, entre 800 et 1800 d'altitude; dans la Sierra Nevad à 2300 m. Croît dans l'intérieur sud-ouest de l'Amérique du No

Bois: léger, tendre et spongieux difficile à travailler. Brun très cla ou presque blanc.

Intérêts: pâte à papier. Les Indie moulaient les graines comestible pour en faire de la farine.

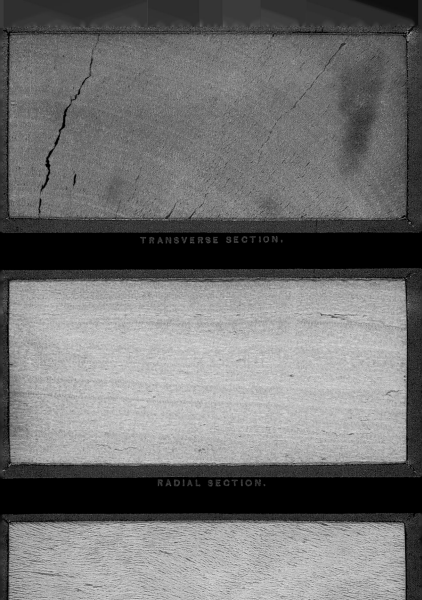

TRANSVERSE SECTION.

RADIAL SECTION.

L | LINDEN FAMILY, LIME FAMILY

American Lime, Basswood

FAMILIE DER LINDENGEWÄCHSE
Amerikanische Linde

FAMILLE DU TILLEUL
Tilleul d'Amérique, Bois blanc (Canada)

TILIACEAE
Tilia americana L.

Description: From 65 to 80 ft. (20 to 24 m) in height and from 3 to 4 ft. (0.9 to 1.2 m) in diameter. Exceptionally from 100 to 160 ft. (30 to 45 m) in height and up to 6 ft. (1.8 m) in diameter. Strongly scented, yellow flowers in clusters.

Habitat: Rich, moist, but well-drained slopes and lowlands of Southern North America, always abundant.

Wood: Light, soft and tough. Heartwood pale brown or often tinged light red; sapwood scarcely distinguishable.

Use: Wood much used for cheap furniture, for cars, articles in turned wood and paper pulp. After maceration fibers of the inner bark are used for mats and wickerwork and as binding material. A food plant for bees.

One of the most common and very useful forest trees of eastern North America and Canada.

Beschreibung: 20 bis 24 m hoch bei 0,9 bis 1,2 m Durchmesser. Ausnahmsweise 30 bis 45 m Höhe bei bis zu 1,8 m Durchmesser. Weithin intensiv duftende, gelbe Blüten in Büscheln.

Vorkommen: Reiche, feuchte, aber gut drainierte Hänge und Flachland im südlichen Nordamerika, stets zahlreich.

Holz: Leicht, weich und zäh. Kern hellbraun oder oft leicht rot getönt; Splint kaum unterscheidbar.

Nutzung: Viel verwendetes Holz für billige Möbel, für Wagen, Drechslerarbeiten und Papier-Pulpe. Nach Mazeration werden Fasern der inneren Rinde zu Bastmatten, Flechtwerk und Bindematerial verarbeitet. Bienenfutterpflanze.

Einer der häufigsten und sehr nützlichen Waldbäume des östlichen Nordamerika und Kanada.

Description : arbre de 20 à 24 m de hauteur, avec un tronc de 0,9 1,2 m de diamètre. Atteint parfo 30 à 45 m de hauteur et 1,8 m d diamètre. Fleurs jaunes disposée en cymes, dégageant un parfum perceptible de loin.

Habitat : plaines et pentes riches humides mais bien drainées, dan le sud de l'Amérique du Nord, toujours en grandes colonies.

Bois : léger, tendre et tenace. Boi de cœur brun clair ou souvent le rement teinté de rouge ; aubier à peine distinct.

Intérêts : bois aux usages multipl meubles bon marché, charronna tournage et pâte à papier. Après macération des fibres de l'écorce interne, fabrication de rabanes, d'ouvrages en treillis et de cordages. Espèce mellifère.

Une des essences forestières les plus répandues et les plus utiles l'est du continent nord-américai

Basswood, American Linden or Lin, Lime-Tree, Bee-Tree.

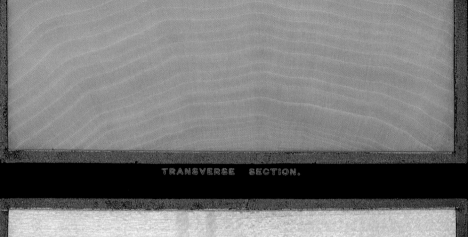

TRANSVERSE SECTION.

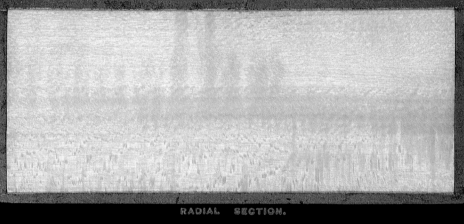

RADIAL SECTION.

TANGENTIAL SECTION.

Ger. Amerikanische Linde.   *Sp.* Tilio Americano.

Fr. Tilluel d'Amerique.

M | MAGNOLIA FAMILY
Cucumber Tree

FAMILIE DER MAGNOLIENGEWÄCHSE
Gurkenmagnolie, Gurkenbaum

FAMILLE DU MAGNOLIA
Magnolia acuminé, Arbre à concombres,
Magnolia à feuilles acuminées (Canada)

MAGNOLIACEAE
Magnolia acuminata (L.) L.

Description: Up to 100 ft. (30 m) high and from 3 to 4 ft. (0.9 to 1.2 m) in diameter. Straight, columnar trunk. Crown large and dense. Bark grayish brown, furrowed. Greenish flowers appear after the leaves.

Habitat: Mountain slopes, valleys, stony riverbanks, in association with indicators of good soils and damp climates, including White Oak, species of hickory, White Ash, Tulip Tree, and Sugar Maple. Does not form homogeneous stands of its own. Develops best in the Carolinas and Tennessee.

Wood: Light, soft, close-grained and with a silky sheen; durable and similar to that of the Tulip Tree, used commercially. Heartwood yellowish brown; sapwood paler, almost white.

Use: In tree nurseries as a stock for Asiatic magnolias.

The hardiest of the American magnolias. A quick-growing park tree in the USA and Europe.

Beschreibung: Bis 30 m hoch bei 0,9 bis 1,2 m Durchmesser. Gerader, säulenförmiger Stamm. Krone groß und dichtlaubig. Borke in graubraunen Riefen. Grünliche Blüten nach dem Laubaustrieb.

Vorkommen: Berghänge, Täler, kiesige Ufer in Gesellschaft von Anzeigern guter Böden und feuchten Klimas: Weiß-Eiche, Hickories, Weiß-Esche, Tulpenbaum, Zucker-Ahorn u. a. Bildet selbst keine Bestände. Optimale Ausbildung in den Karolinas und Tennessee.

Holz: Leicht, weich, dicht und seidig; dauerhaft und dem des Tulpenbaums ähnlich (Handel!). Kern gelblich braun; Splint heller, fast weiß.

Nutzung: In Baumschulen Pfropfunterlage für asiatische Magnolien.

Winterhärteste der amerikanischen Magnolien. Schnellwüchsiger Parkbaum in den USA und Europa.

Description : arbre pouvant attein 30 m de hauteur, avec un tronc droit, columnaire de 0,9 à 1,2 m diamètre et des branches étalées, constituant un houppier très den Ecorce parcourue de fissures bru gris. Fleurs blanc verdâtre après l feuillaison.

Habitat : versants montagneux, vallées, berges caillouteuses, en compagnie d'espèces indicatrices de sols fertiles et de climat humide : chêne blanc, caryers, frêne blanc, tulipier, érable à suc etc. Ne constitue lui-même aucu peuplement forestier. Développe ment optimum en Caroline et au Tennessee.

Bois : léger, tendre, dense, satiné durable et similaire à celui du tulipier (commerce !). Bois de cœur jaune brunâtre ; aubier plus clair, presque blanc.

Intérêts : utilisé en pépinière comme porte-greffe des magnolia asiatiques.

Le plus rustique des magnolias américains. Arbre de parcs à croi sance rapide aux Etats-Unis et en Europe.

Cucumber-Tree, Mountain Magnolia.

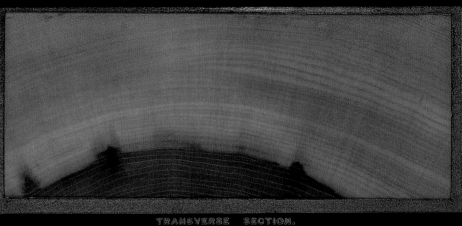

TRANSVERSE SECTION.

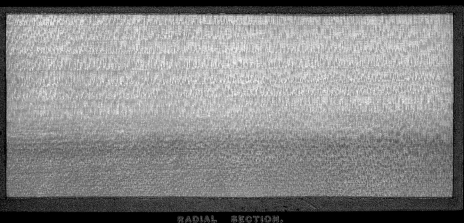

RADIAL SECTION.

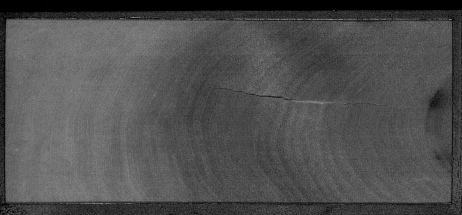

TANGENTIAL SECTION.

Ger. Langgespitzte Magnolia.   *Sp.* Magnolia acuminada.

Fr. Magnolia à feuilles pointes.

M | MAGNOLIA FAMILY

Evergreen Magnolia, Southern Magnolia, Bull Bay,
Loblolly Magnolia

FAMILIE DER MAGNOLIENGEWÄCHSE
Großblütige Magnolie

FAMILLE DU MAGNOLIA
Magnolia à grandes fleurs, Laurier tulipier

MAGNOLIACEAE
Magnolia grandiflora L.

Description: 70 to 85 ft. (20 to 25 m) high and up to 4 ft. (1.2 m) in diameter, with a straight, smooth, columnar trunk. Flowers delicately scented, creamy white, and as large as a dessert plate.

Habitat: Rich, moist soils in association with Water and Willow Oak, tupelos, and Sweet Gum. Coastal areas on the Gulf of Mexico, and the North American Atlantic coast. Develops best by the lower Mississippi.

Wood: Heavier, harder, and more valuable than the wood of other American magnolias.

Use: Used only for firewood.

Grown in Europe as an ornamental tree since the early 18th century, but only survives without winter protection in the milder regions, e.g. England and the island of Mainau on Lake Constance. It is the least hardy of all the American magnolias.

Beschreibung: 20 bis 25 m hoch bei bis zu 1,2 m Durchmesser. Gerader, säulenförmiger Stamm, glatt. Delikat duftende, schüsselgroße, cremeweiße Blüten.

Vorkommen: Reiche, feuchte Böden in Gesellschaft von Wasser- und Weiden-Eiche, den Tupelos und dem Amberbaum. Küstennahe Gebiete am Golf von Mexiko und der Atlantikküste Nordamerikas. Beste Entwicklung am unteren Mississippi.

Holz: Schwerer, härter und wertvoller als das der anderen amerikanischen Magnolien.

Nutzung: Nur als Brennholz genutzt.

In Europa seit dem frühen 18. Jh. als Zierbaum, überlebt aber selbst in milden Gebieten (England, Insel Mainau) nur mit Winterschutz. Die kälteempfindlichste der amerikanischen Magnolien.

Description: arbre de 20 à 25 m de hauteur, avec un tronc droit, columnaire et lisse pouvant atteindre 1,2 m de diamètre. Gra[n] fleurs blanc crème, délicatemen[t] parfumées.

Habitat: sols riches, humides e[n] association avec le chêne noir e[t] le chêne saule, les nyssas et le liquidambar. Croît dans les régi[ons] côtières du golfe du Mexique et la côte atlantique de l'Amériqu[e] Nord. Développement optimal [dans] la vallée inférieure du Mississip[pi]

Bois: plus lourd, plus dur et plu[s] précieux que celui des autres magnolias américains.

Intérêts: utilisé seulement comm[e] bois de chauffage.

Introduite en Europe au début d[u] 18ᵉ siècle, cette espèce ornemen[tale] a besoin d'une protection contre le froid même dans les régions [à] climat doux (Angleterre, île de Mainau sur le lac de Constance) Le plus rustique des magnolias américains.

Big Laurel, Bull Bay, Magnolia.

TRANSVERSE SECTION.

RADIAL SECTION.

TANGENTIAL SECTION

Ger. Grossblumige Magnolia.   Fr. Grand Magnolier.

Sp. Magnolia floregrande.

M | MAGNOLIA FAMILY

Fraser Magnolia, Mountain Magnolia,
Long-leaved Cucumber Tree

FAMILIE DER MAGNOLIENGEWÄCHSE
Berg-Magnolie, Frasers Magnolie

FAMILLE DU MAGNOLIA
Magnolia de Fraser

MAGNOLIACEAE
Magnolia fraseri Walter

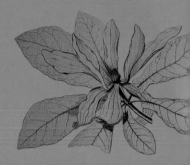

Description: 30 to 40 ft. (9 to 12 m) high and 12 to 18 in. (0.3 to 0.45 m) in diameter. Often without branches in its lower half, or shrublike and branching near the ground into several stems. Bark thin, dark brown, smooth, also fissured on older specimens. Leaves ovate-spatulate, 12 to 14 in. (25 to 28 cm) long and 6 to 7 in. (15 to 18 cm) broad (on young trees even twice as large), bright green, membranous, and auriculate at the base. Flowers large, 8 to 9 in. (20 to 23 cm) in diameter, cream in color. Tips of the carpels long and recurved.

Habitat: On rich soils in southeastern North America in a mild, damp climate.

Wood: Light, soft, not strong, and close-grained. Heartwood light brown, only apparent in older specimens; sapwood thick (30 to 40 annual rings), cream-colored.

Use: Only rarely cultivated, as it is very demanding in regard to climate and soils, and is difficult to propagate.

Beschreibung: 9 bis 12 m hoch bei 0,3 bis 0,45 m Durchmesser. Oft bis zur halben Höhe unverzweigt oder bereits in der Nähe des Bodens buschartig in mehrere Stämmchen verzweigt. Borke dünn, dunkelbraun, glatt, an alten Exemplaren auch rissig. Blätter oval-spatelförmig, 25 bis 28 cm lang und 15 bis 18 cm breit (an jungen Bäumen auch doppelt so groß) und von hellgrüner Farbe, membranös, Basis geöhrt. Blüten groß, 20 bis 23 cm im Durchmesser, cremefarben. Spitzen der Fruchtblätter lang und zurückgebogen.

Vorkommen: Auf reichen Böden des südöstlichen Nordamerika in mildem, feuchtem Klima.

Holz: Leicht, weich, nicht stabil und mit feiner Textur. Kern hellbraun, erst bei alten Exemplaren auftretend; Splint dick (30 bis 40 Jahresringe), cremefarben.

Nutzung: Nur selten kultiviert, da sehr hohe Ansprüche an Klima und Böden stellend und schwierig zu vermehren.

Description: 9 à 12 m de hauteur sur un diamètre de 0,3 à 0,45 m. Tronc souvent non ramifié jusqu mi-hauteur ou divisé dès la base en plusieurs tiges. Ecorce mince, brun sombre, lisse, se fissurant s les vieux spécimens. Feuilles de à 28 cm de long et de 15 à 18 cm de large (pouvant être deux fois plus grandes sur les jeunes arbre obovales, vert clair, membraneus munies de deux oreillettes à la base. Grandes fleurs (20 à 23 cm de large) blanc crème. Follicules fruit recourbées et à apex pointu

Habitat: sols riches du sud-est de l'Amérique du Nord, sous climat doux et humide.

Bois: léger, tendre, non stable et grain fin. Bois de cœur brun clai apparaissant seulement sur les vieux arbres; aubier épais (30 à cernes annuels), couleur crème.

Intérêts: rarement cultivé car exigeant en matière de sols et de climat et difficile à multiplier.

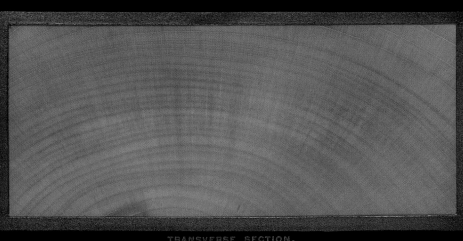

TRANSVERSE SECTION.

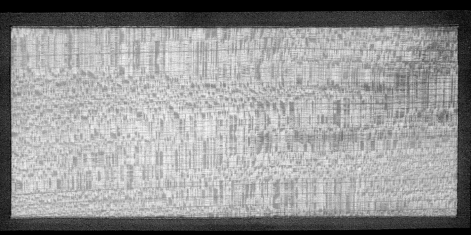

RADIAL SECTION.

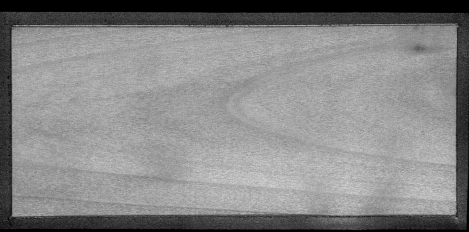

TANGENTIAL SECTION.

Ger. Fraser-Magnolie. *Fr.* Magnolier de Fraser.
Sp. Magnolia de Fraser.

M | MAGNOLIA FAMILY

Sweet Bay, Swamp Bay, Swamp Laurel

FAMILIE DER MAGNOLIENGEWÄCHSE
Süße Magnolie, Blaugrüne Magnolie

FAMILLE DU MAGNOLIA
Magnolia de Virginie, Arbre de castor (Canada)

MAGNOLIACEAE
Magnolia virginiana L.
Syn.: *M. glauca* L.

Description: 70 to 80 ft. (20 to 23 m) high and 2 to 3 ft. (0.6 to 0.9 m) in diameter. Bark brownish gray and uniformly smooth or with bumpy excrescences. Uppersides of leaves dark green, blotched, undersides white. Flowers creamy white, very fragrant, appearing in early summer.

Habitat: Low-lying swampy soils and the edges of ponds in association with Loblolly Pine, Swamp Bay, Wild Olive, Evergreen Magnolia, Swamp Holly, Yaupon, Red Maple, etc., in eastern North America.

Wood: Light, soft; resembling that of other magnolias.

Use: For fine cabinetwork. Once used medicinally for the bitter substances it contains.

An ornamental park tree, mainly in the Gulf and southern states of the USA. Not sufficiently hardy for central Europe.

Beschreibung: 20 bis 23 m hoch bei 0,6 bis 0,9 m Durchmesser. Borke bräunlich grau und durchweg glatt oder besetzt mit buckligen Auswüchsen. Dunkelgrüne Oberseiten der Blätter scheckig, Unterseiten weiß. Beinweiße, herrlich duftende Blüten im Frühsommer.

Vorkommen: Tief liegende Sumpfböden und die Säume von Teichen in Gesellschaft von Loblolly-Kiefer, Sumpf-Avocado, Wildem Ölbaum, Immergrüner Magnolie, Sumpf-Stechhülse, Yaupon-Stechhülse, Rot-Ahorn u. a. im östlichen Nordamerika.

Holz: Leicht, weich; mit dem anderer Magnolien übereinstimmend.

Nutzung: Kunstschreinerei. Wegen seiner Bitterstoffe früher auch medizinisch genutzt.

Vor allem in den Golf- und Südstaaten der USA als Ziergehölz in Parks und Gärten. Für Mitteleuropa nicht robust genug.

Description : 20 à 23 m de hauteu sur un diamètre de 0,6 à 0,9 m. Ecorce gris brun, lisse ou couver de protubérances squamiformes. Feuilles vert sombre tachetées dessus, blanchâtres dessous. Fle d'un blanc pur, délicieusement parfumées au début de l'été.

Habitat : sols des bas-marais et abords des étangs de l'est de l'Amérique du Nord, en associat avec le pin à l'encens, l'avocatier palustre, l'olivier sauvage, le magnolia toujours vert, le houx décidu, le houx vomitif, l'érable rouge etc.

Bois : léger, tendre ; identique à celui des autres magnolias.

Intérêts : utilisé en ébénisterie, autrefois en médecine pour sa teneur en substances amères.

Arbre de parcs ornemental, surt aux Etats-Unis dans les Etats du sud et du Golfe. Essence manqu de rusticité et de plasticité pour l'Europe moyenne.

Sweet Bay, Small Magnolia, White or Swamp Laurel, Beaver-Tree.

TRANSVERSE SECTION.

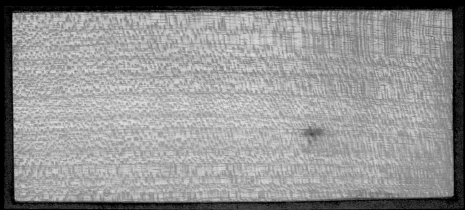

RADIAL SECTION.

TANGENTIAL SECTION

Ger. Grauer Biberbaum.   *Sp.* Laurel dulce.,

Fr. Magnolier glauque.

M | MAGNOLIA FAMILY
Tulip Tree, Whitewood, Yellow Poplar

FAMILIE DER MAGNOLIENGEWÄCHSE
Tulpenbaum

FAMILLE DU MAGNOLIA
Tulipier de Virginie, Bois jaune (Canada)

MAGNOLIACEAE
Liriodendron tulipifera L.

Description: Up to 130 ft. (60 m) high and from 6 to 13 ft. (2 to 4 m) in diameter. Straight, columnar trunk, free of branches for the first 80 to 100 ft. (25 to 30 m). In that respect, not surpassed by any deciduous tree in Atlantic America. Leaves lyre-shaped. Flowers tulip-like.

Habitat: Deep, rich, well-drained soils, always in association with similarly demanding species. Develops best on the western slopes of the Allegheny Mountains in Tennessee and North Carolina.

Wood: Light and soft; easy to work (used by the eastern, forest-dwelling Indians for their 50-man canoes) and similar to the wood of magnolias. Heartwood pale yellow or brown; sapwood thin, almost white.

Use: A valuable wood, much used in commerce for buildings, interior construction, roof shingles, boat-building, water pumps, etc.

A common park tree in the USA and Europe.

Beschreibung: Bis zu 60 m hoch bei 2 bis 4 m Durchmesser. Gerader, säulenförmiger Stamm, astfrei bis in 25, 30 m Höhe. Darin von keinem Laubbaum des atlantischen Amerika übertroffen. Leierschnittiges Laub. Tulpenartige Blüten.

Vorkommen: Tiefgründige, reiche, gut drainierte Böden, immer in Gesellschaft ähnlich anspruchsvoller Arten. Optimale Ausbildung an den Westhängen der Alleghenies in Tennessee und Nordkarolina.

Holz: Leicht und weich; leicht zu bearbeiten (50-Mann-Kanus der östlichen Waldindianer) und dem Magnolienholz ähnlich. Kern hellgelb oder braun; Splint dünn, fast weiß.

Nutzung: Wertvolles, viel verwendetes Nutzholz für Bauten, Innenausbau, Dachschindeln, Bootsbau, Wasserpumpen usw.

Häufiger Parkbaum in USA und Europa.

Description : arbre s'élevant jusqu' 60 m, avec un tronc droit, colur naire de 2 à 4 m de diamètre et dénudé jusqu'à 25 ou 30 m de hauteur. Le plus grand des feuil de la façade atlantique de l'Amé rique. Feuille de forme particuli formée de quatre lobes dont les deux terminaux font paraître la pointe tronquée. Fleur rappelant celle de la tulipe.

Habitat : sols profonds, riches et bien drainés, toujours en associa tion avec des espèces similaires. Surtout sur les versants ouest de Alleghanys dans le Tennessee et la Caroline du Nord.

Bois : léger, tendre ; se travaillan bien (canots pour 50 personnes des Indiens habitant les forêts orientales) et semblable à celui du magnolia. Bois de cœur jaun clair ou brunâtre ; aubier mince quasiment blanc.

Intérêts : bois très recherché en m nuiserie intérieure et extérieure pour fabriquer bateaux, bardeau pompes à eau etc.

Arbre de parcs commun aux Eta Unis et en Europe.

Tulip-Tree,   White-Wood,   White or Yellow "Poplar," Canoe-Wood.

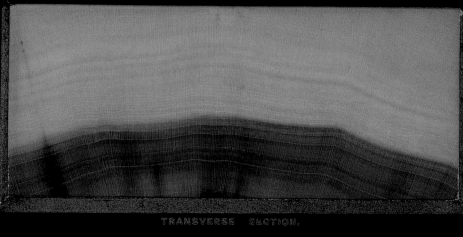

TRANSVERSE SECTION.

RADIAL SECTION.

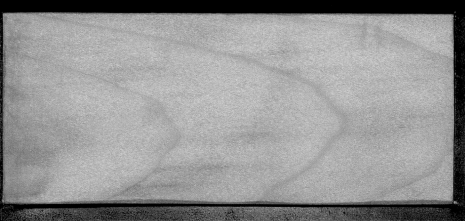

TANGENTIAL SECTION.

Ger. Tulpenbaum.   Fr. Tulipier.   Sp. Tulipifera.

M | MAHOGANY FAMILY
Bead Tree, China Tree, Indian (Cape) Lilac

FAMILIE DER MAHAGONIGEWÄCHSE
Glatter Zedrach, Indischer Zedrach

FAMILLE DE L'ACAJOU AMER
Margousier, Lilas des Indes

MELIACEAE
Melia azedarach L.

Description: To about 70 ft. (20 m) high and 2 ft. (0.6 m) in diameter. Leaves pinnate; sweet-smelling, lilac-scented, purple flowers. Very fast-growing.

Habitat: In North America, only under cultivation. Naturalized in Africa, Central America, and South America.

Wood: Hard, raised areas smooth; polishable, difficult to dry, tends to snap and warp, weathers well and resistant to insect attack. Heartwood and sapwood clearly separated: heartwood yellow-gray with a weak reddish tinge; sapwood yellowish white. Resembles ash wood. Dark markings, very decorative.

Use: For furniture. Imported to Europe, mainly from Asia, rarely used in turnery (roundwood) (Begemann, 1988).

Beschreibung: Bis etwa 20 m hoch bei 0,6 m Durchmesser. Gefiedertes Laub, fliederartig duftende, violette Blüten. Sehr schnellwüchsig.

Vorkommen: In Nordamerika wahrscheinlich nur auf Kulturboden angepflanzt. Verwildert in Afrika, Mittel- und Südamerika.

Holz: Hart und gehobelte Flächen glatt; polierfähig, schlecht zu trocknen, neigt dazu zu reißen und sich zu verziehen, witterungsbeständig und resistent gegen Insekten. Kern und Splint scharf getrennt, Kern gelbgrau mit schwach rötlichem Ton; Splint gelblich weiß. Dem Holz der Esche ähnlich. Dunkel gezeichnet, sehr dekorativ.

Nutzung: Für Möbel. Importe nach Europa vor allem aus Asien, selten als Rundholz verwendet. (Begemann 1988).

Description : arbre atteignant 20 environ de hauteur sur un diam de 0,6 m. Feuilles composées e bipennées; fleurs violettes à od de lilas. Croissance très rapide.

Habitat : en Amérique du Nord, probablement uniquement sou une forme cultivée sur un sol d culture. Subspontané en Afriqu en Amérique centrale et en Am rique du Sud.

Bois : dur et lisse sur surfaces rabotées; susceptible de prend un beau poli, difficile à sécher, tendance au gauchissement et à l'éclatement; résistant aux inte péries et aux insectes. Bois de et aubier très distincts; cœur g jaune, teinté très légèrement de rouge; aubier blanc jaunâtre. B similaire à celui du frêne. Prése des dessins foncés, très décorat

Intérêts : utilisé pour la fabricati de meubles. Importé essentielle ment d'Asie; rarement employ comme bois de grume (Begema 1988).

TRANSVERSE SECTION.

RADIAL SECTION.

TANGENTIAL SECTION

M | MAHOGANY FAMILY
Mahogany

FAMILIE DER MAHAGONIGEWÄCHSE
Mahagoni

FAMILLE DE L'ACAJOU AMER
Acajou des Antilles

MELIACEAE
Swietenia mahagoni (L.) Jacq.

Description: 40 to 50 ft. (12 to 15 m) high and 2 (8) ft. (0.6 (2.4) m) in diameter. Very stout, spreading branches. Bark moderately thick, with a dark red-brown outer surface that is broken up into rather thick scales. Leaves evergreen, divided into three or four pairs of ovate-lanceolate leaflets that taper irregularly towards the base. Leaflets entire, pale yellow-green, occasionally slightly reddish beneath. Flowers white, with an erect tube formed from the united bases of the stamens. Fruit long-stalked, with a dark brown surface.

Habitat: Occurs from the Caribbean across Central America and in South America as far as Peru. In North America in southern Florida.

Wood: Heavy, extremely hard, strong, and close-grained; very durable, darkens with age and on exposure to the air. Heartwood rich red-brown; sapwood 20 annual rings, yellow.

Use: For cabinets and for ship and boat-building, because of the combination of being easy to work and very durable. Bitter and astringent bark used in the treatment of fevers. Practically disappeared.

Beschreibung: 12 bis 15 m hoch bei 0,6 (bis 2,4) m Durchmesser. Sehr kräftige, ausladende Äste. Borke mäßig dick, mit dunkel rotbrauner Oberfläche, die sich in relativ dicken Schuppen abschält. Blätter immergrün, in 3 bis 4 oval-lanzettliche, an der Basis in ungleichmäßig verschmälerte Fiederpaare geteilt. Fiedern ganzrandig, blass gelbgrün, Unterseite gelegentlich leicht rötlich. Blüten weiß, mit aufrechtem Tubus aus den verschmolzenen Basalteilen der Staubblätter. Frucht lang gestielt, dunkelbraune Hülle.

Vorkommen: Von der Karibik über Mittelamerika bis nach Peru verbreitet. In Nordamerika nur in Süd-Florida.

Holz: Schwer, außerordentlich hart, fest und mit feiner Textur; sehr haltbar, dunkelt im Alter und an der Luft nach. Kern tief rotbraun; Splint 20 Jahresringe (maximal 2,5 cm) dick, gelb.

Nutzung: Vermutlich das begehrteste Holz für Tischlerarbeiten, im Schiff- und Bootsbau wegen der Kombination aus Leichtigkeit und Haltbarkeit sehr gefragt. Bittere und adstringierende Borke als Fiebermittel. Praktisch ausgestorben.

Description: arbre de 12 à 15 m de hauteur, avec un tronc de 0,6 de diamètre (2,4 m max.), armé de branches vigoureuses et étalé Ecorce moyennement épaisse, brun rouge sombre, s'exfoliant en écailles relativement épaisses Feuilles persistantes, composées 3 ou 4 folioles ovales-lancéolées, les basales plus ou moins aminc entières, vert jaune pâle, parfois rougeâtres à la face inférieure. L fleurs blanches ont des étamines soudées formant un tube dressé Fruit longuement pédonculé, avec un épicarpe brun sombre.

Habitat: des Caraïbes au Pérou e passant par l'Amérique centrale. En Amérique du Nord, seuleme dans le sud de la Floride.

Bois: lourd, extrêmement dur, résistant et à grain fin; très dura fonçant avec l'âge et à l'air. Bois de cœur d'un brun rouge profor aubier constitué de 20 cernes annuels (2,5 cm max.), jaune.

Intérêts: très recherché en ébéni terie, aussi dans la construction navale car il associe légèreté et durabilité. Ecorce amère et astri gente employée comme fébrifug Pratiquement disparu.

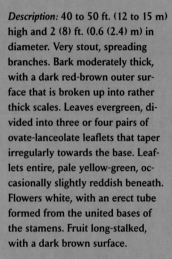

MAHOGANY. MADEIRA-WOOD.

TRANSVERSE SECTION.

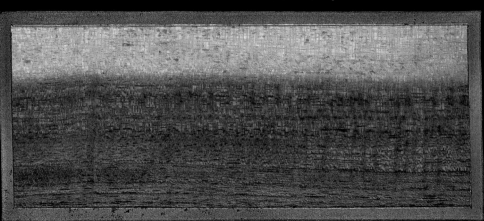

RADIAL SECTION.

TANGENTIAL SECTION.

M | MALPIGHIA FAMILY
Locust Berry, Candle Berry

FAMILIE DER MALPIGHIENGEWÄCHSE
–

FAMILLE DU MALPIGHIAE
Byrsonima, Bois-d'Arrada

MALPIGHIACEAE
Byrsonima lucida (Mill.) DC.

Description: Usually only a shrub, occasionally reaching the dimensions of a tree, but even then only rarely higher than 20 ft. (6.5 m) with a trunk 10 in. (0.25 m) in diameter. The bark is pale. The opposite, leathery leaves are spatulate to ovate and have thickened margins. The upper surface is lustrous and dark green, but the lower surface is pale. The flowers are white at first, but later turn yellow or pink, and are arranged in erect clusters. The roundish fruits are greenish in color. They have a thin, dry flesh.

Habitat: On sandy, not too moist soils of southern Florida.

Wood: The pale red heartwood is light, soft, and weak. It is surrounded by sapwood that is only slightly paler.

Use: The bark is rich in tannins and was used in tanning and dying. The wood is suitable for building purposes, but plants of adequate size are rare. The fruits are edible. In addition, the leaves and bark have been used for medicinal purposes because of their tannin content.

Beschreibung: Meist nur als Strauch, gelegentlich auch die Dimensionen eines Baumes erreichend, aber nur selten höher als 6,5 m mit 0,25 m dickem Stamm. Die Borke ist blass. Die gegenständigen ledrigen Blätter sind spatelförmig bis oval und besitzen verdickte Blattränder. Die glänzende Oberseite ist dunkelgrün, die Unterseite bleibt blass. Die anfangs weißen, später gelb oder rosa gefärbten Blüten stehen in aufrechten Trauben. Die rundlichen Früchte sind grünlich. Sie besitzen ein dünnes, trockenes Fleisch.

Vorkommen: Auf sandigen und nicht zu feuchten Böden Süd-Floridas.

Holz: Das hellrote Kernholz ist leicht, weich und schwach. Es wird von einem nur etwas helleren Saftholz umgeben.

Nutzung: Die Borke ist reich an Tanninen und wurde in der Gerberei und Färberei genutzt. Das Holz eignet sich für Bauzwecke, hinreichend große Exemplare der Pflanze sind aber selten. Die Früchte sind essbar. Blätter und Borke wurden wegen ihres Tannin-Gehaltes auch medizinisch genutzt.

Description : ne se présente d'ordinaire que sous forme de buisson mais peut prendre quelquefois un port arborescent ; dans ce cas, sa taille dépasse rarement 6,5 m et son tronc, 0,25 m de diamètre. Ecorce pâle. Feuilles opposées, coriaces, spatulées à ovales, à marge épaissie, vert foncé et brillantes dessus, restant pâles dessous. Fleurs blanches puis jaunes ou roses, en grappes dressées. Fruits globuleux verdâtres, à chair mince et sèche.

Habitat : sols sablonneux et pas trop humides du sud de la Floride.

Bois : léger, tendre et faible. Bois de cœur rouge clair ; aubier légèrement plus clair.

Intérêts : l'écorce, riche en tannin, était employée en teinture et en tannage. Le bois a les qualités nécessaires pour être employé en construction mais les spécimens assez grands pour cela sont rares. Les fruits sont comestibles. Les feuilles et l'écorce étaient utilisées autrefois en médecine pour leurs tannins.

TALLOW BERRY. GLAMBERRY.

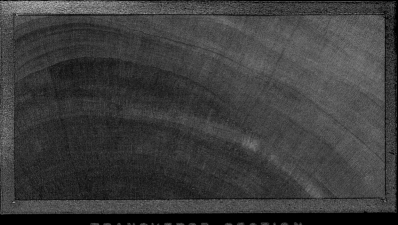

TRANSVERSE SECTION.

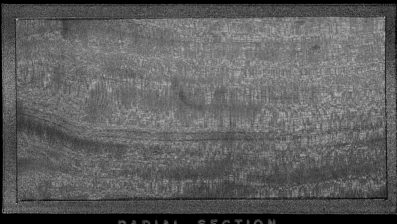

RADIAL SECTION.

TANGENTIAL SECTION.

Fr. Bois d'arrada. Sp. Palo de Doncello (Porto Rico)

M | MANGROVE FAMILY
Mangrove, Red Mangrove

FAMILIE DER MANGROVENGEWÄCHSE
Amerikanische Mangrove

FAMILLE DU PALÉTUVIER
Rhizophore manglier, Mangle rouge, Palétuvier d'Amérique

RHIZOPHORACEAE
Rhizophora mangle L.

Description: A shrubby tree, mostly 16 to 20 ft. (4.5 to 6 m) high. Forms numerous aerial roots. Rarely 75 to 85 ft. (21 to 24 m) high, free of branches up to half its height, and with a narrow crown. Bark thin to moderately thick, gray with a red tinge. Leaves ovate to elliptic, rounded at the tip. Flowering throughout the year. Fruit soon enlarging to a length of 15 in. (35 cm) through the emergence of the radicle from the seed. When this falls from the branch, the primary root either bores into the mud or the seed is carried away by the water.

Habitat: Muddy shores. Only developing into a tree above high-water mark. Widespread on the coasts of Central and South America, and West Africa. In North America confined to the islands of Florida; also in Lower California and other parts of Mexico.

Wood: Heavy, hard, strong, with a silky sheen, and close-grained; polishable, and resistant to attack by shipworm or drill shell (*Teredo*).

Use: Firewood and wharf piles. An important habitat for mussels, crustaceans, and young fish. Provides coastal protection.

Beschreibung: Meist 4,5 bis 6 m hoher, buschiger Baum. Zahlreiche Luftwurzeln ausbildend. Selten 21 bis 24 m hoch, bis zur Hälfte astfreier Stamm und schmale Krone. Borke dünn bis mäßig dick, grau mit roter Färbung. Blätter oval bis elliptisch, an der Spitze abgerundet. Blüten während des ganzen Jahres. Frucht durch auswachsende Primärwurzel schnell auf 35 cm Länge vergrößert. Von den Ästen herabfallend bohrt sie sich in den Schlamm oder wird vom Wasser fortgetragen.

Vorkommen: Schlammige Strände. Nur außerhalb der Gezeitenzone baumförmig. Weit verbreitet an den Küsten Mittel- und Südamerikas sowie Westafrikas. In Nordamerika nur auf den Inseln vor Florida, in Niederkalifornien und in Teilen Mexikos heimisch.

Holz: Außerordentlich schwer, hart, fest, seidig und mit feiner Textur; polierfähig und resistent gegen Schiffsbohrwurmbefall (*Teredo*).

Nutzung: Feuerholz und Kaipfähle. Wichtiger Lebensraum für Muscheln, Krebstiere und Jungfische. Küstenschutz.

Description : arbuste buissonnant de 4,5 à 6,0 m, et qui développe de nombreuses racines aérienne Atteint rarement 21 à 24 m de haut; tronc non ramifié jusqu'à hauteur et une couronne étroite Ecorce mince à peu épaisse, gris teintée de rouge. Feuilles ovales à elliptiques, arrondies à l'apex. Fleurs toute l'année. Fruits de 35 cm grâce aux racines primair qui se développent; ils tombent ensuite et disséminés par l'eau.

Habitat : vasières littorales. Port arborescent uniquement en deh des zones soumises à l'influence des marées. Très répandu sur les côtes de l'Amérique centrale et du Sud ainsi que de l'Afrique de l'Ouest. En Amérique du Nord, seulement sur les îles en face de Floride, en Basse-Californie et d d'autres régions du Mexique.

Bois : très lourd, dur, résistant, satiné et à grain fin; peut être p et insensible aux attaques de mo lusques xylophages (*teredo*).

Intérêts : combustible et pilotis d quais. Constitue un milieu biotiq très important pour les mollusqu les crabes et les poissons. Espèce protectrice du littoral.

MANGROVE. RED MANGROVE.

TRANSVERSE SECTION.

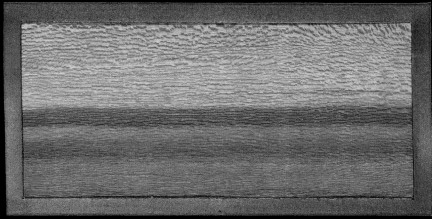

RADIAL SECTION.

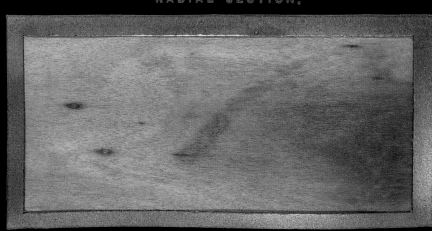

TANGENTIAL SECTION.

M | MAPLE FAMILY

Bigleaf Maple, Broadleaf Maple

FAMILIE DER AHORNGEWÄCHSE
Großblättriger Ahorn

FAMILLE DE L'ÉRABLE
Érable à grandes feuilles, Érable de l'Oregon

ACERACEAE
Acer macrophyllum Pursh

Description: 85 to 100 ft. (25 to 30 m) high, 2 to 3 ft. (0.6 to 0.9 m) in diameter. Trunk tall and straight. Bark brown, weakly tinged red or a deep russet. Deeply furrowed and broken into small, square scales.

Habitat: Banks of streams and rich, low-lying soils in the coastal regions of Pacific North America. In California sometimes on rocky slopes in humid areas.

Wood: Light, soft, not very strong, close-grained, often attractively rippled and twisted, sometimes in concentric circles as in Sugar Maple; polishable and easy to work. Heartwood dark brown, with red tinge; sapwood thick, paler or almost white.

Use: The most valuable wood from a deciduous tree in the forests of western North America. Also source of sugar in California. Young twigs have milky sap. In Washington and Oregon for interior construction, furniture, ax handles, and broomsticks. Shade and street-tree in the Pacific states.

Beschreibung: 25 bis 30 m hoch bei 0,6 bis 0,9 m Durchmesser. Hoher, gerader Stamm. Borke braun, schwach rot getönt oder kräftig rötlich braun. Tief gefurcht und in kleine, quadratische Schuppen gebrochen.

Vorkommen: Bachufer und reiche Niederungsböden in den Küstenregionen des pazifischen Nordamerika. In Kalifornien manchmal auf felsigen Hängen in humidem Klima.

Holz: Leicht, weich, nicht sehr fest und mit feiner Textur, diese oft schön gekräuselt und gedreht und manchmal in konzentrischen Ringen wie beim Zucker-Ahorn; polierfähig und leicht zu bearbeiten. Kern kräftig braun, rot getönt; Splint dick, heller oder fast weiß.

Nutzung: Das wertvollste Holz eines laubabwerfenden Baums in den Wäldern des westlichen Nordamerika. Wird in Kalifornien zu Zucker verarbeitet. In Washington und Oregon zum Innenausbau, für Möbel, Axtschäfte und Besenstiele. Schatten- und Straßenbaum in den pazifischen Staaten.

Description: 25 à 30 m de hauteu sur un diamètre de 0,6 à 0,9 m. Tronc droit, au port élancé. Ecorc brune, légèrement teintée de rou ou brun rougeâtre vif, profondément fissurée et se craquelant en petites écailles carrées.

Habitat: berges de ruisseaux et sols riches de secteurs déprimés des régions côtières de la façade pacifique de l'Amérique du Nord. Quelquefois sur les pentes rocheuses de la Californie humie

Bois: léger, tendre, pas très résistant et à texture fine, souvent fri et sinueuse, parfois en anneaux concentriques comme chez l'éral à sucre; susceptible d'un beau p et facile à travailler. Bois de cœu brun vif, teinté de rouge; aubier épais, plus clair ou presque blan

Intérêts: le plus précieux des feui lus des forêts pacifiques nord-an ricaines; sucre de bonne qualité en Californie. Jeunes rameaux secrétant un suc laiteux. Utilisé menuiserie intérieure et d'ameul ment, pour fabriquer des manch de haches et à balais. Arbre d'alignement et d'ombrage dans Etats de la façade pacifique.

Oregon Maple, Broad or Big-leaved Maple, Maple.

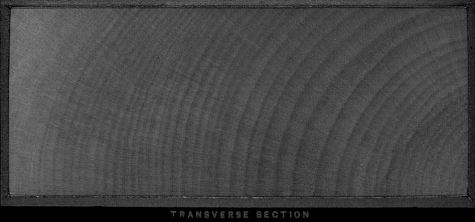

TRANSVERSE SECTION

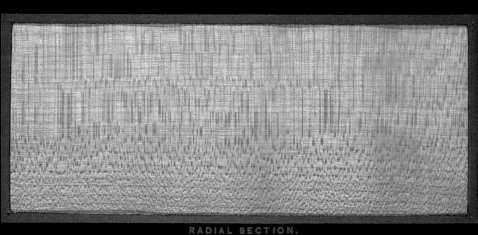

RADIAL SECTION.

TANGENTIAL SECTION

Ger. Grossblättriger Ahorn. Fr. Erable à grandes feuilles
Sp. Arce de hoja grande.

M | MAPLE FAMILY

Box Elder, Ash-leaved Maple

FAMILIE DER AHORNGEWÄCHSE
Eschen-Ahorn

FAMILLE DE L'ÉRABLE
Érable négundo typique

ACERACEAE
Acer negundo L.

Description: 50 to 85 ft. (16 to 25 m) high and 2 to 4 ft. (0.6 to 1.2 m) in diameter. Very quick-growing. Attractive foliage. Unusually resistant to drought, and therefore a popular and much planted street tree in the interior of the USA and in Europe.

Habitat: Riverbanks, lake shores, and low-lying ground, damp localities (though has a high resistance to drought). Widespread in the interior of North America.

Wood: Light, soft, and close-grained; easy to work.

Use: Interior construction, paper pulp. Sap sometimes used to produce sugar.

Introduced into Germany (at Mannheim) in the 18th century, and spread naturally from there into the meadows by the Rhine.

Beschreibung: 16 bis 25 m hoch bei 0,6 bis 1,2 m Durchmesser. Sehr schnellwüchsig. Gefälliges Laub. Ungewöhnlich widerstandsfähig gegen Trockenheit. Daher gern und viel gepflanzter Straßenbaum im Inneren der USA und in Europa.

Vorkommen: Flussufer, Seeufer und Tiefland, feuchte Wuchsorte (bei hoher Trockenresistenz) in weiten Teilen des inneren Nordamerika.

Holz: Leicht, weich und von feiner Textur; gut zu bearbeiten.

Nutzung: Innenausbau, Papier-Pulpe. Saft manchmal zu Zucker verarbeitet.

In Deutschland (Mannheim) im 18. Jh. eingeführt, breitet sich in der Rheinaue von selbst aus.

Description: 16 à 25 m de haute sur un diamètre de 0,6 à 1,2 m. Croissance très rapide. Beau feuillage abondant. Supporte bien la sécheresse. Arbre d'alignement apprécié pour cette raison à l'intérieur des Etats-Unis et en Europe.

Habitat: rives des cours d'eau et des lacs, secteurs déprimés, milieux humides (malgré une grande résistance à la sécheresse) dans régions étendues dans l'intérieur de l'Amérique du Nord.

Bois: léger, tendre et à grain fin, se travaillant bien.

Intérêts: utilisé pour la pâte à papier et en menuiserie intérieur. On extrait quelquefois la sève pour en faire du sucre.

Introduit en Allemagne (Mannheim) au 18e siècle ; se répand spontanément dans la plaine du Rhin.

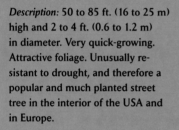

Box-Elder, Ash-leaved Maple, Negundo.

TRANSVERSE SECTION.

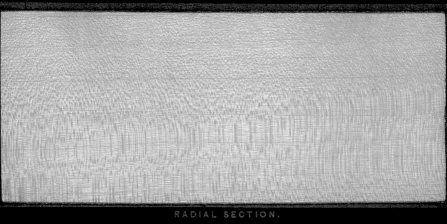

RADIAL SECTION.

TANGENTIAL SECTION

Ger. Eschenblättriger Ahorn. *Sp.* Negundo de Arce.

Fr. Erable à feuilles de frène.

M | MAPLE FAMILY
Red Maple, Scarlet Maple, Swamp Maple

FAMILIE DER AHORNGEWÄCHSE
Rot-Ahorn

FAMILLE DE L'ÉRABLE
Érable rouge, Plaine rouge (Canada)

ACERACEAE
Acer rubrum L. var. *trilobum* Torr. & A. Gray ex K. Koch
Syn.: *A. carolinianum* Walt.

Description: Up to 100 ft. (30 m) or more, and 3 to 4 ft. (0.9 to 1.2 m) in diameter. One of the first trees to show its brilliant colors in the fall. The swelling of the buds in late winter is an indication of an early spring. An early-flowering food plant for bees. The sugary sap is less sweet than that of the Sugar Maple. The indehiscent fruits are winged, and ornamental because of their red color; also they remain on the tree until new leaves are produced.

Habitat: Mainly lowland, margins of rivers and swamps, in the north in association with Black and Red Ash, American Arbor-vitae, American Larch, etc., in eastern North America.

Wood: Characteristics and color like those of the Silver Maple (see pp. 384/385).

Use: A considerable proportion of "Quilted Maple" (so called on account of its markings) comes from the Red Maple.

Much planted in Europe in parks and streets.

Beschreibung: Bis zu 30 m und höher bei 0,9 bis 1,2 m Durchmesser. Im Herbst einer der ersten Bäume, die ihre brillante Färbung zeigen. Das Anschwellen seiner Knospen im Spätwinter gilt als Vorfrühlingsbote. Früh blühende Bienenfutterpflanze. Zuckersaft weniger süß als der des Zucker-Ahorns. Geflügelte Schließfrüchte, zierend durch ihr Rot und ihr Verbleiben am Baum bis zum nächsten Laubaustrieb.

Vorkommen: Hauptsächlich Flachland, Ufer von Flüssen und Sümpfen, im Norden in Gesellschaft von Schwarz- und Rot-Esche, Lebensbaum, Amerikanischer Lärche u. a. im östlichen Nordamerika.

Holz: Eigenschaften und Farbe wie beim Silber-Ahorn (S. 384/385).

Nutzung: Ein beträchtlicher Teil des „Kräusel-Ahorns" (Zeichnung) stammt vom Rot-Ahorn.

In Europa vielfach als Park- und Straßenbaum.

Description: peut atteindre 30 m de hauteur, voire davantage, su un diamètre de 0,9 à 1,2 m. Un premiers arbres à arborer sa par flamboyante en automne. Le go flement de ses bourgeons à la s tie de l'hiver est considéré comr l'annonce du printemps. Espèce mellifère à floraison précoce. Sè moins sucrée que celle de l'érab sucre. Fruits décoratifs, sous form de disamares rouges à ailes form un U et persistant sur les ramea jusqu'à la prochaine feuillaison.

Habitat: essentiellement en plai bords des cours d'eau et des marécages; au nord, en compag des frênes rouge et noir, du thu géant, du mélèze laricin etc. On le trouve dans l'est de l'Amériqu du Nord.

Bois: mêmes qualités et couleur que celui de l'érable argenté (p. 384/385).

Intérêts: l'érable dit «frisé» (des doit une bonne partie de ses car tères spécifiques à l'érable roug

Arbre de parcs et de rues répan en Europe.

Red Maple, Swamp Maple, Soft Maple, Red-flowered Maple.

TRANSVERSE SECTION.

RADIAL SECTION.

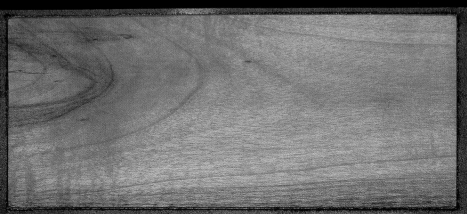

TANGENTIAL SECTION.

Ger. Rother Ahorn.   Sp. Arce rojo.

M | MAPLE FAMILY

Silver Maple, Silver-leaf Maple, White Maple

FAMILIE DER AHORNGEWÄCHSE
Silber-Ahorn

FAMILLE DE L'ÉRABLE
Érable argenté, Plaine de France (Canada)

ACERACEAE
Acer saccharinum L.
Syn.: *A. dasycarpum* Ehrh.

Description: 100 to 130 ft. (30–40 m) high and 3 to 5 ft. (1 to 1.70 m) in diameter. Leaves deeply indented, the upper-side green, the underside silvery white, decorative. Quick-growing. Sugary sap of high quality but smaller in quantity compared with *A. saccharum.*

Habitat: Low-lying ground, occasionally flooded water meadows in association with willows, River Birch, Red and Black Maple, Swamp White Oak, etc. Edged the banks of navigable rivers in the interior of eastern North America.

Wood: Fairly hard, strong, and very close-grained; easy to work.

Use: Wood for furniture. The so-called Quilted Maple (see p. 386) belongs to this species.

A park tree in the USA and Europe, also available in many cultivars.

Beschreibung: 30 bis 40 m hoch bei 1 bis 1,70 m Durchmesser. Laub tief eingeschnitten, Oberseite grün, Unterseite silberweiß, zierend. Schnellwüchsig. Führt einen qualitativ hochwertigen Zuckersaft, jedoch in geringerer Menge als *A. saccharum.*

Vorkommen: Tief gelegenes Flachland, gelegentlich überschwemmte Flussauen in Gesellschaft von Weiden, Flussbirke, Rot- und Schwarzem Ahorn, Sumpf-Weiß-Eiche u. a. Säumte die Ufer schiffbarer Flüsse im Inneren des östlichen Nordamerika.

Holz: Ziemlich hart, fest und mit sehr feiner Textur; gut zu bearbeiten.

Nutzung: Möbelholz. Der sog. Kräusel-Ahorn (vgl. S. 386) gehört zu dieser Art.

Parkbaum in den USA und in Europa, auch in vielen Kulturvarietäten.

Description: 30 à 40 m de haute sur un diamètre de 1 à 1,70 m. Feuilles découpées en lobes sépa par de profonds sinus étroits. Fac supérieure verte, face inférieure blanc argenté, finement pubesce et décorative. Arbre à croissance rapide. Sève de grande qualité m produite en moindre quantité qu chez *A. saccharum.*

Habitat: plaines basses, lits de hautes eaux inondés occasionnel lement, en compagnie des saules du bouleau des rivières, des érab rouge et noir, du chêne bicolore. Autrefois, arbre d'alignement en bordure des rivières navigables dans les régions de l'intérieur es de l'Amérique du Nord.

Bois: assez dur, résistant et au grain très fin; se travaillant bien.

Intérêts: utilisé comme bois d'ameublement. Espèce proche: l'érable frisé (voir p. 386).

Arbre de parcs aux Etats-Unis et Europe, ayant donné naissance à nombreux cultivars.

Silver-leaved Maple, White or Silver Maple, Soft Maple.

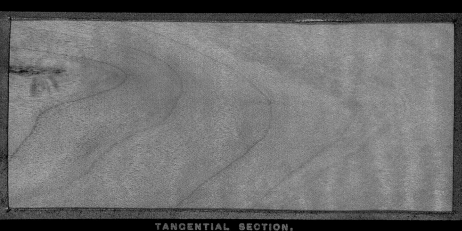

TRANSVERSE SECTION.

RADIAL SECTION.

TANGENTIAL SECTION.

Ger. Silberblätteriger Ahorn.   **Fr.** Erable blanc.

Sp. Arce con hojas plateadas.

M | MAPLE FAMILY

Silver Maple, Silver-leaf Maple, White Maple,
Quilted Maple

FAMILIE DER AHORNGEWÄCHSE

Silber-Ahorn, Kräusel-Ahorn

FAMILLE DE L'ÉRABLE

Érable argenté, Érable frisé, Plaine de France (Canada)

ACERACEAE

Acer saccharinum L.
Syn.: *A. dasycarpum* Ehrh.

Description: 100 to 130 ft.
(30 to 40 m) high and 3 to 5 ft.
(1 to 1.70 m) in diameter. Leaves
deeply indented, the upperside green, the underside silvery
white, decorative. Quick growing. Sugary sap of high quality but
smaller in quantity compared with
A. saccharum.

Habitat: Low-lying ground, occasionally flooded water meadows
in association with willows, River Birch, Red and Black Maple,
Swamp White Oak, etc. Edged
the banks of navigable rivers in the
interior of eastern North America.

Wood: Fairly hard, strong, and
very close-grained; easy to work.

Use: Wood for furniture.

A park tree in the USA and Europe,
also available in many cultivars.

Beschreibung: 30 bis 40 m hoch bei
1 bis 1,70 m Durchmesser. Laub
tief eingeschnitten, Oberseite grün,
Unterseite silberweiß, zierend.
Schnellwüchsig. Führt einen
qualitativ hochwertigen Zuckersaft,
jedoch in geringerer Menge als *A.*
saccharum.

Vorkommen: Tief gelegenes Flachland, gelegentlich überschwemmte
Flussauen in Gesellschaft von Weiden, Flussbirke, Rot- und Schwarzem Ahorn, Sumpf-Weiß-Eiche
u. a. Säumte die Ufer schiffbarer
Flüsse im Inneren des östlichen
Nordamerikas.

Holz: Ziemlich hart, fest und
mit sehr feiner Textur; gut zu
bearbeiten.

Nutzung: Möbelholz.

Parkbaum in den USA und in Europa, auch in vielen Kulturvarietäten.

Description: 30 à 40 m de haute
sur un diamètre de 1 à 1,70 m.
Feuilles découpées en lobes sép
par de profonds sinus étroits. Fa
supérieure verte, face inférieure
blanc argenté, finement pubesce
et décorative. Arbre à croissance
rapide. Sève de grande qualité r
produite en moindre quantité q
chez *A. saccharum.*

Habitat: plaines basses, lits de
hautes eaux inondés occasionne
lement, en compagnie des saule
du bouleau des rivières, des éra
rouge et noir, du chêne bicolore
Autrefois, arbre d'alignement er
bordure des rivières navigables
dans les régions de l'intérieur
est de l'Amérique du Nord.

Bois: assez dur, résistant et au
grain très fin; se travaillant bien

Intérêts: utilisé comme bois
d'ameublement.

Arbre de parcs aux Etats-Unis e
en Europe, ayant donné naissan
à de nombreux cultivars.

Silver Maple – Curly Figure, Curly Maple.

TRANSVERSE SECTION.

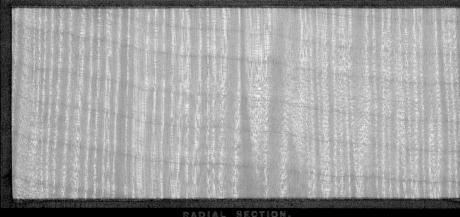

RADIAL SECTION.

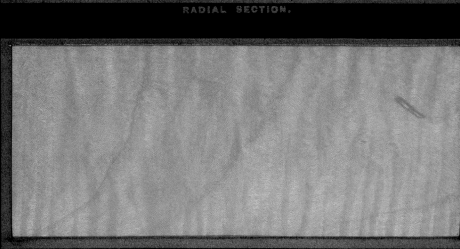

TANGENTIAL SECTION.

Ger. Gekränselter Ahorn.   Fr. Erable frisé.

M | MAPLE FAMILY

Striped Maple, Moosewood, Moose-bark

FAMILIE DER AHORNGEWÄCHSE
Amerikanischer Streifen-Ahorn

FAMILLE DE L'ÉRABLE
Erable de Pennsylvanie, Bois barré (Canada)

ACERACEAE
Acer pensylvanicum L.

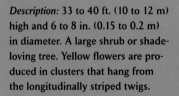

Description: 33 to 40 ft. (10 to 12 m) high and 6 to 8 in. (0.15 to 0.2 m) in diameter. A large shrub or shade-loving tree. Yellow flowers are produced in clusters that hang from the longitudinally striped twigs.

Habitat: Grows here and there in forests of Sugar Maple, Yellow Birch, Beech, Hemlock, Red Spruce, Butternut, etc., in north-eastern North America, particularly by mountain streams.

Wood: Rather light. Heartwood deep reddish brown; sapwood substantially paler, usually marked with stripes.

Use: In the north the tree may lose its branches because of the avid browsing of moose, hence some of its English names.

Beschreibung: 10 bis 12 m hoch bei 0,15 bis 0,2 m Durchmesser. Großstrauch, schattenliebender Baum. Gelbe Blüten in überhängenden Blütenständen an längs gestreiften Zweigen.

Vorkommen: Eingesprengt in Wälder von Zucker-Ahorn, Gelb-Birke, Buche, Hemlock, Rot-Fichte, Butternuss u. a. im nordöstlichen Nordamerika, besonders an Gebirgsbächen.

Holz: Eher leicht. Kern kräftig rosabraun; Splint reichlich heller, gewöhnlich mit Markflecken.

Nutzung: Im Norden dient der Baum als bevorzugte Nahrungsquelle der Elche und wird daher auch als „Elchbaum" bezeichnet.

Description: 10 à 12 m de hauteu sur un diamètre de 0,15 à 0,2 m Arbre d'ombre au port buissonnant. Fleurs jaunes groupées en grappes pendantes sur des rame striés de veines blanches.

Habitat: disséminé dans les forê d'érables à sucre, de bouleaux jaunes, de hêtres, de pruches, d'épicéas rouges, de noyers cend etc., surtout au bord des torrent de montagne dans le nord-est de l'Amérique du Nord.

Bois: plutôt léger. Bois de cœur d'un brun rose intense; aubier nettement plus clair, généraleme marqué de taches médullaires.

Intérêts: Dans le Nord, les élans broutent très volontiers, d'où so nom vernaculaire d'«arbre à éla»

TRANSVERSE SECTION.

RADIAL SECTION.

TANGENTIAL SECTION.

M | MAPLE FAMILY
Sugar Maple, Rock Maple, Blister Maple

FAMILIE DER AHORNGEWÄCHSE
Echter Zucker-Ahorn, Bläschen-Ahorn

FAMILLE DE L'ÉRABLE
Érable à sucre, Érable à fruits cotonneux, Érable du Canada,
Érable pustuleux, Érable moiré (Canada)

ACERACEAE
Acer saccharum Marshall

Description: 100 ft. (30 m) high or more, and 3 to 5 ft. (1 to 1.5 m) in diameter.

Habitat: Well-drained soils, especially northwards in association with Beech, Yellow Birch, Hemlock, Black Cherry, etc., sometimes forming woods itself. Eastern North America.

Wood: Heavy, hard, and strong.

Use: Excellent for all kinds of applications. A valuable wood in itself, a particular form such as the Blister Maple (illustrated here) produces an even more valuable, high-quality wood for veneers. Formerly also a first-rate firewood. Rich in potash.

Without doubt the most valuable broad-leaved tree of North America, both because of its wood and also as the producer of a sugary juice when the sap rises from the end of February to the beginning of April. Obtained by tapping the tree without killing it. Thickened into syrup and crystallized out.

Beschreibung: 30 m hoch und mehr bei 1 bis 1,5 m Durchmesser.

Vorkommen: Gut drainierte Böden, besonders nordwärts in Gesellschaft von Buche, Gelb-Birke, Hemlock, Schwarzkirsche u. a., bildet manchmal eigene Wälder. Östliches Nordamerika.

Holz: Schwer, hart und fest.

Nutzung: Ausgezeichnet für alle Arten der Verarbeitung und Verwendung. Das an sich schon wertvolle Holz wird durch die Besonderheiten seiner Unterarten – z. B. des hier abgebildeten Bläschen-Ahorns – noch kostbarer; hochwertiges Furnierholz. Früher auch vorzügliches Brennholz. Reich an Pottasche.

Zweifellos der wertvollste Laubbaum Nordamerikas durch sein Holz und als Lieferant seines Zuckersaftes beim „Saftsteigen" (Ende Februar bis Anfang April). Gewinnung durch Anzapfen ohne Abtötung des Baumes möglich. Eindicken zu Sirup und Auskristallisation.

Description: arbre de 30 m ou pl[u]s de hauteur, avec un tronc de 1 à 1,5 m de diamètre.

Habitat: sols frais et bien drainés en particulier vers le nord de son aire, en compagnie d'autres espèces comme le hêtre, le boule[au] jaune, la pruche, le cerisier tardi[f] etc. Constitue parfois des peuple[ments] forestiers. On le trouve da[ns] l'est de l'Amérique du Nord.

Bois: dur, lourd et résistant.

Intérêts: se prête à toutes sortes de traitements et d'utilisations. C[e] bois de grande valeur est représe[n]té ici sous une forme particulière[,] l'érable pustuleux de plus grand[e] valeur encore. Excellent bois de placage, encore exporté de nos jours. Autrefois, très bon bois de chauffage. Riche en potasse.

Sans conteste le plus précieux de[s] feuillus d'Amérique du Nord po[ur] l'exploitation de son bois et de s[a] sève sucrée, acheminée vers les rameaux à la sortie de l'hiver (fi[n] février-début avril). On récolte la sève en incisant l'écorce, sans conduire l'arbre à la mort. Epaiss[ie] et concentrée elle donne le sirop d'érable.

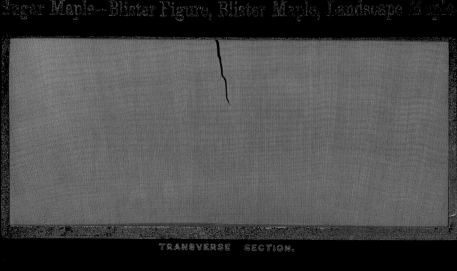

TRANSVERSE SECTION.

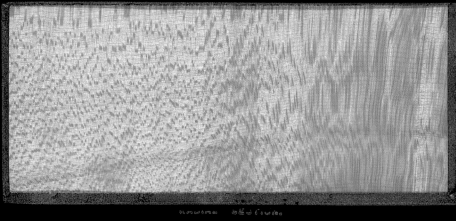

RADIAL SECTION.

TANGENTIAL SECTION.

M | MAPLE FAMILY

Sugar Maple, Rock Maple, Hard Maple

FAMILIE DER AHORNGEWÄCHSE
Echter Zucker-Ahorn

FAMILLE DE L'ÉRABLE
Érable à sucre, Érable à fruits cotonneux, Érable du Canada,
Érable moiré (Canada)

ACERACEAE
Acer saccharum Marshall

Description: 100 ft. (30 m) high or more, and 3 to 5 ft. (1 to 1.5 m) in diameter.

Habitat: Well-drained soils, especially northwards in association with Beech, Yellow Birch, Hemlock, Black Cherry, etc., sometimes forming woods itself. Eastern North America.

Wood: Heavy, hard, and strong.

Use: Excellent for all kinds of applications. A valuable wood in itself, particular forms such as the Blister Maple (see p. 390) and the Birdseye Maple (see p. 394) produce an even more valuable, high-quality wood for veneers. Formerly also a first-rate firewood. Rich in potash.

Without doubt the most valuable broad-leaved tree of North America, both because of its wood and also as the producer of a sugary juice when the sap rises from the end of February to the beginning of April. Obtained by tapping the tree without killing it. Thickened into syrup and crystallized out.

Beschreibung: 30 m hoch und mehr bei 1 bis 1,5 m Durchmesser.

Vorkommen: Gut drainierte Böden, besonders nordwärts in Gesellschaft von Buche, Gelb-Birke, Hemlock, Schwarzkirsche u. a., bildet manchmal eigene Wälder. Östliches Nordamerika.

Holz: Schwer, hart und fest.

Nutzung: Ausgezeichnet für alle Arten der Verarbeitung und Verwendung. Das an sich schon wertvolle Holz wird durch die Besonderheiten seiner Unterarten – z. B. des hier abgebildeten Bläschen-Ahorns (vgl. S. 390) und Vogelaugen-Ahorns (vgl. S. 394) – noch kostbarer; hochwertiges Furnierholz. Früher auch vorzügliches Brennholz. Reich an Pottasche.

Zweifellos der wertvollste Laubbaum Nordamerikas durch sein Holz und als Lieferant seines Zuckersaftes beim „Saftsteigen" (Ende Februar bis Anfang April). Gewinnung durch Anzapfen ohne Abtötung des Baumes möglich. Eindicken zu Sirup und Auskristallisation.

Description: arbre de 30 m ou plus de hauteur, avec un tronc de 1 à 1,5 m de diamètre.

Habitat: sols frais et bien drainés en particulier vers le nord de son aire, en compagnie d'autres espèces comme le hêtre, le boul jaune, la pruche, le cerisier tardi etc. Constitue parfois des peuplements forestiers. On le trouv dans l'est de l'Amérique du Nor

Bois: dur, lourd et résistant.

Intérêts: se prête à toutes sortes de traitements et d'utilisations. bois de grande valeur se rencon sous d'autres formes particulière telles que l'érable pustuleux (vo p. 390) et l'érable œil d'oiseau (p. 394). Excellent bois de placag encore exporté de nos jours. Au fois, très bon bois de chauffage. Riche en potasse.

Le plus précieux des feuillus d'Amérique du Nord pour l'exp tation de son bois et de sa sève s crée, acheminée vers les rameau la sortie de l'hiver (fin février-dé avril). On récolte la sève en incis l'écorce, sans conduire l'arbre à mort. Epaissie et concentrée elle donne le sirop d'érable.

Sugar Maple, Hard Maple, Rock Maple, Sugar-Tree.

TRANSVERSE SECTION.

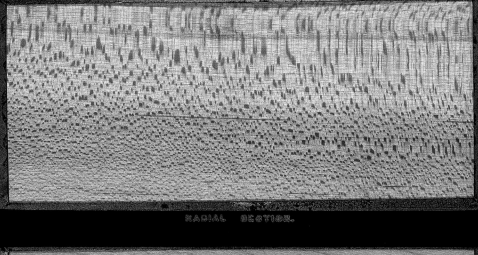

RADIAL SECTION.

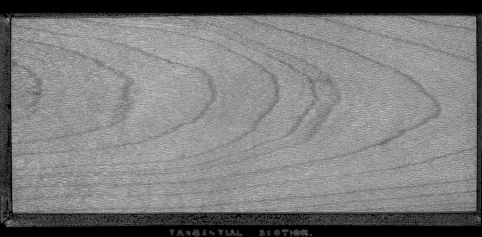

TANGENTIAL SECTION.

Ger. Zucker Ahorn. Fr. Erable à Sucre. Sp. Arce de azucar.

M | MAPLE FAMILY

Sugar Maple, Rock Maple, Hard Maple, Birds-eye Maple

FAMILIE DER AHORNGEWÄCHSE

Echter Zucker-Ahorn, Vogelaugen-Ahorn

FAMILLE DE L'ÉRABLE

*Érable à sucre, Érable à fruits cotonneux, Érable du Canada,
Érable œil d'oiseau, Érable moiré (Canada)*

ACERACEAE

Acer saccharum Marshall

Description: 100 ft. (30 m) high or more, and 3 to 5 ft. (1 to 1.5 m) in diameter.

Habitat: Well-drained soils, especially northwards in association with Beech, Yellow Birch, Hemlock, Black Cherry, etc., sometimes forming woods itself. Eastern North America.

Wood: Heavy, hard, and strong.

Use: Excellent for all kinds of applications. A valuable wood in itself, a particular form such as the Birdseye Maple produces an even more valuable, high-quality wood for veneers. Formerly also a first-rate firewood. Rich in potash.

Without doubt the most valuable broad-leaved tree of North America, both because of its wood and also as the producer of a sugary juice when the sap rises from the end of February to the beginning of April. Obtained by tapping the tree without killing it. Thickened into syrup and crystallized out.

Beschreibung: 30 m hoch und mehr bei 1 bis 1,5 m Durchmesser.

Vorkommen: Gut drainierte Böden, besonders nordwärts in Gesellschaft von Buche, Gelb-Birke, Hemlock, Schwarzkirsche u. a., bildet manchmal eigene Wälder. Östliches Nordamerika.

Holz: Schwer, hart und fest.

Nutzung: Ausgezeichnet für alle Arten der Verarbeitung und Verwendung. Das an sich schon wertvolle Holz wird durch die Besonderheiten seiner Unterarten – z. B. des hier abgebildeten Vogelaugen-Ahorns – noch kostbarer; hochwertiges Furnierholz. Früher auch vorzügliches Brennholz. Reich an Pottasche.

Zweifellos der wertvollste Laubbaum Nordamerikas durch sein Holz und als Lieferant seines Zuckersaftes beim „Saftsteigen" (Ende Februar bis Anfang April). Gewinnung durch Anzapfen ohne Abtötung des Baumes möglich. Eindicken zu Sirup und Auskristallisation.

Description: arbre de 30 m ou plus de hauteur, avec un tronc de 1 à 1,5 m de diamètre.

Habitat: sols frais et bien drainés en particulier vers le nord de son aire, en compagnie d'autres espèces comme le hêtre, le boule jaune, la pruche, le cerisier tardi etc. Constitue parfois des peuplements forestiers. On le trouve dans l'est de l'Amérique du Nor

Bois: dur, lourd et résistant.

Intérêts: se prête à toutes sortes de traitements et d'utilisations. O bois de grande valeur est représe té ici sous une forme particulière l'érable œil d'oiseau de plus gra valeur encore. Excellent bois de placage, encore exporté de nos jours. Autrefois, très bon bois de chauffage. Riche en potasse.

Sans conteste le plus précieux de feuillus d'Amérique du Nord po l'exploitation de son bois et de s sève sucrée, acheminée vers les rameaux à la sortie de l'hiver (fi février-début avril). On récolte la sève en incisant l'écorce, sans conduire l'arbre à la mort. Epaiss et concentrée elle donne le sirop d'érable.

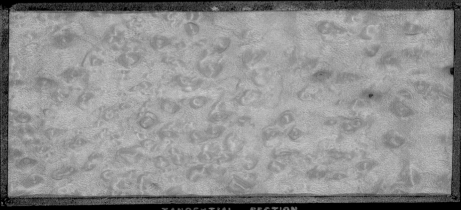

TRANSVERSE SECTION.

RADIAL SECTION.

TANGENTIAL SECTION.

M | MAPLE FAMILY
Vine Maple

FAMILIE DER AHORNGEWÄCHSE
Wein-Ahorn

FAMILLE DE L'ÉRABLE
Erable circiné, Erable à feuilles de vigne

ACERACEAE
Acer circinatum Pursh

Description: Rarely 30 to 40 ft. (9 to 12 m) high and to maximum of 10 or 12 in. (0.25 to 0.3 m) in diameter. Often only stocky, bushy or creeping. Bark pale red-brown with many shallow fissures. Leaves almost round, seven to nine-lobed. Lobes pointed, with doubly serrate margin, dark green. Flowers in umbel-like terminal clusters. Petals markedly shorter than sepals. Fruits almost at right angles to stalk, with thin wings.

Habitat: Alongside fast-flowing waters in the coastal regions of Pacific North America, forms extensive thickets in the north.

Wood: Heavy, hard, not very strong, close-grained. Heartwood pale brown to almost white; sapwood thick and pale. Many thin medullary rays.

Use: Firewood, ax and other tool handles. American Indians of the north-west use it to make fishing net frames.

Beschreibung: Selten 9 bis 12 m hoch bei maximal 0,25 bis 0,3 m Durchmesser. Oft nur rebstockartig, als Busch oder kriechend. Borke hell rotbraun mit zahlreichen flachen Rissen. Blätter fast rund, sieben- bis neunlappig. Lappen zugespitzt, mit doppelt gesägtem Blattrand, dunkelgrün. Blüten endständig und doldenähnlich. Kronblätter deutlich kürzer als die Kelchblätter. Früchte fast rechtwinklig zum Fruchtstiel ansetzend, mit dünnen Flügeln.

Vorkommen: Entlang von Fließgewässern in den Küstenregionen des pazifischen Nordamerika, im Norden ausgedehnte Dickichte bildend.

Holz: Schwer, hart, nicht sehr fest und mit feiner Textur. Kern hellbraun bis fast weiß; Splint dick und heller. Viele dünne Markstrahlen.

Nutzung: Feuerholz, Axt- und Werkzeuggriffe, die Indianer im Nordwesten fertigten damit auch die Rahmen ihrer Fischkescher.

Description : rarement 9 à 12 m de hauteur sur un diamètre ne dépassant pas de 0,25 à 0,3 m. se réduisant souvent à celui du de vigne, buissonnant ou rampa Ecorce brun rouge clair parcouru de nombreuses fissures plates. Feuilles vert sombre, presque rondes, formées de 7 à 9 lobes pointus, à contour doublement denté. Fleurs en corymbes, avec des pétales nettement plus cour que les sépales. Disamares à aile fines, insérées à angle presque droit sur le pédoncule.

Habitat : le long des eaux couran des régions côtières du Pacifiqu de l'Amérique du Nord ; en vas fourrés dans le nord.

Bois : lourd, dur, pas très résista et à grain fin. Bois de cœur brun clair à presque blanc ; aubier ép. et plus clair. Rayons médullaires nombreux et fins.

Intérêts : utilisé comme bois de feu, pour fabriquer des manches de haches et d'outils. Les Indien du nord-ouest s'en servaient po confectionner des cerceaux de trubles.

Vine Maple.

TRANSVERSE SECTION.

RADIAL SECTION.

TANGENTIAL SECTION.

Ger. Rebenahorn.   *Fr*. Erable de vigne.

Sp. Arce de vid.

M | MULBERRY FAMILY
Common Fig

FAMILIE DER MAULBEERBAUMGEWÄCHSE
Gemeiner Feigenbaum

FAMILLE DU MÛRIER DES FICUS
Figuier commun, Figuier de Carie

MORACEAE
Ficus carica L.

Description: Mostly a shrub, but occasionally taking on the dimensions of a tree, and then reaching a height of 35 ft. (10 m) The pale green leaves are five to seven-lobed, and have toothed margins. Their surface is roughly hairy. The bark is pale gray. The flowers are insignificant, sessile or shortly stalked within the large, roundish receptacles (which later become figs). Male flowers are extremely rare in cultivated plants. Receptacle subovoid. Fruits ovoid to pear-shaped, hidden in the fleshy receptacle.

Habitat: It is considered to have originated in the Mediterranean region, from where it has been introduced into large parts of the warmer regions of the world. It is extensively cultivated in southern North America.

Wood: Light and soft, but nevertheless strong and flexible.

Use: The Common Fig is cultivated for its edible receptacles (figs), but it is also used as a decorative plant and as a shade-tree. It is the only edible species of fig.

Beschreibung: Meist als Strauch wachsend, gelegentlich aber auch die Ausmaße eines Baumes annehmend und dann bis zu 10 m hoch. Die hellgrünen Blätter sind fünf- bis siebenlappig mit gezähnten Rändern. Ihre Oberfläche ist rau behaart. Die Borke ist blassgrau. Die Blüten sind unauffällig, sitzend oder kurz gestielt im Innern großer, rundlicher Rezeptakuli (der späteren „Feige"). Männliche Blüten sind bei der Kulturform äußerst selten. Rezeptakulum kurz und oval. Früchte oval bis birnenförmig, im fleischigen Rezeptakulum verborgen.

Vorkommen: Seine Heimat wird im Mittelmeerraum vermutet, von wo aus er in weite Teile der wärmeren Regionen der Erde eingeführt wurde. Im südlichen Nordamerika in großem Umfang angebaut.

Holz: Leicht und weich, aber dennoch fest und elastisch.

Nutzung: Der Feigenbaum wird wegen seiner essbaren Rezeptakuli („Feige") angebaut, daneben aber auch als Zierpflanze und Schattenspender verwendet. Es handelt sich um die einzige essbare Feigenart.

Description: buisson le plus sou~ mais prenant quelquefois un po arborescent, dans ce cas, jusqu'à 10 m de hauteur. Les feuilles ve clair sont palmatilobées, avec 5 7 lobes et dentées. Le limbe est rugueux et pubescent. L'écorce gris pâle. Les fleurs sont insigni fiantes, sessiles ou brièvement pédonculées, incluses dans un grand réceptacle arrondi (la futu «figue»). Les fleurs mâles sont extrêmement rares chez les vari tés cultivées. Le réceptacle est o et court. Les fruits drupacés son ovales à piriformes, cachés dans le réceptacle charnu.

Habitat: on pense qu'il est origi- naire du Bassin méditerranéen; introduit un peu partout dans le régions chaudes et très largeme cultivé dans le sud de l'Amériqu du Nord.

Bois: léger, tendre, poreux mais résistant et élastique.

Intérêts: cultivé pour ses fruits n aussi comme arbre d'ornement d'ombrage. C'est la seule espèce figuier à fruits comestibles.

COMMON FIG.

TRANSVERSE SECTION.

RADIAL SECTION.

TANGENTIAL SECTION.

M | MULBERRY FAMILY
Fig, Wild Fig

FAMILIE DER MAULBEERBAUMGEWÄCHSE

–

FAMILLE DU MÛRIER DES FICUS
Figuier blanc, Aralie-cerise

MORACEAE
Ficus citrifolia Mill.
Syn.: *F. brevifolia* Nutt.

Description: Rarely 40 to 50 ft. (12 to 15 m) high and 12 to 18 in. (0.3 to 0.45 m) in diameter, with an open, irregular crown. Aerial roots grow down to the ground from the branches. Bark moderately thick, smooth, light brown with an orange-colored tinge. Leaves broadly ovate to heart-shaped, dark green. Flowers insignificant, sessile or shortly stalked, with large, roundish receptacles inside (which later become figs). Receptacle subovoid. Fruits ovoid, almost completely enclosed by the calyx, and hidden in the receptacle.

Habitat: Slightly elevated coral rocks on shores. Occasionally as an epiphyte on other trees. Widespread in the Caribbean. In North America confined to southern Florida.

Wood: Light, soft, and close-grained. Heartwood orange-brown or yellow; sapwood thick, hardly distinguishable. With many distinct medullary rays and scattered, large, open pores in concentric circles.

Use: Small boats, guitars, boxes, small buildings, fences, telegraph poles (Begemann, 1988).

Beschreibung: Selten 12 bis 15 m hoch bei 0,3 bis 0,45 m Durchmesser. Offene, unregelmäßige Krone. Von den Ästen können Luftwurzeln zur Erde wachsen. Borke mäßig dick, glatt, hellbraun mit orangefarbener Tönung. Blätter breit und oval bis herzförmig, dunkelgrün. Blüten unauffällig, sitzend oder kurz gestielt im Innern großer rundlicher Rezeptakuli (der späteren „Feige"). Rezeptakulum kurz und oval. Früchte oval, fast völlig vom Kelch umhüllt, im Rezeptakulum verborgen.

Vorkommen: Leicht erhöht liegende Korallenfelsen am Strand. Gelegentlich als Aufsitzer auf anderen Bäumen. Verbreitet auf den Westindischen Inseln. In Nordamerika nur in Süd-Florida heimisch.

Holz: Leicht, weich und von feiner Textur. Kern orangebraun oder gelb; Splint dick, kaum unterscheidbar. Mit vielen deutlichen Markstrahlen und großen offenen, zerstreuten Gefäßen in konzentrischen Kreisen.

Nutzung: Leichte Boote, Gitarren, Kisten, leichte Bauten, Zäune, Telegrafenmasten (Begemann 1988).

Description: arbre atteignant rarement 12 à 15 m de hauteur, avec tronc de 0,3 à 0,45 m de diamètre et un houppier ouvert et irrégulier. Des racines aériennes peuvent naître des branches et pousser jusqu'au sol. Ecorce moyennement épaisse, lisse, brun clair teinté d'orange. Feuilles larges-ovales à cordiformes, vert foncé. Fleurs insignifiantes, sessiles ou brièvement pédonculées, tapissant l'intérieur d'un grand réceptacle court-ovale (la future «figue»). Petits fruits espacés ovales, presque inclus dans calice, cachés dans le réceptacle.

Habitat: écueils coralliens exhaussés sur les plages. Quelquefois aussi sous forme épiphyte. Répandu aux Caraïbes. En Amérique du Nord, dans le sud de la Floride.

Bois: léger, tendre et à grain fin. Pas de contraste entre le bois de cœur brun orange ou jaune et l'aubier épais. Nombreux rayons médullaires distincts et grands vaisseaux ouverts disséminés en cercles concentriques.

Intérêts: bateaux légers, guitares, caisses, constructions légères, clôtures, poteaux télégraphiques (Begemann 1988).

TRANSVERSE SECTION.

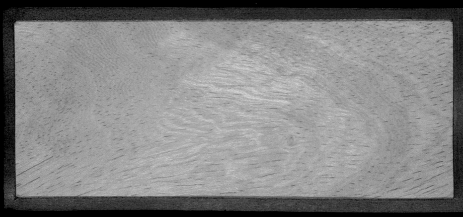

RADIAL SECTION.

TANGENTIAL SECTION.

M | MULBERRY FAMILY

Osage Orange, Bow Wood

FAMILIE DER MAULBEERBAUMGEWÄCHSE
Bogenholz, Gelbholz

FAMILLE DU MÛRIER DES FICUS
*Maclure épineuse, Bois-d'arc, Oranger des Osages,
Mûrier des Osages*

MORACEAE
Maclura pomifera (Raf.) C.K. Schneid.
Syn.: *Toxylon pomiferum* Raf. ex Sarg.

Description: 50 to 60 ft. (15 to 18 m) high and 2 to 3 ft. (0.6 to 0.9 m) in diameter. Branches stout, erect, forming an open, rather irregular, rounded crown. Bark moderately thick, deeply furrowed. Leaves oblong, broad at the base, tapering gradually to a long point, green. Male flowers in globular clusters; the female in round heads. Fruits spherical, 4 to 5 in. across.

Habitat: Rich, low-lying soils in the interior of south-eastern North America.

Wood: Heavy, not particularly tough, very strong, very flexible; shrinks considerably as it dries, durable. Heartwood brown, with longitudinal yellow markings (Begemann, 1987).

Use: Veneers, railroad sleepers, wagon-building, wheel rims, street surfaces, gymnastic apparatus. Of great importance as a source of dye in the tanning of leather, etc. Its rapid and dense growth, and its resistance to diseases and insect infestation renders it one of the most popular plants for "living fences". Its abundant, foliage and large, attractive fruits make it also valuable as an ornamental plant.

Beschreibung: 15 bis 18 m hoch bei 0,6 bis 0,9 m Durchmesser. Äste kräftig, aufrecht, eine offene, unregelmäßige, rundliche Krone bildend. Borke mäßig dick, tief rissig. Blätter länglich, aus breiter Basis allmählich lang zugespitzt, grün. Männliche Blüten in kugeligen Trauben, weibliche in runden Köpfchen. Früchte rund, 10 bis 13 cm groß.

Vorkommen: Reiche Tieflandböden. Im Inneren des südöstlichen Nordamerika.

Holz: Schwer, nicht besonders zäh, sehr fest, sehr elastisch; beim Trocknen beträchtlich schwindend und dauerhaft. Kern braun, längs gelb (Begemann 1987).

Nutzung: Furniere, Eisenbahnschwellen, Waggonbau, Radfelgen, Straßenpflaster, Turngeräte. Hauptbedeutung als Färbeholz, Ledergerbung etc. Ihr rascher und dichter Wuchs und die Resistenz gegen Krankheiten und Insekten ließen sie zu einer der beliebtesten Pflanzen für „lebende Zäune" werden. Ihr reiches Laub und die attraktiven großen Früchte machen sie auch als Zierpflanze wertvoll.

Description: arbre de 15 à 18 m de hauteur, avec un tronc de 0, à 0,9 m de diamètre. Branches vigoureuses, dressées, constitua un houppier rond, plutôt irrégu et ouvert. Ecorce épaisse, à text profondément fissurée. Feuilles ovales-lancéolées, terminées en longue pointe fine, vertes. Fleu mâles en petites grappes sphériques, fleurs femelles en globu Fruits ronds, de 10 à 13 cm.

Habitat: sols riches de plaine d le sud-est de l'Amérique du No

Bois: lourd, pas particulièreme tenace, très résistant et élastiqu fort retrait au séchage et durab Bois de cœur brun, jaune dans sens de la longueur (Begemann 1987).

Intérêts: utilisé en déroulage, e vage, pour fabriquer des traver de chemin de fer, des wagons, jantes et des agrès, mais surtou en teinture et en tannage. Une des espèces les plus appréciées haies vives à cause de sa croissance rapide et vigoureuse et d sa résistance aux parasites. Esp ornementale également par so feuillage dense et ses gros fruit attrayants.

Osage Orange.

TRANSVERSE SECTION

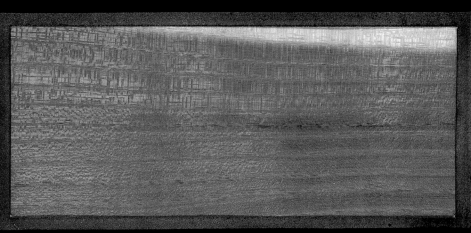

RADIAL SECTION.

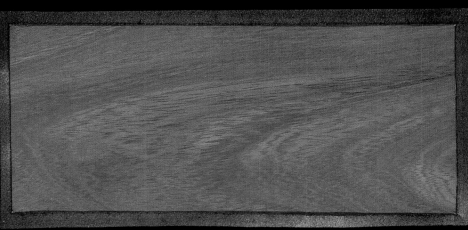

TANGENTIAL SECTION.

Ger. Bogen-Holz.   *Fr.* Bois d'Arc.

M | MULBERRY FAMILY
Paper Mulberry

FAMILIE DER MAULBEERBAUMGEWÄCHSE
Papiermaulbeerbaum

FAMILLE DU MÛRIER DES FICUS
Mûrier à papier

MORACEAE
Broussonetia papyrifera (L.) L'Hér. ex Vent.

Description: In North America a low tree not more than 45 ft. (14 m) high, often with a stout, curved, or twisted trunk, which can be up to 3 ft. (1 m) in diameter. Old trees in an isolated position are usually somewhat broader than high. The bark of young trees is smooth and marked with pale yellow lines; later it becomes rough and more irregular. The leaves are usually ovate, and can be without lobes, three to five-lobed, or asymmetrical with one side enlarged, all on the same tree. The flowers are arranged in spikes.

Habitat: Endemic to eastern Asia and on the Pacific islands. Since its introduction into south-eastern North America, it has spread and now grows wild in many parts of eastern North America.

Wood: The deep pink heartwood is light and rather soft. It is surrounded by yellowish white sapwood.

Use: The tree was introduced as an ornamental plant. Its name refers to the use of its inner bark for the manufacture of paper.

Beschreibung: In Nordamerika als niedriger Baum von nicht mehr als 14 m Höhe und oft mit einem gebogenen und verdrehten kräftigen Stamm, der bis zu 1 m dick werden kann. Freistehende alte Bäume sind meist etwas breiter als hoch. Die Borke junger Bäume ist glatt und wird von blassgelben Linien geziert, später wird sie rau und unregelmäßiger. Die Blätter sind meist oval, neben ungelappten können aber auch drei- bis fünflappige oder asymmetrisch einseitig vergrößerte Blätter am selben Baum gefunden werden. Die Blüten sind in Ähren angeordnet.

Vorkommen: Seine Heimat liegt in Ostasien und auf den pazifischen Inseln. Nach seiner Einführung im südöstlichen Nordamerika verwilderte er in weiten Teilen des östlichen Nordamerika.

Holz: Das kräftig rosafarbene Kernholz ist leicht und eher weich. Es wird von gelblich weißem Saftholz umgeben.

Nutzung: Der Baum wurde als Ziergehölz eingeführt. Sein Name verweist auf die Nutzung der inneren Rinde zur Papierherstellung.

Description: arbre ne dépassant 14 m de hauteur en Amérique Nord et doté d'un tronc souvent tortueux, vigoureux, pouvant atteindre 1 m de diamètre. Les spécimens âgés sont d'ordinaire un peu plus larges que hauts en situation isolée. Ecorce lisse, par courue de lignes jaune pâle à l'é juvénile, devenant rugueuse et irrégulière avec l'âge. Feuilles ov les le plus souvent mais adoptar aussi d'autre formes sur le mêm arbre : elles peuvent être découp en 3 à 5 lobes ou présenter un é gissement latéral. Fleurs disposé en épis.

Habitat: originaire de l'Asie orie tale et des îles du Pacifique. Intr duit dans le sud-est de l'Amériq du Nord puis devenu subsponta un peu partout à l'est du sous-continent.

Bois: léger et plutôt tendre. Bois de cœur d'un rose intense ; aubi blanc jaunâtre.

Intérêts: introduit comme arbre d'ornement. L'écorce interne ser à la fabrication de papier dans le pays dont elle est originaire, d'o son nom.

TRANSVERSE SECTION.

RADIAL SECTION.

TANGENTIAL SECTION.

Papier-Maulbeerbaum. *Fr.* Murier á papier.

M | MULBERRY FAMILY
Red Mulberry

FAMILIE DER MAULBEERBAUMGEWÄCHSE
Rote Maulbeere

FAMILLE DU MÛRIER DES FICUS
Mûrier rouge, Mûrier rouge d'Amérique

MORACEAE
Morus rubra L.

Description: 65 to 85 ft. (20 to 25 m) high and 3 to 4 ft. (0.9 to 1.2 m) in diameter. Leaves variously shaped, mostly large, casting dense shade. When fully ripe, the dark purple mulberries (a "false fruit") are juicy and delicious, and strongly staining.

Habitat: Rich lowland and lower slopes of hills in eastern North America.

Wood: Fairly heavy and moderately soft; very durable.

Use: In cooperage, in boatbuilding, and for fences.

Peter Kalm (1715–1779), a pupil of Linnaeus, was sent to North America in order to test the suitability of the Red Mulberry for its introduction to Sweden as a food plant for silkworms. This part of his mission failed.

Beschreibung: 20 bis 25 m hoch bei 0,9 bis 1,2 m Durchmesser. Vielgestaltige Blätter, die meisten groß, dichten Schatten werfend. Die schwärzlich purpurne Maulbeere (eine Scheinfrucht) ist, wenn vollreif, saftig, delikat und stark färbend.

Vorkommen: Nährstoffreiche Flachlandböden und niedrige Hangfüße im östlichen Nordamerika.

Holz: Ziemlich schwer und mäßig weich; sehr haltbar.

Nutzung: In der Böttcherei, im Bootsbau, für Zäune.

Peter Kalm (1715–1779), ein Schüler Linnés, wurde nach Nordamerika entsandt, um die Eignung und Einführung der Roten Maulbeere als Futterpflanze für Seidenraupen in Schweden zu prüfen. Dieser Teil seiner Mission misslang.

Description: 20 à 25 m de hauteu sur un diamètre de 0,9 à 1,2 m. Feuilles variables, généralement grande taille, offrant un ombrage dense. Les fruits, appelés « mûres sont pourpres presque noirs et possèdent une chair juteuse et savoureuse à maturité mais très tachante.

Habitat: sols de plaine riches et pieds de pentes bas dans l'est de l'Amérique du Nord.

Bois: assez lourd et moyenneme tendre; très durable.

Intérêts: utilisé en tonnellerie, dans la construction navale et pour fabriquer des clôtures.

Peter Kalm (1715–1779), un élèv de Linné, fut envoyé en Amériqu pour étudier les possibilités d'acc mater le mûrier rouge en Suède de le cultiver en tant qu'arbre du ver à soie. Mais sa mission devai échouer sur ce dernier point.

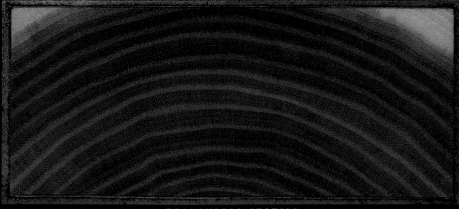

TRANSVERSE SECTION.

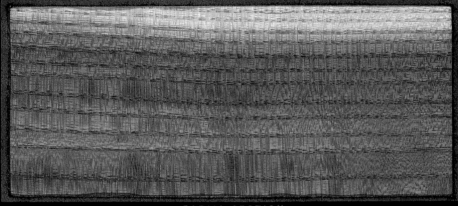

RADIAL SECTION.

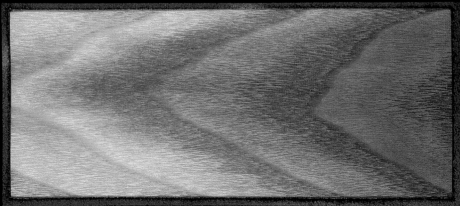

TANGENTIAL SECTION

Ger. Rother Maulbeerbaum.   Sp. Moral colorado.

Fr. Murier rouge.

M | MULBERRY FAMILY
Strangler Fig, Golden Fig

FAMILIE DER MAULBEERBAUMGEWÄCHSE
Wilder Feigenbaum

FAMILLE DU MÛRIER DES FICUS
Figuier-étrangleur, Figuier blanc, Figuier étrangleur de Floride

MORACEAE
Ficus aurea Nutt.

Description: 50 to 60 ft. (15 to 18 m) high, a broad-crowned plant rising above the other trees. It germinates on the trunk or branches of the host tree, and sends down aerial roots into the ground that fuse together and "strangle" the host. After reaching the ground, the roots are able to develop into new stems and in this way a single tree can extend over large areas. Bark moderately thick, smooth, ash-gray or light brown. The surface separates into flat scales, which, when they fall, expose the almost black inner bark. Leaves oblong-lanceolate, pointed at both ends. Flowers reddish purple. Fruit half-enclosed by the sepals.

Habitat: Wooded hills on the coasts and islands of southern Florida, and on the Bahamas.

Wood: Very light, soft, very weak, and coarse-grained; rapidly decaying in the ground.

Use: Occasionally used on Key West as a shade tree. Only lately introduced into the botanic gardens of the USA and Europe.

Beschreibung: 15 bis 18 m hoch, breitkronige „Überpflanze" auf anderen Bäumen. Keimt an den Stämmen oder Ästen des Wirtsbaumes, von wo aus Luftwurzeln zum Boden wachsen, miteinander verschmelzen und den Wirt „erwürgen" („Würgefeige"). Wurzeln können sich nach Erreichen des Bodens zu neuen Stämmen umwandeln und so aus einem Baum heraus große Flächen überziehen. Borke mäßig dick, glatt, aschgrau oder hellbraun. Auf der Oberfläche flache Schuppen abschälend, die nach dem Abfallen die fast schwarze innere Rinde entblößen. Blätter länglich-lanzettlich, an beiden Enden zugespitzt. Blüten rötlich violett. Frucht zur Hälfte von Kelchblättern umhüllt.

Vorkommen: Bewaldete Hügel der Strände und Inseln Süd-Floridas und der Bahamas.

Holz: Sehr leicht, weich, sehr schwach und mit grober Textur; im Boden sehr schnell faulend.

Nutzung: Auf Key West gelegentlich als Schattenbaum verwendet. Erst spät in die botanischen Gärten der USA und Europas eingeführt.

Description : arbre épiphyte de 15 à 18 m de hauteur, à couronne ample. Ses graines germent sur le tronc ou les branches de l'arbre hôte, et les racines aériennes formées croissent vers le sol en constituant un réseau de plus en plus dense autour du tronc du support (racines-lianes, ou racines étrangleuses). Après avoir atteint le sol, elles peuvent se transformer en de nouveaux troncs et occuper ainsi de vastes surfaces. L'écorce, moyennement épaisse, lisse, gris cendré ou brun clair, s'exfolie en écailles fines qui révèlent un endoderme presque noir. Feuilles allongées-lancéolées, terminées en pointe à l'apex et à la base. Fleurs violettes. Sépales recouvrant à moitié le fruit.

Habitat : buttes boisées des plages et des îles du sud de la Floride et des Bahamas.

Bois : très léger, tendre, très faible et à grain grossier ; se décompose très rapidement sur pied.

Intérêts : planté quelquefois comme arbre d'ombrage à Key West en Floride. Introduction tardive dans les jardins botaniques des Etats-Unis et d'Europe.

Golden Fig.   Wild Rubber-tree.   Strangle-tree.

TRANSVERSE SECTION.

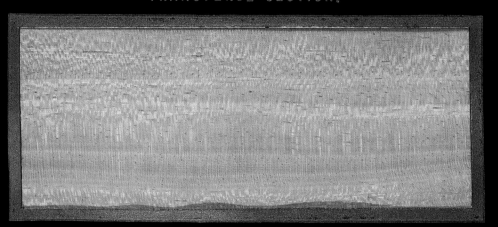

RADIAL SECTION.

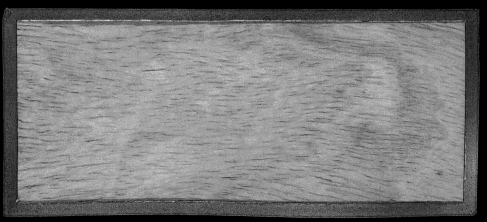

TANGENTIAL SECTION.

Ger. Wilder Feigenbaum.   Fr. Figuier doré.

Sp. Metapolo.

M | MULBERRY FAMILY
White Mulberry

FAMILIE DER MAULBEERBAUMGEWÄCHSE
Weiße Maulbeere

FAMILLE DU MÛRIER DES FICUS
Mûrier blanc

MORACEAE
Morus alba L.

Description: In North America a tree only rarely higher than 27 ft. (8 m), with a trunk up to 3 ft. (1 m) in diameter. The bark is rough and yellowish brown in color. The predominantly ovate, shining dark green leaves are toothed along the margins. The small flowers, lacking a corolla, are arranged in spikes. The composite fruits resemble blackberries. The individual fruits have at their tips the remains of the style.

Habitat: Native to north-eastern Asia, but after its introduction into North America it spread and is now found growing wild in many places in the eastern part of the continent.

Wood: Very hard, heavy, and durable.

Use: The foliage of the Mulberry is food for the larvae of the silk moth (silkworms), so attempts to cultivate this food plant have been made in all the warmer parts of the world. However, because of the high labor cost in harvesting the silk, its use in North America has not become established. Nor is its wood much used.

Beschreibung: In Nordamerika als Baum nur selten höher als 8 m mit bis zu 1 m dickem Stamm. Die Borke ist rau und gelblich braun gefärbt. Die überwiegend ovalen, glänzend dunkelgrünen Blätter sind am Rande gezähnt. Die kleinen Blüten, denen eine Blütenkrone fehlt, sind in Ähren angeordnet. Die Sammelfrüchte ähneln Brombeeren. Die Einzelfrüchte tragen an ihrer Spitze die Reste des Griffels.

Vorkommen: Ihre Heimat ist Nordostasien. Nach ihrer Einführung in Nordamerika verwilderte sie im östlichen Teil des Kontinents vielerorts.

Holz: Sehr hart, schwer und haltbar.

Nutzung: Das Laub des Maulbeerbaumes dient der Raupe des Seidenspinners als Nahrung. Versuche zur Kultivierung des Futterbaumes wurden daher in allen wärmeren Teilen der Welt unternommen. Wegen des hohen manuellen Aufwandes bei der Seidenernte hat sich die Nutzung in Nordamerika aber nicht durchgesetzt. Auch das Holz wird nur wenig genutzt.

Description: arbuste s'élevant rarement à plus de 8 m en Amérique du Nord, avec un tronc pouvant atteindre 1 m de diamètre. Écorce rugueuse, brun jaunâtre. Feuilles généralement ovales, à limbe vert foncé brillant et denté. Petites fleurs sans corolle, disposées en épis. Fruits ressemblant à des mûres. Chaque fruit est surmonté des restes du style.

Habitat: originaire du nord-est de l'Asie; introduit en Amérique du Nord et retourné à l'état sauvage un peu partout dans l'est du sous-continent.

Bois: très dur, lourd et durable.

Intérêts: les feuilles sont consommées par les chenilles du bombyx du mûrier. L'arbre fut introduit pour la sériciculture dans toutes les régions chaudes du globe mais l'élevage du ver à soie a été abandonné en Amérique du Nord car il nécessite une main-d'œuvre abondante. Le bois est également peu utilisé.

White Mulberry.

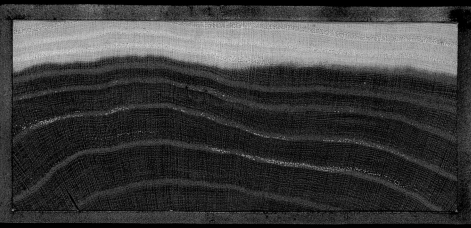

TRANSVERSE SECTION

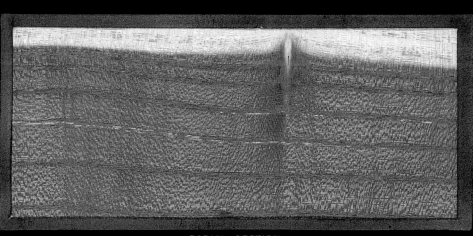

RADIAL SECTION.

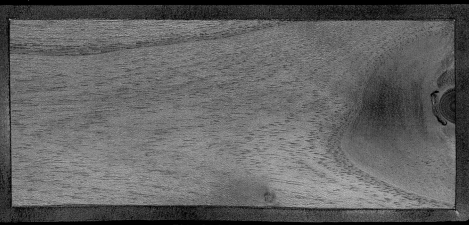

TANGENTIAL SECTION.

Ger. Weisze Maulbeerbaum. *Fr.* Mûrier blanc.

Sp. Moral blanco.

M | MYRRH FAMILY
Gumbo-limbo, West Indian Birch

FAMILIE DER WEISSGUMMIBAUMGEWÄCHSE
Gumbo Limbo

FAMILLE DE LA MYRRHE
Gommier rouge

BURSERACEAE
Bursera simaruba (L.) Sarg.

Description: 50 to 60 ft. (15 to 18 m) high and 2 ft. 6 in. to 3 ft. (0.75 to 0.9 m) in diameter. Main branches stout, growing out at right angles to the trunk. Bark thick, with glandular dots. Outer surface bright red-brown, dark brown or gray beneath. Leaves only at the ends of the branches, green on both sides. Flowering occurs before or while the leaves are unfolding. Flowers in long, slender panicles, with five calyx-lobes and petals. Fruit surrounded by dark red valves.

Habitat: Throughout the Caribbean. In North America in southern Florida.

Wood: Very light, extremely soft, spongy, and weak. Heartwood light brown; sapwood thick, similarly colored, soon discolored by decay.

Use: The wood is so light and weak that it is of little use. If pieces of the wood are cut off, they root easily and grow rapidly, so are suitable as hedge plants. The aromatic resin ("caranna") obtained by wounding the tree was formerly used in the treatment of gout, and on the West Indies was processed to form a varnish. In Florida the leaves are occasionally used as a tea substitute.

Beschreibung: 15 bis 18 m hoch bei 0,75 bis 0,9 m Durchmesser. Hauptäste kräftig, rechtwinklig vom Stamm abgehend. Borke dick mit drüsigen Punkten. Helle, rotbraune Oberfläche, darunter dunkelbraun oder grau. Blätter nur an den Enden der Äste, beiderseits grün. Blüte vor oder während der Laubentfaltung. Blüten in langen, schmalen Rispen mit 5 Kelch- und Kronblättern. Frucht von dunkelroten Valven umhüllt.

Vorkommen: In der Karibik verbreitet, in Nordamerika nur in Süd-Florida.

Holz: Sehr leicht, außerordentlich weich, schwammig und schwach. Kern hellbraun; Splint dick, von gleicher Farbe, rasch entfärbend durch Zerfall.

Nutzung: Das Holz ist so leicht und vergänglich, dass sich die Nutzung nicht lohnt. Abgeschnittene Triebe leicht wurzelnd und raschwüchsig, daher zur Bildung „lebender Zäune" geeignet. Aromatisches Harz frischer Wunden früher als „Caranna" zur Behandlung von Gicht verwendet, auf den Westindischen Inseln zu Lack verarbeitet. Blätter in Florida als Tee-Ersatz.

Description: 15 à 18 m de hauteu[r] sur un diamètre de 0,75 à 0,9 m. Branches maîtresses insérées à angle droit sur le tronc. Ecorce épaisse, marquée de points glan[du]laires, claire et brun rouge en su[r]face, brun foncé ou grise dessou[s]. Feuilles terminales uniquement, vertes sur les deux faces. Floraiso[n] avant ou pendant la feuillaison. Fleurs comptant 5 pétales et 5 sépales et réunies en longues panicules étroites. Fruit à péricar[pe] valvé rouge sombre.

Habitat: répandu dans les Caraïb[es] en Amérique du Nord seulemen[t] dans le sud de la Floride.

Bois: très léger, extrêmement tendre, spongieux et faible. Bois [de] cœur brun clair et aubier épais, d[e] même couleur; déteinte rapide p[ar] décomposition.

Intérêts: bois léger et éphémère, peu employé. Il convient par con[séquent] pour les haies vives car les ramea[ux] coupés prennent facilement racin[e] et croissent vite. La résine aroma[-] tique qui s'écoule des blessures fraîches était utilisée jadis contre la goutte et, aux Antilles, pour fabriquer un vernis. En Floride, l[es] feuilles sont un succédané du th[é].

GUMBO LIMBO.   WEST INDIAN BIRCH.

TRANSVERSE SECTION.

RADIAL SECTION.

TANGENTIAL SECTION.

Ger. Gummitragender Bursere. Fr.Gomart d'Amerique. Gommier (Fr.W.I.).

Sp. Almácigo, Carano (Sp.W.I., Mex., etc.), Jinocuave (Costa Rica).

M | MYRTLE FAMILY
Blue Gum

FAMILIE DER MYRTENGEWÄCHSE
Eukalyptus, Blauer Gummibaum

FAMILLE DE L'EUCALYPTUS ET DU GOYAVIER
Gommier bleu, Gommier globuleux

MYRTACEAE
Eucalyptus globulus Labill.

Description: Tree with thin, grayish brown, mottled bark, which separates in long strips. The leaves are extremely variable: ovate on young trees and in the lower part of the crown on older trees, lanceolate to sickle-shaped on older trees. Branches distinctly quadrangular in section. The flowers are large and mostly arise individually in the axils of the leaves. The fruits are oblong capsules, angular in appearance because of four ridges.

Habitat: Native in South Australia and Tasmania, often introduced into other warm parts of the world, including North America, California, where it sometimes escapes cultivation and grows wild.

Wood: Very heavy, hard, difficult to split and strong, grayish brown in color. The wood is very durable even when in contact with salt water.

Use: As firewood, as a decorative plant, and for protection from wind. Its scent drives away malaria-carrying insects. The leaves are used for medicinal purposes, and occasionally smoked.

Beschreibung: Baumförmig mit dünner, graubraun gefleckter Borke, die sich in langen Streifen ablöst. Die Blätter sind extrem variabel: oval auf jungen Bäumen und im unteren Teil der Krone alter Bäume, lanzettlich bis sichelförmig auf älteren Bäumen. Äste deutlich vierkantig. Die Blüten sind groß und stehen meist einzeln in den Blattachseln. Die Früchte sind rundliche Kapseln, die durch vier Grate eckig wirken.

Vorkommen: Heimisch in Südaustralien und Tasmanien, vielfach in warme Gegenden anderer Erdteile eingeführt; in Nordamerika besonders nach Kalifornien, wo er teilweise verwildert.

Holz: Sehr schwer, hart, schlecht spaltbar und fest, von graubrauner Farbe. Das Holz ist sehr widerstandsfähig, selbst im Kontakt mit Salzwasser.

Nutzung: Als Brenn- und Zierholz und als Windschutz. Sein Geruch vertreibt Anopheles-Mücken. Die Blätter werden für medizinische Zwecke genutzt, mitunter auch geraucht.

Description: arbre à écorce minc tachée de brun gris et se détachant en longues lanières. Feuil extrêmement variables: ovales chez les jeunes arbres et dans la partie inférieure de la couronne chez les sujets adultes, lancéolé à lunulaires sur les spécimens â Branches de section nettement quadrangulaire. Grandes fleurs taires, à l'aisselle des feuilles. Fr en capsule globuleuse marquée quatre arêtes qui lui donnent u apparence anguleuse.

Habitat: originaire du sud de l'A tralie et de la Tasmanie, souven introduit dans les régions chaud d'autres continents; en Amériqu du Nord, on le trouve surtout e Californie où il est devenu subs pontané à certains endroits.

Bois: très lourd, dur, difficile à fendre et résistant; très durable même au contact de l'eau; brun gris.

Intérêts: utilisé comme combustible, en ornement et en brise-ve Son odeur éloigne les moustiqu vecteurs du paludisme. Les feui sont employées en médecine po soigner certaines affections resp ratoires.

Eucalyptus, Blue Gum, Gum-tree.

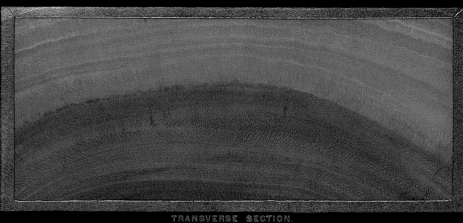

TRANSVERSE SECTION.

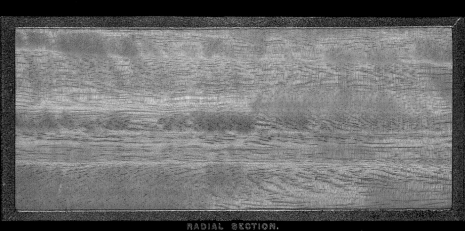

RADIAL SECTION.

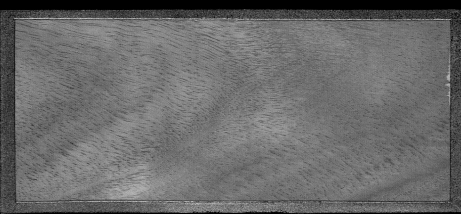

TANGENTIAL SECTION.

Ger. Eucalyptus. *Fr.* Eucalyptus. *Sp.* Eucalyptus.

M | MYRTLE FAMILY
Guava

FAMILIE DER MYRTENGEWÄCHSE
Guayava

FAMILLE DE L'EUCALYPTUS ET DU GOYAVIER
Goyavier

MYRTACEAE
Psidium guajava L.

Description: This small tree grows to a height of only 18 ft. (5 m) in North America. It has a trunk 8 in. (0.2 m) in diameter covered with a papery bark rich red-brown in color. The bark peels off in large flakes, revealing the brownish white inner bark. The oblong leaves are dark green on the upper side and covered with pale hairs beneath. The yellow or pink-tinged fruits are roundish to pear-shaped, and have a pleasant sharp-tasting flesh.

Habitat: Endemic to Central and South America. Since its introduction into North America it has escaped from cultivation and grows wild in the southern regions.

Wood: The pale reddish brown heartwood is heavy, hard, strong, and flexible. It is surrounded by even paler sapwood.

Use: The tree was introduced into North America because of its pleasant-tasting fruit.

Beschreibung: Der kleine Baum wird in Nordamerika nur 5 Meter hoch mit einem 0,2 m dicken Stamm, der von einer papierartigen, kräftig rotbraun gefärbten Borke bedeckt ist. Die Borke schält sich in großen Fetzen ab und legt dabei die bräunlich weiße innere Rinde frei. Die oberseits dunkelgrünen, länglichen Blätter sind auf der Unterseite blass behaart. Die gelben oder rosa getönten Früchte sind rundlich bis birnenförmig und besitzen ein angenehm säuerlich schmeckendes Fruchtfleisch.

Vorkommen: Seine Heimat ist das tropische Mittel- und Südamerika. Nach seiner Einführung in Nordamerika verwilderte er in den südlichen Regionen.

Holz: Das hell rötlich braune Kernholz ist schwer, hart, fest und elastisch. Es wird von noch hellerem Saftholz umgeben.

Nutzung: Der Baum wurde wegen seiner wohlschmeckenden Früchte nach Nordamerika eingeführt.

Description : arbuste n'atteignant que 5 m de hauteur en Amériqu du Nord, avec un tronc de 0,2 m diamètre. L'écorce papyracée d' brun rouge intense pèle en gran lambeaux et laisse apparaître dessous une écorce d'un blanc b nâtre. Les feuilles sont allongées vert foncé à la face supérieure e couvertes d'une pubescence pâl à la face inférieure. Les fruits, jaunes ou roses, sont globuleux piriformes et ont une saveur agr blement acide.

Habitat : originaire des zones tro cales de l'Amérique centrale et Sud. Introduit en Amérique du Nord et retourné à l'état sauvag dans les régions méridionales.

Bois : lourd, dur, résistant et élastique. Bois de cœur d'un bru rougeâtre clair ; aubier plus clair

Intérêts : l'arbre fut introduit en Amérique du Nord pour son fru savoureux.

GUAVA.

TRANSVERSE SECTION.

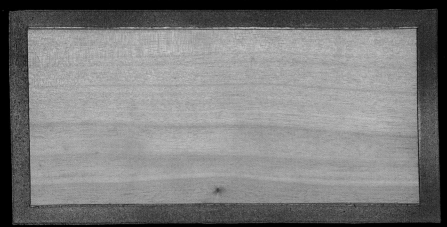

RADIAL SECTION.

TANGENTIAL SECTION.

Ger. GUAJAVA.   Fr. GOYAVIER

Sp. GUAJABA.

M | MYRTLE FAMILY
Naked Stopper, Nakedwood

FAMILIE DER MYRTENGEWÄCHSE
–

FAMILLE DE L'EUCALYPTUS ET DU GOYAVIER
Myrcianthe bois-d'Inde, Bois créole

MYRTACEAE
Myrcianthes fragrans (Sw.) McVaugh
Syn.: *Eugenia dicrana* Berg

Description: This small tree is rarely more than 30 ft. (8 m) high or 8 in. (0.2 m) in diameter. The red-brown bark is smooth, but its surface separates into small scales. The somewhat leathery, ovate leaves are covered with small black dots. The white flowers are arranged in groups of three in the axils of young leaves near the ends of the branches. The reddish-brown fruit is roundish to oblong, and has a thin, dry flesh. The brown, bean-shaped seeds are edible and aromatic.

Habitat: On sandy or stony soils in the coastal region of southern Florida.

Wood: The light brown or red heartwood is heavy and hard. It is surrounded by yellow sapwood.

Use: The fruits are edible but only rarely used.

Beschreibung: Der kleine Baum wird selten höher als 8 m und dicker als 0,2 m. Die rotbraune Borke ist glatt, schuppt sich aber an der Oberfläche. Die etwas ledrigen ovalen Blätter sind fein schwarz gepunktet. Die weißen Blüten stehen jeweils zu dritt in den Achseln junger Blätter in der Nähe der Astenden. Die rotbraune Frucht ist rundlich bis länglich. Sie besitzt ein dünnes und trockenes Fleisch. Die bohnenförmigen braunen Samen sind essbar und aromatisch.

Vorkommen: Auf sandigen oder steinigen Böden der Küstenregion Süd-Floridas.

Holz: Das hellbraune oder rote Kernholz ist schwer und hart. Es wird von gelbem Saftholz umgeben.

Nutzung: Die Früchte sind essbar, werden aber nur selten genutzt.

Description: arbuste atteignant rarement plus de 8 m de hauteu et 0,2 m de diamètre. Ecorce bru rouge, lisse mais se desquamant surface. Feuilles ovales, légèrem coriaces, finement ponctuées de noir. Fleurs blanches, groupées trois à l'aisselle de feuilles juvén subterminales. Fruit brun rouge, globuleux à allongé, avec une ch sèche et mince. Les graines brun réniformes sont comestibles et aromatiques.

Habitat: sols sablonneux ou caill teux des zones côtières du sud la Floride.

Bois: lourd et dur. Bois de cœur brun clair ou rouge; aubier jaun

Intérêts: les fruits sont comestib mais rarement consommés.

NAKED WOOD. NAKED STOPPER.

TRANSVERSE SECTION.

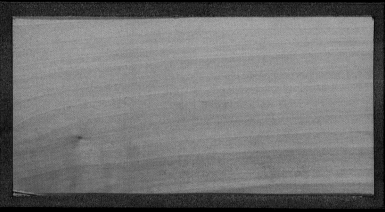

RADIAL SECTION.

TANGENTIAL SECTION.

M | MYRTLE FAMILY
Red Gum

FAMILIE DER MYRTENGEWÄCHSE
Rotgummi

FAMILLE DE L'EUCALYPTUS ET DU GOYAVIER
Gommier rouge

MYRTACEAE
Eucalyptus camaldulensis Dehnh.
Syn.: *E. rostrata* Schlecht.

Description: In its native country it grows very tall, with a trunk up to 15 ft. (4 m) in diameter and ash-gray or brownish bark that separates into plates. The leaves are lanceolate and both sides are uniformly colored. The flowers are arranged in umbels. Fruits cup-shaped with a broadened rim.

Habitat: Its native country is Australia, where it is widely distributed, mainly on damp soils. It was first introduced into North America in California, and rapidly became naturalized there.

Wood: Dense, hard, heavy, and strong. Polishable, extremely durable (even in contact with the ground) and is hardly ever attacked by wood-destroying insects or by the shipworm (Teredo).

Use: One of the most valuable of Eucalyptus timbers, good firewood and used everywhere where durability is required. The bark contains large quantities of tannins, which are also of medicinal use.

Beschreibung: In seiner Heimat von sehr großem Wuchs, mit bis zu 4 m dickem Stamm und aschgrauer oder bräunlicher Borke, die sich in Platten ablöst. Die Blätter sind lanzettlich und beiderseits gleich gefärbt. Die Blüten sind in Dolden angeordnet. Frucht tassenförmig mit verbreitertem Rand.

Vorkommen: Seine Heimat ist Australien, wo er vor allem auf feuchten Böden weit verbreitet ist. Er wurde in Nordamerika zuerst in Kalifornien eingeführt, wo er sich bald einbürgerte.

Holz: Dicht, hart, schwer und fest. Es ist gut polierbar, außerordentlich dauerhaft (selbst bei Bodenkontakt) und wird kaum von holzzerstörenden Insekten und dem Schiffsbohrwurm (*Teredo*) angegriffen.

Nutzung: Eines der wertvollsten Eukalyptus-Hölzer, gut brennbar und überall dort einsetzbar, wo Haltbarkeit wichtig ist. Die Rinde enthält große Mengen Tannine, die auch medizinisch genutzt werden.

Description: arbre de très grande taille dans son aire d'origine, ave un tronc pouvant atteindre 4 m diamètre. Ecorce gris cendré, se desquamant en plaques. Feuilles lancéolées, de même couleur su les deux faces. Fleurs groupées e ombelles. Fruits globuleux, tronqués et à bord évasé.

Habitat: très répandu dans son p d'origine l'Australie, surtout dan les sols humides. Introduit d'abo en Californie, où il s'est naturalis puis dans d'autres régions d'Am rique du Nord.

Bois: dense, dur, lourd et résista pouvant se polir et extrêmemen durable (même au contact du sol); quasiment inattaqué par le insectes xylophages (*teredo*).

Intérêts: un des bois d'eucalyptu les plus précieux; c'est également un bon combustible et il e recommandé en plantation pour durabilité. L'écorce est employée en thérapeutique car elle est rich en tannins.

Red Gum, Biall.

TRANSVERSE SECTION.

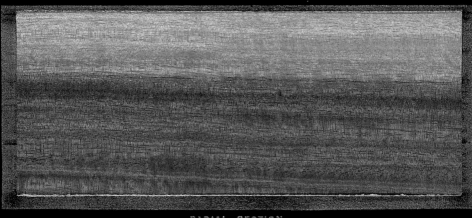

RADIAL SECTION.

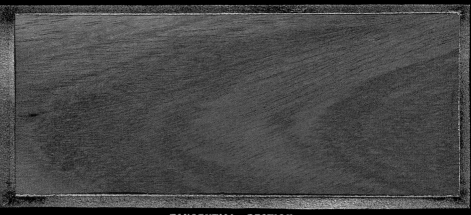

TANGENTIAL SECTION.

Ger. Rothgummi.   *Fr*. Gommier rouge.

Sp. Goma colorada.

M | MYRTLE FAMILY
Red Stopper, Garber's Stopper

FAMILIE DER MYRTENGEWÄCHSE
–

FAMILLE DE L'EUCALYPTUS ET DU GOYAVIER
Eugénie à fruits rouges, Bois ciceron, Boisiette,
Merisier jaune

MYRTACEAE
Eugenia confusa DC.

Description: The tree can reach a height of 70 ft. (20 m) and it has a trunk 20 in. (0.5 m) in diameter. The bark of the straight, columnar trunk is bright brownish red, but that of the slender branchlets is gray. The stout, erect main branches form a narrow, dense crown. The upper surface of the oblong-ovate leaves is a lustrous dark green. The lower surface is paler and marked with black dots. The white flowers are grouped in many-flowered clusters. The roundish fruit is bright scarlet. It has a thin, dry flesh that surrounds the almost globular seed.

Habitat: On small hills in southern Florida.

Wood: The rich red-brown heartwood is surrounded by a thick ring of dark-colored sapwood. It is hard, heavy, and uncommonly strong.

Use: There is no information about the uses of this tree.

Beschreibung: Der Baum kann eine Höhe von 20 m bei einem Stammdurchmesser von 0,5 m erreichen. Die Borke des säulenförmig geraden Stammes ist hell bräunlich rot, jene der schlanken Zweige dagegen grau. Die kräftigen, aufrechten Hauptäste formen eine schmale und dichte Krone. Die Oberseite der länglich ovalen Blätter ist glänzend dunkelgrün. Die Unterseite ist blasser und schwarz gepunktet. Die weißen Blüten sind in vielblütigen Büscheln vereinigt. Die rundliche Frucht ist scharlachrot. Sie besitzt ein dünnes, trockenes Fleisch, das den fast runden Samen umgibt.

Vorkommen: Auf kleinen Hügeln Süd-Floridas.

Holz: Das kräftig rotbraune Kernholz wird von einem dicken Ring des dunklen Saftholzes umgeben. Es ist hart, schwer und ungewöhnlich fest.

Nutzung: Über die Nutzung des Baumes liegen keine Informationen vor.

Description : arbre pouvant attein 20 m de hauteur sur un diamètr de 0,5 m. Le tronc est droit, columnaire, et recouvert d'une écorce rouge brunâtre clair ; cell des rameaux fins, par contre, es grise. Les branches maîtresses s vigoureuses, dressées et constituent un houppier étroit et dens Les feuilles allongées-ovales son d'un vert sombre brillant à la fac supérieure, plus pâles et tachées noir à la face inférieure. Les fleu blanches sont groupées en touff denses. Le fruit de couleur pour possède une chair mince et sèch enveloppant une graine presque sphérique.

Habitat : petites collines du sud de la Floride.

Bois : dur, lourd, extrêmement résistant. Bois de cœur d'un brun rouge intense ; aubier épais sombre.

Intérêts : aucun renseignement sur les utilisations éventuelles de cet arbre.

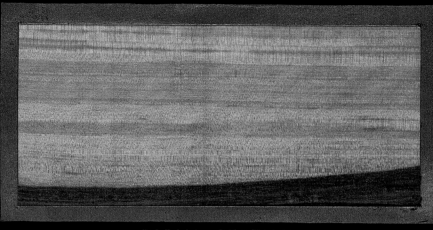

RADIAL SECTION.

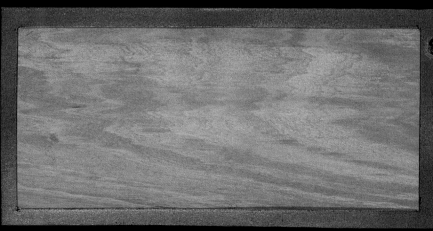

TANGENTIAL SECTION.

Sp. Sieneguillo (Porto Rico).

N | NIGHTSHADE FAMILY

Tree Tobacco

FAMILIE DER NACHTSCHATTENGEWÄCHSE
Baum-Tabak

FAMILLE DE LA TOMATE
Tabac glauque

SOLANACEAE
Nicotiana glauca Graham

Description: A small tree with a slender crown, only rarely reaching a height of up to 20 ft. (6 m), with a trunk 10 in. (0.25 m) in diameter. Bark at first green, later brown. Leaves broadly ovate to oblong with long stalks. Both the young branches and the leaves are covered with a bluish-green bloom. Flowers greenish yellow, in slender panicles at the ends of the branches. The fruits are oblong, two-valved capsules, tightly enclosed by the persistent calyx.

Habitat: The tree originates from around Buenos Aires, Argentina. It was introduced into southern California, where it frequently escapes cultivation and grows wild, especially in the coastal region, and can be found, for example, on riverbanks and on the edges of towns.

Wood: Light, soft, and brittle. The heartwood is brownish yellow in color and is surrounded by paler sapwood.

Use: A decorative plant. In spite of its name, it is in no way suitable as a substitute for tobacco.

Beschreibung: Kleiner Baum mit schlanker Krone, der nur selten eine Höhe von bis zu 6 m bei 0,25 m Stammdurchmesser erreicht. Borke anfangs grün, später braun. Blätter breit und oval bis länglich mit langen Blattstielen. Sowohl junge Äste als auch die Blätter sind blaugrün bereift. Blüten grünlich gelb, in schlanken Rispen am Ende der Äste. Die Früchte sind längliche, zweiklappige Kapseln, die von dem ausdauernden Kelch dicht umschlossen bleiben.

Vorkommen: Der Baum stammt aus der Gegend von Buenos Aires, Argentinien. Er wurde nach Südkalifornien eingeführt, wo er vor allem in der Küstenregion häufig verwildert und z.B. an Flussufern und an den Stadträndern zu finden ist.

Holz: Leicht, weich und brüchig. Das Kernholz hat eine bräunlich gelbe Farbe und ist von hellerem Saftholz umgeben.

Nutzung: Zierpflanze. Trotz des Namens in keiner Weise als Tabakersatz geeignet.

Description : petit arbre à couron étroite, atteignant rarement 6 m hauteur, avec un tronc de 0,25 m de diamètre. Ecorce verte deven brune avec l'âge. Feuilles larges-ovales à allongées, longuement pétiolées. Les jeunes rameaux, a que les feuilles, sont recouverts d'une pruine vert bleu. Fleurs d jaune franc, groupées en panicu terminales. Fruits en capsules allongées, bivalves et incluses d le calice persistant.

Habitat : originaire de la région d Buenos Aires, en Argentine. Intr duit en Californie où il est souv retourné à l'état sauvage, surtou dans les régions côtières, poussa par exemple sur les berges et au abords des villes.

Bois : léger, tendre et cassant. Bo de cœur jaune brunâtre ; aubier plus clair.

Intérêts : espèce ornementale. M gré son nom, il ne peut rempla en aucune manière le vrai tabac

Tree Tobacco, Wild Tobacco.

TRANSVERSE SECTION.

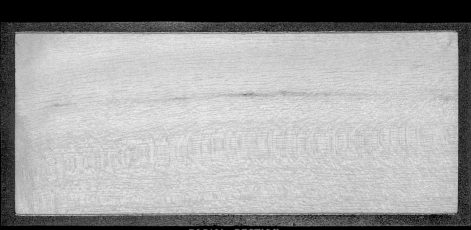

RADIAL SECTION.

TANGENTIAL SECTION.

Ger. Baumischer Tabak. *Fr.* Tabac d'arbre.

O | OLIVE FAMILY
Biltmore Ash

FAMILIE DER ÖLBAUMGEWÄCHSE
Biltmores Weißesche

FAMILLE DU FRÊNE ET DE L'OLIVIER
Frêne blanc, Frêne blanc d'Amérique, Franc-frêne (Canada)

OLEACEAE
Fraxinus americana L.
Syn.: *F. biltmoreana* Beadle

Description: 40 to 50 ft. (12 to 15 m) high and 1 ft. (0.3 m) in diameter. Branches erect or spreading, forming an open symmetrical crown. Bark dark gray, slightly furrowed. Leaves with seven to nine leaflets, the leaflets long-stalked, lanceolate, pointed, the upper surface at first yellow-bronze in color, later thick in texture, dark green, and somewhat lustrous, the lower surface pale, with long white hairs along the midrib. Stalks of leaves and leaflets hairy, leaf margins with rounded indentations or weakly serrate. Flowers dioecious, appearing at the same time as the leaves, in relatively compact, glabrous or hairy panicles. Fruits only slightly narrowing towards the end, and with a terminal wing, notched at the tip, and two to three times longer than the elliptic, many-veined "body" of the seed.

Habitat: Riverbanks in eastern North America.

Wood: The wood of this tree does not yet seem to have been analyzed.

Use: Occasionally planted as a park tree (Schmeil, 1994).

Beschreibung: 12 bis 15 m hoch bei 0,3 m Durchmesser. Äste aufrecht oder ausladend, eine offene, symmetrische Krone bildend. Borke dunkelgrau, schwach rissig. Blätter mit 7 bis 9 Fiedern, Blättchen lang gestielt, lanzettlich, zugespitzt, Oberseite zunächst gelbbronzefarben, später dick, dunkelgrün und etwas glänzend, Unterseite blass und längs der Mittelrippe mit langen weißen Haaren. Blatt- und Blättchenstiele behaart, Blattrand eingebuchtet bis schwach gesägt. Blüten zur Zeit der Laubentfaltung, zweihäusig, in relativ kompakten, unbehaarten oder behaarten Rispen. Früchte mit vorne nur wenig verschmälerten und an der Spitze eingebuchteten Flügeln, die zwei- bis dreimal länger als der kurzelliptische, vielnervige „Körper" des Samens sind.

Vorkommen: Flussufer im östlichen Nordamerika.

Holz: Das Holz des Baumes scheint bis heute noch nicht ausführlich untersucht worden zu sein.

Nutzung: Gelegentlich als Parkbaum (Schmeil 1994).

Description : arbre de 12 à 15 m d hauteur avec un tronc de 0,3 m e diamètre et des branches dressée ou étalées, constituant un houpp symétrique et ouvert. Ecorce gris sombre, peu fissurée. Feuilles à 7 ou 9 folioles longuement pétiolé lancéolées, acuminées à l'apex ; face supérieure d'abord bronze jaunâtre puis épaisse, vert foncé un peu lustrée, face inférieure pâ à nervure médiane garnie de lon poils blancs ; pétioles pubescents et marge échancrée à faiblement dentée. Fleurs dioïques au mom de la feuillaison, groupées en panicules relativement compacte glabres ou veloutées. Samare à péricarpe légèrement rétréci à la base, échancré au sommet, deux trois fois plus long que le « corps elliptique court et nervé de la graine.

Habitat : rives des cours d'eau de l'est de l'Amérique du Nord.

Bois : semble n'avoir fait jusqu'à présent l'objet d'aucune étude approfondie.

Intérêts : planté de temps à autre dans les parcs (Schmeil 1994).

TRANSVERSE SECTION

RADIAL SECTION

O | OLIVE FAMILY
Black Ash, Hoop Ash (burl wood)

FAMILIE DER ÖLBAUMGEWÄCHSE
Schwarz-Esche (Knotenholz)

FAMILLE DU FRÊNE ET DE L'OLIVIER
Frêne noir, Frêne gras (Canada) (Bois des nœuds)

OLEACEAE
Fraxinus nigra Marshall
Syn.: *F. sambucifolia* Lam.

Description: 85 to 100 ft. (25 to 30 m) high and 3 to 4 ft. (0.9 to 1.2 m) in diameter. Trunk straight, columnar, with gray, scaly bark and here and there knotty outgrowths ("ash burls").

Habitat: Low riverbanks and cold swamps in association with American Arbor-vitae, Balsam Poplar, larches, Silver Maple, Black Spruce, etc., in north-eastern North America, extending far into the north. Sometimes forming forests entirely by itself.

Wood: Fairly heavy, moderately hard and strong.

Use: Interior construction, furniture, strips for basket-making (as the wood splits easily). The wood of the knots is used for veneers.

Beschreibung: 25 bis 30 m hoch bei 0,9 bis 1,2 m Durchmesser. Gerader, säulenförmiger Stamm mit grauer, schuppiger Borke und hier und da knotigen Auswüchsen („ash burls").

Vorkommen: Niedrige Flussufer und kalte Sümpfe in Gesellschaft von Lebensbaum, Balsam-Pappel, Lärche, Silber-Ahorn, Schwarz-Fichte u. a. im nordöstlichen Nordamerika, dabei weit nach Norden reichend. Manchmal eigene Waldbestände bildend.

Holz: Ziemlich schwer, mäßig hart und fest.

Nutzung: Innenausbau, Möbel, Späne für Körbe (da es sich gut spalten lässt). Das Holz der Knoten für Furniere.

Description: arbre de 25 à 30 m de hauteur, avec un tronc droit, columnaire de 0,9 à 1,2 m de diamètre. Ecorce grise et écailleuse, présentant des nœuds («a burls») à certains endroits.

Habitat: berges basses et maréca froids dans le nord-est de l'Amé rique du Nord, s'étend jusque tr loin dans le nord, en association avec le thuya géant, le peuplier baumier, le mélèze, l'érable argenté, l'épicéa noir etc. Consti parfois des peuplements forestie

Bois: assez lourd, moyennement dur et résistant.

Intérêts: utilisé en menuiserie intérieure et d'ameublement; les copeaux (le bois se fend bien servent à faire des cageots et les nœuds, du bois de placage.

Black Ash Burl.

TRANSVERSE SECTION.

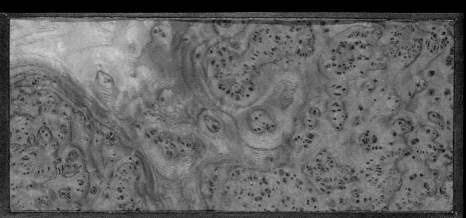

RADIAL SECTION.

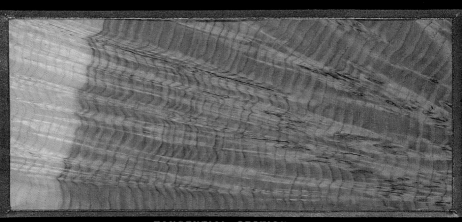

TANGENTIAL SECTION.

Schwarze Esche Knoten.　　Sp. Fresno negro batanado nud

Fr. Frêne noneux.

O | OLIVE FAMILY

Black Ash, Hoop Ash (stem wood)

FAMILIE DER ÖLBAUMGEWÄCHSE
Schwarz-Esche (Stammholz)

FAMILLE DU FRÊNE ET DE L'OLIVIER
Frêne noir, Frêne gras (Canada) (Bois du tronc)

OLEACEAE
Fraxinus nigra Marshall
Syn.: *F. sambucifolia* Lam.

Description: 85 to 100 ft. (25 to 30 m) high and 3 to 4 ft. (0.9 to 1.2 m) in diameter. Trunk straight, columnar, with gray, scaly bark and here and there knotty outgrowths ("ash burls").

Habitat: Low riverbanks and cold swamps in association with American Arbor-vitae, Balsam Poplar, larches, Silver Maple, Black Spruce, etc., in north-eastern North America, extending far into the north. Sometimes forming forests entirely by itself.

Wood: Fairly heavy, moderately hard and strong.

Use: Interior construction, furniture, strips for basket-making (as the wood splits easily). The wood of the knots ("ash burls") (see pp. 428/429) is used for veneers.

Beschreibung: 25 bis 30 m hoch bei 0,9 bis 1,2 m Durchmesser. Gerader, säulenförmiger Stamm mit grauer, schuppiger Borke und hier und da knotigen Auswüchsen („ash burls").

Vorkommen: Niedrige Flussufer und kalte Sümpfe in Gesellschaft von Lebensbaum, Balsam-Pappel, Lärche, Silber-Ahorn, Schwarz-Fichte u. a. im nordöstlichen Nordamerika, dabei weit nach Norden reichend. Manchmal eigene Waldbestände bildend.

Holz: Ziemlich schwer, mäßig hart und fest.

Nutzung: Innenausbau, Möbel, Späne für Körbe (da es gut spaltet). Das Holz der Knoten („ash burls") (s. S. 428/429) für Furniere.

Description: arbre de 25 à 30 m de hauteur, avec un tronc droit, columnaire de 0,9 à 1,2 m de diamètre. Ecorce grise et écailleuse, présentant des nœuds («a burls») à certains endroits.

Habitat: berges basses et maréca froids dans le nord-est de l'Amérique du Nord, s'étend jusque tr loin dans le nord, en association avec le thuya géant, le peuplier baumier, le mélèze, l'érable argenté, l'épicéa noir etc. Consti parfois des peuplements forestie

Bois: assez lourd, moyennement dur et résistant.

Intérêts: utilisé en menuiserie in rieure et d'ameublement; les copeaux (le bois se fend bien) serv à faire des cageots et les nœuds («ash burls») (v. p. 428/429), du bois de placage.

TRANSVERSE SECTION.

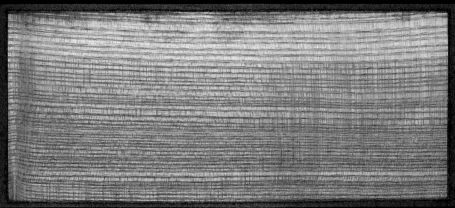

RADIAL SECTION.

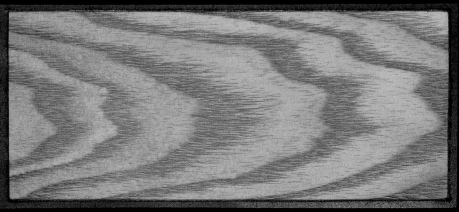

TANGENTIAL SECTION

Ger. Schwarze Esche. *Sp.* Fresno negro.

O | OLIVE FAMILY
Blue Ash

FAMILIE DER ÖLBAUMGEWÄCHSE
Blau-Esche

FAMILLE DU FRÊNE ET DE L'OLIVIER
Frêne bleu, Frêne anguleux (Canada)

OLEACEAE
Fraxinus quadrangulata Michx.

Description: A slender tree up to 120 ft. (37 m) high and 2 to 3 ft. (0.6 to 0.9 m) in diameter, but usually much smaller and only 60 to 70 ft. (18 to 21 m) high. Bark moderately thick, irregularly furrowed, light gray with a red tinge. Branches quadrangular. Leaves composed of five to nine, usually seven leaflets, margins coarsely serrate. When the leaves begin to unfold, they are covered with a thick, brown felt, but later they become yellow-green, glabrous on the upper surface, but occasionally with tufts of hairs on the lower surface. Flowers in panicles. Fruits oblong.

Habitat: Rich soils on limestone hills, rarely also on low ground in river valleys. In the Interior of eastern North America.

Wood: Heavy, hard, rather brittle, and close-grained. Heartwood pale yellow streaked with brown.

Use: Much used for flooring and carriage-building. A blue dye can be obtained from the inner bark, hence its name. Because of its rapid growth, beautiful appearance, and great resistance to diseases and insect pests it is an excellent park tree.

Beschreibung: Schlanker Baum von bis zu 37 m Höhe bei 0,6 bis 0,9 m Durchmesser, meist aber viel kleiner und nur 18 bis 21 m hoch. Borke mäßig dick, von unregelmäßigen Rissen durchzogen, hellgrau mit roter Tönung. Äste vierkantig. Blätter in 5 bis 9, meist aber 7 Fiedern geteilt, Blattränder grob gesägt. Nach der Laubentfaltung zunächst mit dickem braunem Filz bedeckt, später gelbgrün und oberseits kahl, auf der Unterseite gelegentlich mit einzelnen Haarbüscheln. Blüten in Rispen. Früchte länglich.

Vorkommen: Reiche Böden auf Kalkhügeln, selten auch im Tiefland von Flusstälern. Im Inneren des östlichen Nordamerika.

Holz: Schwer, hart, eher brüchig und mit feiner Textur. Kern hellgelb mit braunen Streifen.

Nutzung: Viel verwendet für Holzfußböden und Kutschen. Aus der inneren Rinde kann eine blaue Farbe gewonnen werden (Name!). Wegen seines raschen Wachstums, schönen Wuchses und seiner weitgehenden Resistenz gegen Krankheiten und Insektenfraß ein ausgezeichneter Parkbaum.

Description: arbre élancé pouvan atteindre 37 m de hauteur, avec tronc de 0,6 à 0,9 m de diamètre mais d'ordinaire bien plus petit à 21 m). Ecorce d'une couleur gr clair teintée de rouge, moyennement épaisse et irrégulièrement fissurée. Branches anguleuses. Feuilles composées de 5 à 9 folioles, 7 le plus souvent, grossi rement dentées, couvertes d'une épaisse pubescence brune à leur naissance puis devenant vert jau et glabres dessus, avec parfois quelques touffes de poils au reve Fleurs groupées en panicules. Samares allongées.

Habitat: sols riches sur collines calcaires, rarement fonds de vall fluviales. Croît à l'est dans l'inté rieur de l'Amérique du Nord.

Bois: lourd, dur, plutôt cassant e grain fin. Bois de cœur jaune cla avec des rubans bruns.

Intérêts: largement utilisé pour fabriquer des parquets et des calèches. Extraction d'une teintu bleue (d'où son nom!) de l'écord interne. Excellent arbre de parcs en raison de sa croissance rapide de sa belle stature et de sa grand résistance aux agents pathogène

Blue Ash,

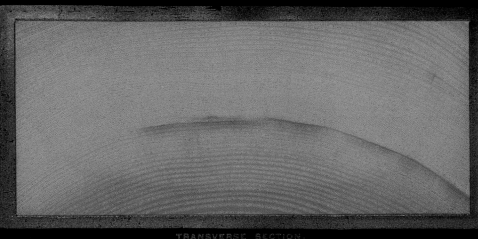

TRANSVERSE SECTION.

RADIAL SECTION.

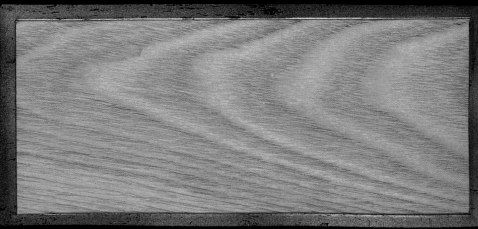

TANGENTIAL SECTION.

O | OLIVE FAMILY
Devilwood

FAMILIE DER ÖLBAUMGEWÄCHSE
Amerikanischer Ölbaum

FAMILLE DU FRÊNE ET DE L'OLIVIER
Osmanthe d'Amérique, Olivier d'Amérique,
Osmanthe américain

OLEACEAE
Osmanthus americanus (L.) Benth. & Hook. f. ex A. Gray

Description: 50 ft. (15 m) high and 4 in. (0.1 m) in diameter. A small tree or shrub. Bark thin, dark gray, or gray tinged with red, rough from small, thin, appressed scales. When this falls, the dark cinnamon-red inner bark becomes visible. The marvelously scented flowers develop before the leaves, the latter falling in their second year. Fruit dark blue.

Habitat: Rich, damp soils, riverbanks, and edges of the (typical) ponds in the "pine-barrens," i.e. poor pine-covered moorland spread throughout the region. Also grows in swamps and occasionally on higher, dry, and sandy soils. Widespread along the Gulf and Atlantic coasts in south-eastern North America.

Wood: Heavy, very hard, very strong, and close-grained.

Use: Very difficult to work.

Beschreibung: 15 m hoch bei 0,1 m Durchmesser. Kleinbaum, auch Strauch. Borke dünn, dunkelgrau oder rötlich grau, rau und mit kleinen, dünnen, anliegenden Schuppen. Fallen diese ab, wird die dunkel zimtrote, innere Rinde sichtbar. Die herrlich duftenden Blüten wachsen vor den Blättern, diese fallen im 2. Jahr. Frucht dunkelblau.

Vorkommen: Reicher, feuchter Boden, Flussufer und Ränder von (typischen) Teichen in den sog. „pine-barrens", das sind regional verbreitete arme Kiefernheiden. Bewohnt auch Sümpfe und gelegentlich höher gelegene, trockensandige Böden. Verbreitet entlang der Golf- und Atlantikküsten im südöstlichen Nordamerika.

Holz: Schwer, sehr hart, sehr fest und mit feiner Textur.

Nutzung: Sehr schwierig zu bearbeiten.

Description: petit arbre de 15 m hauteur sur un diamètre de 0,1 également sous forme d'arbriss buissonnant. Ecorce mince, gris foncé ou grise teintée de rouge couverte de petites écailles fine qui en se détachant révèlent un écorce interne rouge cannelle sombre. Les fleurs, très parfum apparaissent avant le feuillage, lequel tombe la deuxième anné Fruits bleu foncé.

Habitat: sols riches et humides, berges et bords des étangs dans ce qu'on appelle les «pine-barrens», landes à pins établies da certaines régions. Croît aussi da les marécages et quelquefois da les sols sablonneux secs situés en hauteur. On le trouve le lon Golfe et de la côte atlantique d le sud-est de l'Amérique du No

Bois: lourd, très dur et résistant à grain fin.

Intérêts: très difficile à travailler

Devil-wood, Wild Olive

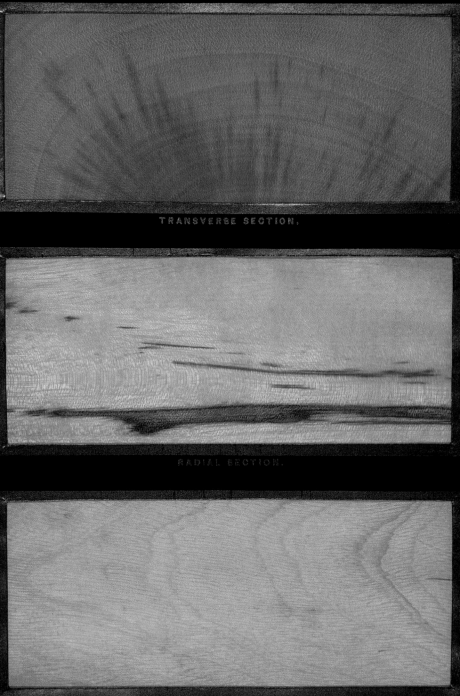

TRANSVERSE SECTION.

RADIAL SECTION.

TANGENTIAL SECTION

Ger. Amerikanischer Oehlbaum.   Fr. Olivier d'Amerique.

Sp. Madera del diablo.

O | OLIVE FAMILY
Fringe-flowered Ash

FAMILIE DER ÖLBAUMGEWÄCHSE
Fransenblühende Esche

FAMILLE DU FRÊNE ET DE L'OLIVIER
Frêne à deux pétales

OLEACEAE
Fraxinus dipetala Hook. & Arn.

Description: A many-stemmed shrub, 10 to 12 ft. (3.0 to 3.6 m) high, rarely a small tree, only under favorable conditions. Branches stout and at first almost rectangular. Bark dark green in the first year, later light brown or gray with a red tinge. Leaves with a short stalk and three to nine leaflets. Corolla white, divided to the base into two petals. Flowers in pyramidal panicles. Fruits linear-oblong to narrowly spatulate.

Habitat: Abundant on dry slopes near rivers in California.

Wood: Growing only rarely into a tree, and so for a long time arousing little interest.

Use: Presumably only suitable for firewood.

Beschreibung: Vielstämmiger Strauch von 3 bis 3,6 m Höhe, selten und nur unter günstigen Bedingungen als kleiner Baum. Äste kräftig und anfangs fast rechteckig. Borke im ersten Jahr dunkelgrün, später hellbraun oder -grau mit roter Tönung. Blätter mit kurzem Blattstiel und 3 bis 9 Blattfiedern. Blütenkrone weiß, bis zur Basis in zwei Blütenblätter zerteilt. Blüten in pyramidalen Rispen. Früchte linear-länglich bis schmal und spatelförmig.

Vorkommen: Häufig an trockenen Hängen in der Nähe von Flüssen in Kalifornien.

Holz: Nur sehr selten als Baum wachsend und daher bislang kaum untersucht.

Nutzung: Vermutlich nur als Brennholz.

Description: buisson à tronc multiple de 3,0 à 3,6 m de hauteur, rarement à port arborescent dans des conditions favorables. Branches vigoureuses, à section quasi quadrangulaire jeunes. Ec vert sombre la première année, e devenant brun ou gris clair tein de rouge avec l'âge. Feuilles bri ment pétiolées, composées de 3 9 folioles. Corolle blanche, divis jusqu'à sa base en deux pétales fleurs en panicules pyramidales Samares linéaires ou oblongues et étroites.

Habitat: souvent sur les pentes sèches à proximité des cours d'eau de Californie.

Bois: peu étudié jusqu'ici car il se présente rarement sous form arborescente.

Intérêts: n'est probablement util que comme bois de chauffage.

Fringe-flower Ash.

TRANSVERSE SECTION.

RADIAL SECTION.

TANGENTIAL SECTION

Ger. Fransenblühende Esche. *Fr.* Frêne à fleurs de frange.
Sp. Fresno de flores de franja.

O | OLIVE FAMILY
Green Ash

FAMILIE DER ÖLBAUMGEWÄCHSE
Grün-Esche

FAMILLE DU FRÊNE ET DE L'OLIVIER
Frêne vert

OLEACEAE
Fraxinus pennsylvanica R. S. Marsh.
Syn.: *F. lanceolata* Borkh.

Description: Rarely more than 60 ft. (18 m) high and 2 ft. (0.6 m) in diameter. A round-topped tree with gray, furrowed bark. Leaves light green on both sides, smooth at least above, somewhat paler beneath and occasionally hairy. Leaf margins sharply serrate. Flowers in panicles. Fruits linear to narrowly spatulate.

Habitat: Riverbanks from eastern North America westward to the eastern slopes of the Rocky Mountains.

Wood: Heavy, hard, strong, brittle, and rather coarse-grained. Heartwood brown; sapwood paler, with numerous medullary rays and rows of open pores along the edges of the annual rings.

Use: The wood is of inferior quality compared with that of *F. americana*, but is sometimes used as a substitute for this in a variety of products. It is distributed in great quantities in the northern USA. Because of its luxuriant, deeply green foliage, easy propagation, and tolerance of irregular supplies of water, the tree is popular for planting as an avenue tree, in parks, and as a shade tree.

Beschreibung: Selten mehr als 18 m hoch bei 0,6 m Durchmesser. Kugelige Krone sowie graue, rissige Borke. Beiderseits hellgrüne, zumindest oberseits glatte Blätter, unterseits etwas blasser und gelegentlich behaart. Blattränder scharf gesägt. Blüten in Rispen. Früchte linealisch bis schmal spatelförmig.

Vorkommen: Flussufer vom östlichen Nordamerika westwärts bis zur Ostabdachung der Rocky Mountains.

Holz: Schwer, hart, fest, brüchig und eher mit grober Textur. Kern braun; Splint heller, mit zahlreichen Markstrahlen sowie einigen Reihen offener Gefäße längs der Grenzen der Jahresringe.

Nutzung: Holz qualitativ weniger wertvoll als jenes von *Fraxinus americana*, gelegentlich als Ersatz von diesem für eine Vielzahl von Produkten verwendet. In großen Mengen in die nördlichen USA exportiert. Wegen des üppigen und kräftig grünen Laubes, der leichten Vermehrung und der Toleranz gegenüber unregelmäßiger Wasserversorgung ist der Baum beliebt für die Pflanzung als Alleebaum, in Parks und als Schattenspender.

Description : rarement plus de 18 m, avec un tronc de 0,6 m de diamètre. Houppier sphérique et écorce grise, fissurée. Feuilles ve clair sur les deux faces, glabres dessus, un peu plus pâles et par pubescentes dessous, bordées d dents aiguës. Fleurs en panicule Samares linéaires à oblongs.

Habitat : rives des cours d'eau depuis l'est de l'Amérique du N en direction de l'ouest jusqu'au versants est des Rocheuses.

Bois : lourd, dur, résistant, cassa et à grain plutôt grossier. Bois de cœur brun ; aubier plus clair, av de nombreux rayons médullaire et quelques faisceaux vasculaire de type ouvert le long des cerne annuels.

Intérêts : bois de moindre qualité que celui de *Fraxinus americana* mais parfois employé comme su titut pour un bon nombre de pro duits. Exporté en grande quantit vers le nord des Etats-Unis. App cié comme arbre d'alignement, c parcs et d'ombrage en raison de couleur et de l'abondance de so feuillage, de sa facilité à se multi plier et de son aptitude à suppor les contraintes hydriques.

TRANSVERSE SECTION.

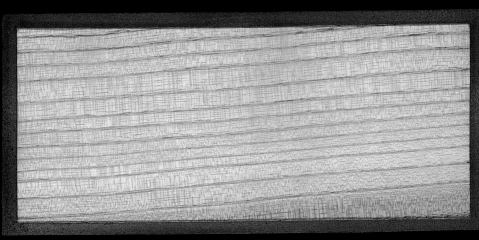

RADIAL SECTION.

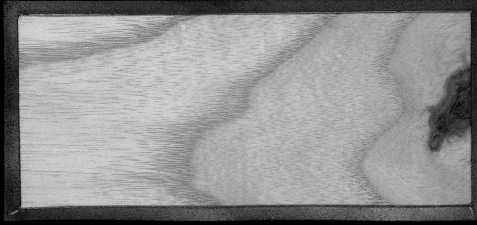

TANGENTIAL SECTION.

Ger. Grüne Esche. *Fr.* Frêne Vert. *Sp.* Fresno verde.

O | OLIVE FAMILY

Olive

FAMILIE DER ÖLBAUMGEWÄCHSE

Olivenbaum

FAMILLE DU FRÊNE ET DE L'OLIVIER

Olivier d'Europe

OLEACEAE

Olea europaea L.

Description: A tree up to 45 ft. (14 m) high with an irregular crown formed from a few stout branches. The trunk is irregularly formed with fissures and hollows, and can be as much as 15 in. (0.4 m) in diameter. The bark is gray and furrowed. Leaves lanceolate, narrowing gradually to the base, stiff and leathery in texture, and with a very short stalk. The upper surface is dark green, the lower surface covered with silvery white scales. The flowers are arranged in panicles. The fruits are stone fruits and rich in oil.

Habitat: Native to the Old World, and introduced into North America first in southern California, where it thrives on poor, loose soils.

Wood: Heavy, hard, and strong, very close-grained, with irregular annual rings and easy to polish.

Use: The wood is much prized in joinery. The olives, and the olive oil obtained from them, are used as foodstuffs.

Beschreibung: Bis zu 14 m hoher Baum mit unregelmäßiger Krone, die aus wenigen kräftigen Ästen gebildet wird. Der Stamm ist unregelmäßig geformt mit Rissen und Höhlen und kann bis 0,4 m dick werden. Die Borke ist grau und rissig. Blätter lanzettlich mit allmählich verschmälerter Basis, ledrig-steifer Konsistenz und sehr kurzem Blattstiel. Die Oberseite ist dunkelgrün, die Unterseite silbrig weiß beschuppt. Die Blüten sind in Rispen angeordnet. Ölreiche Kernfrüchte.

Vorkommen: Heimisch in der Alten Welt, in Nordamerika zuerst nach Südkalifornien eingeführt, wo er auf nährstoffarmen, lockeren Böden gut gedeiht.

Holz: Schwer, hart und fest, sehr feinkörnig mit unregelmäßigen Jahresringen und gut polierbar.

Nutzung: Das Holz ist sehr begehrt in der Tischlerei. Oliven und Olivenöl als Nahrungsmittel.

Description: arbre atteignant jusqu' 14 m de hauteur, avec un houpp irrégulier, composé de quelques branches vigoureuses. Tronc irré lier, présentant des fissures et de cavités et pouvant atteindre 0,4 de diamètre. Ecorce blanc grisâtr et crevassée. Feuilles lancéolées, s'allongeant à la base, d'une text rigide et très brièvement pétiolé Face supérieure vert foncé, rever blanc argenté couvert d'écailles. Petites fleurs réunies en grappes. Drupes oléagineuses (olives).

Habitat: originaire du Vieux Monde, importé en Amérique d Nord par le sud de la Californie il prospère sur des sols pauvres é éléments nutritifs mais légers.

Bois: lourd, dur et résistant, à gr très fin avec des cernes annuels irréguliers; prenant un beau poli

Intérêts: le bois est très recherch en ébénisterie. Cultivé pour ses fruits et pour l'huile que l'on en tire.

Olive.

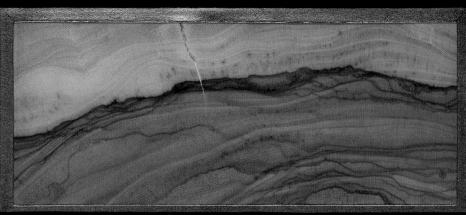

TRANSVERSE SECTION.

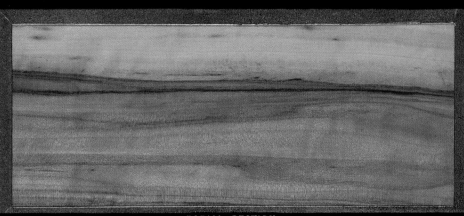

RADIAL SECTION.

TANGENTIAL SECTION.

Ger. Olivenholz.   *Fr.* Olivier.   *Sp.* Olivo.

O | OLIVE FAMILY
Oregon Ash

FAMILIE DER ÖLBAUMGEWÄCHSE
Oregon-Esche

FAMILLE DU FRÊNE ET DE L'OLIVIER
Frêne de l'Oregon

OLEACEAE
Fraxinus latifolia Benth.
Syn.: *F. oregona* Nutt.

Description: 80 to 90 ft. (23 to 26 m) high and 4 ft. (1.2 m) in diameter. Trunk tall. Bark dark gray or brown with reddish tinge. Deeply divided into broken furrows by broad, flat ridges. Surface has papery scales.

Habitat: Widespread along the North American Pacific coast. Good river valley soils in southwest Oregon, with alder and Bigleaf Maple. In southern California on good, moist soils close to streams.

Wood: Light, hard, brittle, and coarse-grained. Heartwood brown; sapwood thick, paler.

Use: Much used for furniture, vehicles and carts, framework, interior construction, cooperage. Firewood.

Street tree in Washington, Oregon, and Victoria, British Columbia. Hardy in Atlantic America and in western and southern Europe.

Beschreibung: 23 bis 26 m hoch bei 1,2 m Durchmesser. Hoher Stamm. Borke dunkelgrau oder braun und leicht rot getönt. Tief geteilt von unterbrochenen Rillen in breite, flache Riefen. Oberfläche aus papierenen Schuppen.

Vorkommen: Verbreitet entlang der nordamerikanischen Pazifikküste. Beste Niederungsböden der Flüsse in Südwest-Oregon mit Erle und Großblättrigem Ahorn. In Süd-Kalifornien auf guten, feuchten Böden in der Nähe von Bächen.

Holz: Leicht, hart, spröde und mit grober Textur. Kern braun; Splint dick, heller.

Nutzung: Viel für Möbel, Wagen und Waggons und deren Rahmen, zum Innenausbau, in der Böttcherei. Als Brennholz.

Straßenbaum in Washington, Oregon und Victoria, Britisch-Kolumbien. Winterhart im atlantischen Amerika und in West- und Südeuropa.

Description : arbre de 23 à 26 m de hauteur, avec un tronc élancé de 1,2 m de diamètre. Ecorce gris foncé ou brune légèrement teinte de rouge, à sillons et à larges côtes plates, se détachant en feuillets écailleux.

Habitat : répandu le long de la côte pacifique nord-américaine. Meilleurs sols des lits de rivières dans le sud-ouest de l'Oregon, en association avec l'aulne et l'érable à grandes feuilles ; sols humides et de bonne qualité, près des ruisseaux dans le sud de la Californie.

Bois : léger, dur, cassant et à grain grossier. Bois de cœur brun ; aubier épais et plus clair.

Intérêts : employé en charronnage, en ameublement, en menuiserie intérieure, en tonnellerie et pour fabriquer des wagons ; bois de chauffage également.

Arbres de routes et d'avenues dans le Washington, l'Oregon, le Victoria et la Colombie-Britannique. Résiste à l'hiver sur la façade atlantique de l'Amérique ainsi qu'en Europe occidentale et méridionale.

Oregon Ash.

TRANSVERSE SECTION

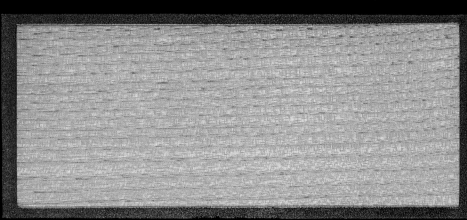

RADIAL SECTION.

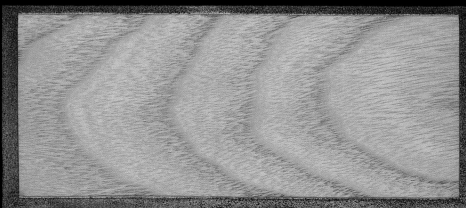

TANGENTIAL SECTION.

Ger. Oregonische Esche.　　　*Fr.* Frêne d'Oregon.

Sp. Fresno de Oregon.

O | OLIVE FAMILY
Red Ash

FAMILIE DER ÖLBAUMGEWÄCHSE
Rot-Esche

FAMILLE DU FRÊNE ET DE L'OLIVIER
Frêne rouge, Frêne de rivage (Canada)

OLEACEAE
Fraxinus pennsylvanica Marshall

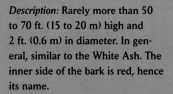

Description: Rarely more than 50 to 70 ft. (15 to 20 m) high and 2 ft. (0.6 m) in diameter. In general, similar to the White Ash. The inner side of the bark is red, hence its name.

Habitat: Rich, low-lying land (in contrast to the White Ash), the margins of swamps and riverbanks in association with Nettle Tree, elms, Swamp and Water Oaks, Bitternut, hickory, Red and Silver Maple, Sweet Gum and Black Gum, etc. Very widespread in eastern North America.

Wood: Heavy, hard, fairly strong, brittle, and coarse-grained. Heartwood light brown; sapwood paler, streaked with yellow.

Use: Sometimes confused with the better wood of the White Ash and used instead.

Beschreibung: Selten mehr als 15 bis 20 m hoch bei 0,6 m Durchmesser. Der Weiß-Esche ähnlich, unterscheidet sich aber in den deutlich gezähnten Blattfiedern (bei der Weiß-Esche ganzrandig oder nur schwach gezähnt). Innenseite der Rinde rot, daher der Name.

Vorkommen: Reiches, tief liegendes Flachland (im Gegensatz zur Weiß-Esche), Ufer von Sümpfen und Flüssen, meist in Gesellschaft von Zürgelbaum, Ulme, Sumpf- und Wasser-Eiche, Bitternuss, Hickory, Rot- und Silber-Ahorn, Amberbaum und Wald-Tupelo u. a. Sehr verbreitet im östlichen Nordamerika.

Holz: Schwer, hart, ziemlich stark, spröde und von grober Textur. Kern hellbraun; Splint heller mit gelbem Einschlag.

Nutzung: Manchmal verwechselt mit dem besseren Holz der Weiß-Esche und wie dieses verwendet.

Description : rarement plus de 15 20 m de hauteur sur un diamètre de 0,6 m. Ressemblant généralement au frêne blanc d'Amérique. L'écorce interne est rouge, d'où son nom.

Habitat : sols riches des plaines basses (contrairement au frêne blanc), bords des marais et des rivières, en compagnie du micocoulier, des ormes, du chêne des marais et du chêne noir, du noyer cendré, du caryer, des érables rouge et argenté, du liquidambar et du nyssa sylvestre. Espèce très répandue dans l'est de l'Amérique du Nord.

Bois : lourd, dur et assez fort, cassant et à grain grossier. Bois de cœur brunâtre ; aubier plus clair, avec une maillure jaune.

Intérêts : confondu quelquefois avec le bois du frêne blanc, de meilleure qualité, et utilisé comme celui-ci

Red Ash,   Gray Ash.

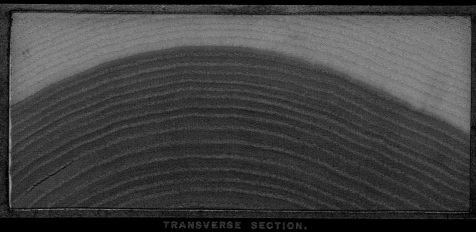

TRANSVERSE SECTION.

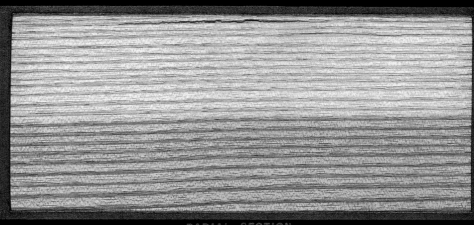

RADIAL SECTION.

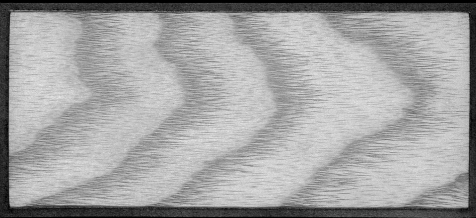

TANGENTIAL SECTION.

Ger. Rothesche.   Fr. Frêne rouge.   Sp. Fresno colorado.

O | OLIVE FAMILY
Swamp Privet

FAMILIE DER ÖLBAUMGEWÄCHSE
Forestiere

FAMILLE DU FRÊNE ET DE L'OLIVIER
Forestiéra acuminé, Troène des marais

OLEACEAE
Forestiera acuminata (Michx). Poir.

Description: Occasionally up to 27 ft. (8 m) high and 12 in. (0.3 m) in diameter. Often only a small, spreading shrub with a few crooked or deformed stems.

Habitat: Low riverbanks, lake margins, and flooded swamps, together with various willows, the American Plane, Water Elm, Swamp Cypress, tupelos, Water Locust, Water Hickory, Deciduous Holly, etc., in (south-)eastern North America.

Wood: Fairly light, hard, strong, and close-grained.

Use: Suitable for turnery.

Beschreibung: Gelegentlich bis 8 m hoch bei 0,3 m Durchmesser. Oft nur ein kleiner, ausgebreiteter Strauch mit mehreren krumm- oder schiefwüchsigen Stämmen.

Vorkommen: Flache Flussufer, Seeufer und überschwemmte Sümpfe, gemeinsam mit verschiedenen Weiden, der Amerikanischen Platane, der Wasser-Ulme, Sumpfzypresse, den Tupelos, der Wasser-Gleditschie, Wasser-Hickory, Laubabwerfender Stechhülse u. a. im (süd-) östlichen Nordamerika.

Holz: Ziemlich leicht, hart, fest und mit feiner Textur.

Nutzung: Geeignet für Drechslerarbeiten.

Description : atteint parfois 8 m de hauteur sur un diamètre de 0,3 m. Se présente souvent sous forme d'un petit buisson au port ample, composé de plusieurs tig tortueuses ou obliques.

Habitat : berges plates, rives des et marais inondés du (sud-) est l'Amérique du Nord, en associat avec divers saules, le platane d'O cident, le planéra des marécages taxode chauve, les nyssas, le fév aquatique, le caryer aquatique, l houx décidu etc.

Bois : assez léger, dur, résistant e à grain fin.

Intérêts : convient aux ouvrages de tournerie.

Swamp Privet.

TRANSVERSE SECTION.

RADIAL SECTION.

O | OLIVE FAMILY
Velvet Ash

FAMILIE DER ÖLBAUMGEWÄCHSE
Lederblättrige Esche

FAMILLE DU FRÊNE ET DE L'OLIVIER
Frêne velouté, Frêne de l'Arizona

OLEACEAE
Fraxinus velutina Torr.

Description: 30 to 40 ft. (9 to 12 m) high and rarely more than 8 in. (0.2 m) in diameter. Branches stout and spreading, usually forming a small spherical crown. Bark fairly thick, deeply grooved. Leaves of three to nine leaflets, entire or serrated, lateral leaflets with short stalks or almost sessile, dark yellow green, occasionally with scattered long hairs along the veins on the underside. Male and female flowers separate in short panicles. Fruits flat and spatulate (spoon-shaped).

Habitat: Mostly near rivers or in canyons, also occasionally in dry mesas in south-western North America.

Wood: Heavy, rather soft, not strong, close-grained. Heartwood pale brown; sapwood thick, paler. Contains many thin medullary rays.

Use: Ax handles and for building wagons. In the states of southern Arizona and northern Mexico often for shade along roads and at edges of irrigation channels.

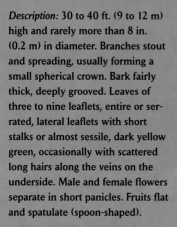

Beschreibung: 9 bis 12 m hoch bei nur selten mehr als 0,2 m Durchmesser. Äste kräftig und ausladend, normalerweise eine kugelige kleine Krone bildend. Borke mäßig dick, tiefrissig. Blätter in 3 bis 9 Blattfiedern geteilt, ganzrandig oder gesägt, die seitlichen Fiedern kurz gestielt oder fast sitzend, von dunkel gelbgrüner Farbe, gelegentlich mit einzelnen langen Haaren längs der Blattrippen auf der Unterseite. Männliche und weibliche Blüten separat in kurzen Rispen. Früchte flach spatelförmig.

Vorkommen: Meist in der Nähe von Flüssen oder in Cañons, gelegentlich aber auch in trockenen Mesas im Südwesten Nordamerikas.

Holz: Schwer, eher weich, nicht fest und mit feiner Textur, Kern hellbraun; Splint dick, heller. Enthält zahlreiche dünne Markstrahlen.

Nutzung: Für Axtschäfte und im Waggonbau. In den Städten von Süd-Arizona und Nord-Mexiko, häufig längs der Straßen und an den Rändern von Bewässerungsgräben als Schattenspender.

Description : 9 à 12 m de hauteur sur un diamètre dépassant rarement 0,2 m. Rameaux vigoureux et étalés, constituant d'ordinaire un petit houppier globuleux. Ecc moyennement épaisse, parcouru de fissures profondes. Feuilles co posées de 3 à 9 folioles, entières ou dentées ; folioles latérales brièvement pétiolées ou subsessiles, vert jaune foncé, garnies parfois de quelques poils longs s les nervures à la face inférieure. Espèce dioïque à fleurs groupées courtes panicules. Samares oblo gues et plates.

Habitat : le plus souvent à proximité des cours d'eau ou dans les canyons, mais parfois aussi sur l mesas sèches dans le sud-ouest de l'Amérique du Nord.

Bois : lourd, plutôt tendre, non résistant et à grain fin. Bois de cœur brun clair ; aubier épais, pl clair. Rayons médullaires nombr et fins.

Intérêts : manches de haches et construction de wagons. Arbre d'ombrage dans les villes du sud l'Arizona et du nord du Mexiqu souvent le long des routes ou de canaux d'irrigation.

Leather-leaf Ash.

TRANSVERSE SECTION.

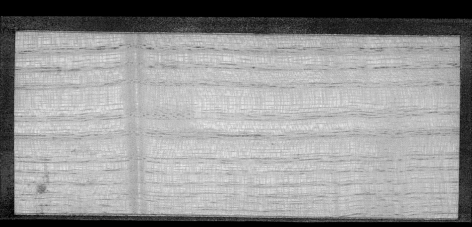

RADIAL SECTION.

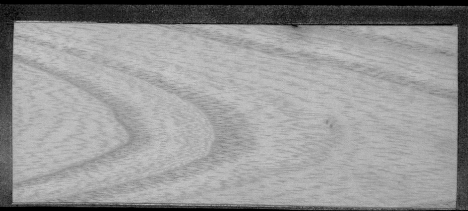

TANGENTIAL SECTION.

Ger. Lederblättrige Esche. *Fr*. Frêne à feuilles de cuir.

Sp. Fresno des hojas de cuero.

O | OLIVE FAMILY
Water Ash, Swamp Ash

FAMILIE DER ÖLBAUMGEWÄCHSE
Wasser-Esche

FAMILLE DU FRÊNE ET DE L'OLIVIER
Frêne de la Caroline, Frêne des marais

OLEACEAE
Fraxinus caroliniana Mill.

Description: A maximum of 40 ft. (12 m) high and 12 in. (0.3 m) diameter. Branches slender, forming a narrow, often round-topped crown. Bark thin, light gray with brown patches, broken on the surface into thin scales. Leaves divided into five to seven leaflets, sharply serrate or entire, glabrous or hairy, dark green on the upper surface, the underside paler or yellow-green. Flowers dioecious, in panicles. Fruits elliptic, oblong-ovoid, or spatulate.

Habitat: In deep swamps by rivers, under the shade of other trees, in the coastal regions of North America along the Atlantic and the Gulf of Mexico, and also in Cuba.

Wood: Light, soft, weak, and close-grained. Heartwood almost white, sometimes tinged red; sapwood thick, paler. Well-separated but indistinct medullary rays and pores.

Use: The smallest of the eastern ashes, and providing wood only of inferior quality. It was introduced into English parks in 1724, but is now once again scarcely found outside its native swamps.

Beschreibung: Maximal 12 m hoch bei 0,3 m Durchmesser. Äste schlank, eine schmale, oft kugelige Krone bildend. Borke dünn, hellgrau mit braunen Flecken, an der Oberfläche in dünne Schuppen zerfallend. Blätter in 5 bis 7 Fiedern zerteilt, scharf gesägt oder ganzrandig, glatt oder behaart, mit dunkelgrüner Ober- und blasser oder gelbgrüner Unterseite. Blüten zweihäusig, in Rispen. Früchte elliptisch, länglich oval oder spatelförmig.

Vorkommen: In tiefgründigen Sümpfen, an Flüssen, im Schatten anderer Bäume in den Küstenregionen Nordamerikas längs des Atlantiks und des Golfes von Mexiko sowie auf Kuba.

Holz: Leicht, weich, schwach und mit feiner Textur. Kern fast weiß, manchmal rot getönt; Splint dick, heller. Gut getrennte, aber undeutliche Markstrahlen und Gefäße.

Nutzung: Unter den östlichen Eschen die kleinste und eine schlechtere Holzqualität bietend. Wurde bereits 1724 in englische Parks eingeführt, ist aber heute kaum noch außerhalb von Sümpfen zu finden.

Description: 12 m de hauteur maximum avec un tronc de 0,3 de diamètre et des branches fin formant souvent un houppier rond. Ecorce mince, d'un gris cla tacheté de brun et se desquama Feuilles en 5 à 7 folioles, à dents aiguës ou entières, lisses ou pub centes, face supérieure vert som face inférieure plus pâle ou vert jaune. Fleurs dioïques, groupées en panicules. Samares elliptique allongées-ovales ou spatulées.

Habitat: marécages profonds, pr des cours d'eau, à l'ombre d'aut espèces dans les régions côtière de l'Amérique du Nord, le long l'Atlantique et du golfe du Mexi ainsi qu'à Cuba.

Bois: léger, tendre, faible et à gr fin. Bois de cœur presque blanc, parfois teinté de rouge; aubier épais, plus clair. Rayons médulla et vaisseaux bien séparés mais indistincts.

Intérêts: le plus petit des frênes orientaux et offrant un bois de moindre qualité. Introduit dans parcs anglais dès 1724, mais il n se rencontre pratiquement plus dehors des lieux marécageux qu affectionne.

Water Ash.

TRANSVERSE SECTION

RADIAL SECTION.

TANGENTIAL SECTION.

O | OLIVE FAMILY
White Ash

FAMILIE DER ÖLBAUMGEWÄCHSE
Weiß-Esche

FAMILLE DU FRÊNE ET DE L'OLIVIER
Frêne blanc, Frêne blanc d'Amérique, Franc-frêne (Canada)

OLEACEAE
Fraxinus americana L.

Description: Exceptionally as much as 135 ft. (40 m) high and 5 to 6 ft. (1.5 to 1.8 m) in diameter. A particularly imposing representative of its genus.

Habitat: Rich soils on slopes and on flat land, if not too damp. One of the main components of the forests of eastern North America and Canada.

Wood: Heavy, hard, and strong.

Use: Used for knife handles, ax-handles, agricultural implements, carts, furniture, etc.

One of the most valuable American forest trees. A popular shade tree in the USA.

Beschreibung: Ausnahmsweise bis 40 m hoch bei 1,5 bis 1,8 m Durchmesser. Ein besonders stattlicher Vertreter seiner Gattung.

Vorkommen: Reiche Böden an Hängen und im Flachland, wenn nicht zu feucht. Einer der Hauptvertreter in den Wäldern des östlichen Nordamerika und Kanada.

Holz: Schwer, hart und fest.

Nutzung: Genutzt für Messergriffe, Axtstiele, landwirtschaftliches Gerät, Wagen, Möbel usw.

Einer der wertvollsten amerikanischen Waldbäume. In den USA beliebt als schattenspendender Baum.

Description : s'élève de façon exceptionnelle à 40 m sur un diamètre de 1,5 m à 1,8 m. Représentant particulièrement imposant de son genre.

Habitat : sols riches de pentes et de plaines, à condition qu'ils ne soient pas trop humides. Une des principales essences forestières de l'est de l'Amérique du Nord et du Canada.

Bois : lourd, dur et résistant.

Intérêts : utilisé en charronnage et pour fabriquer des manches de couteaux et de haches, des outils agricoles, des meubles etc.

Une des essences forestières les plus précieuses d'Amérique du Nord. Très apprécié aux Etats-Unis pour son bel ombrage.

White Ash.

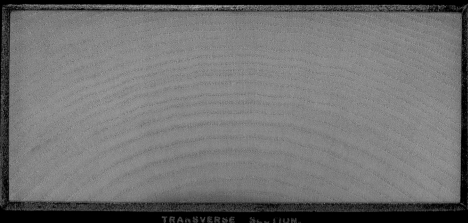

TRANSVERSE SECTION.

RADIAL SECTION.

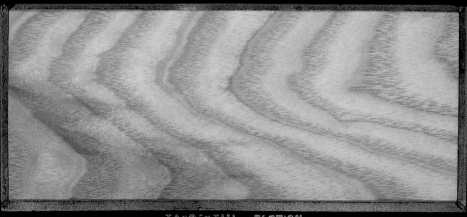

TANGENTIAL SECTION.

P | PALM FAMILY

Desert Palm, Fan Palm

FAMILIE DER PALMENGEWÄCHSE

Washingtonia

FAMILLE DU COCOTIER

Washingtonie de Californie

ARECACEAE

Washingtonia filifera (L. Linden) H. Wendl.
Syn.: *W. filamentosa* (Fenzi) Kuntze

Description: 85 ft. (25 m) high and 2 to 3 ft. (0.6 to 0.9 m) in diameter. Stem (not trunk) about 85 to 100 ft. (25 to 30 m) tall. Unbranched. Bark thick, pale red-brown, lightly scaly. Covered with layer upon layer of dead, hanging leaves, from the growing head right down to the soil.

Habitat: Damp, alkaline soil and canyons. Desert margins in south-western North America from Colorado to Lower California.

Wood: Light, soft, with many attractive fibrovascular bundles.

Use: Fruits gathered by American Indians and milled into flour.

First grown by Jesuit missionaries in southern California. Fast and strong-growing: may also be transplanted. South-west California, southern Europe, South America.

Beschreibung: 25 m hoch bei 0,6 bis 0,9 m Durchmesser. Der Strunk (nicht Stamm) etwa 25 bis 30 m hoch. Unverzweigt. Rinde dick, hell rotbraun, leicht schuppig. Darüber Schicht um Schicht ein Mantel abgestorbener, herunterhängender Blätter vom lebenden Schopf bis nahe an den Boden.

Vorkommen: Nasse Alkaliböden und Cañons. Ränder von Wüsten im südwestlichen Nordamerika von der Colorado-Wüste bis Niederkalifornien.

Holz: Leicht, weich, mit zahllosen auffallenden Faserbündeln.

Nutzung: Früchte von Indianern gesammelt und zu Mehl verarbeitet.

Zuerst von Jesuitenmissionaren in Süd-Kalifornien kultiviert. Schnell- und starkwüchsig, auch zum Verpflanzen. In Südwest-Kalifornien, Südeuropa, Südamerika.

Description: 25 m de hauteur sur un diamètre de 0,6 à 0,9 m. Le stipe (ce n'est pas un tronc mais une base foliaire persistant atteint 25 à 30 m de hauteur et est dépourvu de rameaux. Epais écorce brun rouge clair, légèrem écailleuse et enveloppée de la ba de la tige jusqu'au bouquet de feuilles vivantes par des feuilles mortes pendantes.

Habitat: sols alcalins mouillés et canyons; en bordure des déserts du sud-ouest de l'Amérique du Nord depuis le désert du Colora jusqu'en Basse-Californie.

Bois: léger, tendre, avec d'innom brables faisceaux fibreux remarquables.

Intérêts: fruits comestibles, moul en farine par les Indiens.

Les jésuites ont été les premiers à le cultiver dans leurs missions du sud de la Californie. Arbre à croissance rapide et vigoureuse se laisse replanter. Cultivé dans le sud-ouest de la Californie, en Europe méridionale et en Amérique du Sud.

TRANSVERSE SECTION

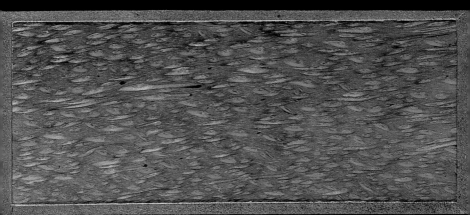

RADIAL SECTION.

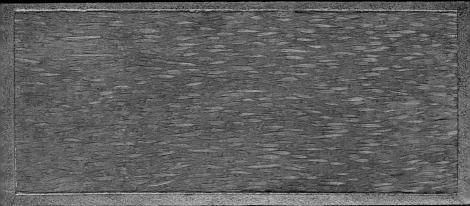

TANGENTIAL SECTION.

Ger Californianische Wedelpalme. *Fr* Palmier d'eventail.

P | PINE FAMILY

American Larch, Tamarack, Eastern Larch

FAMILIE DER KIEFERNGEWÄCHSE
Westamerikanische Lärche

FAMILLE DES PINS
Mélèze laricin, Tamarack (Canada)

PINACEAE
Larix laricina (Du Roi) K. Koch
Syn.: *L. americana* Michx.

Description: Usually not more than 70 ft. (20 m) high and 2 ft. (0.6 m) in diameter. Trunk straight, columnar, bearing a head of small branches when young.

Habitat: In northern North America on swampy soils, low-lying lake shores, also peat bogs. At times forms open forests by itself.

Wood: Heavy, fairly hard and strong; very durable in the ground. Heartwood has an orange-brown tinge; sapwood thin, somewhat paler.

Use: For railroad sleepers, telegraph poles, planks, and interior construction.

Together with the Black Spruce forms the vanguard of the forests of the subarctic region.

Beschreibung: Meist nicht über 20 m hoch bei 0,6 m Durchmesser. Gerader, säulenförmiger oder sich leicht verjüngender Stamm.

Vorkommen: Im nördlichen Nordamerika auf entschieden sumpfigen Böden, an niedrigen Seeufern, auch in Torfmooren. Bildet bisweilen offene Reinbestände.

Holz: Schwer, ziemlich hart und fest; sehr haltbar im Boden. Kern leicht orangebraun; Splint, dünn, etwas heller.

Nutzung: Für Eisenbahnschwellen, Telegrafenmasten, Planken und den Innenausbau.

Bildet mit der Schwarz-Fichte die Vorhut der Wälder der subarktischen Region.

Description : arbre atteignant jusc 20 m de hauteur, avec un tronc columnaire ou légèrement pyram dal de 0,6 m de diamètre.

Habitat : dans le nord de l'Amérique du Nord sur des sols très marécageux et les rives basses de lacs et tourbières. Constitue parf des peuplements purs d'arbres n jointifs.

Bois : lourd, assez dur et résistan très durable sur pied. Bois de cœ teinté de brun-orange ; aubier mince, un peu plus clair.

Intérêts : utilisé pour fabriquer de traverses de chemin de fer, des p teaux télégraphiques, des bordag et en menuiserie intérieure.

Constitue avec l'épicéa noir l'ava garde des forêts boréales.

Tamarack, American or Black Larch, Hackmatack.

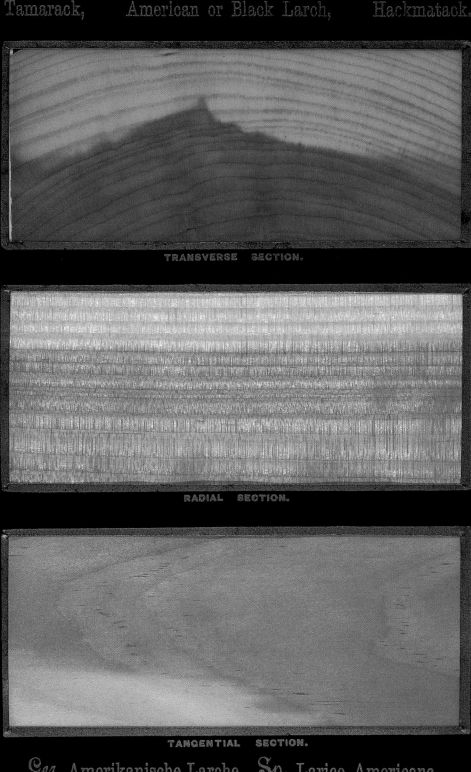

TRANSVERSE SECTION.

RADIAL SECTION.

TANGENTIAL SECTION.

Ger. Amerikanische Larche. *Sp.* Larice Americana. *Fr.* Meleze Americain.

P | PINE FAMILY
Balsam Fir

FAMILIE DER KIEFERNGEWÄCHSE
Balsam-Tanne

FAMILLE DES PINS
Sapin baumier, Baumier du Canada

PINACEAE
Abies balsamea (L.) Mill.

Description: Rarely more than 85 ft. (25 m) high and 2 ft. (0.7 m) in diameter. Bark everywhere, except on old trunks, containing pockets of resin (Canada balsam). Twigs soft and aromatic.

Habitat: Swampy soils in association with Tamarack, Black Ash, White Cedar, etc., less common on higher ground with Beech, Hemlock, etc. Northern North America.

Wood: Light.

Use: Occasionally used for making wooden crates, and recently for paper. Canada balsam is used as an embedding agent in microscopy.

Beschreibung: Selten höher als 25 m bei 0,7 m Durchmesser. Rinde durchweg mit Harz-Nestern (Kanada-Balsam), ausgenommen alte Stämme. Weiche duftende Zweige.

Vorkommen: Sumpfige Böden in Gesellschaft von Tamarack, Schwarz-Esche, Lebensbaum u. a., weniger verbreitet in Hochlagen mit Buche, Hemlock u. a. Nördliches Nordamerika.

Holz: Leicht.

Nutzung: Gelegentlich als Kistenholz. Neuerdings zur Papiergewinnung. Kanada-Balsam, ein Einbettungsmittel in der Mikrokospie.

Description: rarement plus de 25 m de hauteur sur un diamètr de 0,7 m. L'écorce est couverte de poches de résine («baume du Canada»), sauf chez les très vieu spécimens. Rameaux souples, exhalant une odeur balsamique agréable.

Habitat: sols marécageux en compagnie du mélèze laricin, du frêne noir, du thuya géant etc. ; moins répandu sur les hauteurs, il est associé au hêtre, à la pruch etc. On le trouve dans le nord d l'Amérique du Nord.

Bois: léger.

Intérêts: employé quelquefois en caisserie, depuis peu dans la fabrication du papier. Le baume du Canada servait de conservate pour les coupes microscopiques.

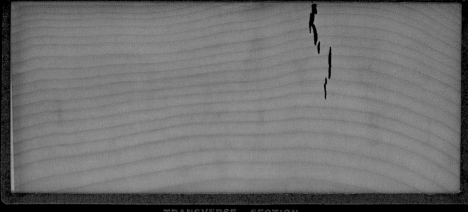

TRANSVERSE SECTION.

RADIAL SECTION.

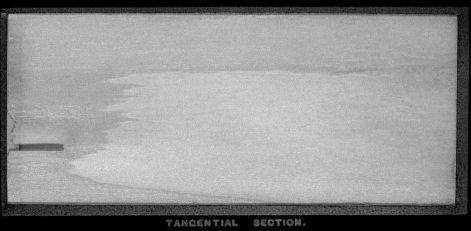

TANGENTIAL SECTION.

Gen. Balsam-Tanne. *Sp.* Abeto balsamico.

Fr. Sapin baumier.

P | PINE FAMILY
Blue Spruce, Colorado Spruce

FAMILIE DER KIEFERNGEWÄCHSE
Schwarzfichte

FAMILLE DES PINS
Epicéa bleu, Epicéa bleu du Colorado,
Epinette bleue (Canada)

PINACEAE
Picea pungens Engelm.

Description: 85 to 100 (150) ft. (25 to 30 (-45) m) high, and a maximum of 3 ft. (0.9 m) in diameter. Growth very variable. At first (up to 40 years old) of pyramidal form, but then becoming more irregular owing to the increased growth of individual branches half way up the trunk, and the withering away of areas situated below this point. Bark of young trees gray with a cinnamon-red tinge, but deeply furrowed and pale gray or more rarely cinnamon-red in color in older trees. Needles stiff, pointed, blue-green or silvery white. Male flowers in oblong-ovoid inflorescences, the female in cylindrical ones. Cones cylindrical.

Habitat: On riverbanks, solitary or in small groups in the interior of North America in Colorado, Utah, and Wyoming.

Wood: Very light, soft, weak, with a silky sheen, and close-grained. Heartwood very light brown or white; sapwood hardly distinguishable.

Use: Owing to the blue-green color of its foliage and its pyramidal shape when young, it is one of the most popular decorative trees in the USA.

Beschreibung: 25 bis 30 (bis 45) m hoch, mit maximal 0,9 m Durchmesser. Wuchs sehr variabel. Anfangs (bis 40 Jahre) von pyramidaler Gestalt, später durch das verstärkte Wachstum einzelner Äste auf halber Höhe des Stammes und das Absterben unterhalb gelegener Bereiche unregelmäßiger werdend. Borke junger Bäume grau mit zimtigem, rötlichem Farbton, im Alter tief rissig und blassgrau. Nadeln steif, spitz, blaugrün oder silbrig weiß. Männliche Blüten in länglich ovalen, weibliche in zylindrischen Blütenständen. Zapfen zylindrisch.

Vorkommen: An Flussufern, einzeln oder in kleinen Gruppen im Inneren Nordamerikas in Colorado, Utah und Wyoming.

Holz: Sehr leicht, weich, schwach, seidig und mit feiner Textur. Kern sehr hell, braun oder weiß; Splint kaum unterscheidbar.

Nutzung: Wegen der blaugrünen Farbe des Laubes und der pyramidalen Wuchsform junger Bäume eines der beliebtesten Ziergehölze in den USA.

Description: 25 à 30 m de hauteur (45 m max.) sur un diamètre de 0,9 m maximum. Port très variable pyramidal à l'état juvénile (jusqu'à 40 ans), devenant plus irrégulier par suite d'une croissance plus for de quelques branches médianes et de l'élagage naturel des basses branches. Ecorce grise teintée de rouille sur les jeunes arbres, se fis rant profondément et devenant gr pâle ou plus rarement rousse avec l'âge. Aiguilles rigides, piquantes, vert bleu ou blanc argenté. Fleurs mâles en inflorescences ovales-allongées, les femelles cylindriques. Cônes cylindriques.

Habitat: rives des cours d'eau, solitaire ou en petites formations dans l'intérieur de l'Amérique du Nord entre le Colorado, l'Utah et le Wyoming.

Bois: très léger, tendre, faible, satiné et à grain fin. Bois de cœur très clair, brun ou blanc; aubier à peine distinct.

Intérêts: une des espèces ornementales préférées aux Etats-Unis à cause de son feuillage vert bleu et de son port pyramidal à l'état juvénile.

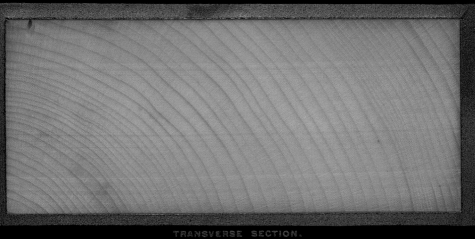

TRANSVERSE SECTION.

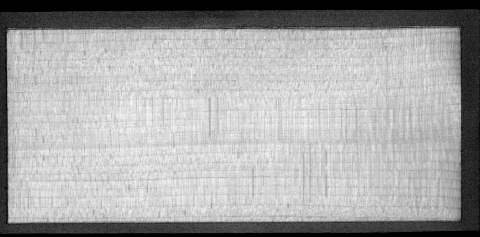

RADIAL SECTION.

TANGENTIAL SECTION.

P | PINE FAMILY

Bristle-cone Fir, Silver Fir, Santa Lucia Fir

FAMILIE DER KIEFERNGEWÄCHSE
Grannen-Tanne, Santa Lucia-Tanne

FAMILLE DES PINS
Sapin bractéifère, Sapin de Santa Lucia

PINACEAE
Abies bracteata (D. Don) D. Don ex Poit.
Syn.: *A. venusta* (Dougl.) K. Koch

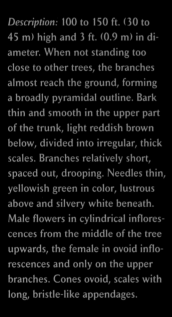

Description: 100 to 150 ft. (30 to 45 m) high and 3 ft. (0.9 m) in diameter. When not standing too close to other trees, the branches almost reach the ground, forming a broadly pyramidal outline. Bark thin and smooth in the upper part of the trunk, light reddish brown below, divided into irregular, thick scales. Branches relatively short, spaced out, drooping. Needles thin, yellowish green in color, lustrous above and silvery white beneath. Male flowers in cylindrical inflorescences from the middle of the tree upwards, the female in ovoid inflorescences and only on the upper branches. Cones ovoid, scales with long, bristle-like appendages.

Habitat: At altitudes around 3,000 ft. (900 m) above sea level on the moist bottoms of canyons, which may dry out completely in summer. Only in the Santa Lucia mountains of California.

Wood: Heavy, not strong, and coarse-grained.

Use: Because of the difficulty in reaching the places where it grows and the scarcity of the trees it is only occasionally used as fuel.

Beschreibung: 30 bis 45 m hoch bei 0,9 m Durchmesser. Wenn nicht allzu dicht stehend, reicht die Beastung fast bis auf den Boden, breite pyramidale Form bildend. Borke dünn und glatt im oberen Stammbereich, unten hell rotbraun, in unregelmäßige, dicke Schuppen zergliedert. Äste relativ kurz, locker stehend, herabhängend. Nadeln dünn, von gelbgrüner Farbe, mit glänzender Ober- und silberweißer Unterseite. Männliche Blüten in zylindrischen Blütenständen ab der halben Baumhöhe aufwärts, weibliche in ovalen Blütenständen und nur auf den oberen Ästen. Zapfen eiförmig, Schuppen mit langen, grannenartigen Anhängseln.

Vorkommen: In Höhen um 900 m am feuchten Grund von Cañons, die im Sommer komplett austrocknen können. Nur in den kalifornischen Santa Lucia Mountains.

Holz: Schwer, nicht stabil und mit grober Textur.

Nutzung: Wegen der schwierigen Erreichbarkeit der Standorte und der Seltenheit der Bäume nur gelegentlich als Feuerholz.

Description: arbre de 30 à 45 m hauteur, avec un tronc de 0,9 m diamètre, ramifié presque jusqu' sol s'il a de la place. Houppier ample et pyramidal. Ecorce minc et lisse dans la partie supérieure tronc, brun rouge clair et à textu craquelée, en écailles épaisses et irrégulières dans la partie basale Branches relativement courtes, lâches et retombantes. Aiguilles fines, brillantes dessus, blanc argenté au revers. Inflorescences mâles cylindriques, vert jaune, d la moitié supérieure de l'arbre, inflorescences femelles ovales, n'apparaissant que sur les branc supérieures. Cônes ovoïdes, avec des écailles se terminant en bart

Habitat: à 900 m environ d'altitude, sur le fond humide de canyons susceptibles de s'asséch l'été. On le trouve seulement da les montagnes californiennes de Santa Lucia.

Bois: lourd, non stable et à grain grossier.

Intérêts: chauffage à l'occasion e. raison de sa rareté et de l'accès difficile des lieux qu'il colonise.

Bristle-cone Fir, Santa Lucia Silver Fir.

TRANSVERSE SECTION.

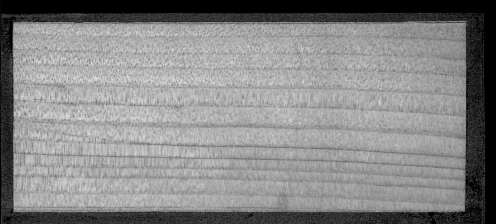

RADIAL SECTION.

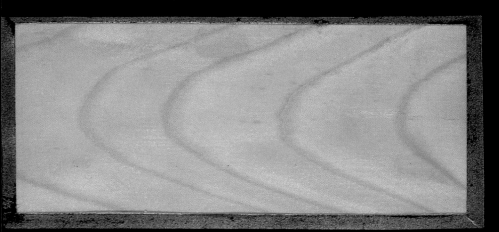

TANGENTIAL SECTION

Ger. Tanne von Santa Lucia. *Fr.* Sapin de Santa Lucia. *Sp.* Abeto de Santa Lucia.

P | PINE FAMILY

Bristle-cone Pine, Foxtail Pine, Hickory Pine

FAMILIE DER KIEFERNGEWÄCHSE
Fuchsschwanz-Kiefer

FAMILLE DES PINS
Pin de Balfour, Pin à queue de renard, Pin foxtail

PINACEAE
Pinus balfouriana Grev. & Balf.

Description: 40 to 50 ft. (12 to 15 m) high and 2 to 3 ft. (0.6 to 0.9 m) in diameter. In higher altitudes it often occurs as a low shrub with almost creeping stems. Trunk short. At first pyramidal growth, but becomes increasingly irregular with age. Young bark thin and smooth, milky white, later reddish brown and irregularly fissured. Needles in fives, stiff, curved, 1 to 1½ in. (2.5 to 4 cm) long, lying close to the branches. Flowers somewhat distant from the ends of the branches: the male in short inflorescences; the female singly or in small groups. Cones oblong-ovoid, with long, curved bristles.

Habitat: Rocky or stony slopes in south-western North America on the upper tree-line between 7,500 and 10,500 ft. (2,400 to 3,700 m) above sea level.

Wood: Light, soft, not strong, close-grained. Heartwood red; sapwood thin, almost white. Indistinct dark bands of summer wood, only occasional resin canals, but numerous distinct medullary rays.

Use: Occasionally as timber for mines and as fuel. As a result it has almost been wiped out in places.

Beschreibung: 12 bis 15 m hoch bei 0,6 bis 0,9 m Durchmesser. In höheren Lagen oft als niedriger Busch mit fast kriechenden Stämmen. Stamm kurz. Anfangs von pyramidalem Wuchs, im Alter aber zunehmend unregelmäßig. Borke jung dünn und glatt, milchig weiß, später rotbraun und unregelmäßig rissig. Nadeln zu 5, steif, gebogen, 2,5 bis 4 cm lang, den Ästen anliegend. Blüten etwas unterhalb der Astenden, die männlichen in kurzen Blütenständen, die weiblichen einzeln oder in kleinen Gruppen. Zapfen länglich oval, mit langen, gebogenen Dornen.

Vorkommen: Felsige oder steinige Hänge im südwestlichen Nordamerika an der oberen Baumgrenze zwischen 2400 bis 3700 m.

Holz: Leicht, weich, nicht fest und mit feiner Textur. Kern rot; Splint dünn, fast weiß. Undeutliche dunkle Bänder von Sommerzellen, wenige Harzgänge und zahlreiche deutliche Markstrahlen.

Nutzung: Gelegentlich als Minen- und Feuerholz. Lokal dadurch fast ausgerottet.

Description: 12 à 15 m de hauteu sur un diamètre de 0,6 à 0,9 m. Souvent de type arbrisseau à l'étage alpin, avec des ramificatic presque rampantes. Tronc court, port pyramidal lorsqu'il est jeun mais devenant plus irrégulier av l'âge. Ecorce d'abord mince, liss et d'un blanc laiteux puis brun rouge et parcourue de fissures irrégulières. Aiguilles par 5, rigic courbes, de 2,5 à 4 cm de long, plaquées sur les rameaux. Fleurs presque terminales, les mâles en inflorescences courtes, les femell solitaires ou en petits groupes. Cônes ovales-allongés, à écailles recourbées en crochets piquants.

Habitat: versants rocheux ou caillouteux du sud-ouest de l'Amérique du Nord, à la limite supérieure des arbres, entre 240 et 3700 m d'altitude.

Bois: léger, tendre, non résistant à grain fin. Bois de cœur rouge; bier mince, presque blanc. Band sombres de cellules d'été, quelqu canaux résinifères et nombreux rayons médullaires.

Intérêts: chauffage, boisage des mines; quasi disparu dans certai régions.

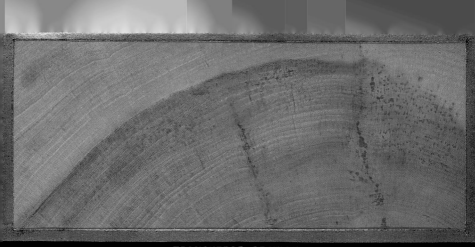

TRANSVERSE SECTION.

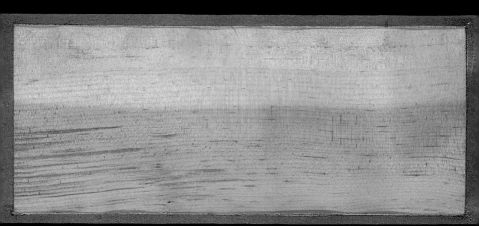

RADIAL SECTION.

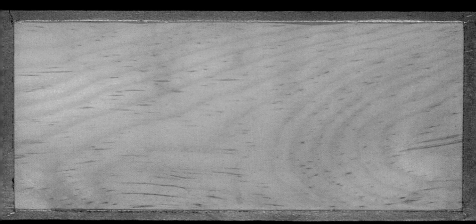

TANGENTIAL SECTION.

Ger. Fuchsschwanzige Fichte   *Fr.* Pin de queue de renard

P | PINE FAMILY
Canadian Hemlock, Eastern Hemlock

FAMILIE DER KIEFERNGEWÄCHSE
Kanadische Hemlock, Hemlocktanne

FAMILLE DES PINS
Pruche du Canada, Haricot (Canada)

PINACEAE
Tsuga canadensis (L.) Carrière

Description: Up to 100 ft. (30 m) high and 2 to 3 ft. (0.6 to 0.9 m) in diameter. Gracefully branched. Foliage of flat needles, the upper side dark green, whitish beneath. Cones at the ends of the branches.

Habitat: Well-drained uplands and the steep banks of gorges, usually in association with Digger Pine, Red Spruce, maples, Beech, Yellow Birch, etc., in north-eastern North America.

Wood: Light, soft, and brittle.

Use: Timber mostly used as rough wood for building. Bark more important.

Formerly one of the commonest forest trees of north-east America. Almost wiped out through over-exploitation of the tannin-rich bark.

Beschreibung: Bis 30 m hoch bei 0,6 bis 0,9 m Durchmesser. Graziöse Verzweigung. Belaubung mit flachen Nadeln, oberseits dunkelgrün, unterseits weißlich. Zapfen an den Zweigspitzen.

Vorkommen: Gut drainierte Hochlagen und die Hänge von Schluchten, gewöhnlich in Gesellschaft von Weiß-Kiefer, Rot-Fichte, Ahornen, Buche, Gelb-Birke u. a. im nordöstlichen Nordamerika.

Holz: Leicht, weich und brüchig.

Nutzung: Meist als grobes Bauholz hauptsächlich jedoch Verwendung der gerbstoffreichen Rinde.

Einst einer der häufigsten Waldbäume des amerikanischen Nordostens. Aufgrund von Übernutzung der gerbstoffreichen Rinde fast ausgerottet.

Description : jusqu'à 30 m de hauteur sur un diamètre de 0,6 à 0,9 m. Elégante et gracieuse ra fication. Feuillage à aiguilles apla ties, vert foncé dessus, blanchâtr au revers. Petits cônes terminau

Habitat : hautes terres bien drain et pentes de ravins, généraleme en compagnie du pin blanc, de l'épicéa rouge, des érables, du hêtre, du bouleau jaune etc. On le trouve dans le nord-est de l'Amérique du Nord.

Bois : tendre, léger et cassant.

Intérêts : bois d'œuvre grossier. S tout employé pour son écorce.

Autrefois, un des arbres forestie les plus répandus du nord-est ar ricain. A pratiquement disparu e raison de l'exploitation excessive de son écorce riche en tannins.

TRANSVERSE SECTION.

RADIAL SECTION.

TANGENTIAL SECTION.

Ger. Schierling-Tanne. Fr. Peruche. Sp. Abeto Canadense.

P | PINE FAMILY
Carolina Hemlock

FAMILIE DER KIEFERNGEWÄCHSE
Karolina-Hemlock

FAMILLE DES PINS
Pruche de la Caroline, Hemlock de Caroline

PINACEAE
Tsuga caroliniana Engelm.

Description: 40 to 50 (70) ft. (12 to 15 (20) m) high and up to 2 ft. (0.6 m) in diameter. Trunk with short, stout, and often drooping branches, which give the tree a compact crown of pyramidal form. Bark red-brown, divided by deep fissures, the surface separating into small, flat scales. Leaves very dark green in color. Flowers in short, oblong inflorescences: the male tinged with purple; the female deep purple in color. Cones ovoid, hanging downwards when mature, the cone-scales spreading almost at right angles at maturity.

Habitat: On stony riverbanks at altitudes of between 2,000 and 3,000 ft. (600 to 900 m), as a solitary tree or in small groups. In south-eastern North America from Virginia to Georgia. Not common anywhere.

Wood: Light, soft, not strong, brittle, and coarse-grained.

Use: As a park tree in damp, cool climates, e.g. New England, with more luxuriant and longer-lasting foliage than in similar species.

Beschreibung: 12 bis 15 (20) m hoch bei bis zu 0,6 m Durchmesser. Stamm mit kurzen, kräftigen und oft herabhängenden Ästen, die dem Baum eine kompakte Krone von pyramidaler Form geben. Borke rotbraun, durch tiefe Risse gefeldert, Oberfläche sich in flachen Schüppchen ablösend. Blätter von sehr dunkler grüner Farbe. Blüten in kurzen, länglichen Blütenständen, männliche mit violettem Farbton, weibliche von kräftig violetter Farbe. Zapfen oval, reif hängend, Zapfenschuppen zur Reifezeit fast rechtwinklig abspreizend.

Vorkommen: An steinigen Flussufern in Höhen zwischen 600 bis 900 m als Einzelbaum oder in kleinen Gruppen. Im südöstlichen Nordamerika von Virginia bis Georgia. Nirgendwo häufig.

Holz: Leicht, weich, nicht stabil, brüchig und mit grober Textur.

Nutzung: Als Parkbaum in feuchtkühlem Klima, z.B. in Neuengland, mit üppigerem und länger ausdauerndem Laub als bei ähnlichen Arten.

Description: 12 à 15 m de hauteu (20 m max.), avec un tronc pouv atteindre 0,6 m de diamètre et des branches courtes, vigoureus souvent retombantes, constituar un houppier compact et pyrami dal. Ecorce brun rouge, à textur profondément fissurée en secteu et s'exfoliant en écailles plates. Feuilles d'un vert très foncé. Fle en courtes inflorescences allong les mâles violacées, les femelles d'un violet vif. Cônes ovales, pe dants à maturité, écailles s'ouvra presque perpendiculairement à l'axe du cône à maturité.

Habitat: rives caillouteuses des cours d'eau à une altitude de 60 à 900 m, solitaire ou en petites formations. Croît dans le sud-es de l'Amérique du Nord de la Vi ginie à la Géorgie; nulle part trè répandu.

Bois: léger, tendre, non stable, cassant et à grain grossier.

Intérêts: arbre de parcs sous un climat frais et humide, comme e Nouvelle-Angleterre par exempl avec un feuillage plus abondant et demeurant plus longtemps su l'arbre que chez d'autres espèce similaires.

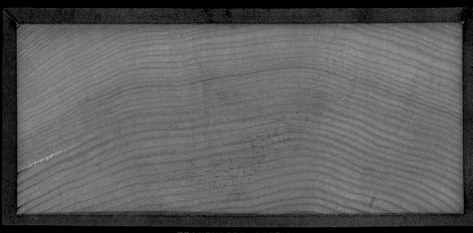

TRANSVERSE SECTION

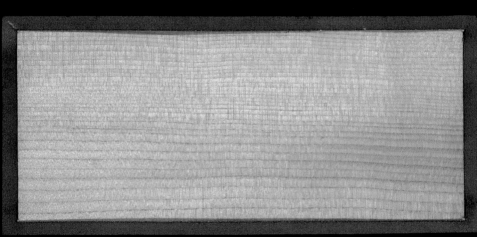

RADIAL SECTION.

TANGENTIAL SECTION.

P | PINE FAMILY
Digger Pine, Bull Pine

FAMILIE DER KIEFERNGEWÄCHSE
Nuss-Kiefer, Sabines Föhre

FAMILLE DES PINS
Pin de Sabine

PINACEAE
Pinus sabiniana Douglas ex Douglas

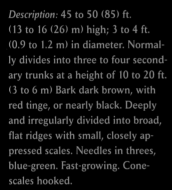

Description: 45 to 50 (85) ft. (13 to 16 (26) m) high; 3 to 4 ft. (0.9 to 1.2 m) in diameter. Normally divides into three to four secondary trunks at a height of 10 to 20 ft. (3 to 6 m) Bark dark brown, with red tinge, or nearly black. Deeply and irregularly divided into broad, flat ridges with small, closely appressed scales. Needles in threes, blue-green. Fast-growing. Cone-scales hooked.

Habitat: Prefers sunny hills in central California, about 2,000 ft. (600 m), with Blue Oak.

Wood: Light, soft, not strong, coarse-grained; not durable. Heartwood pale brown or red; sapwood thick, yellow or almost white. Summer wood has very resin-rich bands and a few large resin canals.

Use: Seeds gathered in large quantities for food by American Indians. Distillation of the seeds produces an hydrocarbon, abietene, of medicinal value.

Of ornamental value for its open foliage with large branches with massive cones.

Beschreibung: 13 bis 16 (26) m bei 0,9 bis 1,2 m Durchmesser. Allgemein in 3 bis 6 m über Grund Teilung in 3 bis 4 kräftige Sekundärstämme. Borke dunkelbraun, rot getönt, oder nahezu schwarz. Tief und unregelmäßig geteilt in große, dicke, abgerundete Riefen mit kleinen, eng anliegenden Schuppen. Nadeln zu dritt, blaugrün. Schnellwüchsig. Zapfenschuppen hakig.

Vorkommen: Vorzugsweise sonnenverbrannte Hügel in Mittel-Kalifornien bei etwa 600 m.

Holz: Leicht, weich, nicht fest und mit grober Textur; nicht dauerhaft. Kern hellbraun oder rot; Splint dick, gelb oder fast weiß. Im Spätholz sehr harzreiche Bänder und einige große Harzgänge.

Nutzung: Samen, in großen Mengen gesammelt, als Nahrungsmittel der Indianer von Bedeutung. Durch Destillation der Samen wird ein Kohlenwasserstoff, Abietin, für medizinische Zwecke gewonnen.

Wegen der dünnen Belaubung treten die großen Äste mit den massiven Zapfen dekorativ hervor.

Description: 13 à 16 m (26 m) de hauteur sur un diamètre de 0,9 à 1,2 m. Tronc divisé d'ordinaire e ou 4 tiges secondaires vigoureus entre 3 et 6 m au-dessus du sol. Ecorce d'un brun foncé teinté de rouge ou quasiment noire, à fissures profondes et à grandes crê épaisses, irrégulières, arrondies et recouvertes de petites écailles appliquées. Aiguilles par deux, v bleu. Croissance rapide. Ecailles des cônes recourbées en crochet piquants.

Habitat: préfère les collines brûle par le soleil en moyenne Californ à 600 m environ d'altitude.

Bois: léger, tendre, non résistant et à grain grossier; non durable. Bois de cœur brun clair ou rouge aubier épais, jaune ou presque blanc. Bandes très résineuses et quelques gros canaux résiniferes dans le bois final.

Intérêts: les graines étaient conso mées par les Indiens. Extraction distillation de l'abiétine (hydroca bure contenu dans les graines), à des fins médicales.

Grandes branches décoratives, rehaussées par de gros cônes.

TRANSVERSE SECTION.

RADIAL SECTION.

P | PINE FAMILY

Douglas Fir, Red Fir

FAMILIE DER KIEFERNGEWÄCHSE
Douglas-Fichte

FAMILLE DES PINS
Douglas de Menzies, Douglas bleu,
Pin de Douglas (Canada)

PINACEAE
Pseudotsuga menziesii (Mirb.) Franco var. *menziesii*
Syn.: *P. taxifolia* (Lamb.) Britton

Description: About 200 to 230 ft. (60 to 70 m) high and 3 to 4 ft. (0.9 to 1.2 m) in diameter, but up to between 13 and 15 ft. (3 to 3.5 m) in diameter. Also shrubby at high altitude. Young bark smooth, thin, a rather shiny dark gray-brown, later divided by irregularly fused, rounded ridges with small, thick, closely appressed dark red-brown scales. In dry areas paler, soft, and spongy.

Habitat: Slopes in western North America (excluding arid regions); forms mountain forests by itself.

Wood: Hardens after felling, becoming difficult to work. Two types: red and yellow, the red coarse-grained and of lesser value; sapwood varies in thickness and density, almost white. Resin canals in summer wood.

Use: Building timber, railroad sleepers, posts, poles, firewood. In Europe as woodland and park tree.

Beschreibung: Um 60 bis 70 m hoch bei 0,9 bis 1,2 m Durchmesser, wenn höher bei 3 bis 3,5 m Durchmesser. In hohen Gebirgslagen auch strauchig. Borke jung glatt, dünn, ziemlich glänzend dunkel graubraun, später in unregelmäßig miteinander verbundene, gerundete Riefen mit kleinen, dicken, dicht anliegenden, dunkel rotbraunen Schuppen geteilt. In trockenen Regionen blasser, weich und schwammig.

Vorkommen: Hanglagen des westlichen Nordamerika (ohne die Trockengebiete), bildet selber Bergwälder.

Holz: Nach dem Fällen härtet das Holz nach und wird schwerer zu bearbeiten. Zwei Arten: rot und gelb, das rote mit grober Textur und weniger wertvoll; Splint variiert in Dicke und Dichte, fast weiß. Harzreiche Bänder im Spätholz.

Nutzung: Bauholz, Eisenbahnschwellen, Pfähle, Masten, Brennholz. In Europa als Wald- und Parkbaum.

Description : arbre d'environ 60 à 70 m de hauteur, avec un tronc d 0,9 à 1,2 m de diamètre, atteignant 3 à 3,5 m sur les spécimen plus grands. Peut prendre aussi un port buissonnant en altitude. Ecorce brun gris, lisse, fine et ass brillante à l'état juvénile ; à textu entrelacée, formée de sillons arro dis et d'écailles adhérentes, petit et épaisses, brun rouge foncé sur les sujets adultes. Plus terne, mo et spongieuse dans les régions sèches.

Habitat : versants de montagnes de l'ouest de l'Amérique du Nor (à l'exception des régions sèches constitue des forêts d'altitude.

Bois : difficile à travailler car il du cit après l'abattage. Deux sortes : rouge et jaune, le rouge est à gra grossier et de moindre valeur ; aubier d'épaisseur et de densité variables, presque blanc. Bandes résineuses dans le bois final.

Intérêts : utilisé en construction, pour la fabrication de traverses d chemin de fer, de piquets, de mâ de marine et comme bois de cha fage. Essence de reboisement et parcs en Europe.

Douglas Spruce, Red Fir, Yellow Fir, Oregon Pine.

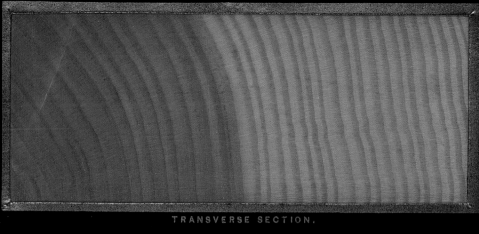

TRANSVERSE SECTION.

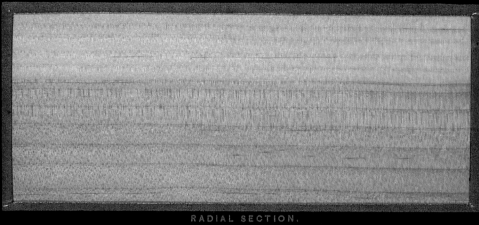

RADIAL SECTION.

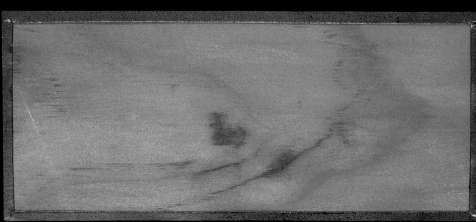

TANGENTIAL SECTION

Ger. Tanne von Douglas.   *Fr.* Sapin de Douglas.

Sp. Abeto de Douglas,

P | PINE FAMILY
Engelmann Spruce, White Spruce

FAMILIE DER KIEFERNGEWÄCHSE
Engelmann-Fichte

FAMILLE DES PINS
*Epicéa d'Engelmann, Epicéa de l'Arizona,
Epinette des montagnes (Canada)*

PINACEAE
Picea engelmannii Parry ex Engelm.

Description: 150 ft. (45 m) high
and 4 to 5 ft. (1.2 to 1.5 m) in diameter. Also occurs as a shrub in
higher places. At first it is of dense,
compact, pyramidal form, free of
branches for much of its height,
and with a narrow, pyramidal
crown. Bark light cinnamon-red,
separating into large, thin, loosely
attached scales. Needles soft and
flexible, blue-green, with the unpleasant smell of a skunk. Male and
female flowers separate, in cylindrical inflorescences. Cones oblong-
ovoid.

Habitat: Mountain forests in western North America, frequently
as the main tree species.

Wood: Very light, soft, not strong,
with a silky sheen, and close-
grained. Heartwood pale yellow
with a red tinge; sapwood thick,
scarcely distinguishable, with numerous medullary rays, occasional
small resin canals, and insignificant
bands of summer wood.

Use: To a large extent used as building timber, firewood, and for the
production of charcoal. Locally the
bark is used for tanning leather.

Beschreibung: 45 m hoch bei 1,2
bis 1,5 m Durchmesser. In hohen
Lagen auch strauchförmig. Anfangs
von dichter, kompakter, pyramidaler Form, weit herauf astfrei und
mit schmaler pyramidaler Krone.
Borke hell zimtfarben rötlich, in
große, dünne und locker befestigte
Schuppen gegliedert. Nadeln weich
und flexibel, blaugrün, mit unangenehmem Geruch nach Stinktier
(Skunk). Männliche und weibliche
Blüten separat, in zylindrischen
Blütenständen. Zapfen länglich und
oval.

Vorkommen: Gebirgswälder im
westlichen Nordamerika, oft als
Hauptbaumart.

Holz: Sehr leicht, weich, nicht
stabil, seidig und mit feiner Textur.
Kern blassgelb mit roter Tönung;
Splint dick, kaum unterscheidbar,
mit zahlreichen Markstrahlen,
wenigen kleinen Harzgängen und
unauffälligen Bändern kleiner
Sommerzellen.

Nutzung: In großem Umfang als
Bauholz, Brennmaterial und zur
Holzkohlegewinnung genutzt.
Lokal wird die Borke zur Lederger-
bung verwendet.

Description : 45 m de hauteur
sur un diamètre de 1,2 à 1,5 m.
Se présente aussi sous forme de
buisson en altitude. Port dense,
pyramidal, compact, haut branchu, avec une couronne étroite et
pyramidale à l'état juvénile. Ecor
rougeâtre clair teintée de cannell
à texture craquelée, en grandes
écailles minces et lâches. Aiguille
molles et souples, vert bleu, déga
geant une odeur nauséabonde de
mouffette. Fleurs mâles et femell
en inflorescences cylindriques di
tinctes. Cônes ovales-allongés

Habitat : forêts montagnardes dar
l'ouest de l'Amérique du Nord,
souvent en élément dominant.

Bois : très léger, tendre, non stabl
satiné et à grain fin. Très peu de
contraste entre le bois de cœur
jaune pâle teinté de rouge et
l'aubier épais, comportant de nor
breux rayons médullaires, quelqu
petits canaux résinifères et des
bandes de petites cellules d'été.

Intérêts : comme bois d'œuvre,
comme combustible et pour le
charbon de bois ; écorce employé
localement pour le tannage des
cuirs.

474

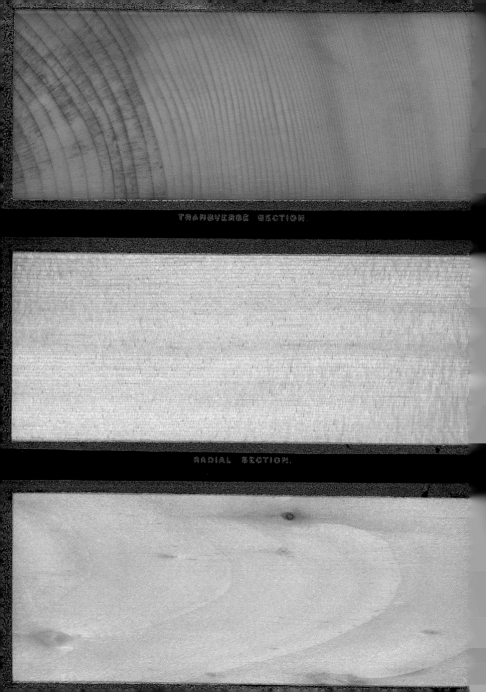

TRANSVERSE SECTION.

RADIAL SECTION.

P | PINE FAMILY

Fraser Fir, Southern Balsam Fir, She Balsam

FAMILIE DER KIEFERNGEWÄCHSE
Frasers Tanne, Balsam-Tanne

FAMILLE DES PINS
Sapin de Fraser

PINACEAE
Abies fraseri (Pursh) Poir.

Description: 30 to 40 (70) ft. (9 to 12 (-21) m) high and a maximum of 2 ft. 6 in. (0.75 m) in diameter. A rapid-growing but short-lived tree. At first pyramidal in form, but soon losing branches from the lower part of the trunk. Branches slender, relatively rigid. Needles deep dark green and shining above, pale beneath. The base of the needles twists so that all point upwards. Male and female flowers in ovoid-cylindrical inflorescences. Cones ovoid.

Habitat: Only at the highest levels of the southern Appalachians in eastern North America, and forming extensive forests from 4,000 to 6,000 ft. (1,200 to 1,800 m) above sea level.

Wood: Very light, soft, not strong, and close-grained. Heartwood light brown; sapwood almost white. Numerous thin medullary rays.

Use: Occasionally used as building timber. Introduced for the first time into England in 1811, but because of its short life and its similarity to the longer-lived *Abies lasiocarpa*, it has almost completely disappeared from European gardens.

Beschreibung: 9 bis 12 (bis 21) m hoch, bei maximal 0,75 m Durchmesser. Schnellwüchsiger, kurzlebiger Baum. Zu Beginn von pyramidaler Form, aber bereits früh von unten her verkahlend. Äste schlank, relativ steif. Nadeln tief dunkelgrün und glänzend auf der Oberseite, Unterseite blass. Durch Verdrehung der Nadelbasis weisen alle Nadeln nach oben. Männliche und weibliche Blüten in oval-zylindrischen Blütenständen. Zapfen oval.

Vorkommen: Nur auf den höchsten Erhebungen der südlichen Appalachen im östlichen Nordamerika. Ausgedehnte Wälder zwischen 1200 bis 1800 m. Höhe bildend.

Holz: Sehr leicht, weich, nicht stabil und mit feiner Textur. Kern hellbraun; Splint fast weiß. Zahlreiche dünne Markstrahlen.

Nutzung: Gelegentlich als Bauholz. 1811 erstmalig nach England eingeführt, jedoch wegen der Kurzlebigkeit und der Ähnlichkeit mit der besser haltbaren *Abies lasiocarpa* aus den europäischen Gärten fast völlig wieder verschwunden.

Description: 9 à 12 m de hauteur (21 m max.) sur un diamètre ne dépassant pas 0,75 m. Arbre à croissance rapide et d'une brève durée de vie, de forme pyramida jeune mais se dénudant très tôt la base. Branches fines, relativement rigides. Aiguilles d'un vert profond et brillantes dessus, pâle dessous, disposées horizontalement et relevées en brosse. Fleur mâles et femelles en inflorescenc ovales-cylindriques. Cônes ovales ovoïdes.

Habitat: uniquement sur les relie les plus élevés des Appalaches dans l'est de l'Amérique du Nord Constitue de vastes peuplements forestiers entre 1200 et 1800 m.

Bois: très léger, tendre, non stabl et à grain fin. Bois de cœur brun clair; aubier presque blanc. Rayo médullaires fins et nombreux.

Intérêts: utilisé de temps à autre pour la construction. L'arbre fut introduit en Angleterre en 1811 mais très vite il disparut presque des jardins européens à cause de sa brève durée de vie et de sa ressemblance avec *Abies lasiocarp* plus durable.

Fraser Fir.

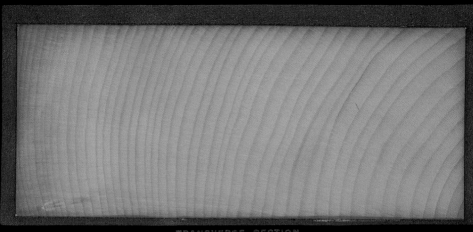

TRANSVERSE SECTION.

RADIAL SECTION.

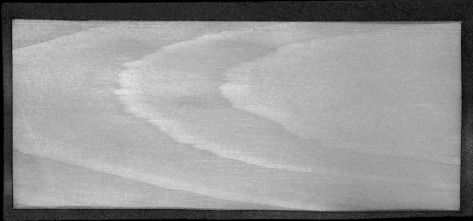

TANGENTIAL SECTION.

Ger. Fraser Tanne.   *Fr.* Sapin de Fraser.

Sp. Abeto de Fraser.

P | PINE FAMILY
Hemlock

FAMILIE DER KIEFERNGEWÄCHSE
Großfrüchtige Douglastanne

FAMILLE DES PINS
Douglas à grands cônes, Douglas de San Bernardino

PINACEAE
Pseudotsuga macrocarpa (Vasey) Mayr

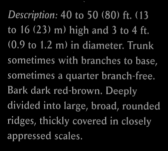

Description: 40 to 50 (80) ft. (13 to 16 (23) m) high and 3 to 4 ft. (0.9 to 1.2 m) in diameter. Trunk sometimes with branches to base, sometimes a quarter branch-free. Bark dark red-brown. Deeply divided into large, broad, rounded ridges, thickly covered in closely appressed scales.

Habitat: Characteristically in sparse forests on the west and south slopes of arid mountains, on steep slopes of narrow gorges between 3,300 and 5,000 ft. (1,000 to 1,600 m), above banks of streams in southern California. Also sometimes in high-altitude woods with oaks and pines (e.g. Jeffrey's Pine).

Wood: Heavy, hard, strong and close-grained, durable. Heartwood dark red; sapwood almost white.

Use: Sometimes for timber, but mostly for firewood.

Beschreibung: 13 bis 16 (23) m hoch bei 0,9 bis 1,2 m Durchmesser. Stamm manchmal bis zum Grunde beastet, manchmal ein Viertel astfrei. Borke dunkel rotbraun. Tief geteilt in große, breite, abgerundete Riefen, dick bedeckt mit enganliegenden Schuppen.

Vorkommen: In Form charakteristischer, spärlicher Wälder, West- und Südhänge arider Berge, steile Hänge enger Schluchten zwischen 1000 und 1600 m, oberhalb von Bachufern in Süd-Kalifornien. Bildet gelegentlich in höheren Lagen Haine mit Eichen und Kiefern (z. B. Jeffreys Westlicher Gelb-Kiefer).

Holz: Schwer, hart, fest und mit feiner Textur; dauerhaft. Kern dunkelrot; Splint fast weiß.

Nutzung: Gelegentlich als Bauholz, aber hauptsächlich als Brennholz.

Description : 13 à 16 m (jusqu'à 23 m) de hauteur sur un diamèt de 0,9 à 1,2 m. Tronc parfois ran fié jusqu'au sol, parfois non rami jusqu'au quart de sa hauteur. Ecorce brun rouge foncé, creusée de profonds et larges sillons aux crêtes arrondies, épaissies par de écailles adhérentes.

Habitat : en forêts clairsemées, caractéristiques sur les versants sud et ouest de montagnes aride sur les pentes escarpées de gorge étroites, entre 1000 et 1600 m au-dessus du niveau de la mer et sur les talus surplombant les rive de torrents du sud de la Californ Constitue de temps à autre, en altitude, des fourrés avec les chê et les pins (pin jaune de l'Ouest exemple).

Bois : lourd, dur, résistant et à gra fin ; durable. Bois de cœur rouge foncé ; aubier presque blanc.

Intérêts : utilisé quelquefois en construction, mais surtout comm bois de chauffage.

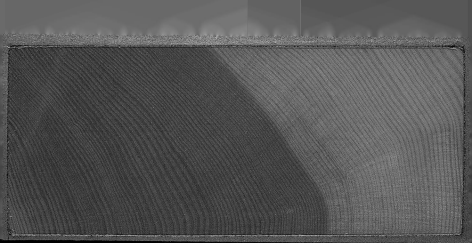

TRANSVERSE SECTION.

RADIAL SECTION.

TANGENTIAL SECTION

Ger. Grossfrüchtige Fichte.　　Fr. Sapin de cones grands.

P | PINE FAMILY
Jack Pine, Gray Pine, Northern Scrub Pine

FAMILIE DER KIEFERNGEWÄCHSE
Banks-Kiefer

FAMILLE DES PINS
Pin gris, Pin du Labrador, Pin de Banks (Canada)

PINACEAE
Pinus banksiaña Lamb.

Description: 75 to 85 ft. (23 to 25 m) high and 2 to 3 ft. (0.6 to 0.9 m) in diameter. Bark dark reddish brown, rough, with irregular scales and plates. Needles less than 2 in. (4 cm) long, curved and twisted lengthwise. The trees develop best in the north-west of their range, forming entire forests, but eastwards they are rarer and weaker, with low-set branches.

Habitat: Even in the poorest soils, especially in the far north from the Atlantic to the Rocky Mountains; to the south, no further than northern Nebraska.

Wood: Light, soft, and not strong. Heartwood reddish brown; sapwood thick and paler.

Use: Occasionally for railroad sleepers, posts, and building wood, also as firewood.

Beschreibung: 23 bis 25 m hoch bei 0,6 bis 0,9 m Durchmesser. Borke dunkel rotbraun, rau, unregelmäßige Schuppen und Platten. Nadeln unter 4 cm lang, gekrümmt und längs gedreht. Im Nordwesten ihres Verbreitungsgebietes am besten ausgebildet und ganze Wälder bildend, ostwärts seltener und schwächer, mit tief ansetzenden Ästen.

Vorkommen: Auch magerste Böden, besonders im hohen Norden, vom Atlantik bis in die Rocky Mountains, südlich nur bis ins nördliche Nebraska.

Holz: Leicht, weich und nicht stark. Kern rotbraun, Splint dick und heller.

Nutzung: Gelegentlich für Eisenbahnschwellen, Pfosten und Bauholz, auch als Brennmaterial.

Description : 23 à 25 m de hauteu sur un diamètre de 0,6 à 0,9 m. Ecorce brun rouge foncé, se craquelant en écailles et en plaques asymétriques. Les aiguilles tordu vrillées mesurent moins de 4 cm de long. Espèce la mieux formée constituant des peuplements for tiers au nord-ouest de son aire d répartition ; vers l'est, en revanch elle est plus rare, moins vigoureu et bas branchue.

Habitat : se contente également d sols les plus maigres, en particuli dans le Grand Nord depuis la cô atlantique jusqu'aux Rocheuses, vers le sud seulement jusqu'au nord du Nebraska.

Bois : léger, tendre et pas fort. Bo de cœur brun rouge ; aubier épai et plus clair.

Intérêts : bois de qualité médiocre utilisé de temps à autre pour fab quer des traverses de chemin de fer, des poteaux et des charpent

Gray Pine, Northern Scrub Pine, Prince's Pine.

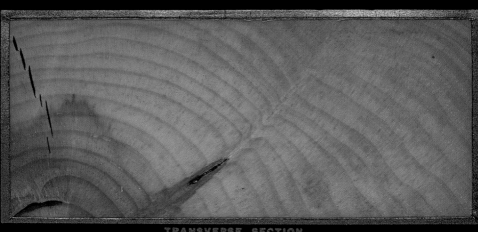

TRANSVERSE SECTION.

RADIAL SECTION.

TANGENTIAL SECTION.

Ger. Banks-Fichte.   Fr. Pin grise.   Sp. Pino pardo.

P | PINE FAMILY
Knobcone Pine

FAMILIE DER KIEFERNGEWÄCHSE
Knaufzapfen-Kiefer

FAMILLE DES PINS
*Pin à cônes pointus, Pin de l'Eldorado,
Pin à cônes en tubercule*

PINACEAE
Pinus attenuata Lemmon

Description: 20 ft. (6 m) high and 12 in. (0.3 m) in diameter, rarely 85 to 100 ft. (25 to 30 m) high and 3 ft. (0.75 m) in diameter. Trunk usually soon branching into several secondary trunks. Often begins to bear fruit when only 4 to 5 ft. (1.2 to 1.5 m) high. Needles in threes, stiff, pale yellow-green, 5 to 7 in. (13 to 18 m) long. Flowers inconspicuous: male flowers in long inflorescences; female flowers in groups of two to four. Cones elongate-conical, growing in groups.

Habitat: Dry, sun-warmed mountain slopes in Pacific North America. Forms extensive forests in the north; mixed with other trees in the south.

Wood: Light, soft, rather weak, brittle, coarse-grained. Heartwood pale brown; sapwood thick, white, sometimes tinged pale red, with very wide but rather indistinct rings of small summer cells and many large resin canals.

Use: Occasionally grown in Europe.

Beschreibung: 6 m hoch bei 0,3 m Durchmesser, selten auch 25 bis 30 m hoch bei 0,75 m Durchmesser. Stamm sich meist bald in mehrere Teilstämme aufspaltend. Oft schon ab einer Größe von 1,2 bis 1,5 m Früchte ausbildend. Nadeln zu dreien, steif, blass gelbgrün, 13 bis 18 cm lang. Blüten unauffällig, männliche in länglichen Blütenständen, weibliche in Gruppen zu 2 bis 4. Zapfen länglich-konisch, in Gruppen stehend.

Vorkommen: Trockene, von der Sonne erwärmte Berghänge im pazifischen Nordamerika. Im Norden ausgedehnte Wälder bildend, im Süden mit anderen Bäumen vermischt.

Holz: Leicht, weich, nicht stabil, brüchig und mit grober Textur. Kern hellbraun; Splint dick, weiß und manchmal mit leicht rotem Farbton, sehr breiten, aber undeutlichen Bändern kleiner Sommerzellen sowie zahlreichen großen Harzgängen.

Nutzung: Gelegentlich in Europa kultiviert.

Description: 6 m de hauteur sur diamètre de 0,3 m, rarement 25 30 m sur un diamètre de 0,75 m Tronc se divisant d'ordinaire à la base en tiges secondaires et donnant des fruits dès qu'il atte 1,2 à 1,5 m de hauteur. Aiguilles par 3, rigides, vert jaune clair, d 13 à 18 cm de long. Fleurs peu marquantes, les mâles en inflore cences allongées, les femelles gr pées par 2, 3 ou 4. Cônes allong coniques, en bouquets.

Habitat: versants montagneux se chauffés par le soleil du Pacifiqu en Amérique du Nord. Constitu dans le nord de vastes forêts, da le sud, élément des forêts mixte

Bois: léger, tendre, non stable, cassant et à grain grossier. Bois de cœur brun clair; aubier épais blanc et parfois légèrement teint de rouge. Faisceaux larges mais visibles de petites cellules d'été nombreux canaux résinifères.

Intérêts: cultivé quelquefois en Europe.

Knob-cone Pine.

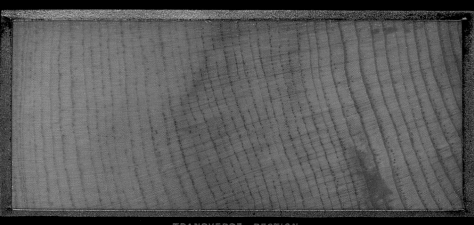

TRANSVERSE SECTION.

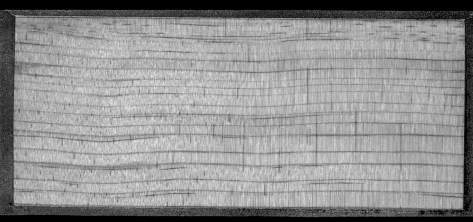

RADIAL SECTION.

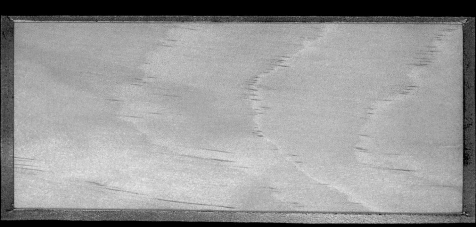

TANGENTIAL SECTION.

P | PINE FAMILY
Limber Pine, White Pine

FAMILIE DER KIEFERNGEWÄCHSE
Biegsame Kiefer

FAMILLE DES PINS
*Pin flexible, Pin des Montagnes Rocheuses,
Pin blanc de l'Ouest (Canada)*

PINACEAE
Pinus flexilis E. James

Description: 40 to 50 (85) ft. (12 to 15 (-24) m) high and 2 to 5 ft. (0.6 to 1.5 m) in diameter. At high altitudes it is found as a low shrub. Trunk thick, short. When young, the tree has an open, pyramidal shape. Later it becomes free of branches at the base. Bark at first thin, smooth, and pale gray or silvery white; later thick, dark brown with a reddish tinge, divided into plates. Needles in fives, dark green, thick, stiff, 1 to 3 in. (4 to 8 cm) long. The male in short inflorescences; the female in small groups. Cones oblong-ovoid.

Habitat: In the interior of western North America, grows from 4,800 to 10,000 ft. above sea level in mountain forests. It is the main tree species in Montana.

Wood: Light, soft, and close-grained, turning red on exposure to air. Heartwood pale yellow; sapwood thin, almost white. Indistinct bands of summer wood, numerous resin-canals, and numerous prominent medullary rays.

Use: Timber. A park tree in the eastern states of the USA., though only slow-growing. Thrives better in European gardens.

Beschreibung: 12 bis 15 (bis 24) m hoch bei 0,6 bis 1,5 m Durchmesser. In hohen Lagen als niedriger Busch. Stamm dick, kurz. Jung mit offen pyramidaler Form. Später basal astfrei. Borke anfangs dünn, glatt und hellgrau oder silberweiß, später dick, dunkelbraun mit rotem Farbton, in Platten zergliedert. Nadeln zu 5, dunkelgrün, dick, steif, 4 bis 8 cm lang. Die männlichen in kurzen Blütenständen, die weiblichen in kleinen Gruppen. Länglich ovale Zapfen.

Vorkommen: Im Inneren des westlichen Nordamerika von 1500 bis 3000 m in Gebirgswäldern mit anderen Koniferen, in Montana auch als Hauptbaumart.

Holz: Leicht, weich und mit feiner Textur; an der Luft rot werdend. Kern blassgelb; Splint dünn, fast weiß. Undeutliche Bänder kleiner Sommerzellen, zahlreiche Harzgänge und zahlreiche hervortretende Markstrahlen.

Nutzung: Bauholz. In den östlichen Staaten der USA als Parkbaum nur langsam wüchsig, gedeiht besser in europäischen Gärten.

Description: 12 à 15 m de hauteur (24 m max.) sur un diamètre de 0,6 à 1,5 m. Buisson bas en altitu Arbre au tronc court et épais, à p pyramidal ouvert à l'état juvénile, dégarni de branches à la base sur les sujets adultes. Ecorce mince, lisse et gris clair ou blanc argenté sur les jeunes arbres, plus épaisse brun sombre teinté de rouge et se craquelant en plaques avec l'âg Aiguilles par 5, vert foncé, épaisse rigides, de 4 à 8 cm de long. Fleu mâles en inflorescences courtes, l femelles en petits groupes. Cônes ovales-allongés.

Habitat: croît dans l'ouest à l'inte rieur de l'Amérique du Nord ent 1500 et 3000 m dans les forêts de conifères, ou comme essence principale dans le Montana.

Bois: léger, tendre et à grain fin, rougissant à l'air. Bois de cœur jaune pâle; aubier mince, presqu blanc. Bandes indistinctes de petites cellules d'été, nombreux canaux résinifères et nombreux rayons médullaires bien visibles.

Intérêts: bois d'œuvre. Arbres de parcs à croissance lente dans l'es des Etats-Unis, réussit mieux dan les jardins européens.

Limber Pine, Rocky Mountain White Pine.

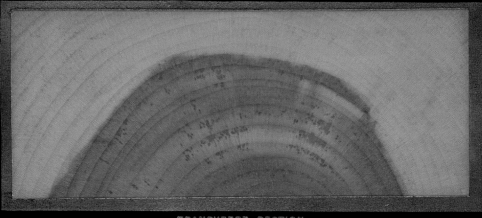

TRANSVERSE SECTION.

RADIAL SECTION.

TANGENTIAL SECTION.

Ger. Biegsame Fichte.　　　*Fr.* Pin souple.

Sp. Pino flexible.

P | PINE FAMILY

Loblolly Pine, Old Field Pine, Frankincense Pine,
Southern Pine

FAMILIE DER KIEFERNGEWÄCHSE
Weihrauchkiefer, Kienbaum

FAMILLE DES PINS
Pin à encens, Pin taeda

PINACEAE
Pinus taeda L.

Description: 80 to 100 ft. (24–30 m) high and 2 ft. (0.6 m) in diameter. Trunk straight, free of branches for much of its height. Rarely up to 165 ft. (50 m) or more high and 5 ft. (1.5 m) in diameter. Branches in the lower part of the crown horizontal, but erect above, forming a compact, rounded crown. Bark thick, light reddish brown, with irregular fissures. Needles in threes, slender, stiff, pale green, 6 to 9 in. (15 to 22 m) long. Flowers in erect spikes, the male slightly curved. Cones oblong. Very quick-growing.

Habitat: From sea level, in soils influenced by the tide, to more than 1,500 ft. (450 m) above sea level in the mountains, often planted on degraded land. Eastern North America.

Wood: Varying considerably according to location, close-grained in the case of slow-growing specimens, and with little sapwood. The wood from plantations not very durable, coarse-grained, with a high proportion of sapwood, and not very tough.

Use: Dependent on the quality of the wood. Formerly used in shipbuilding, especially for masts. Occasionally cultivated in Europe since 1713.

Beschreibung: 24 bis 30 m hoch bei 0,6 m Durchmesser. Gerader, weit herauf astfreier Stamm. Selten bis über 50 m hoch bei 1,5 m Durchmesser. Äste im unteren Teil der Krone horizontal, oben aufrecht und eine kompakte, runde Krone bildend. Borke dick, hell rotbraun, mit unregelmäßigen Rissen. Nadeln zu dreien, schlank, steif, blassgrün, 15 bis 22 cm lang. Blüten in aufrechten Kätzchen, die männlichen leicht gebogen. Zapfen länglich. Sehr schnellwüchsig.

Vorkommen: Von Meereshöhe in von der Tide beeinflußten Böden bis ins Gebirge auf über 450 m, oft auf degradierten Böden angepflanzt. Im östlichen Nordamerika.

Holz: Je nach Standort sehr verschieden, langsam gewachsene Exemplare mit feiner Textur, mit wenig Splint, Plantagenholz nicht sehr widerstandsfähig, mit grober Textur und mit hohem Splintanteil, wenig belastbar.

Nutzung: Abhängig von der Qualität des Holzes. Früher zum Schiffbau, besonders als Masten. In Europa seit 1713 gelegentlich kultiviert.

Description : 24 à 30 m de hauteu tronc droit et haut branchu de 0,6 m de diamètre. Rarement pl de 50 m sur un diamètre de 1,5 Branches horizontales à la base houppier, dressées dans sa parti supérieure, formant une couronne compacte et ronde. Ecorce épaisse, brun rouge clair, à fissu irrégulières. Aiguilles par 3, fine et rigides, vert pâle, de 15 à 22 c de long. Fleurs en chatons dress les mâles légèrement recourbées Cônes allongés. Essence à croissance très rapide.

Habitat : depuis le niveau de la mer, sur des sols soumis à l'actio des marées, jusqu'à plus de 450 m ; souvent cultivé sur des sols dégradés. Dans l'est de l'Améric du Nord.

Bois : très différent selon les lieu à grain fin et aubier peu dévelop chez les spécimens à croissance lente ; pas très solide, à grain gr sier et aubier épais, peu résistan mécaniquement chez les arbres plantation.

Intérêts : autrefois dans la constru tion navale, pour les mâts. Culti de temps à autre en Europe dep 1713.

TRANSVERSE SECTION.

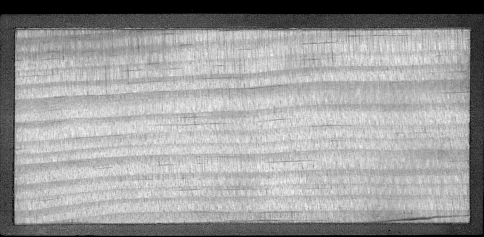

RADIAL SECTION.

TANGENTIAL SECTION.

Ger. Rosmarin-Fichte. *Fr.* Pin de Romarin. *Sp.* Romero en pino

The Alaskan Indians made the
inner bark into a kind of bread.
Germinates after fire (pyrophyte).

Die Alaskaindianer verarbeiteten
die innere Rinde zu einer Art Brot.
Feuerkeimer (Pyrophyt).

de fer et les etais de mines).

Avec l'écorce interne, les Indiens
d'Alaska confectionnaient, une
sorte de pain. Essence pyrophyte

California Scrub Pine.

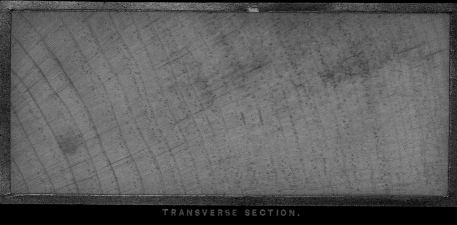

TRANSVERSE SECTION.

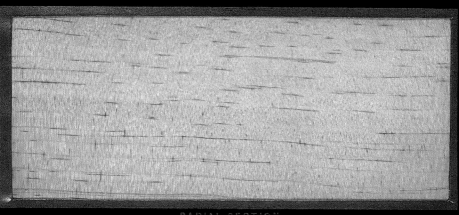

RADIAL SECTION.

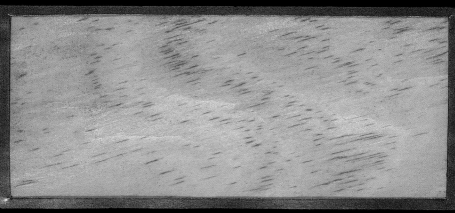

TANGENTIAL SECTION

Californianische schlechte Fichte. *Fr.* Pin tordu de Califo

Sp. Pino torcido de California.

P | PINE FAMILY
Longleaf Pine, Southern Pine

FAMILIE DER KIEFERNGEWÄCHSE
Gelb-Kiefer, Parkett-Kiefer, Sumpf-Kiefer

FAMILLE DES PINS
Pin des marais, Pin à longues feuilles, Pitchpin du Sud

PINACEAE
Pinus palustris Mill.

Description: About 100 to 130 ft. (30 to 40 m) high and 2 to 3 ft. (0.6 to 0.7 (0.9) m) in diameter. Trunk tall, straight, and with few branches. Bark pale orange-brown. In large, thin scales, or thick elongate plates. Deep-rooted.

Habitat: Belt of late-Tertiary sands and gravels along the Atlantic and Gulf coasts.

Wood: Heavy, unusually hard, tough, very strong, coarse-grained, resinous; durable. Heartwood pale red or orange; sapwood thin, almost white.

Use: The most valuable timber of all the 20 or so "pitch pine" species. One of North America's most important timber trees. Masts, spars, bridges, viaducts, scaffolding, railroad carriages, railroad sleepers, also for fencing, floorboards, and interior construction, firewood and charcoal. Turpentine and tar production. Worldwide for shipbuilding and ship furniture.

Beschreibung: Um 30 bis 40 m hoch bei 0,6 oder 0,7 (0,9) m Durchmesser. Hoher, gerader, wenig abholziger Stamm. Borke hell orangebraun. In großen, dünnen Schuppen oder dickeren, durch Längs- und Querrisse gebildeten länglichen Platten. Tiefwurzler.

Vorkommen: Der Gürtel spättertiärer Sande und Kiese längs der Atlantik- und der Golfküste.

Holz: Schwer, außerordentlich hart, zäh, sehr fest, von grober Textur und harzreich; dauerhaft. Kern hellrot oder orangefarben; Splint dünn, fast weiß.

Nutzung: Das wertvollste Holz der etwa 20 zu den „pitch pines" gerechneten Arten. Eines der wichtigsten Hölzer Nordamerikas. Masten und Rahen, Brücken, Viadukte, Gerüste, Eisenbahnwaggons, Eisenbahnschwellen, ferner für Zäune, Dielen und Innenausbau, Brennholz und Holzkohle. Terpentin- und Teergewinnung. Für Schiffbauer und Schiffsausrüster in aller Welt.

Description: environ 30 à 40 m d hauteur sur un diamètre de 0,6 o 0,7 m (0,9 m). Tronc élancé, droi et peu ligneux. Ecorce brun oran clair se craquelant en minces écailles ou en longues plaques épaisses, formées par des sillons longitudinaux et transversaux. Racines profondes.

Habitat: ceinture de sables et de galets du Tertiaire supérieur le lo de l'Atlantique et sur le pourtou du golfe du Mexique.

Bois: lourd, extrêmement dur, tenace, très résistant, à grain gros sier et résineux; durable. Bois de cœur rouge clair ou orangé; aubi mince, presque blanc.

Intérêts: bois de très bonne quali le plus précieux de la vingtaine d «pitchpins» dénombrée. L'un des principaux bois utiles d'Amériqu du Nord. Fabrication de mâts, ve gues, ponts, viaducs, échafaudage wagons et traverses de chemin de fer; utilisé en outre pour les pieux de clôtures, les parquets et en menuiserie intérieure, comme bois de chauffage et charbon de bois. Production de goudron et de térébenthine. Bois des architectes navals et des armateurs.

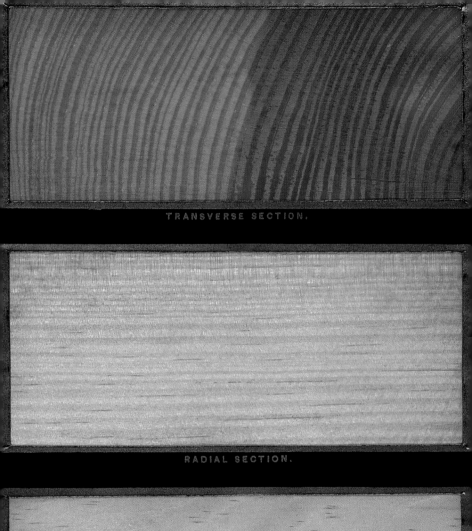

TRANSVERSE SECTION.

RADIAL SECTION.

P | PINE FAMILY
Monterey Pine

FAMILIE DER KIEFERNGEWÄCHSE
Monterey-Kiefer

FAMILLE DES PINS
Pin de Monterey

PINACEAE
Pinus radiata D. Don

Description: 85 to 100 ft. (26 to 30 m) high and 2 to 3 (5 to 6) ft. (0.6 to 0.9 (1.5 to 1.8) m) in diameter. Trunk tall. Bark dark brown, deeply divided by broad, low ridges with thick, appressed scales.

Habitat: Sea cliffs and margins of dunes and beaches in California.

Wood: Light, soft, not strong, brittle, close-grained. Heartwood pale brown; sapwood thick, almost white. Some narrow resin canals in summer wood.

Use: Occasionally as firewood.

Ornamental. Flourishes well. In North America, planted more often than any other from Vancouver Island southwards. Also in southeast USA, Mexico, Australia, and New Zealand. Since 1833 in England. Favored in parks of western and southern Europe.

Beschreibung: 26 bis 30 m hoch bei 0,6 bis 0,9 (1,5 bis 1,8) m Durchmesser. Langer Stamm. Borke dunkelbraun, tief geteilt in breite, flache Riefen mit dicken, anliegenden, plattigen Schuppen.

Vorkommen: Seeklippen und Ränder von Dünen und Stränden Kaliforniens.

Holz: Leicht, weich, nicht fest, spröde und mit feiner Textur. Kern hellbraun. Splint dick, fast weiß. Einige schmale, harzreiche Bänder im Spätholz.

Nutzung: Gelegentlich als Brennmaterial.

Ornamental. Gedeiht sehr gut. In Nordamerika von Vancouver Island südwärts häufiger gepflanzt als irgendeine andere. Auch im Südosten der USA, in Mexiko, Australien und Neuseeland. Seit 1833 in England. In West- und Südeuropa bevorzugt als Parkbaum gepflanzt.

Description : arbre de 26 à 30 m de hauteur, avec un tronc élancé de 0,6 à 0,9 m (1,5 à 1,8 m) de diamètre. Ecorce brun sombre à sillons profonds et à crêtes plates se desquamant en plaques épaisses et adhérentes.

Habitat : brisants, en bordure des dunes et des rivages en Californie.

Bois : léger, tendre, non résistant, cassant et à grain fin. Bois de cœur brun clair ; aubier épais, presque blanc. Quelques bandes résinifères étroites dans le bois final.

Intérêts : utilisé de temps à autre comme bois de chauffage.

Espèce ornementale se développant bien. La plus couramment plantée en Amérique du Nord, de l'île de Vancouver au sud de la façade pacifique, mais aussi dans le sud-est des Etats-Unis, au Mexique, en Australie et en Nouvelle-Zélande. Introduite en Angleterre depuis 1833. Arbre de parcs préféré en Europe occidentale et méridionale.

TRANSVERSE SECTION.

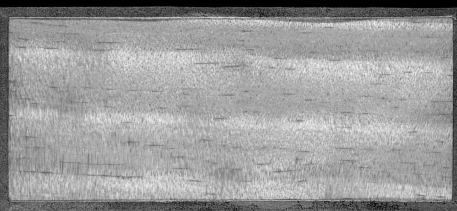

RADIAL SECTION.

P | PINE FAMILY
Mountain Hemlock

FAMILIE DER KIEFERNGEWÄCHSE
Berg-Hemlocktanne

FAMILLE DES PINS
Pruche subalpine, Hemlock de Mertens,
Pruche de Patton (Canada)

PINACEAE
Tsuga mertensiana (Bong.) Carrière

Description: 80 to 100 (165) ft. (23 to 30 (50) m) high and 4 to 5 ft. (1.2 to 1.5 m) in diameter. When freestanding, has full covering of branches and twigs, and a pyramidal shape; trees in a stand are branch-free for up to two-thirds of their height. Shrubby and creeping at high altitudes. Bark deeply furrowed. Close, rounded ridges with thin, appressed scales. Dark cinnamon red, tinged blue or purple.

Habitat: Exposed ridges and slopes in western North America, at upper tree line under snow for months each year. There forms low thickets, creeping. The tough, supple branches survive the fiercest storms for centuries.

Wood: Light, soft, not strong, close-grained; polishable. Heartwood pale brown or red; sapwood thin, almost white.

Use: Occasionally as timber, but little used because it grows in exposed, inaccesible sites.

Beschreibung: 23 bis 30 (50) m hoch bei 1,2 bis 1,5 m Durchmesser. Nahezu vollholzig. Im Freistand für 100 bis 200 Jahre von Ästen und Zweigen pyramidal eingeschlossen. Im Bestandesschluss auf 2/3 der Länge astlos. In großen Gebirgshöhen fast ohne aufrechten Stamm, kriechend. Borke tief geteilt. Verbundene, abgerundete Riefen mit dünnen, anliegenden Schuppen. Dunkel zimtrot mit blauen oder purpurnen Tönen.

Vorkommen: Exponierte Höhenrücken und Hänge im westlichen Nordamerika, an der oberen Waldgrenze alljährlich monatelang unter Schnee. Dort als niedrige, dichte Dickichte, am Boden kriechend. Die zähen, biegsamen Äste trotzen jahrhundertelang den wildesten Stürmen.

Holz: Leicht, weich, nicht fest und mit feiner Textur; polierfähig. Kern blassbraun oder rot; Splint dünn, fast weiß.

Nutzung: Gelegentlich als Bauholz, aber aufgrund der exponierten, schwer zugänglichen Lage kaum genutzt.

Description: 23 à 30 m (jusqu'à 50 m) de hauteur, avec un tronc presque cylindrique de 1,2 à 1,5 de diamètre. Houppier pyramida pendant 100 à 200 ans en situati isolée; en peuplements, tronc no ramifié jusqu'aux 2/3 de sa hauteur. En haute altitude, ne préser presque jamais de tronc droit ma un port rampant. Ecorce à textur entrelacée, constituée de fissures profondes et de crêtes arrondies aux minces écailles adhérentes; rouge cannelle sombre teinté de bleu ou de pourpre.

Habitat: à la limite supérieure de arbres dans l'ouest de l'Amériqu du Nord, enneigée plusieurs moi de l'année, sur les croupes et les versants de montagnes exposés, où il forme un épais fourré à por rampant. Les branches coriaces, flexibles résistent plusieurs siècle aux tempêtes les plus violentes.

Bois: léger, tendre, non résistant à grain fin; susceptible d'être pol Bois de cœur brun pâle ou rouge aubier mince, presque blanc.

Intérêts: utilisé de temps à autre comme bois d'œuvre mais sans exploitation véritable car les zone où il croît sont difficiles d'accès.

TRANSVERSE SECTION.

RADIAL SECTION.

TANGENTIAL SECTION

Ger. Alpische Tanne. Fr. Peruche alpestre. Sp. Abeto alpino.

P | PINE FAMILY

Noble Fir (sometimes sold as "larch")

FAMILIE DER KIEFERNGEWÄCHSE
Edle Tanne

FAMILLE DES PINS
Sapin noble, Sapin bleu, Sapin de l'Oregon

PINACEAE
Abies procera Rehder
Syn.: *A. nobilis* (Dougl. ex D. Don) Lindl.

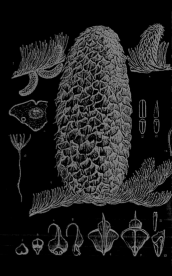

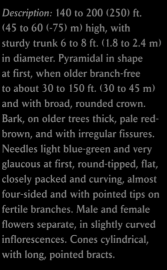

Description: 140 to 200 (250) ft. (45 to 60 (-75) m) high, with sturdy trunk 6 to 8 ft. (1.8 to 2.4 m) in diameter. Pyramidal in shape at first, when older branch-free to about 30 to 150 ft. (30 to 45 m) and with broad, rounded crown. Bark, on older trees thick, pale red-brown, and with irregular fissures. Needles light blue-green and very glaucous at first, round-tipped, flat, closely packed and curving, almost four-sided and with pointed tips on fertile branches. Male and female flowers separate, in slightly curved inflorescences. Cones cylindrical, with long, pointed bracts.

Habitat: Mountain forests in north-western North America.

Wood: Light, hard, strong, and rather close-grained. Heartwood pale brown with red streaks; sap-wood darker! Prominent dark and resinous bands of cells in summer wood and thin, inconspicuous medullary rays.

Use: Building timber, often going by the name "larch," (also used for other conifers) for interior construction and as packing material.

Beschreibung: 45 bis 60 (bis 75) m hoch, mit kräftigem Stamm von 1,8 bis 2,4 m Durchmesser. Anfangs von pyramidaler Wuchs-form, im Alter astfrei bis in etwa 30 bis 45 m Höhe und mit breiter, abgerundeter Krone. Borke an älteren Bäumen dick hell rotbraun und durch unregelmäßige Risse zergliedert. Nadeln hell blaugrün und anfangs stark bereift, abgerundet und flach, dichter stehend und gebogen, fast vierseitig und mit spitzen Enden an fertilen Ästen. Männliche und weibliche Blüten, in leicht gebogenen Blütenständen. Zapfen länglich-zylindrisch mit langen Dornen.

Vorkommen: Gebirgswälder im nordwestlichen Nordamerika.

Holz: Leicht, hart, stabil und mit ziemlich feiner Textur. Kern blass-braun mit roten Streifen; Splint dunkler! Deutliche dunkle und har-zige Bänder von Sommerzellen und dünne, undeutliche Markstrahlen.

Nutzung: Bauholz; außerdem unter dem auch für andere Nadelhöl-zer verwendeten Handelsnamen „larch" zum Innenausbau und als Verpackungsmaterial verwendet.

Description: arbre s'élevant entre 45 et 60 m (moins de 75 m) sur un diamètre de 1,8 à 2,4 m. Tron vigoureux, pyramidal dans le jeune âge; non ramifié jusqu'à une hauteur de 30 à 45 m, avec un houppier ample et arrondi su les sujets adultes. Ecorce épaisse, brun rouge et creusée de fissures irrégulières sur les spécimens âg Aiguilles vert bleu clair, rayées de bandes pruineuses blanches à l'état juvénile, à apex arrondi et plates; plus rapprochées, incurvé presque quadrilatérales et à apex aigu sur les rameaux fertiles. Fle mâles et femelles distinctes, en inflorescences légèrement courbe Cônes cylindriques, allongés, à bractées épineuses.

Habitat: forêts de montagnes du nord-ouest de l'Amérique du N

Bois: léger, dur, stable et à grain assez fin. Bois de cœur brun pâle avec des bandes rouges; aubier plus foncé! Faisceaux de cellules d'été sombres, résineux, et rayon médullaires fins, indistincts.

Intérêts: bois d'œuvre, menuiseri intérieure et caisserie sous le no commercial de «larch», commun à d'autres conifères.

Noble Fir   Oregon "Larch".

TRANSVERSE SECTION

RADIAL SECTION.

TANGENTIAL SECTION.

Ger. Erlauchte Tanne. *Fr.* Sapin noble. *Sp.* Abeto noble.

P | PINE FAMILY

Nut Pine, Pinyon

FAMILIE DER KIEFERNGEWÄCHSE
Einblättrige Nusskiefer

FAMILLE DES PINS
Pin pinyon

PINACEAE
Pinus monophylla Torr. & Frém.

Description: 16 to 20 (sometimes 45 to 50) ft. (5 to 6 (sometimes 13 to 16) m)high; rarely more than 12 in. (0.3 m) in diameter. Often dividing into several spreading branches from close to the base. Bark with deep, irregular furrows and narrow, irregular, flat ridges with thin, closely appressed light or dark-brown scales tinged red or orange. Needles single or paired.

Habitat: Gravel slopes and high ground, sometimes with Ponderosa Pine, Bristle-cone Pine, and California Juniper. Many discontinuous locations in the arid areas of western North America.

Wood: Light, soft, brittle, and close-grained. Heartwood yellow or pale brown; sapwood thick and almost white.

Use: Firewood. Produces the best charcoal for smelting. Seeds used by American Indians in desert areas, raw or roasted.

Since 1847 in Europe.

Beschreibung: 5 bis 6 m hoch, gelegentlich auch 13 bis 16 m, bei selten mehr als 0,3 m Durchmesser. Oft schon nahe am Boden in mehrere seitwärts ausgebreitete starke Stämme geteilt. Borke mit tiefen, unregelmäßigen Rillen und engen, unregelmäßigen, flachen Riefen mit dünnen, eng anliegenden hell- oder dunkelbraunen Schuppen, rot oder orange getönt. Nadeln einzeln oder zu zweit.

Vorkommen: Kiesige Hänge und Hochflächen, teilweise mit der Westlichen Gelb-Kiefer, der Grannen-Kiefer und dem Kalifornischen Wacholder. Viele disjunkte Areale in den Trockengebieten des westlichen Nordamerika.

Holz: Leicht, weich, spröde und mit feiner Textur. Kern gelb oder hellbraun; Splint dick und fast weiß.

Nutzung: Als Brennholz. Bringt im Großen Becken die beste Holzkohle für das Verhütten. Samen als Nahrungsmittel der Indianer in Wüstengebieten, roh oder geröstet.

Seit 1847 in Europa.

Description : 5 à 6 m (parfois 13 16 m) de hauteur sur un diamè atteignant rarement plus de 0,3 Tronc souvent divisé dès la base en plusieurs grosses tiges étalée latéralement. Ecorce creusée de sillons irréguliers aux côtes étro plates et irrégulières, recouverte d'écailles minces et appliquées, d'un brun sombre ou clair teint rouge ou d'orange. Aiguilles iso ou groupées par deux.

Habitat : pentes caillouteuses et hautes plaines, seul ou en assoc tion avec le pin jaune de l'Oues sapin bractéifère et le genévrier Californie. On le trouve de faço discontinue dans les régions ari de l'ouest de l'Amérique du No

Bois : léger, tendre, fragile et à grain fin. Bois de cœur brun cla ou jaune ; aubier épais et presqu blanc.

Intérêts : utilisé comme combust Produit le meilleur charbon de pour la fonte dans le Grand Bas Les graines, ou pignons, étaient consommées telles quelles ou grillées par les Indiens des régic désertiques.

Acclimaté en Europe depuis 18

Single-leaf Piñon Pine, Nut Pine, Piñon.

TRANSVERSE SECTION.

RADIAL SECTION.

TANGENTIAL SECTION.

Ger. Einzigblättrige Fichte. *Fr.* Pin monofeuillier.

P | PINE FAMILY
Pacific Silver Fir, Red Silver Fir, White Fir

FAMILIE DER KIEFERNGEWÄCHSE
Purpur-Tanne, Silber-Tanne

FAMILLE DES PINS
Sapin gracieux, Sapin argenté, Sapin amabilis (Canada)

PINACEAE
Abies amabilis (Douglas ex Loudon) Douglas ex J. Forbes

Description: Up to 250 ft. (75 m) high and 4 to 6 ft. (1.2 to 1.8 m) in diameter. At higher altitudes and in the north only 70 to 85 ft. (20 to 25 m). Trunk in dense forests often free of branches for the lower 150 ft. (45 m). Isolated trees however may have relatively short branches down to the ground. Bark on trees up to 150 years old thin, smooth, and pale or silvery white, later becoming darker. Needles deep dark green and flattened, strongly lustrous on the upper surface, silvery white beneath. Male and female flowers separate in oblong-cylindrical inflorescences. Cones oblong-ovoid, a rich purple color when young.

Habitat: On well-drained slopes in mountain forests of Pacific North America almost to the tree line, more rarely also on riverbanks.

Wood: Light, hard, not strong, and close-grained.

Use: Occasionally used in Washington for interior construction. An impressive tree on account of the snow-white bark, rich green foliage, and shining purple cones, but one that does not do well in cultivation.

Beschreibung: Bis 75 m hoch bei 1,2 bis 1,8 m Durchmesser. In höheren Lagen und im Norden nur 20 bis 25 m. Stamm in dichten Wäldern oft bis in 45 m Höhe astfrei. Einzelbaum dagegen bis zum Grund von relativ kurzen Ästen eingehüllt. Borke an bis zu 150 Jahre alten Bäumen dünn und glatt mit blasser oder silbriger Farbe. Borke später dunkler werdend. Nadeln tief dunkelgrün und abgeflacht. Stark glänzend an der Oberseite, Unterseite silbrig weiß. Männliche und weibliche Blüten separat in länglich-zylindrischen Blütenständen. Zapfen länglich oval, jung von kräftig violetter Farbe.

Vorkommen: An gut drainierten Hängen der Gebirgswälder des pazifischen Nordamerika fast bis zur Baumgrenze, seltener auch an Flussufern.

Holz: Leicht, hart, nicht stabil und mit feiner Textur.

Nutzung: In Washington gelegentlich zum Innenausbau. Wegen der schneeweißen Borke, des kräftig grünen Laubs und der leuchtend violetten Zapfen ein eindrucksvoller Baum, der in Kultur aber nur schlecht gedeiht.

Description : peut atteindre 75 m hauteur sur un diamètre de 1,2 à 1,8 m. Sa taille ne dépasse pas 2 ou 25 m en altitude et dans le no Tronc souvent non ramifié jusqu 45 m de haut dans les forêts der mais enveloppé jusqu'au sol par des branches relativement courte en situation isolée. Ecorce mince lisse, pâle ou argentée sur les su de moins de 150 ans, plus sombr par la suite. Aiguilles d'un vert profond et aplaties, très brillante dessus, blanc argenté dessous. Inflorescences mâles et femelles cylindriques, distinctes sur le mê arbre. Cônes ovoïdes-longs, d'un violet intense jeunes.

Habitat : pentes bien drainées de forêts montagnardes du Pacifiqu en Amérique du Nord, presque jusqu'à la limite des arbres, plus rarement au bord des cours d'ea

Bois : léger, dur, non stable et à grain fin.

Intérêts : utilisé parfois en menuis rie intérieure dans l'Etat de Wash ington. Cet arbre est splendide à son écorce blanche comme neige son feuillage vert intense et ses cônes d'un violet lumineux, mai réussit mal en culture.

Amabilis Fir, Red Silver Fir.

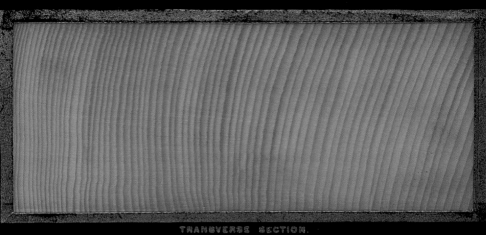

TRANSVERSE SECTION.

RADIAL SECTION.

TANGENTIAL SECTION

P | PINE FAMILY

Parry Pinyon Pine

FAMILIE DER KIEFERNGEWÄCHSE
Parrys Föhre

FAMILLE DES PINS
Pin à quatre aiguilles

PINACEAE
Pinus quadrifolia Paul. ex Sudw.

Description: 30 to 40 ft. (9 to 12 m) high and 15 in. (0.45 m) in diameter. Crown compact and pyramidal. Bark dark brown with a red tinge, divided by shallow fissures into plates covered with flat scales. Needles in clusters of three to five (mostly four), glaucous, green, 1 to 1½ in. (3 to 4 cm) long. Male flowers in long, cylindrical inflorescences at the ends of the branches, the female somewhat distant from the ends, singly or in groups. Cones almost globular. Probably very slow-growing.

Habitat: Forms open forests in arid regions and on the mountain slopes of Lower California.

Wood: Light, soft, and close-grained. Heartwood light brown or yellow, sapwood considerably paler. Light-colored zones of summer wood, and distinct resin canals.

Use: The seeds are eaten raw or roasted by the Indians and represent an important source of food. The tree is occasionally cultivated in gardens in California.

Beschreibung: 9 bis 12 m hoch bei 0,45 m Durchmesser. Krone kompakt pyramidal. Borke dunkelbraun mit rotem Farbton, durch flache Risse in Schollen zerteilt, deren Oberfläche sich in flachen Schuppen schält. Nadeln zu 3 bis 5 (meist zu viert), bereift, grün. 3 bis 4 cm lang. Männliche Blüten in lang zylindrischen Blütenständen an den Enden der Äste, weibliche etwas unterhalb der Enden, einzeln oder in Gruppen. Zapfen fast kugelig. Wahrscheinlich sehr langsamwüchsig.

Vorkommen: Bildet lichte Wälder in wüstenartigen Gegenden und an Berghängen Niederkaliforniens.

Holz: Leicht, weich und mit feiner Textur. Kern blassbraun oder gelb, Splint erheblich heller. Helle Zonen von Sommerzellen und deutliche Harzgänge.

Nutzung: Die Samen werden von den Indianern roh oder geröstet gegessen und stellen eine wichtige Nahrungsquelle dar. Der Baum wird in Kalifornien gelegentlich in Gärten kultiviert.

Description : 9 à 12 m de hauteur sur un diamètre de 0,45 m. Houppier pyramidal et compact Ecorce brun sombre teinté de rouge, creusée de fissures plates aux crêtes s'exfoliant en écailles plates. Faisceaux de 3 à 5 aiguill (4 le plus souvent) marquées de blanc pruineux, vertes, 3 à 4 cm Inflorescences mâles cylindrique longues, terminales ; fleurs feme solitaires ou groupées un peu pl bas sur les rameaux. Cônes pres sphériques. Croissance très lent probablement.

Habitat : constitue des forêts clai dans les zones désertiques et su les versants montagneux de la Basse-Californie.

Bois : léger, tendre et à grain fin. Bois de cœur brun pâle ; aubier beaucoup plus clair. Plages clair de cellules d'été et canaux résin fères distincts.

Intérêts : les graines, crues ou grillées, jouent un rôle importar dans l'alimentation des Indiens. Planté quelquefois dans les jard en Californie.

Parry Pine, Mexican Piñon or Nut Pine.

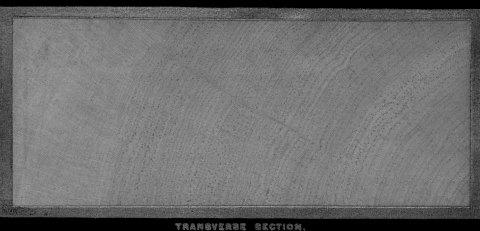

TRANSVERSE SECTION.

RADIAL SECTION.

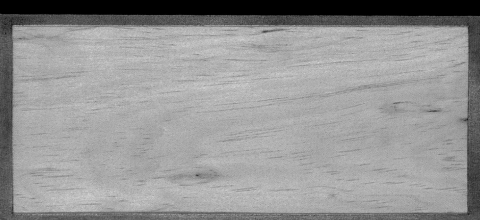

TANGENTIAL SECTION.

Ger. Vierblättrige Fichte.   *Fr.* Pin quadrifeuillier.

Sp. Pino de quatro hojas.

P | PINE FAMILY
Pitch Pine

FAMILIE DER KIEFERNGEWÄCHSE
Kiefer

FAMILLE DES PINS
Pin de Coulter, Pins à gros cônes

PINACEAE
Pinus coulteri D. Don

Description: 50 to 60 ft. (16 to 18 m) high and 4 ft. (1.2 m) in diameter, although usually smaller. Bark dark brown or nearly black; deeply divided into broad, rounded ridges, with thin, appressed scales. Needles in threes, blue-green.

Habitat: Wooded ridges of the Californian mountain ranges, with Ponderosa Pine, Sugar Pine, and White Fir. Smaller on southern slopes, and often with Knobcone Pine.

Wood: Light, soft, not strong, brittle, and coarse-grained. Heartwood pale red; sapwood thick and almost white. Handsome resin bands in summer wood, with fewer large resin canals.

Use: Occasionally as firewood. Seeds collected and eaten by American Indians in southern California.

Used as an ornamental tree owing to its huge cones, heavier than those of any other pine. Hardy and cultivated in western and central Europe.

Beschreibung: 16 bis 18 m hoch bei 1,2 m Durchmesser, obwohl im allgemeinen kleiner. Borke dunkelbraun oder fast schwarz. Tief geteilt in breite, abgerundete, verbundene Riefen. Mit dünnen, anliegenden Schuppen besetzt. Nadeln zu dritt, blaugrün.

Vorkommen: Waldbedeckte Höhenrücken der kalifornischen Küstengebirge mit Westlicher Gelb-Kiefer, der Zucker-Kiefer und der Kalifornischen Tanne. Auf südseitigen Hängen kleiner, aber häufiger mit der Knaufzapfen-Kiefer.

Holz: Leicht, weich, nicht fest, spröde und mit grober Textur. Kern hellrot; Splint dick und fast weiß. Ansehnliche Harzbänder im Spätholz und wenige, große Harzgänge.

Nutzung: Gelegentlich als Brennholz. Samen in Süd-Kalifornien von Indianern gesammelt und gegessen.

Ornamental durch die riesigen Zapfen, schwerer als die aller anderen Kiefern. Winterhart in West- und Mitteleuropa und dort in Kultur.

Description: 16 à 18 m de hauteu sur un diamètre de 1,2 m, bien d'ordinaire plus petit. Ecorce bru foncé ou presque noire, à fissure profondes et larges crêtes arrondies, recouvertes de fines écaille adhérentes. Aiguilles vert bleuté groupées par trois.

Habitat: croupes de montagnes boisées proches de la côte californienne, en compagnie du pin jaune, du pin à sucre et du sapir blanc du Colorado. Sur les versa d'adret, plus petit mais plus fréquent et associé au Knaufzapfen kiefer.

Bois: léger, tendre, pas résistant cassant et à grain grossier. Bois de cœur rouge clair; aubier épai presque blanc. Belles bandes résineuses dans le bois final et quelques grands canaux résinifè

Intérêts: utilisé à l'occasion comr bois de chauffage. Graines come tibles, récoltées par les Indiens c sud de la Californie.

Essence ornementale par ses énormes cônes, plus lourds que ceux des autres pins. Résiste à l'hiver en Europe occidentale et méridionale, où il est cultivé.

Big-cone Pine.

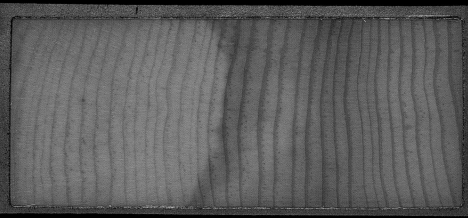

TRANSVERSE SECTION.

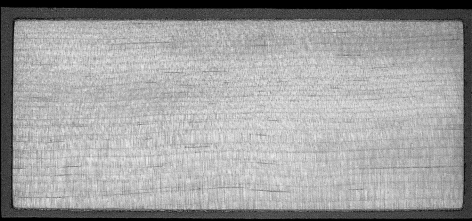

RADIAL SECTION.

TANGENTIAL SECTION

Ger. Grossfrüchtige Fichte. *Fr.* Pin de cones grands.
Sp. Pino de conos grandes.

P | PINE FAMILY

Pitch Pine, Northern Pitch Pine

FAMILIE DER KIEFERNGEWÄCHSE
Pech-Kiefer

FAMILLE DES PINS
Pin rigide, Pin à aubier, Pin des corbeaux (Canada)

PINACEAE
Pinus rigida Mill.

Description: 50 to 70 (15 to 20 m), rarely 85 ft. (25 m) high and 2 to 3 ft. (0.6 to 0.9 m) in diameter. Trunk short, with outgrowths of young shoots on the old wood, on the stumps after felling, and after fire. Bark thick, dark brown, fissured, separating in irregular plates. Needles in threes.

Habitat: Poor sandy soils in the plain and gravely uplands, infertile stretches of land, as for example the "Barrens" of Kentucky. Eastern North America.

Wood: Moderately heavy and hard, brittle, and coarse-grained; very durable. Heartwood light brown or red; sapwood thick, yellow to almost white.

Use: Rough building timber, floorboards, thresholds for doorways, firewood, charcoal. Formerly a source of tar, turpentine, and lampblack (soot). A commodity in New England until the Revolution.

Beschreibung: 15 bis 20, selten 25 m hoch bei 0,6 bis 0,9 Durchmesser. Stamm kurz mit Ausschlag junger Triebe am alten Holz, am Stubben nach Abholzen und nach Feuer. Borke dick, dunkelbraun, rissig, in unregelmäßigen Platten abblätternd. Nadeln zu dritt.

Vorkommen: Sandige, nährstoffarme Böden der Ebene und kiesiger Anhöhen, unfruchtbare Landstriche („barren"). Im östlichen Nordamerika.

Holz: Mittelschwer, mittelhart, spröde und grobe Textur; sehr haltbar. Kern hellbraun oder rot; Splint dick, gelb bis fast weiß.

Nutzung: Grobes Bauholz, Dielen, Schwellen, Brennholz, Holzkohle. Früher zur Teergewinnung, Terpentin und Lampenschwarz (Ruß). Bis zur Revolution Handelsware in Neuengland.

Description : 15 à 20 m de hauteu rarement 25 m, sur un diamètre 0,6 à 0,9 m. Tige courte, donnan naissance à de nombreuses ramu sur le bois ancien, et sur la souch après une coupe ou le passage d' feu. Ecorce épaisse brun sombre, creusée de sillons et se desquam en plaques irrégulières. Aiguilles courtes groupées par trois.

Habitat : sols sablonneux, maigre en plaine et sols graveleux des hauteurs, terres arides («barren» dans l'est de l'Amérique du Norc

Bois : moyennement lourd et dur, cassant et à grain grossier ; très durable. Bois de cœur brun clair ou rouge ; aubier épais, jaune à presque blanc.

Intérêts : gros bois d'œuvre, lames de parquet, traverses, bois de chauffage, charbon de bois. Utilis autrefois pour la production de goudron, de térébenthine et de suie. Marchandise en Nouvelle-Angleterre jusqu'à la Révolution.

Pitch Pine.

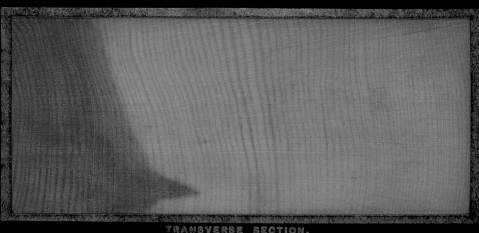

TRANSVERSE SECTION.

RADIAL SECTION.

TANGENTIAL SECTION.

Ger. Pechkiefer. *Fr.* Pin de goudron. *Sp.* Pino de pez

P | PINE FAMILY
Pond Pine, Marsh Pine

FAMILIE DER KIEFERNGEWÄCHSE
Teich-Fichte

FAMILLE DES PINS
Pin sérotineux, Pin des mares, Pin tardif

PINACEAE
Pinus serotina Michx.

Description: 40 to 50 (12 to 15 m), sometimes 75 to 80 ft. (23 to 26 m) high and 2 to 3 ft. (0.6 to 0.9 m) in diameter. Trunk short. Bark dark red-brown, divided into small plates by irregular, narrow, flat furrows, with thin, closely appressed scales on the surface. Three-needled. Sprouts well after fire or from stump.

Habitat: Wet valleys, sandy or peaty swamps with magnolias, Swamp Bay and gums (*Nyssa* spp.), either as the only pine or with Loblolly Pine in the coastal regions of the eastern Gulf states as well as the Atlantic states as far as New Jersey.

Wood: Heavy, soft, splintering, coarse-grained, and very resinous. Heartwood dark orange; sapwood thick, pale yellow. Conspicuous dark resin canals.

Use: For building in North Carolina. Formerly for masts of small boats. Turpentine source.

Beschreibung: 12 bis 15, gelegentlich 23 bis 26 m hoch bei 0,6 bis 0,9 m Durchmesser. Stamm kurz. Borke dunkel rotbraun, durch unregelmäßige enge, flache Rillen in kleine Platten geteilt, die an ihrer Oberfläche dünne, eng anliegende Schuppen abgeben. Dreinadelig. Nach Feuer oder aus dem Stubben kräftig austreibend.

Vorkommen: Nasse Niederungen, sandige oder torfige Sümpfe mit Magnolien, Sumpf-Avocado und Tupelos, entweder als einzige Kiefer oder mit der Loblolly-Kiefer in den Küstenregionen der östlichen Golfstaaten sowie der Atlantikstaaten bis zur Höhe von New Jersey.

Holz: Schwer, weich, splitterig, mit grober Textur und sehr harzreich. Kern dunkel orangefarben; Splint dick, blassgelb. Auffallende, dunkle Harzgänge.

Nutzung: In Nordkarolina als Bauholz. Früher zu Masten kleiner Schiffe verarbeitet. Terpentinegewinnung.

Description: arbre de 12 à 15 m de hauteur, atteignant quelquefois 2 à 26 m, avec un tronc court de 0, à 0,9 m de diamètre. L'écorce bru rouge foncé, creusée d'étroites fissures plates et irrégulières, se découpe en petites plaques aux écailles minces et appliquées. Aiguilles groupées par trois. Reje vigoureusement de souche ou après le passage d'un feu.

Habitat: zones basses détrempée marécages sablonneux ou tourbe en compagnie des magnolias, de l'avocatier palustre et des nyssas, comme seule espèce de pin ou av le pin à l'encens. Croît dans les régions côtières des Etats du Golf à l'est ainsi que dans les Etats de côte atlantique jusqu'à hauteur d New Jersey.

Bois: lourd, tendre, tendant à se fendre, à grain grossier et très rés neux. Bois de cœur orangé foncé aubier épais, jaune pâle. Canaux résinifères sombres et bien visible

Intérêts: bois d'œuvre en Caroline du Nord. Utilisé autrefois pour fa des mâts de navires de faible tonnage. Extraction de la térébenthir

Pond Pine.

TRANSVERSE SECTION.

RADIAL SECTION.

TANGENTIAL SECTION

P | PINE FAMILY
Ponderosa Pine

FAMILIE DER KIEFERNGEWÄCHSE
Felsenföhre, Gelb-Kiefer

FAMILLE DES PINS
Pin ponderosa, Pin à bois lourd, Pin jaune de l'Ouest

PINACEAE
Pinus ponderosa Lawson & C. Lawson

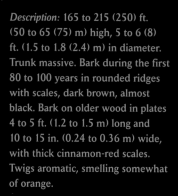

Description: 165 to 215 (250) ft. (50 to 65 (75) m) high, 5 to 6 (8) ft. (1.5 to 1.8 (2.4) m) in diameter. Trunk massive. Bark during the first 80 to 100 years in rounded ridges with scales, dark brown, almost black. Bark on older wood in plates 4 to 5 ft. (1.2 to 1.5 m) long and 10 to 15 in. (0.24 to 0.36 m) wide, with thick cinnamon-red scales. Twigs aromatic, smelling somewhat of orange.

Habitat: Mountain slopes, high plateaus, volcanic tableland, rarely on boggy ground, when dwarf. North America's most extensive pine forests (Colorado Plateau). Exists in several forms in western North America.

Wood: Heavy, hard, strong, rather brittle, comparatively close-grained may be more or less resinous; not durable in the ground. Heartwood pale red; sapwood almost white.

Use: Railroad sleepers, fencing, firewood, and all kinds of timber. American Indians scratched off the inner bark in spring to eat the slimy cambium.

Much variation in form and vitality.

Beschreibung: 50 bis 65 (75) m hoch bei 1,5 bis 1,8 (2,4) m Durchmesser. Massiver Stamm. Borke während der ersten 80 bis 100 Jahre in gerundeten Riefen mit anliegenden Schuppen, dunkelbraun, fast schwarz. An älteren Stämmen Borke in 1,2 bis 1,5 m langen, 0,24 bis 0,36 m breiten Platten mit dicken, zimtroten Schuppen. Zweige aromatisch, etwa wie Orange riechend.

Vorkommen: Berghänge, Hochebenen, vulkanisches Tafelland, selten Sümpfe, dann kümmerlich. Die ausgedehntesten Kiefernwälder Nordamerikas (Colorado-Plateau). In mehreren Formen und Varietäten im westlichen Nordamerika verbreitet.

Holz: Schwer, hart, fest, zuletzt spröde, vergleichsweise feine Textur. Nicht dauerhaft im Boden. Kern hellrot; Splint fast weiß.

Nutzung: Für Eisenbahnschwellen, Zäune, Brennholz und alle Arten von Holzbau. Indianer kratzten die innere Rinde ab und aßen das schleimige Kambium.

Größte Variationsbreite und Vitalität.

Description: 50 à 65 m (75 m) de hauteur, tronc massif de 1,5 à 1,8 m (2,4 m) de diamètre. Ecorc fissurée en crêtes écailleuses arrc dies, brun presque noir pendant les 80 ou 100 premières années, se craquelant en plaques de 1,2 à 1,5 m de long et de 0,24 à 0,36 r de large, avec d'épaisses écailles cannelle, sur les vieux spécimens Rameaux à odeur de mandarine.

Habitat: versants montagneux, hauts plateaux, plateaux volcaniques; rarement marécages, où présente un port rabougri. Les pl vastes forêts de conifères d'Amérique du Nord (plateau du Colorado). Très répandu dans l'ouest continent nord-américain sous de formes et variétés différentes.

Bois: lourd, dur, résistant, à la fir cassant, à grain fin; non durable sur pied. Bois de cœur rouge clai aubier presque blanc.

Intérêts: traverses de chemin de f clôtures, en construction et bois de chauffage. Les Indiens conson maient le cambium mucilagineux

Eventail le plus étendu des variations individuelles et la plus gran plasticité.

California Yellow Pine, Bull Pine.

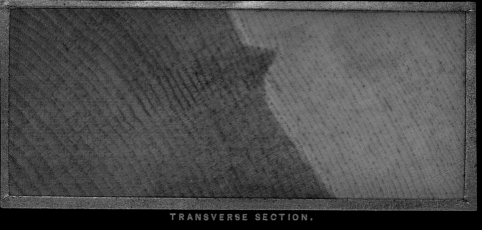

TRANSVERSE SECTION.

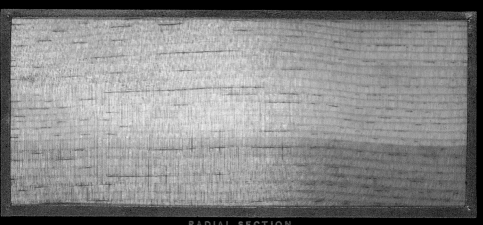

RADIAL SECTION.

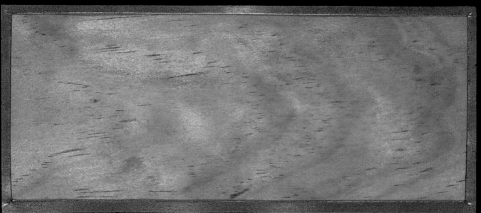

TANGENTIAL SECTION

Californianische Gelbe. Fichte. *Fr.* Pin jaune de Californ

Sp. Pino amarillo de California.

P | PINE FAMILY
Prickle-cone Pine

FAMILIE DER KIEFERNGEWÄCHSE
Bischofs-Kiefer, Stachelkiefer

FAMILLE DES PINS
Pin muriqué, Pin de Bishop

PINACEAE
Pinus muricata D. Don

Description: 40 to 50 (100!) ft. (13 to 16 (30!) m) high and 2 to 3 ft. (0.6 to 0.9 m) in diameter. Lower bark deeply furrowed. Long, narrow, rounded ridges, with rough, appressed dark purple-brown scales. Needles paired, in deep, dark green clusters toward the tips of branches. Cone scales armed with strong, inwardly curved points.

Habitat: The characteristic pine of the Mendocino coast. Steep slopes and bare upland away from the sea, on sandy soils. Sometimes on better soils. Widespread from San Francisco Bay to Lower California.

Wood: Light, hard, very strong and rather coarse-grained. Heartwood pale brown; sapwood thick and almost white. Broad bands of resin cells in summer wood; few resin canals.

Use: Used as timber in the Mendocino area.

Introduced into Europe in 1846. Ornamental in parks, owing to its broad cones.

Beschreibung: Höhe 13 bis 16 (30!) m hoch bei 0,6 bis 0,9 m Durchmesser. Untere Borke tief geteilt. Lange, schmale, abgerundete Riefen mit rauen, anliegenden, dunkel purpurbraunen Schuppen. Nadeln zu zweit in tief dunkelgrünen Büscheln gehäuft an den Zweigenden. Zapfenschuppen mit je starker, eingekrümmter Spitze bewaffnet.

Vorkommen: Charakteristische Kiefer der Mendocino-Küste. Steile Uferklippen und kahles, dem Ozean voll ausgesetztes Oberland, auf Sandflächen. Verbreitet von der San Francisco-Bay bis Niederkalifornien.

Holz: Leicht, hart, sehr fest und mit ziemlich grober Textur. Kern hellbraun; Splint dick und fast weiß. Breite Bänder von Harzzellen im Spätholz, wenige Harzgänge.

Nutzung: In der Gegend von Mendocino als Bauholz.

Seit 1846 in Europa. Ornamentale Wirkung in Parks durch die Zapfenfülle.

Description : 13 à 16 m (jusqu'à 30 m!) de hauteur sur un diamè de 0,6 à 0,9 m. Partie basale du tronc profondément fissurée en longues et étroites crêtes arrondies aux écailles brun pourpre sombre, rugueuses et adhérentes Aiguilles d'un vert sombre profo groupées par deux et disposées e touffes à l'extrémité des rameau. Cônes à écailles fortement recou bées en crochets piquants.

Habitat : pin typique du cap Men docino. Ecueils escarpés et arrièr côte dénudée, soumise à l'action de l'océan, étendues de sable. Cr aussi parfois sur de meilleurs sol Répandu depuis la baie de San Francisco jusqu'en Basse-Californ

Bois : léger, dur, très résistant et à grain assez grossier. Bois de cœu brun clair ; aubier épais, presque blanc. Larges bandes de cellules résinifères dans le bois final, peu de canaux résinifères.

Intérêts : bois de chauffage dans la région de Mendocino.

Introduit en Europe depuis 1846 Bel arbre de parcs par ses cônes abondants.

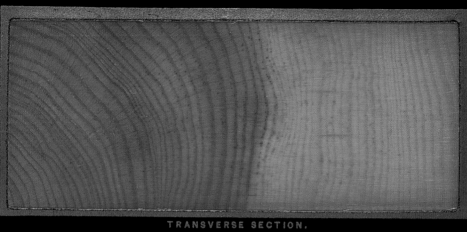

TRANSVERSE SECTION.

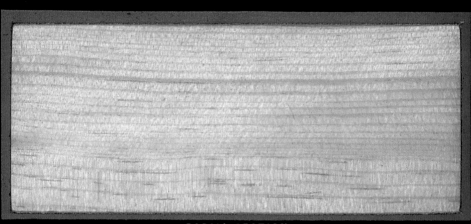

RADIAL SECTION.

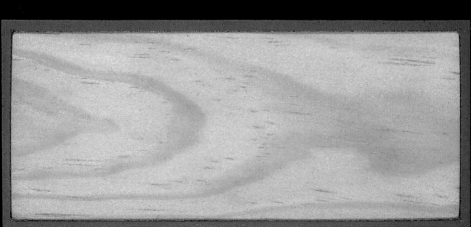

TANGENTIAL SECTION

Ger. Californische Sumpffichte. Fr. Pin de marias.

P | PINE FAMILY
Red Fir

FAMILIE DER KIEFERNGEWÄCHSE
Pracht-Tanne

FAMILLE DES PINS
Sapin rouge, Sapin magnifique, Sapin majestueux

PINACEAE
Abies magnifica A. Murray

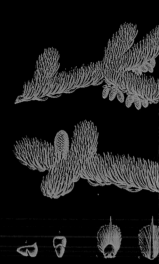

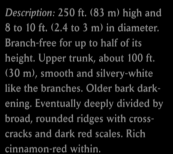

Description: 250 ft. (83 m) high and 8 to 10 ft. (2.4 to 3 m) in diameter. Branch-free for up to half of its height. Upper trunk, about 100 ft. (30 m), smooth and silvery-white like the branches. Older bark darkening. Eventually deeply divided by broad, rounded ridges with cross-cracks and dark red scales. Rich cinnamon-red within.

Habitat: Forms homogeneous forests on fine, moraine soils in mountainous regions in south-western North America.

Wood: Light, soft, not strong, silky-sheened; polishable, hard to store, and comparatively durable in the ground. Heartwood pale red-brown; sapwood thick, darker!

Use: Mainly as firewood. In California as coarse timber.

Discovered by Frémont in 1845. Around 1851 to England, France, and northern Italy. Noblest of the firs and, along with Bigtree, Sugar Pine, Ponderosa Pine, Incense Cedar, and Douglas Fir, the pride of the Sierra Nevada forests.

Beschreibung: 83 m hoch bei 2,4 bis 3 m Durchmesser. Bis auf halbe Höhe astfrei. Oberer Stamm, etwa 30 m, wie die Zweige glatt und silbrig weiß. Ältere Borke dunkelt. Zuletzt tief geteilt in breite, abgerundete Riefen mit Querrissen und dunkelroten Schuppen. Innen kräftig zimtrot.

Vorkommen: Auf feinen Moränenböden des südwestlichen Nordamerika reine Wälder bildend. Gebirge.

Holz: Leicht, weich, nicht fest und seidig; polierfähig, schwierig zu lagern und vergleichsweise dauerhaft im Boden. Kern hell rotbraun; Splint dick, dunkler!

Nutzung: Meist als Brennholz. In Kalifornien als grobes Bauholz.

1845 von Frémont entdeckt. Um 1851 nach England, Frankreich und Norditalien eingeführt. Edelste Art der Tannen und mit dem Mammutbaum, der Zucker-Kiefer, der Westlichen Gelb-Kiefer, der Flusszeder und der Douglasie der Stolz der Wälder der Sierra Nevada.

Description: 83 m de hauteur, tr de 2,4 à 3 m de diamètre, non ramifié jusqu'à mi-hauteur. La p tie supérieure du tronc, à 30 m haut environ, ainsi que les rame sont lisses et blanc argenté. L'éc devient sombre sur les vieux su et se creuse de profondes fissur arrondies qui se craquellent en quettes recouvertes d'écailles ro foncé. L'endoderme est d'un rou cannelle vif.

Habitat: constitue des forêts pur dans les sols morainiques fins d l'étage montagnard du sud-oues de l'Amérique du Nord.

Bois: léger, tendre, non résistant et satiné; pouvant se polir, diffi à stocker et durable sur pied. Bc de cœur rouge brun clair; aubie épais, plus foncé!

Intérêts: surtout bois de chauffa Bois d'œuvre grossier en Califor

Découvert par Frémont en 1845 Introduit en Angleterre, en Fran et en Italie vers 1851. L'orgueil forêts de la Sierra Nevada avec l séquoia géant, le pin à sucre, le pin jaune de l'Ouest, le cèdre d rivières et le douglas.

TRANSVERSE SECTION.

RADIAL SECTION.

TANGENTIAL SECTION

Ger. Californische rothe Tanne.   Fr. Sapin rouge de Californie
S.   Abete colorado de California

P | PINE FAMILY
Red Pine, Norway Pine

FAMILIE DER KIEFERNGEWÄCHSE
Amerikanische Rot-Kiefer

FAMILLE DES PINS
Pin rouge, Pin résineux (Canada)

PINACEAE
Pinus resinosa Sol.

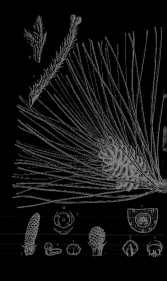

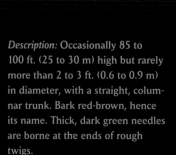

Description: Occasionally 85 to 100 ft. (25 to 30 m) high but rarely more than 2 to 3 ft. (0.6 to 0.9 m) in diameter, with a straight, columnar trunk. Bark red-brown, hence its name. Thick, dark green needles are borne at the ends of rough twigs.

Habitat: Grows on dry, sandy soils on upland slopes and ridges in north-eastern North America, but does not form homogeneous stands.

Wood: Moderately heavy and hard.

Use: Valuable for ships' spars, piles, thresholds of doorways, and wooden structures of all kinds.

Beschreibung: Gelegentlich 25 bis 30 m hoch bei selten mehr als 0,6 bis 0,9 m Durchmesser. Gerader, säulenförmiger Stamm. Rotbraune Borke (bestimmend für die Namensgebung). Dicke, dunkelgrüne Nadeln in Büscheln an den Enden rauer Zweige.

Vorkommen: Bewohnt, ohne je geschlossene Bestände zu bilden, trockene, sandige, hoch gelegene Böden der Hänge und Kämme im nordöstlichen Nordamerika.

Holz: Mäßig schwer und hart.

Nutzung: Geschätzt für Schiffsspieren, Pfahlgründungen, Türschwellen und Holzbauten aller Art.

Description : parfois jusqu'à 25 c 30 m de hauteur, avec un tronc columnaire dépassant rarement à 0,9 m de diamètre. Ecorce bru rouge, qui lui a donné son nom Rameaux rugueux portant à leu extrémité d'épaisses aiguilles ve sombre, groupées en bouquets.

Habitat : occupe, mais sans jama constituer de forêts fermées, les sols secs et sablonneux des hau de versants et des crêtes au nor est de l'Amérique du Nord.

Bois : moyennement lourd et du

Intérêts : recherché pour la fabri tion d'espars, de pilotis, de seui de portes et pour toute sortes d constructions.

Red Pine,   "Norway" Pine.

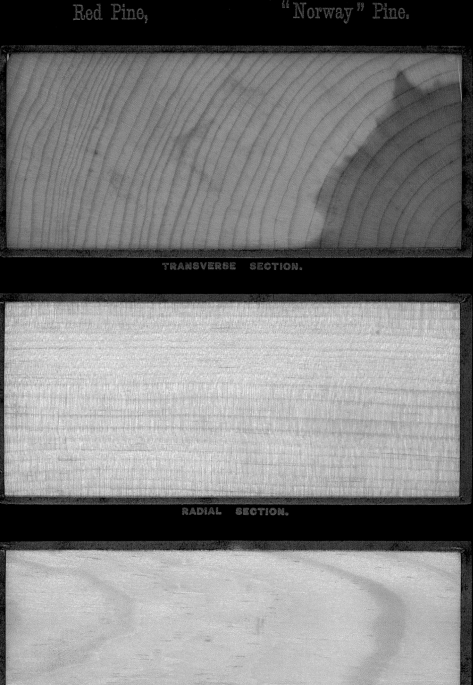

TRANSVERSE SECTION.

RADIAL SECTION.

TANGENTIAL SECTION.

Ger. Harzige Fichte. *Fr.* Pin rouge. *Sp.* Pino rizado.

P | PINE FAMILY
Red Spruce

FAMILIE DER KIEFERNGEWÄCHSE
Rot-Fichte

FAMILLE DES PINS
Epicéa rouge, Prusqueur rouge (Canada)

PINACEAE
Picea rubens Sarg.

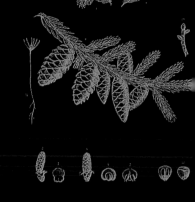

Description: Occasionally up to 100 ft. (30 m) high and 2 to 4 ft. (0.6 to 1.2 m) in diameter, usually considerably smaller. Has a straight, columnar trunk with reddish brown bark, and red cones.

Habitat: Forests in the Allegheny Mountains and the north in association with Hemlock, Beech, Yellow Birch, Sugar Maple, Butternut, etc. Also forms homogeneous woods.

Wood: Light, moderately soft, but strong and flexible.

Use: Building timber, floorboards, sounding boards of musical instruments, and extensively for paper pulp. The exuded resin is sold commercially as "spruce gum."

One of the commonest forest trees in the north of New York State and New England.

Beschreibung: Gelegentlich bis 30 m hoch bei 0,6 bis 1,2 m Durchmesser, meist erheblich kleiner. Gerader, säulenförmiger Stamm mit rotbrauner Borke, roter Zapfen.

Vorkommen: Wälder der Alleghenies und des Nordens in Gesellschaft von Hemlock, Buche, Gelb-Birke, Zucker-Ahorn, Butternuss u. a. Tritt auch waldbildend auf.

Holz: Leicht, mäßig weich, aber fest und elastisch.

Nutzung: Bauholz, Dielenbretter, Resonanzböden von Musikinstrumenten und in großem Umfang für Papier-Pulpe. Die harzigen Ausschwitzungen kommen als Fichtenharz („spruce gum") in den Handel.

Einer der häufigsten Waldbäume im Norden des Staates New York und in Neuengland.

Description: parfois jusqu'à 30 m hauteur sur un diamètre de 0,6 1,2 m, mais d'ordinaire netteme plus petit. Tronc droit, columnai avec une écorce rougeâtre. Côn rouges.

Habitat: forêts des Alleghanys e Nord en compagnie de la pruch du hêtre, du bouleau jaune, de l'érable à sucre, du noyer cendr etc. Peut constituer aussi des pe plements forestiers.

Bois: léger, moyennement tendr mais résistant et élastique.

Intérêts: utilisé en construction, lutherie, dans l'industrie du pap et pour la fabrication de madrie Exsude une matière résineuse appelée dans le commerce résin d'épicéa («spruce gum»).

Une des espèces forestières les f courantes du nord de l'Etat de N York et de la Nouvelle-Angleterr

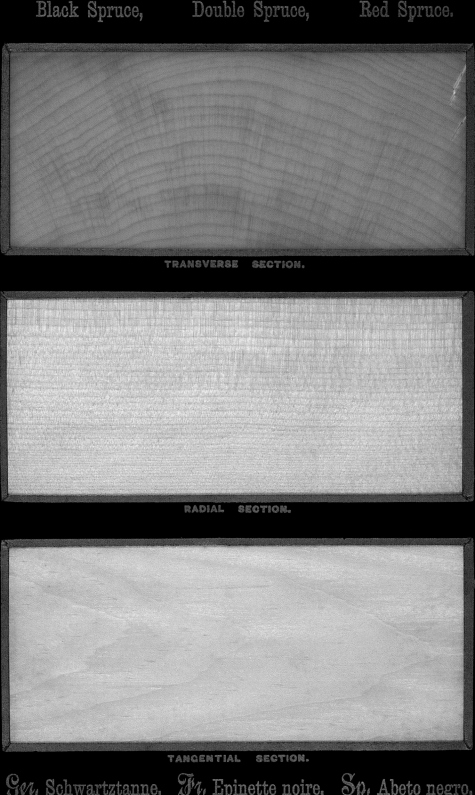

TRANSVERSE   SECTION.

RADIAL   SECTION.

TANGENTIAL   SECTION.

P | PINE FAMILY
Sand Pine, Spruce Pine

FAMILIE DER KIEFERNGEWÄCHSE
Sand-Kiefer

FAMILLE DES PINS
Pin des sables

PINACEAE
Pinus clausa (Chapm. ex Engelm.) Vasey ex Sarg.

Description: Only 15 to 20 or 24 ft. (4.5 to 6 or 7 m) high and scarcely 12 in. (0.3 m) in diameter, but on best sites 75 to 85 ft. (22 to 25 m) and 2 ft. (0.6 m) in diameter. Covered in thin branches down to the ground. Lower bark deeply divided by narrow furrows into irregular long plates with thickly appressed, pale red-brown scales. Upper trunk has thinner, smooth, ash-gray bark. Needles 2 to 3 in. (4 to 7 cm), paired. Very fast-growing.

Habitat: Sand dunes and pure white coastal sand. Inland on dry hills. On better quality soils with magnolias, hickories, Live Oak, Post Oak, etc., in the coastal regions of southeastern North America from Alabama to Florida.

Wood: Light, soft, not strong, brittle, and resinous. Heartwood pale orange-red or yellow; sapwood thick, almost white.

Use: For masts of small boats.
Main Use: soil stabilizer (e.g. on drifting dunes).

Beschreibung: Nur 4,5 bis 6 oder 7 m hoch bei kaum 0,3 m Durchmesser, jedoch bei bestem Standort 22 bis 25 m und 0,6 m Durchmesser. Bis zum Grund in schlanke Äste eingehüllt. Borke am unteren Stamm durch enge Rillen tief in unregelmäßig längliche Platten geteilt, die an ihrer Oberfläche zunächst dicht anliegende, hell rotbraune Schuppen abgeben. Oberer Teil des Stammes mit dünner, glatter, aschgrauer Borke. Nadeln 4 bis 7 cm, zu zweit. Sehr schnell wachsend.

Vorkommen: Sanddünen und reiner, weißer Sand an der Küste. Im Inland trockene Höhenrücken. Auf Böden besserer Qualität mit Magnolien, Hickories, Virginia-Eiche, Pfahl-Eiche u. a. in den Küstenregionen des südöstlichen Nordamerika von Alabama bis Florida.

Holz: Leicht, weich, nicht fest, spröde und harzreich. Kern hell orangerot oder gelb; Splint dick, fast weiß.

Nutzung: Zu Masten kleiner Schiffe verarbeitet. Hauptnutzen: Bodenbefestiger (z. B. von Wanderdünen).

Description : arbuste de 4,5 à 6 ou 7 m de hauteur seulement sur u diamètre d'à peine 0,3 m. Peut atteindre cependant 22 à 25 m de haut et 0,6 m de diamètre. Enveloppé jusqu'au sol de branc fines. L'écorce se creuse à la base du tronc de fissures étroites et p fondes, séparées par des plaquet allongées et asymétriques aux écailles plaquées brun rouge clai puis se desquamant ; elle est gris cendré, mince et lisse dans la pa supérieure du tronc. Aiguilles de 4 à 7 cm de long, groupées par deux. Croissance très rapide.

Habitat : dunes, sable blanc et pu sur la côte. À l'intérieur, sommit de collines sèches. Sur des sols d meilleure qualité en association avec les magnolias, les caryers, le chêne de Virginie, le chêne étoilé etc. Croît dans les régions côtière du sud-est de l'Amérique du No depuis l'Alabama jusqu'à la Flori

Bois : léger, tendre, peu résistant, cassant et résineux. Bois de cœu rouge orangé clair ou jaune ; aub épais, presque blanc.

Intérêts : pour les mâts de petits navires. Principal usage : fixation des dunes.

Sand Pine, Scrub Pine, Upland Spruce Pine.

TRANSVERSE SECTION.

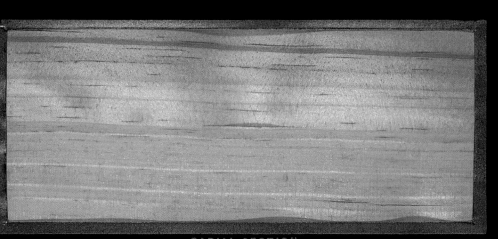

RADIAL SECTION.

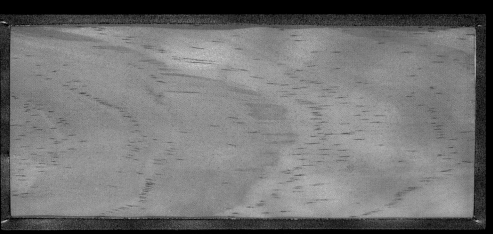

TANGENTIAL SECTION

Sand-Fichte.   Fr. Pin sablonneaux.   Sp. Pino de arena

P | PINE FAMILY
Scrub Pine, Jersey Pine

FAMILIE DER KIEFERNGEWÄCHSE
Virginische Kiefer

FAMILLE DES PINS
Pin de Virginie, Pin de Jersey

PINACEAE
Pinus virginiana Mill.

Description: 40 to 50 ft. (12 to 15 m) high and 15 in. (0.35 m) in diameter. In the west of its range, in southern Indiana, it is twice the size. Bark dark reddish brown, rough, with irregular scales and plates. Needles in pairs, only 1 to 2 in. (3 to 5 cm) long, twisted.

Habitat: Sandy soils. A pioneer tree on wasteland. (North-)eastern North America

Wood: Light, soft, not strong; brittle. Heartwood light reddish brown; sapwood very much paler.

Use: Mostly used as firewood, occasionally for building work.

Beschreibung: 12 bis 15 m hoch bei 0,35 m Durchmesser. Im Westen ihres Verbreitungsgebietes, im südlichen Indiana, doppelt so groß. Borke dunkel rotbraun, rau, mit unregelmäßigen Schuppen und Platten. Nadeln paarweise, nur 3 bis 5 cm lang, gedreht.

Vorkommen: Sandige Böden. Pioniergehölz auf Ödland. (Nord-) Östliches Nordamerika.

Holz: Leicht, weich, nicht stark; brüchig. Kern hell rotbraun; Splint sehr viel heller.

Nutzung: Meist nur als Brennmaterial, gelegentlich als Bauholz.

Description: 12 à 15 m de hauteur sur un diamètre de 0,35 m. Croissance en hauteur et en diamètre deux fois plus grande dans la partie occidentale de son aire de distribution, le sud de l'Indiana. Ecorce brun rouge foncé et rugueuse, à texture en plaques et écailles irrégulières. Aiguilles par deux, de 3 à 5 cm de long seulement et vrillées.

Habitat: sols sablonneux. Espèce pionnière sur les terres incultes. On le trouve dans le (nord-) est de l'Amérique du Nord.

Bois: léger, tendre, pas fort; cassant. Bois de cœur brun rouge clair; aubier beaucoup plus clair.

Intérêts: bois de chauffage mais occasionnellement aussi bois de charpente.

Jersey Pine, Scrub Pine.

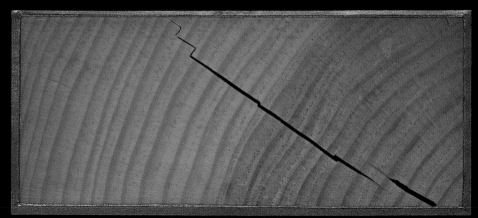

TRANSVERSE SECTION.

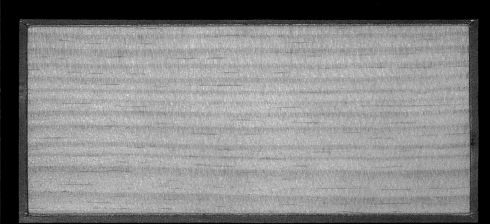

RADIAL SECTION.

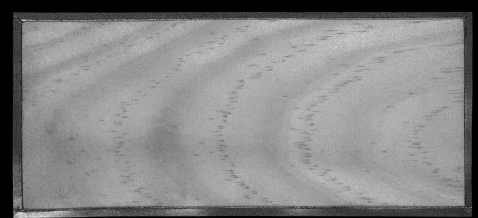

TANGENTIAL SECTION.

Jersey-Fichte. *Fr.* Pin de New Jersey. *Sp.* Pino de New Jersey

P | PINE FAMILY
Sitka Spruce

FAMILIE DER KIEFERNGEWÄCHSE
Sitka-Fichte

FAMILLE DES PINS
Epicéa de Sitka, Epinette de Sitka (Canada)

PINACEAE
Picea sitchensis (Bong.) Carrière

Description: 100 ft. (30 m) high, 3 to 4 ft. (0.9 to 1.2 m) in diameter, sometimes to almost 230 ft. (70 m) high, 15 ft. (4.5 m) in diameter. Shrubby in extreme north-west. Trunk with powerful root buttresses. Bark with thin, loose red-brown scales.

Habitat: Damp sandy soils, boggy soils, wet fields, rocks, wet floodplains with Western Red Cedar and Western Hemlock in the coastal regions of Pacific North America from the Kodiak Islands to California. In California with Redwood and Silver Fir. Optimum: south-east Alaska.

Wood: Light, soft, not strong, straight-grained, and silky-sheened; splits well. Heartwood pale brown, with red tinge; sapwood thick, almost white.

Use: Main timber tree in Alaska. Interior construction, fencing, boatbuilding, cooperage, boxes.

Largest of all the spruces. Fastest-growing of all North American trees.

Beschreibung: 30 m hoch bei 0,9 bis 1,2 m Durchmesser, gelegentlich nahe an 70 m hoch bei 4,5 m Durchmesser. Im extremen Nordwesten strauchig. Stamm stark abholzig mit mächtigen Wurzelanläufen. Borke in dünnen, lose anliegenden, rotbraunen Schuppen.

Vorkommen: Feuchter Sandboden, Sumpfboden, nasse Felder, Felsen, Schwemmland in Flussmündungen zusammen mit dem Westlichen Lebensbaum und der Westlichen Schierlingstanne in den Küstenregionen des pazifischen Nordamerika von den Kodiak-Inseln bis Kalifornien. In Kalifornien mit dem Küstensequoie und der Weiß-Tanne. Optimum: im Südosten Alaskas.

Holz: Leicht, weich, nicht fest, gerade faserig und seidig; spaltet gut. Kern hellbraun, rot getönt; Splint dick, fast weiß.

Nutzung: Hauptbauholz in Alaska. Für Innenausbau, Zäune, Bootsbau, Böttcherei, Kisten.

Größte aller Fichten. In Nordamerika an Starkwüchsigkeit von keinem Baum übertroffen.

Description : 30 m de hauteur sur un diamètre de 0,9 à 1,2 m, parfc près de 70 m sur un diamètre de 4,5 m ; buissonnant dans les conditions extrêmes du nord-oue Tronc très renflé, muni d'énorme contreforts. Ecorce s'exfoliant en minces écailles brun rouge.

Habitat : sols sablonneux frais, sols marécageux, prairies humide rochers, sols alluviaux dans les embouchures dans les régions côtières de la façade pacifique de l'Amérique du Nord des îles Kodiak à la Californie, avec le thuya occidental et la pruche ; en Californie, avec le séquoia géant et le sapin blanc. Prospère dans le sud-est de l'Alaska.

Bois : léger, tendre, non résistant, fil droit et satiné ; se fendant bien Bois de cœur brun clair, teinté de rouge ; aubier épais, presque blan

Intérêts : principal bois de constru tion en Alaska. Menuiserie intérieure, tonnellerie, caisserie, palissages et bateaux.

Le plus grand des épicéas. Sa croissance vigoureuse n'est égalée en Amérique du Nord par aucun autre arbre.

Tide-land Spruce.

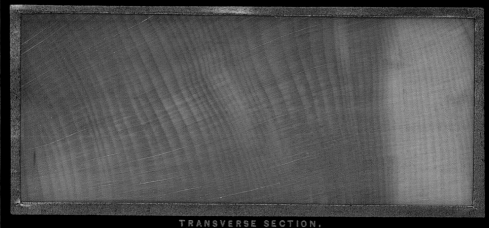

TRANSVERSE SECTION.

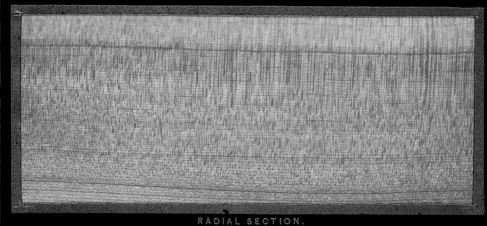

RADIAL SECTION.

TANGENTIAL SECTION

Ger. Fluthland-Tanne.　　　*Fr.* Sapin du rivage de la mer.

Sp. Abeto de la ribera del mar.

P | PINE FAMILY

Slash Pine, Swamp Pine, Bastard Pine

FAMILIE DER KIEFERNGEWÄCHSE
Kuba-Kiefer

FAMILLE DES PINS
Pin de Cuba

PINACEAE
Pinus elliottii Engelm. var. *elliottii*
Syn.: *P. heterophylla* (Elliot) Sudw.

Description: 100 to 120 ft. (30 to 35 m) high and 2 to 3 ft. (0.7 to 0.9 m) in diameter. Branch-free to 70 ft. (20 m). Bark in plates of thin, red-brown scales, which reveal orangey red-brown areas on removal. Needles in twos or threes.

Habitat: The "littoral pine flats" of south-eastern North America, in scattered groves on bays and lagoons, with Longleaf Pine and Loblolly Pine. In Florida the only pine found throughout, from the east to the west coasts and forming homogeneous woods. In Georgia found up to 150 km from the sea. On the Gulf coast penetrating to 75 to 90 km inland along watercourses, with Live Oak and evergreens.

Wood: Heavy, unusually hard, tough, very strong, coarse-grained, and resinous; durable. Heartwood a deep dark orange; sapwood thick and almost white.

Use: Similar quality and market as for Longleaf Pine.

Beschreibung: 30 bis 35 m hoch bei 0,7 bis 0,9 m Durchmesser. Astfrei bis in 20 m Höhe. Borke in Platten aus dünnen, rotbraunen Schuppen. Bei deren Abtrennung treten orangerotbraune Partien auf. Nadeln zu zweit oder dritt.

Vorkommen: Im maritimen Kieferngürtel des südöstlichen Nordamerika, den sog. küstennahen Kiefernebenen („littoral pine flats"), in zerstreuten Hainen an Buchten und Lagunen mit der Sumpf-Kiefer und der Weihrauchkiefer. In Florida als einzige Kiefer durchgehend von der Ost- zur Westküste und als reiner Wald. In Georgia aufsteigend bis zu 150 km von der See. An der Golfküste an Wasserläufen, 75 bis 90 km landeinwärts vordringend, in Gesellschaft von Virginia-Eiche und Immergrünen.

Holz: Schwer, außerordentlich hart, zäh, sehr fest, grobe Textur und harzreich; dauerhaft. Kern kräftig dunkelorange; Splint dick und fast weiß.

Nutzung: Ähnliche Qualität und Verkäufe wie bei der Sumpf-Kiefer.

Description : arbre de 30 à 35 m de hauteur, avec un tronc de 0,7 à 0,9 m de diamètre, non ramifié jusqu'à 20 m. L'écorce se craquel en fines plaques écailleuses brun rouge, révèlant des parties de l'endoderme brun orangé. Aiguil groupées par deux ou trois.

Habitat : occupe la ceinture des pinèdes maritimes du sud-est de l'Amérique du Nord, les «plaines aux pins» (littoral pine flats) près la côte, les bosquets dispersés dar les criques et les lagunes, avec le p des marais et le pin à l'encens. Se pin de Floride à occuper une aire continue de la côte orientale à la côte occidentale et en peuplemen purs. À l'intérieur de la Géorgie, se rencontre jusqu'à 150 km de la mer. Sur le pourtour du Golfe, jusqu'à 75 ou 90 km en amont de cours d'eau, avec le chêne de Virg nie et des espèces sempervirentes

Bois : lourd, très dur, tenace, très résistant, à grain grossier et résineux ; durable. Bois de cœur orange sombre intense ; aubier épais, presque blanc.

Intérêts : qualités et valeur économique similaires à celui du pin des marais.

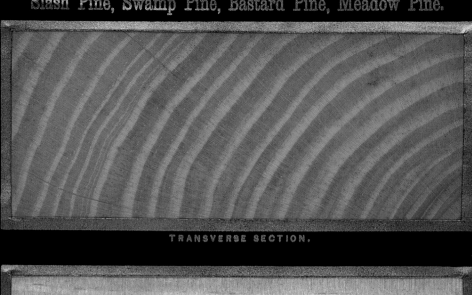

TRANSVERSE SECTION.

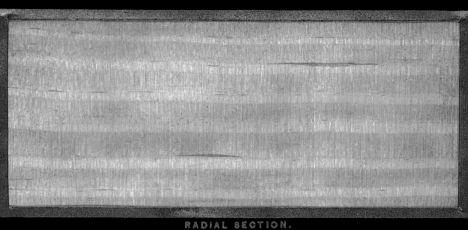

RADIAL SECTION.

TANGENTIAL SECTION

Ger. Cubanische Fichte.   *Fr.* Pin taillade.

Sp. Pino recortado.

P | PINE FAMILY

Spruce Pine, Cedar Pine

FAMILIE DER KIEFERNGEWÄCHSE
Kahle Kiefer

FAMILLE DES PINS
Pin glabre, Pin épinette

PINACEAE
Pinus glabra Walt.

Description: 85 to 100 ft. (25 to 30 m), occasionally 130 ft. (40 m) high and 2 ft. (0.6 to 0.7 m) in diameter. Branch-free to 50, 70 ft. (15 to 20 m). Bark: low ridges and shallow grooves. Pale red-brown with closely appressed scales. Needles in clusters of two.

Habitat: Individually or in small groves on low terraces above river marshes near the coast in southeastern North America. Sometimes in association with magnolias, gums (*Nyssa* spp.), hickories, Beech, and Loblolly Pine. Colonizes abandoned cultivated land. Best developed in north-west Florida.

Wood: Light, soft, not strong, brittle, close-grained; not durable. Heartwood pale brown; sapwood thick, almost white.

Use: Firewood.

One of the largest pines in the eastern USA, but of little economic value.

Beschreibung: 25 bis 30 m, gelegentlich 40 m hoch bei 0,6 bis 0,7 (bis 1) m Durchmesser. Astfrei bis 15, 20 m über dem Boden. Borke: flache Riefen, flache Rillen. Hell rötlichbraun mit dicht anliegenden Schuppen. Zweinadelig.

Vorkommen: Bewohnt immer küstennah einzeln oder in kleinen Hainen niedrige Terrassen oberhalb von Flusssümpfen im südöstlichen Nordamerika. Wenn in Gesellschaft, dann mit Magnolien, Tupelos, Hickories, Buchen und der Weihrauchkiefer. Besiedler von aufgegebenem Kulturland. Beste Entwicklung im Nordwesten Floridas.

Holz: Leicht, weich, nicht fest, spröde und von sehr feiner Textur; nicht dauerhaft. Kern hellbraun; Splint dick, nahezu weiß.

Nutzung: Brennholz.

Eine der größten Kiefern im Osten der USA, aber von geringem ökonomischen Wert.

Description: arbre de 25 à 30 m de hauteur, parfois 40 m, avec u tronc de 0,6 à 0,7 m (1,0 m) de diamètre, non ramifié jusqu'à 15 20 m au-dessus du sol. Ecorce br rougeâtre clair, parcourue de sill plats et de crêtes plates aux écail attachées. Aiguilles groupées par deux.

Habitat: croît toujours à proximité du littoral, en solitaire ou e bosquet, les terrasses peu élevée surplombant les marécages de rivières dans le sud-est de l'Amé rique du Nord. En forêts mixtes avec les magnolias, les nyssas, le caryers, les bouleaux et le pin à l'encens. Colonise les terres arab abandonnées. Optimum de déve loppement dans le nord-ouest de la Floride.

Bois: léger, tendre, peu résistant, cassant et à grain très fin; non durable. Bois de cœur brun clair aubier épais, quasiment blanc.

Intérêts: utilisé comme bois de chauffage.

L'un des plus grands pins de l'es des Etats-Unis mais de faible vale économique.

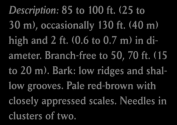

TRANSVERSE SECTION.

RADIAL SECTION.

P | PINE FAMILY
Sugar Pine

FAMILIE DER KIEFERNGEWÄCHSE
Zucker-Kiefer

FAMILLE DES PINS
Pin à sucre, Pin de Lambert, Pin géant

PINACEAE
Pinus lambertiana Dougl.

Description: 100 to 220 ft. (30 to 66 m) high, 6 to 8 (10 to 13) ft. (1.8 to 2.4 or 3 to 4 m) in diameter. Crown 65 to 75 ft. (20 to 23 m) across. Upper branches drooping under the weight of heavy cones. Older bark thick and divided into long, thick, irregular plates with large, loose purple-brown or cinnamon-red scales; younger bark smooth and dark gray.

Habitat: Mountain slopes and sides of canyons and gorges in the coastal states of Pacific North America from California to Oregon.

Wood: Light, soft, straight-grained, silky-sheened, and with a pleasant smell; easy to work. Sapwood pale red-brown with narrow resinous bands in the summer wood and many large resin canals; sapwood thin, almost white.

Use: As timber, for interior construction, roof shingles, doors, cooperage. The American name comes from the sweet substance exuding after fire or after cutting with an ax.

Noblest of the pines. Its majestic aspect comparable only with the Bigtree of the Sierra Nevada.

Beschreibung: 30 bis 66 m hoch bei 1,8 bis 2,4 (3 bis 4) m Durchmesser. Krone misst 20 bis 23 m in der Breite. Obere Äste von schweren Zapfen überhängend. Borke alter Stämme tief und unregelmäßig in lange dicke Platten mit großen, losen, purpurbraunen oder zimtroten Schuppen geteilt, junge Borke glatt und dunkelgrau.

Vorkommen: Berghänge und die Wände von Cañons und Schluchten in den Küstenstaaten des pazifischen Nordamerika von Kalifornien bis Oregon.

Holz: Leicht, weich, gerade faserig, seidig und sehr angenehm riechend; leicht zu bearbeiten. Kern hell rötlichbraun mit schmalen, harzreichen Bändern im Spätholz und zahlreichen großen Harzgängen; Splint dünn, fast weiß.

Nutzung: Als Bauholz, zum Innenausbau, für Dachschindeln, Türen, Böttcherei. Ein süßer Stoff, durch Feuer oder Axthiebe ausgeschwitzt, gab den amerikanischen Namen.

Edelste der Kiefern. Ihr majestätischer Anblick ist nur mit dem des Mammutbaums der Sierra Nevada vergleichbar.

Description: 30 à 66 m de hauteu sur un diamètre de 1,8 à 2,4 m (à 4 m). Houppier de 20 à 23 m d large, branches terminales «coif fantes» sous le poids des cônes. Ecorce à sillons profonds et longu plaques épaisses et irrégulières, se craquelant en grandes écailles brun-pourpre ou cannelle sur les vieux sujets; lisse et gris foncé s les jeunes arbres.

Habitat: versants de montagnes, canyons et de ravins dans les Eta côtiers de la façade pacifique de Californie à l'Oregon.

Bois: léger, tendre, à fil droit, sat et très odorant; facile à travailler Bois de cœur brun rougeâtre clai avec d'étroites bandes résineuse dans le bois final et de nombreu canaux résinifères; aubier mince presque blanc.

Intérêts: construction, menuiseri intérieure, tonnellerie, fabricatio de bardeaux et de portes. Doit s nom à une substance sucrée qu'i exsude à la suite d'un incendie d'un coup de hache.

Le plus beau des pins; son port est comparable à celui du séquoi géant de la Sierra Nevada.

TRANSVERSE SECTION.

RADIAL SECTION.

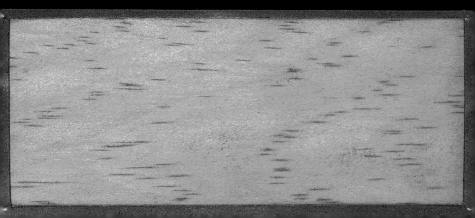

P | PINE FAMILY
Table-mountain Pine, Hickory Pine

FAMILIE DER KIEFERNGEWÄCHSE
Stech-Kiefer

FAMILLE DES PINS
Pin à cônes piquants, Pin du mont de la Table, Pin piquant

PINACEAE
Pinus pungens Lamb.

Description: Up to 60 ft. (18 m) high and 2 to 3 ft. (0.6 to 0.9 m) in diameter with a narrow, round crown when growing close to other trees, but only 20 to 30 ft. (6 to 9 m) high with a short, thick trunk and a broad, irregular and open crown when freestanding. Bark on the lower part of the trunk broken into irregularly shaped plates, separating on the dark brown surface into thin scales. Bark on the upper trunk and the branches dark brown, separating into loose plates. Needles in pairs, short, blue-green, twisted, 1 to 2 in. (3 to 6 cm) long. Flowers insignificant: the male in elongated inflorescences; the female in groups. Cones oblong-conical, armed with stout, sharp spines.

Habitat: On dry, stony slopes and cliffs in eastern North America, especially the Appalachians.

Wood: Light, soft, not strong, brittle, and very close-grained.

Use: As firewood, in Pennsylvania also used to produce charcoal. Planted occasionally as a decorative tree in the USA. Introduced into England for the first time in 1804.

Beschreibung: Im Bestand bis zu 18 m hoch bei 0,6 bis 0,9 m Durchmesser und schmaler, runder Krone, freistehend nur 6 bis 9 m hoch, mit kurzem, dickem Stamm und breiter, unregelmäßiger und offener Krone. Borke am unteren Teil des Stammes in unregelmäßige Felder zergliedert, deren dunkelbraune Oberflächen sich in dünnen Schuppen ablösen. Borke am oberen Stamm und an den Ästen dunkelbraun und in lose Platten aufgelöst. Nadeln zu zweien, kurz, blaugrün, in sich verdreht, 3 bis 6 cm lang. Blüten unauffällig, männliche in länglichen Blütenständen, weibliche in Gruppen. Zapfen länglich-konisch, mit einem kräftigen, stechenden Dorn.

Vorkommen: Auf trockenen, steinigen Hängen und Felsen im östlichen Nordamerika, vor allem der Appalachen.

Holz: Leicht, weich, nicht stabil, brüchig und mit sehr feiner Textur.

Nutzung: Als Brennholz, in Pennsylvania auch zur Holzkohlegewinnung. In den USA gelegentlich als Zierbaum gepflanzt. 1804 erstmals nach England eingeführt.

Description: en peuplement, jusc 18 m de hauteur, avec un tronc de 0,6 à 0,9 m de diamètre et ur houppier étroit, rond; en situati isolée, pas plus de 6 à 9 m de haut, avec un tronc court, épais un houppier ample, irrégulier et ouvert. Dans la partie inférieure du tronc, l'écorce est craquelée e secteurs irréguliers brun sombre s'exfoliant en fines écailles; dan la partie inférieure et sur les branches, elle est brun sombre et se détache en plaques. Aiguill groupées par 2, courtes, vert ble vrillées, de 3 à 6 cm de long. Fle insignifiantes, les mâles en inflo rescences allongées, les femelles groupes. Cônes allongés-conique armés d'une grosse épine acérée

Habitat: pentes caillouteuses sèc et rochers de l'est de l'Amérique du Nord, surtout dans les Appalaches.

Bois: léger, tendre, non stable, cassant et à grain très fin.

Intérêts: utilisé comme bois de chauffage et, en Pennsylvanie, p la fabrication du charbon de boi Planté çà et là en ornement aux Etats-Unis. Introduit en Angleter en 1804.

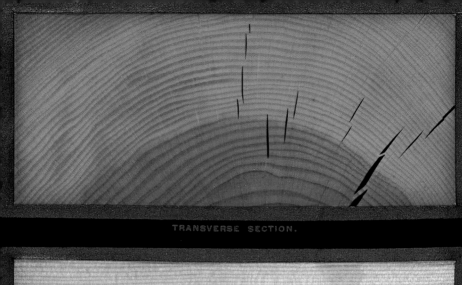

TRANSVERSE SECTION.

RADIAL SECTION.

P | PINE FAMILY
Torrey Pine

FAMILIE DER KIEFERNGEWÄCHSE
Torrey Föhre

FAMILLE DES PINS
Pin de Torrey

PINACEAE
Pinus torreyana Parry ex Carrière

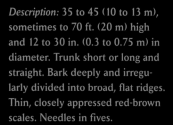

Description: 35 to 45 (10 to 13 m), sometimes to 70 ft. (20 m) high and 12 to 30 in. (0.3 to 0.75 m) in diameter. Trunk short or long and straight. Bark deeply and irregularly divided into broad, flat ridges. Thin, closely appressed red-brown scales. Needles in fives.

Habitat: Coastal cliffs close to the mouth of the Soledad River in southern California and on the off-shore island of Santa Rosa.

Wood: Light, soft, not strong, coarse-grained. Heartwood pale yellow; sapwood thick, yellow or nearly white.

Use: Occasionally as firewood. Seeds edible, raw or roasted.

Park tree in San Diego, California, and in New Zealand. Fast-growing; tall in cultivation.

Beschreibung: 10 bis 13, gelegentlich bis 20 m hoch bei 0,3 bis 0,75 m Durchmesser. Kurzer oder gerader, langer Stamm. Borke tief und unregelmäßig in breite, flache Riefen geteilt. Dünne, dicht anliegende, rotbraune Schuppen. Nadeln zu fünf.

Vorkommen: Klippen an der Küste nahe der Mündung des Soledad Rivers in Süd-Kalifornien sowie auf der vorgelagerten Insel Santa Rosa.

Holz: Leicht, weich, nicht fest und mit grober Textur. Kern hellgelb; Splint dick, gelb oder fast weiß.

Nutzung: Gelegentlich als Brennholz. Samen als Nahrungsmittel roh oder geröstet.

Als Parkbaum in San Diego, Kalifornien und auf Neuseeland. Schnellwüchsig und in Kultur höher.

Description: arbre de 10 à 13 m (parfois jusqu'à 20 m) de hauteu avec un tronc court ou élancé et droit de 0,3 à 0,75 m de diamètr Ecorce se fissurant en larges côte plates et irrégulières, recouverte de fines écailles plaquées brun rouge. Aiguilles groupées par de

Habitat: brisants de la côte pacifique près de l'embouchure de la rivière Soledad dans le sud de la Californie ainsi que sur l'île de Santa Rosa.

Bois: léger, tendre, non résistant et à grain grossier. Bois de cœur jaune clair; aubier épais, jaune o presque blanc.

Intérêts: utilisé de temps à autre pour le chauffage. Graines come tibles, consommées telles quelle ou grillées.

Arbre de parcs à San Diego, Cali fornie, et en Nouvelle-Zélande. Croissance rapide et taille plus élevée en culture.

Torrey Pine, Del Mar Pine, Soledad Pine.

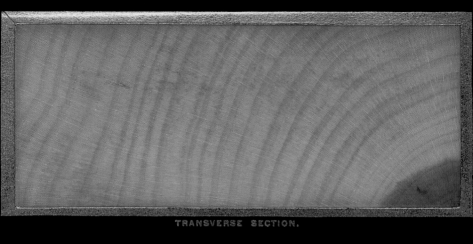

TRANSVERSE SECTION.

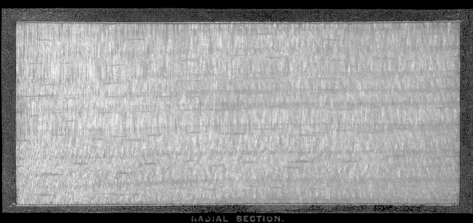

RADIAL SECTION.

TANGENTIAL SECTION.

Ger. Fichte von Torrey.   *Fr.* Pin de Torrey.

Sp. Pino de Torrey.

P | PINE FAMILY
Western Hemlock

FAMILIE DER KIEFERNGEWÄCHSE
Mertens Hemlocktanne

FAMILLE DES PINS
*Pruche de l'Ouest, Hemlock de l'Ouest,
Tsuga de l'Ouest (Canada)*

PINACEAE
Tsuga heterophylla (Raf.) Sarg.

Description: Often 200 ft. (60 m) high and 6 to 9 ft. (1.8 to 3.0 m) in diameter. Branches short, slender, and usually drooping, giving the crown a narrow pyramidal shape. Bark of young trees thin, dark orange-brown with fine fissures, later thick and deeply furrowed, brown with cinnamon tints. Leaves dark green and very glossy. Flowers in short ovoid inflorescences: male flowers yellow; female flowers purple. Cones oblong-ovoid.

Habitat: Damp, rainy forests in western North America.

Wood: Light, hard, and strong; long lasting and easier to work than that of other hemlocks. Heartwood pale brown with yellow tinge; sapwood thin and almost white. It has thin, inconspicuous bands of small summer cells and many distinct medullary rays.

Use: Building timber. The bark is used in large quantities for production of tannin. American Indians make long-lasting cakes from the rising spring sap. Planted as an attractive, fast-growing park tree in Europe since 1851.

Beschreibung: Oft 60 m hoch bei 1,8 bis 3 m Durchmesser. Äste kurz, schlank und normalerweise herabhängend, der Krone eine schmale, pyramidale Form gebend. Borke im Alter dick und tiefrissig, von brauner Farbe mit zimtfarbener Tönung. Blätter dunkelgrün und stark glänzend. Blüten in kurz ovalen Blütenständen, männliche von gelber, weibliche von violetter Farbe. Zapfen oval-länglich.

Vorkommen: Feuchte, niederschlagsreiche Wälder im westlichen Nordamerika.

Holz: Weich, hart und fest; stabiler, länger haltbar und leichter zu bearbeiten als Holz der übrigen Hemlocks. Kern blassbraun mit gelber Tönung; Splint dünn und fast weiß. Dünne, undeutliche Bänder kleiner Sommerzellen und zahlreiche hervortretende Markstrahlen.

Nutzung: Bauholz. Die Rinde wird in großem Umfang zur Gerbstoffgewinnung verwendet. Die Indianer verarbeiten den Saft zu lange haltbaren Kuchen. Seit 1851 in Europa als attraktiver, schnellwüchsiger Parkbaum kultiviert.

Description : souvent 60 m de hauteur sur un diamètre de 1,8 à 3,0 m. Port pyramidal, à cime étroite et branches courtes, minc retombantes en général. Ecorce juvénile mince, brun orangé sombre, finement fissurée ; épais et se creusant de fissures d'un brun teinté de cannelle avec l'âge Feuilles vert foncé, très brillantes Inflorescences mâles courtes, ova et jaunes, inflorescences femelles violettes. Cônes ovoïdes.

Habitat : forêts océaniques très h mides dans l'ouest de l'Amériqu du Nord.

Bois : léger, tendre et résistant ; stable, plus durable et facile à travailler que celui des autres pruches. Bois de cœur brun pâle teinté de jaune ; aubier mince et presque blanc. Faisceaux de petit cellules d'été, minces et indistinc et rayons médullaires nombreux.

Intérêts : bois d'œuvre ; écorce exploitée pour son tannin. Les Indiens utilisent la sève pour confectionner des gâteaux qui se conservent longtemps. Cultivé er Europe depuis 1851 comme arbre de parcs ornemental, à croissance rapide.

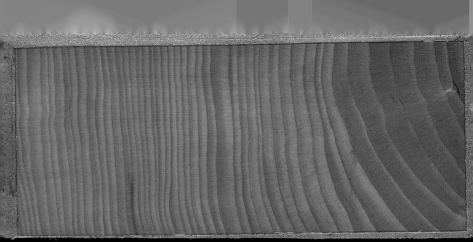

TRANSVERSE SECTION.

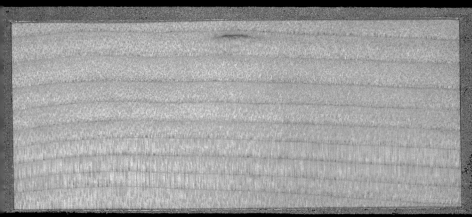

RADIAL SECTION.

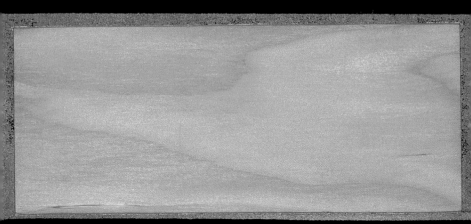

TANGENTIAL SECTION.

Ger. Westliche Tanne.   *Fr.* Peruche occidental.

P | PINE FAMILY
Western Larch, West American Larch

FAMILIE DER KIEFERNGEWÄCHSE
Westamerikanische Lärche

FAMILLE DES PINS
Mélèze de l'Oregon, Mélèze d'Occident,
Mélèze de l'Ouest (Canada)

PINACEAE
Larix occidentalis Nutt.

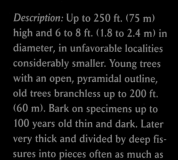

Description: Up to 250 ft. (75 m) high and 6 to 8 ft. (1.8 to 2.4 m) in diameter, in unfavorable localities considerably smaller. Young trees with an open, pyramidal outline, old trees branchless up to 200 ft. (60 m). Bark on specimens up to 100 years old thin and dark. Later very thick and divided by deep fissures into pieces often as much as 2 ft. (60 cm) in length, with a pale cinnamon-red surface. Needles triangular, in clusters, bright green. Male and female inflorescences shortly ovoid. Cones oblong.

Habitat: Solitary or in small groups on damp, low-lying soils in forests, especially by rivers, in north-western North America.

Wood: Very heavy, extremely hard and strong, and close-grained; polishable and very resistant to rot in the ground.

Use: Railroad sleepers, fence posts, building timber, cabinets, and interior construction. In southern British Columbia the Indians eat the sweet substance exuded by the tree.

Thanks to its thick bark it quite often survives the increasing occurrence of forest fires.

Beschreibung: Bis 75 m hoch bei 1,8 bis 2,4 m Durchmesser. An schlechten Standorten bedeutend kleiner. Junge Bäume mit fast pyramidalem, offenem Umriss, im Alter bis zu 60 m astfrei. Borke bei den bis 100-jährigen Exemplaren dünn und dunkel. Später sehr dick und durch tiefe Risse in oft bis zu 60 cm lange Stücke mit hell zimtfarben-rötlicher Oberfläche zerteilt. Nadeln dreikantig, in Büscheln, hellgrün. Männliche und weibliche Blütenstände kurz oval. Zapfen länglich.

Vorkommen: Als Einzelbaum oder in kleinen Gruppen auf feuchten, tiefgründigen Böden in Wäldern, gerne an Flüssen im nordwestlichen Nordamerika.

Holz: Sehr schwer, extrem hart und stabil, mit feiner Textur; polierfähig und im Boden sehr fäulnisresistent.

Nutzung: Eisenbahnschwellen, Zaunpfähle, Bauholz, Kabinette und Innenausbau. Die Indianer im südlichen Britisch-Kolumbien sammeln und verzehren süße Ausscheidungen des Baumes.

Dank ihrer dicken Borke überlebt sie die zunehmenden Waldbrände häufig problemlos.

Description : jusqu'à 75 m de haute sur un diamètre de 1,8 à 2,4 m, nettement plus petit en condition défavorable. Jeunes arbres à port quasi pyramidal et ouvert; tronc non ramifié jusqu'à 60 m de hauteur. Écorce mince et sombre sur les arbres de moins de 100 ans; à texture très épaisse et fissurée en plaques rouge clair, atteignant souvent 60 cm de long sur les vieux sujets. Aiguilles vert clair, à section triangulaire, en rosette sur les rameaux. Inflorescences mâles et femelles courtes, ovales. Cônes allongés.

Habitat : solitaire ou en petites formations sur les sols humides et p fonds dans les forêts; affectionne cours d'eau dans le nord-ouest de l'Amérique du Nord.

Bois : très lourd, extrêmement du et stable, à grain fin; peut être po et résiste à la décomposition.

Intérêts : traverses de chemin de fe pieux de clôtures, lambris; constr tion et menuiserie intérieure. Les Indiens du sud de la Colombie-Br tannique consomment les substances sucrées excrétées.

Se répand malgré les incendies grâce à son écorce épaisse.

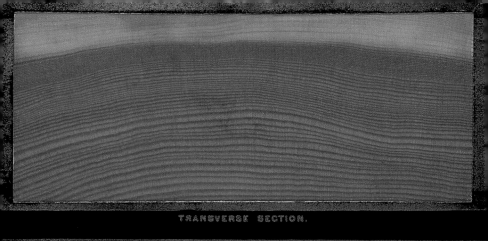

TRANSVERSE SECTION.

RADIAL SECTION.

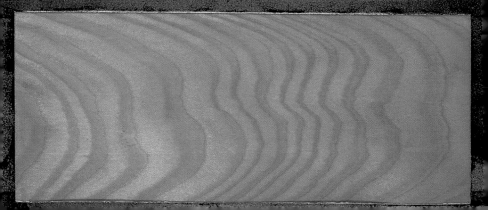

P | PINE FAMILY

Western Red Cedar, Western or Giant Arbor-vitae,
Canoe Cedar

FAMILIE DER KIEFERNGEWÄCHSE
Riesen-Lebensbaum, Riesenthuja

FAMILLE DES PINS
Thuya géant, Cèdre rouge de l'Ouest,
Cèdre de l'Ouest (Canada)

PINACEAE
Thuja plicata Donn ex D. Don

Description: Up to over 200 ft. (60 m) high and 15 ft. (4.5 m) in basal diameter. Branches hanging down from the edge of the crown and almost encircling the trunk to the ground. In the open, forms a closed, pyramidal shape; a shorter, narrower crown in stands. Bark pale cinnamon-brown. Cracks into loose, fibrous scales. Leaves scale-like, closely appressed. Flowers inconspicuous. Fruit large.

Habitat: On rather damp soils and river banks, but also on dry cliffs and mountain slopes. Widespread in western North America.

Wood: Very light, soft, not strong, brittle, coarse-grained; very resistant to rotting in the ground. Heartwood pale brown tinged red and with many medullary rays; sapwood thin, almost white.

Use: Interior furnishings, doors, window frames, fences, roof shingles, cabinets, etc. Used for huts, totem poles, and war canoes by American Indians. Inner bark fibers for ropes, cloth, shawls, etc. Common park tree. Much reduced through forest fires.

Beschreibung: Bis über 60 m hoch bei 4,5 m Durchmesser am Boden. Äste am Rande der Krone herabhängend und den Stamm fast bis zur Erde einhüllend. Freistehend eine geschlossene pyramidale Form ausbildend, im Bestand mit kurzer, schmaler Korne. Borke hell zimtbraun. Durch Risse in lockere, faserige Schuppen zergliedert. Blätter schuppenartig, eng anliegend. Blüten unauffällig. Frucht groß.

Vorkommen: Auf etwas feuchten Böden und Flussufern, aber auch an trockenen Klippen und Berghängen. Im westlichen Nordamerika.

Holz: Sehr leicht, weich, nicht fest, brüchig und mit grober Textur; im Boden sehr fäulnisresistent. Kern blassbraun mit rotem Farbton und zahlreichen Markstrahlen; Splint dünn, fast weiß.

Nutzung: Inneneinrichtung von Häusern, Türen, Fensterrahmen, Zäune, Schindeln, Tischlerarbeiten, etc. Von Indianern für den Hausbau, Totempfähle und Kriegskanus verwendet. Fasern der inneren Borke für Seile, Decken, Umhänge, etc. Häufiger Parkbaum. Bestand durch Waldbrände stark gefährdet.

Description : atteint 60 m de hauteur sur un diamètre de 4,5 m à la base. Branches retombant sur marge de la couronne et enveloppant le tronc presque jusqu'a sol. Forme pyramidale fermée er situation isolée, couronne étroite et courte en forêt. Ecorce brun cannelle clair, fissurée et craquelant en écailles fibreuses et lâche Feuilles squamiformes plaquées les rameaux. Fleurs insignifiantes Cônes de grande taille.

Habitat : sols frais et berges ; écue secs et versants montagneux. Da l'ouest de l'Amérique du Nord.

Bois : très léger, tendre, non résis tant, cassant et à grain grossier ; quasi imputrescible sur pied. Boi de cœur brun pâle teinté de roug avec de nombreux rayons médul laires ; aubier mince, presque bla

Intérêts : menuiserie intérieure, portes, châssis, clôtures, bardeau lambris etc. Utilisé par les Indien pour les habitations, les totems et les canots de guerre. Fibres de l'écorce interne pour faire des cordages, des couvertures, des pè rines etc. Arbre de parcs fréquen Forte diminution des peuplemen due aux incendies.

Giant Cedar, Northwestern Red Cedar, Shingle-wood.

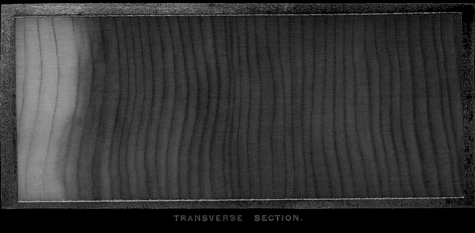

TRANSVERSE SECTION.

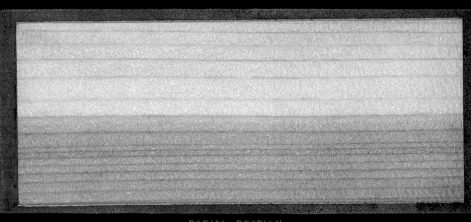

RADIAL SECTION

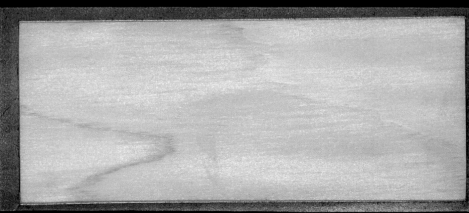

TANGENTIAL SECTION.

Ger. Gigantische Zeder.　　*Fr.* Thuya gigantesque.

Sp. Cedro giganteo.

P | PINE FAMILY
Western White Pine

FAMILIE DER KIEFERNGEWÄCHSE
Amerikanische Kiefer

FAMILLE DES PINS
Pin argenté, Pin blanc de l'Ouest (Canada)

PINACEAE
Pinus monticola Dougl. ex D. Don

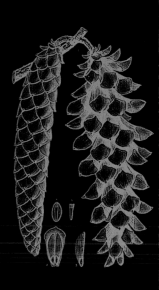

Description: 100 ft. (30 m) high and 4 to 5 ft. (1.2 to 1.5 m) in diameter. Rarely to 150 ft. (45 m) high and then to 8 ft. (2.4 m) in diameter. Trunk large and straight. Young trees pyramidal in shape, becoming somewhat asymmetrical later owing to unequal branch growth. Bark of young trees smooth, later cracked into almost square sections, the purple surface flaking off to reveal cinnamon-brown inner bark. Needles in fives, thick, stiff, 2 to 4 in. (4 to 10 cm) long, blue-green, slightly glaucous. Flowers inconspicuous: male flowers in ovoid inflorescences; female flowers in small groups. Cones long and slender.

Habitat: Mainly mountain forests in Pacific North America, in Washington also in forests to sea level.

Wood: Very light, soft, dense, and rather weak. Heartwood pale brown or red; sapwood thin, white. Numerous medullary rays.

Use: As building timber, especially in north Idaho and Montana. Introduced successfully into Europe.

Beschreibung: 30 m hoch bei 1,2 bis 1,5 m Durchmesser. Selten bis zu 45 m hoch und dann bis zu 2,4 m Durchmesser. Stamm groß, gerade. Jung von pyramidaler Form, im Alter aber leicht asymmetrisch durch das ungleiche Wachstum einzelner Äste. Borke junger Bäume glatt, später durch Risse in fast quadratische Felder zerteilt, deren purpurfarbene Oberflächen sich in flachen Schuppen ablösen und die zimtbraune innere Rinde freilegen. Nadeln zu 5, dick, steif, 4 bis 10 cm lang, blaugrün, leicht bereift. Blüten unauffällig, männliche in ovalen Blütenständen, weibliche in kleinen Gruppen. Zapfen lang und schlank.

Vorkommen: Vorzugsweise Gebirgswälder im pazifischen Nordamerika, in Washington auch Wälder bis auf Meereshöhe.

Holz: Sehr leicht, weich, dicht und wenig stabil. Kern hellbraun oder rot; Splint dünn, weiß. Zahlreiche Markstrahlen.

Nutzung: Als Bauholz besonders im Norden Idahos und Montanas. In Europa erfolgreich eingeführt.

Description: 30 m de hauteur sur un diamètre de 1,2 à 1,5 m, rarement jusqu'à 45 m sur un diamè de 2,4 m. Tronc élancé, droit. Por pyramidal chez les jeunes arbres, légèrement asymétrique chez les sujets adultes. L'écorce est lisse à l'état juvénile puis elle se fissure en carrés de couleur pourpre, se fragmentant et se desquamant en plaquettes laissant apparaître dessous une écorce brun cannellé Aiguilles groupées par 5, épaisses rigides, de 4 à 10 cm de long, d'un vert bleuté légèrement rayé de blanc. Fleurs insignifiantes, les mâles en inflorescences ovales, le femelles en petits groupes. Longs cônes minces.

Habitat: forêts montagnardes du Pacifique mais aussi dans l'Etat d Washington jusqu'au niveau de la mer.

Bois: très léger, tendre, dense et peu stable. Bois de cœur brun cla ou rouge; aubier mince, blanc. Rayons médullaires nombreux.

Intérêts: bois d'œuvre, en particu lier dans le nord de l'Idaho et le Montana. Introduit avec succès e Europe.

Mountain White Pine.

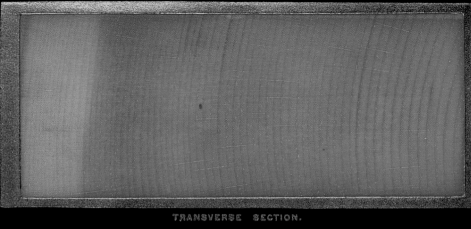

TRANSVERSE SECTION.

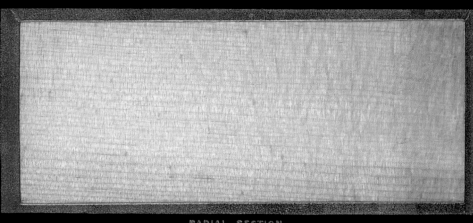

RADIAL SECTION.

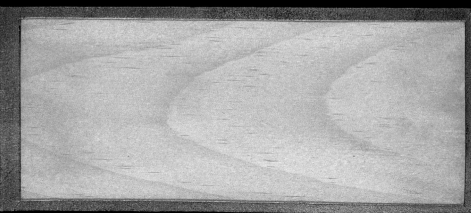

TANGENTIAL SECTION.

Ger. Gebirgige weisze Fichte. *Fr.* Pin blanc de la montagne.
Sp. Pino blanco de los montes.

P | PINE FAMILY
Weymouth Pine, Eastern White Pine

FAMILIE DER KIEFERNGEWÄCHSE
Weymouths-Kiefer, Strobe

FAMILLE DES PINS
Pin blanc, Pin baliveau (Canada)

PINACEAE
Pinus strobus L.

Description: 190 (330) ft. (65 (100) m) high and 3 to 5 (6) ft. (0.9 to 1.5 (1.8) m) in diameter. The largest conifer in north-eastern America. Present-day height less, probably owing to excessive use. Easy to distinguish by the fine, bluish green needles in groups of five and by the long, narrow, pendent cones. Of great vitality, quickgrowing and durable. Rapidly recolonizes abandoned agricultural land.

Habitat: On sandy soils, forming homogeneous woods, but more often in groups in deciduous forest on rich, well-drained soils in northeastern North America.

Wood: Light, soft, not strong, closegrained, and very resinous; polishable and easy to work. Heartwood pale brown, light red; sapwood thin and almost white.

Use: The most common building wood in the north-east (used for whole towns in the times of the early settlers), laths, cupboards, interior construction, matches, window shutters, and ships' masts, ribs, and yards.

Beschreibung: 65 (100) m bei 0,9 bis 1,5 (1,8) m Durchmesser. Größter Nadelbaum im Nordosten Amerikas. Infolge exzessiver Übernutzung heute nur kleinere Exemplare. Durch die feine, blaugrüne Benadelung (zu je 5) und die schmalen, langen, hängenden Zapfen gut zu unterscheiden. Von starker Vitalität, schnellwüchsig und dauerhaft. Erobert brachliegendes Ackerland schnell zurück.

Vorkommen: Auf Sandböden; reine Wälder bildend, öfter noch in Gruppen in Laubwäldern auf nährstoffreichen, gut drainierten Böden im nordöstlichen Nordamerika.

Holz: Leicht, weich, nicht fest, feine Textur und sehr harzig; polierfähig und gut zu bearbeiten. Kern hellbraun, leicht rot; Splint dünn und fast weiß.

Nutzung: Häufigstes Bauholz im Nordosten (ganze Städte der frühen Siedlerzeit), Latten, Schränke, Innenausbau, Streichhölzer, Fensterläden, Schiffsmasten, Spanten und Rahen.

Description: 65 m de hauteur (100 m max.) sur un diamètre de 0,9 à 1,5 m (1,8 m max.). Le plus grand conifère de l'est de l'Amérique du Nord. Sa taille est inférieure à ce qu'elle était autrefois raison d'une exploitation excessi Se distingue par des aiguilles gro pées par cinq, très fines et bleuté et de longs cônes minces, pendants. Arbre d'une grande vigueu à croissance rapide, et durable. Reconquiert rapidement les terre arables abandonnées.

Habitat: sols sablonneux; constit des peuplements purs, plus souv encore en touffes dispersées; for de feuillus sur des sols riches et bien drainés du nord-est de l'Am rique du Nord.

Bois: léger, tendre, pas résistant, à grain fin et très résineux; susceptible de prendre un beau poli et facile à travailler. Bois de cœur brun clair, légèrement roug aubier mince et quasiment blanc

Intérêts: bois d'œuvre le plus util dans le nord-est (construction de villes entières durant la colonisation); lattes, armoires, menuiseri intérieure, allumettes, volets, mâ de marine, couples et vergues.

White Pine, Weymouth Pine.

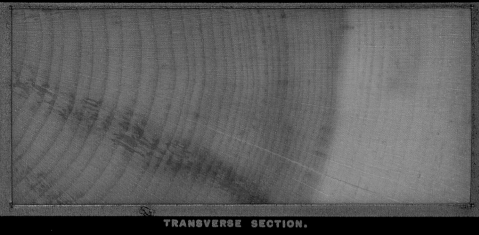

TRANSVERSE SECTION.

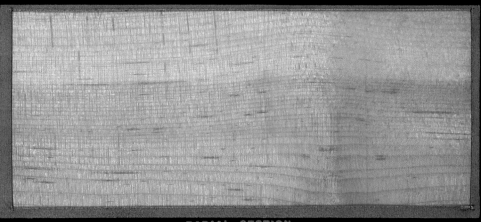

RADIAL SECTION.

TANGENTIAL SECTION.

Ger. Weimuthskiefer. *Fr.* Pin blanc. *Sp.* Pino blanco.

P | PINE FAMILY
White Bark Pine

FAMILIE DER KIEFERNGEWÄCHSE
Weißstämmige Kiefer

FAMILLE DES PINS
Pin à écorce blanche, Pin albicaule (Canada)

PINACEAE
Pinus albicaulis Engelm.

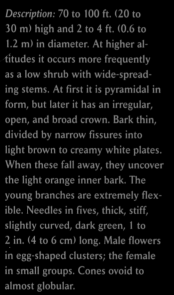

Description: 70 to 100 ft. (20 to 30 m) high and 2 to 4 ft. (0.6 to 1.2 m) in diameter. At higher altitudes it occurs more frequently as a low shrub with wide-spreading stems. At first it is pyramidal in form, but later it has an irregular, open, and broad crown. Bark thin, divided by narrow fissures into light brown to creamy white plates. When these fall away, they uncover the light orange inner bark. The young branches are extremely flexible. Needles in fives, thick, stiff, slightly curved, dark green, 1 to 2 in. (4 to 6 cm) long. Male flowers in egg-shaped clusters; the female in small groups. Cones ovoid to almost globular.

Habitat: Alpine slopes in western North America, on exposed ridges at heights of 4,800 to 10,500 ft. (1,500 to 3,700 m) above sea level. Often marks the tree line.

Wood: Light, soft, brittle, and close-grained. Heartwood light brown; sapwood thin, almost white. Numerous indistinct resin canals and indistinct medullary rays.

Use: The seeds have a sweet taste and used to be gathered by the Indians for food

Beschreibung: 20 bis 30 m hoch bei 0,6 bis 1,2 m Durchmesser. In höheren Lagen häufiger als niedriger Strauch mit weit ausgreifenden Stämmen. Anfangs von pyramidaler Form, später mit unregelmäßiger offener und breiter Krone. Borke dünn, durch schmale Risse in hellbraune bis cremefarben-weißliche Schollen zergliedert. Diese legen mit dem Abfallen die hellorange innere Rinde frei. Junge Äste sind extrem flexibel, sie können sogar verknotet werden. Nadeln zu 5, dick, steif, leicht gebogen, dunkelgrün, 4 bis 6 cm lang. Männliche Blüten in ovalen Blütenständen, weibliche in kleinen Gruppen. Zapfen oval bis fast kugelig.

Vorkommen: Alpine Hänge im westlichen Nordamerika, an exponierten Graten in Höhen von 1500 bis 3700 m. Oft die Baumgrenze markierend.

Holz: Leicht, weich, brüchig und mit feiner Textur. Kern hellbraun; Splint dünn, fast weiß. Zahlreiche undeutliche Harzgänge sowie undeutliche Markstrahlen.

Nutzung: Die süßen Samen dienten den Indianern als Nahrung

Description : 20 à 30 m de haute sur un diamètre de 0,6 à 1,2 m. Plus souvent sous forme de buis bas avec des tiges étalées dans le zones les plus élevées de son air Port pyramidal au début puis ho pier ample, irrégulier et ouvert. Ecorce mince, à texture profondément fissurée et fragmentée e grosses plaques brun clair à blan crème qui en se détachant laisse apparaitre l'écorce interne orang clair. Les jeunes rameaux sont s souples qu'ils peuvent se nouer. Aiguilles par 5, épaisses, rigides, un peu courbes, vert sombre, de 4 à 6 cm de long. Fleurs mâles e inflorescences ovales, les femelle en petits groupes. Cônes ovales quasi sphériques.

Habitat : versants alpins de l'oue: de l'Amérique du Nord, arêtes exposées entre 1500 et 3700 m d'altitude. Marque souvent la lin des arbres.

Bois : léger, tendre, cassant et à grain fin. Bois de cœur brun clai aubier mince, presque blanc. No breux canaux résinifères et rayo médullaires indistincts.

Intérêts : les graines douces étaie

White-bark Pine.

TRANSVERSE SECTION.

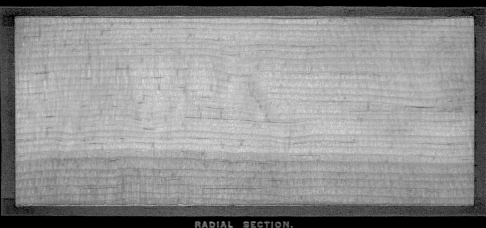

RADIAL SECTION.

TANGENTIAL SECTION.

Ger. Weissborke Fichte.   *Fr.* Pin de écorce blanc.
Sp. Pino de corteza blanca.

P | PINE FAMILY
White Fir

FAMILIE DER KIEFERNGEWÄCHSE
Gleichfarbige Tanne, Kalifornische Tanne

FAMILLE DES PINS
Sapin concolore, Sapin argenté, Sapin blanc du Colorado

PINACEAE
Abies concolor (Gordon & Glend.) Lindl. ex Hildebr.

Description: 200 to 250 ft. (60 to 75 m) high, 5 ft. (1.6 m) in diameter. Often branch-free for up to 100 ft. (30 m). Outside the Sierra Nevada often only 100 to 130 ft. (30 to 40 m) high and 3 ft. (0.9 m) in diameter. Bark of older trunks deeply divided. Broad, rounded ridges with irregular, plate-like scales, ash-gray above, red-brown below. Dull orange within.

Habitat: Varied, rich and poor, dry-rocky soils in (south-) western North America. Withstands heat and drought better than any other fir.

Wood: Very light, soft, not strong, coarse-grained; not durable.

Use: In north California for crates and butter barrels.

Since 1852 in Europe, from Scandinavia to northern Italy. Also as a park tree on the east coast of USA.

Beschreibung: 60 bis 75 m hoch bei 1,6 m Durchmesser. Oft astfrei bis über 30 m. Außerhalb der Sierra Nevada oft nur 30 bis 40 m hoch bei 0,9 m Durchmesser. Borke alter Stämme tief geteilt. Breite, abgerundete Riefen mit unregelmäßig plattigen Schuppen, oben aschgrau, unten rötlich braun. Inneres trüborange.

Vorkommen: Unterschiedlich gute und arme, trocken-felsige Böden im (süd-) westlichen Nordamerika, hält von allen Tannen Hitze und Trockenheit am besten aus.

Holz: Sehr leicht, weich, nicht fest und mit grober Textur; nicht dauerhaft.

Nutzung: In Nord-Kalifornien für Kisten und Butterfässer.

Seit 1852 in Europa von Skandinavien bis Norditalien. Auch an der Ostküste der USA als Parkbaum.

Description : arbre de 60 à 75 m de hauteur, avec un tronc de 1,6 m de diamètre, souvent non ramifié jusqu'à 30 m ou plus. Espèce n'atteignant souvent, en dehors de la Sierra Nevada, que 30 à 40 m de hauteur sur un diamètre de 0,9 m. L'écorce des vieux troncs est profondément fissurée en larges crêtes arrondies, recouvertes de plaques écailleuses irrégulières, gris cendré dessus, brun rougeâtre dessous. L'endoderme est d'un orange terne.

Habitat : croît indifféremment dans les sols caillouteux et secs du (sud-) ouest de l'Amérique du Nord, qu'ils soient de bonne ou de mauvaise qualité, et supporte mieux la chaleur et la sécheresse que tous les autres sapins.

Bois : très léger, tendre, non résistant et à grain grossier ; non durable.

Intérêts : utilisé dans le nord de la Californie en caisserie et pour fabriquer des barattes.

Introduit en Europe depuis 1852, répandu de la Scandinavie à l'Italie du Nord. Arbre de parcs sur la

Cal. White Fir.

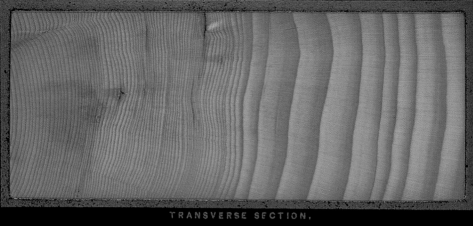

TRANSVERSE SECTION.

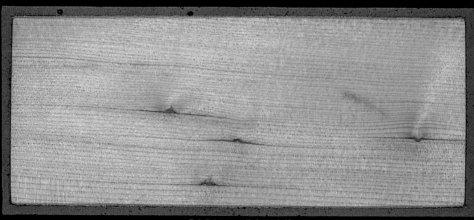

RADIAL SECTION.

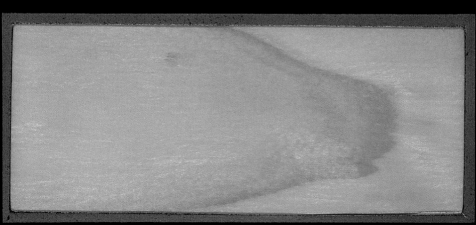

TANGENTIAL SECTION

Ger. Californische Weisse Tanne. *Fr.* Sapin blanc de Californ

Sp. Abeto blanco de California.

P | PINE FAMILY
White Fir, Giant Fir

FAMILIE DER KIEFERNGEWÄCHSE
Riesentanne, Küstentanne

FAMILLE DES PINS
Sapin grandissime, Sapin blanc, Sapin de Vancouver (Canada)

PINACEAE
Abies grandis (Dougl. ex D. Don) Lindl.

Description: 230 to 280 ft. (75 to 90 m) high (on the coast); 6 ft. (1.2 m) in diameter. Markedly smaller on inland ranges, rarely more than 100 ft. (30 m) high. Trunk gradually narrowing, branches protruding, bark often somewhat peeling, thick, gray-brown or red-brown, divided into flat segments by shallow furrows. Needles thin and flexible, standing well out from twigs, very dark green above, silvery white beneath. Flowers in catkin-like inflorescences. Cones pale green, cylindrical, slender.

Habitat: Damp soils in forests with other conifers in western North America, in the south only by the coast in the Pacific sea-mist region.

Wood: Very light, soft, not strong, coarse-grained, not durable. Heartwood pale brown; sapwood thin, paler. Broad dark bands of summer cells and many distinct medullary rays.

Use: Building timber, packaging, etc. Occasionally seen in European parks and gardens, where it is one of the fastest-growing of all conifers.

Beschreibung: 75 bis 90 m hoch (an der Küste) bei 1,2 m Durchmesser. In den Inlandsgebirgen deutlich kleiner, selten höher als 30 m. Sich nach oben verjüngender Stamm, Äste ausladend, häufig etwas herabhängend Borke dick, graubraun oder rotbraun, durch flache Risse in flache Schollen zerteilt. Nadeln dünn und flexibel, vom Ast weit abstehend, oberseits sehr dunkelgrün, unterseits silbrig weiß. Blüten in walzenförmigen Blütenständen. Zapfen hellgrün, zylindrisch, schlank.

Vorkommen: Feuchte Böden in Wäldern anderer Koniferen im westlichen Nordamerika, im Süden nur an der Küste, im Bereich der vom Pazifik ins Land ziehenden Nebel.

Holz: Sehr leicht, weich, nicht fest und mit grober Textur; nicht haltbar. Kern hellbraun; Splint dünn, heller. Breite dunkle Bänder von Sommerzellen und zahlreiche deutliche Markstrahlen.

Nutzung: Bauholz, für Verpackungen etc. Gelegentlich in europäischen Gärten und Parks zu sehen, wo er einer der schnellwüchsigsten Nadelbäume ist.

Description: 75 à 90 m de hauteu. (côte pacifique) sur un diamètre de 1,2 m. Nettement plus petit et s'élevant rarement à plus de 30 m dans les montagnes de l'intérieur. Tronc pyramidal avec des branche étalées, souvent un peu retombantes. Ecorce épaisse, brun gris ou brun rouge, à fissures plates et ar liégeux plats. Aiguilles distiques, bien dressées sur les rameaux, fir et souples, vert très foncé dessus, blanc argenté dessous. Inflorescences cylindriques. Cônes vert clair, cylindriques, minces.

Habitat: sols humides dans les forêts de conifères de l'ouest de l'Amérique du Nord; au sud seu ment près de la côte, dans la zon de brumes et de brouillard.

Bois: très léger, tendre, non résis tant et à texture grossière; non durable. Bois de cœur brun clair aubier mince, plus clair. Marqué sombres et larges faisceaux de ce lules d'été et de rayons médullai nombreux et distincts.

Intérêts: utilisé comme bois d'œuvre, en caisserie etc. Quelqu fois arbre de parcs et de jardins e Europe, où il est l'un des conifère à croître le plus vite.

Great Silver Fir.

TRANSVERSE SECTION.

RADIAL SECTION.

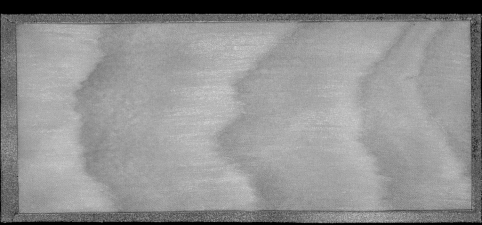

TANGENTIAL SECTION.

Ger. Grosze Tanne. *Fr.* Sapin grand. *Sp.* Abeto grande.

P | PINE FAMILY
White Spruce

FAMILIE DER KIEFERNGEWÄCHSE
Kanadische Fichte, Schimmel-Fichte

FAMILLE DES PINS
*Epicéa blanc, Sapinette blanche du Canada,
Epinette blanche (Canada)*

PINACEAE
Picea glauca (Moench) Voss
Syn.: *P. canadensis* (Mill.) B. S. P.

Description: 50 to 70 ft. (15 to 20 m) high when standing alone, but up to 165 ft. (50 m) in height in dense forest, and 3 to 4 ft. (0.9 to 1.2 m) in diameter. Trunk straight. Bark with small, irregular, reddish brown scales. Needles with a repulsive, polecat-like smell.

Habitat: On the banks of rivers and lakes, in forests across the continent, and as a pioneer on the subarctic forest frontier in association with Aspen, Paper Birch, Balsam Poplar, and Black Spruce. Can only fully develop in the far north, but extends into the northern states of the USA.

Wood: Like that of the Red Spruce.

Use: The long, fine, tough roots of the White Spruce were used by the Indians to sew together the pieces of bark from the Paper Birch to make their canoes, and also to weave watertight bottles.

Beschreibung: Einzeln 15 bis 20 m, in dichtem Bestand im Wald bis 50 m hoch bei 0,9 bis 1,2 m Durchmesser. Gerader Stamm. Borke mit kleinen, unregelmäßigen, rötlich braunen Schuppen. Nadeln mit widerlichem („Iltis"-)Geruch.

Vorkommen: An Fluss- und Seeufern, im transkontinentalen Wald vom Atlantik bis zum Pazifik und als Vorposten an der subarktischen Waldgrenze in Gesellschaft von Zitter-Pappel, Papier-Birke, Balsam-Pappel und Schwarz-Fichte. Kann sich erst im Hohen Norden voll entwickeln, stößt aber auch bis in die nördlichen Staaten der USA vor.

Holz: So wie jenes der Rot-Fichte.

Nutzung: Mit den langen, feinen, zähen Wurzeln der Kanadischen Fichte nähten die Indianer die Rindenteile der Papier-Birke für ihre Kanus zusammen und woben daraus auch wasserdichte Beutel.

Description: 15 à 20 m de hauteu en situation isolée, jusqu'à 50 m en forêt dense, sur un diamètre c 0,9 à 1,2 m. Tronc droit. Ecorce s craquelant en petites écailles irré gulières brun rougeâtre. Aiguilles dégageant une odeur très désagréable quand elles sont cassées.

Habitat: berges des cours d'eau et des lacs, dans la forêt transcon nentale de l'Atlantique au Pacific et comme «avant-poste» de la fo sub-arctique en association avec l peuplier tremble, le bouleau blar le peuplier baumier et l'épicéa nc Développement optimal dans le Grand Nord seulement, mais peu également se rencontrer jusque dans les Etats du nord des Etats-Unis.

Bois: identique à celui de l'épicéa rouge.

Intérêts: les longues et fines racin de l'épicéa blanc étant très solide les Indiens s'en servaient pour coudre les morceaux d'écorce de bouleau blanc avec laquelle ils fabriquaient leurs canots et pour tisser des sacs étanches.

White Spruce.

TRANSVERSE SECTION.

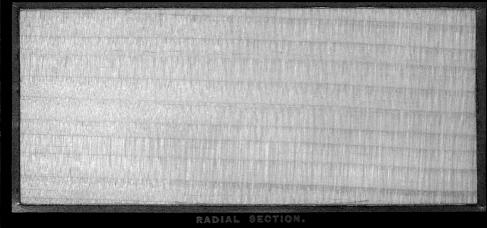

RADIAL SECTION.

TANGENTIAL SECTION.

Ger. Schimmel-Fichte.   *Fr.* Sapin blanc.   *Sp.* Abeto blan

P | PINE FAMILY
Yellow Pine, Southern Pine, Shortleaf Pine,
North Carolina Pine

FAMILIE DER KIEFERNGEWÄCHSE
Amerikanische Kiefer, Fichtenkiefer, Gelb-Kiefer,
Igel-Kiefer

FAMILLE DES PINS
Pin jaune, Pin doux, Pin épineux

PINACEAE
Pinus echinata Mill.

Description: 100 ft. (30 m) or some-what more in height and 3 to 4 ft. (0.9 to 1.2 m) in diameter, with a straight, columnar trunk. Bark red-dish brown, divided by furrows into broad, irregular, scaly plates. Needles in clusters of two or three.

Habitat: Abundant in the forests of the lower Mississippi basin.

Wood: Fairly heavy and hard. Heart-wood reddish yellow; sapwood thick and white.

Use: Only exceeded in value by the Longleaf Pine, and for many pur-poses preferred because it is less hard. The most common timber in the Midwest of the USA.

Beschreibung: 30 m oder etwas höher bei 0,9 bis 1,2 m Durch-messer. Gerader, säulenförmiger Stamm. Borke rotbraun in breiten, unregelmäßig schuppigen Platten und Riefen. Nadeln zu 2 oder 3.

Vorkommen: Wälder im Mississippi-Becken, häufig.

Holz: Ziemlich schwer und hart. Kern rotgelb; Splint dick und weiß.

Nutzung: An Wert nur von dem der Sumpf-Kiefer übertroffen, für manche Zwecke aber beliebter, weil weniger hart. Häufigstes Nutz-holz im Mittleren Westen der USA.

Description : arbre de 30 m de hau-teur, ou un peu plus, avec un tro droit, columnaire de 0,9 à 1,2 m de diamètre. Ecorce brun rouge, se desquamant en larges plaques irrégulières. Aiguilles groupées p deux ou par trois.

Habitat : forêts du bassin du Miss sippi; espèce courante.

Bois : assez lourd et dur. Bois de cœur jaune rougeâtre; aubier épais et blanc.

Intérêts : seul le pin muriqué a un valeur économique plus grande encore, mais le bois du pin jaune moins dur, est plus recherché po certains usages. Bois utile commu dans le Middle West.

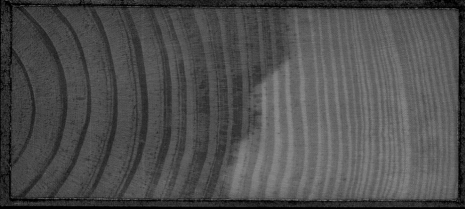

TRANSVERSE SECTION.

RADIAL SECTION.

TANGENTIAL SECTION

Ger. Kurznadelige Fichte.   *Sp.* Pino con hojas certas.

Fr. Pin de feuilles courtes.

P | PLANE-TREE FAMILY
Arizona Sycamore

FAMILIE DER PLATANENGEWÄCHSE
Arizona-Sycamore

FAMILLE DU PLATANE
Platane de Wright

PLATANACEAE
Platanus wrightii S. Watson

Description: 60 to 70 ft. (18 to 21 m) high and 4 to 5 ft. (1.2 to 1.5 m) in diameter. Trunk straight, unbranched to a height of 20 to 30 ft., (6 to 9 m) divided into two or three large stems at ground level or just above, which at first grow out close to the surface. Leaves deeply cut into three to seven slender and elongated lobes. The margins of the leaves are either entire or finely toothed. Flowers and fruits are in globular clusters.

Habitat: Riverbanks in the mountain canyons of southern North America.

Wood: Moderately heavy, fairly hard, strong, flexible, slightly shiny, and close-grained; shrinking considerably as it dries, not durable, suffers from changes of humidity, and is difficult to split. Heartwood light brown to reddish brown; sapwood white to ivory-colored, clearly distinguishable from the heartwood. Annual rings of varying widths and clearly recognizable, medullary rays only visible with a lens.

Beschreibung: 18 bis 21 m hoch bei 1,2 bis 1,5 m Durchmesser. Stamm gerade, bis in eine Höhe von 6 bis 9 m astfrei, vom Boden an oder direkt darüber in 2 bis 3 große Stämme zerteilt, die zunächst niederliegenden Wuchs zeigen. Blätter tief eingeschnitten, drei- bis siebenlappig. Lappen schlank und verlängert, Blattränder glatt oder fein gezähnelt. Blüten und Früchte in kugeligen Trauben.

Vorkommen: Flussufer in Gebirgscañons des südlichen Nordamerika.

Holz: Mittelschwer, ziemlich hart, fest, elastisch, matt glänzend und von feiner Textur; beim Trocknen stark schwindend, nicht dauerhaft, bei Feuchtigkeitsschwankungen leidend und schwer spaltbar. Kern blassbraun bis rötlich braun; Splint weiß bis elfenbeinfarben, deutlich vom Kern getrennt. Jahresringe unterschiedlich breit und deutlich zu erkennen, Markstrahlen nicht ohne Lupe zu sehen.

Nutzung: Wie bei anderen Platanen

Description : 18 à 21 m de hauteu sur un diamètre de 1,2 à 1,5 m. Tronc droit, dénudé entre 6 à 9 r de haut, divisé dans sa partie bas en 2 ou 3 grandes tiges, rampant en début de croissance. Feuilles profondément découpées ; limbe composé de 3 à 7 lobes dentés ou entiers, minces et séparés par des sinus aigus. Fleurs et fruits e grappes globuleuses.

Habitat : berges dans les canyons montagnes du sud de l'Amérique du Nord.

Bois : moyennement lourd, assez dur, résistant, élastique, mat et à grain fin ; fort retrait au séchage, non durable, supportant mal les variations hygrométriques et diff. cile à fendre. Fort contraste entre le bois de cœur brun pâle à brun rougeâtre et l'aubier blanc à ivoir Cernes annuels plus ou moins larges et très distincts ; rayons médullaires invisibles à l'œil nu.

Intérêts : identiques à ceux des autres platanes

TRANSVERSE SECTION.

RADIAL SECTION.

P | PLANE-TREE FAMILY
California Sycamore

FAMILIE DER PLATANENGEWÄCHSE
Kalifornische Platane

FAMILLE DU PLATANE
Platane de Californie

PLATANACEAE
Platanus racemosa Nutt.

Description: 100 to 130 ft. (30 to 40 m) high and 9 ft. (2.7 m) in diameter. Sometimes upright and branch-free for up to half its height. More commonly with secondary trunks from the base. These may be upright, inclined, or creeping to a breadth of some 25 to 35 ft. (6 to 10 m). Old bark dark brown, deeply furrowed, with broad, rounded ridges. Surface with thin scales. Young bark thinner, smooth, and pale, almost white. Leaves deeply three to five-lobed. Inflorescence branched (racemose).

Habitat: Generally on riverbanks in California. Well developed between Monterey and the southern state border. At San Bernardino (about the level of Los Angeles) climbing to 3,300 ft. (1,000 m).

Wood: Comparable with that of the other *Platanus* species.

Use: Decorative park tree in California.

Beschreibung: 30 bis 40 m hoch bei 2,7 m Durchmesser. Manchmal aufrecht und bis zur halben Höhe ohne Äste. Häufiger nahe der Stammbasis Sekundärstämme. Diese aufrecht, schräg oder etwa 6 bis 10 m weit kriechend. Alte Borke dunkelbraun, tief gefurcht, mit breiten, runden Riefen. An deren Oberfläche dünne Schuppen. Junge Borke dünner, glatt und bleich, fast weiß. Blätter tief, drei- bis fünffingrig. Fruchtstände zu mehreren, verzweigt (racemös).

Vorkommen : Allgemein an Flussufern in Kalifornien. Gut entwickelt zwischen Monterey und der Südgrenze des Staates. Am San Bernardino (etwa auf der Höhe von Los Angeles) aufsteigend bis 1000 m.

Holz: Mit dem der anderen Platanen vergleichbar.

Nutzung: Dekorativer Parkbaum in Kalifornien.

Description : 30 à 40 m de hauteu sur un diamètre de 2,7 m. Tronc quelquefois droit et non ramifié jusqu'à mi-hauteur, plus souvent avec des tiges secondaires dans sa partie basale, droites, obliques ou prostrées sur 6 à 10 m. Ecorce brun foncé, creusée de profondes fissures à larges crêtes arrondies, finement écailleuses sur les vieux arbres ; mince, lisse et pâle, presq blanche à l'état juvénile. Feuilles digitées, composées de 3 à 5 lobe Fruits groupés en infrutescences racémeuses (en grappes).

Habitat : berges de Californie en général. Bon développement entr Monterey et la marge méridional de la Californie. Se rencontre jusqu'à une altitude de 1000 m a San Bernardino (env. la hauteur d Los Angeles).

Bois : comparable à celui des autres platanes.

Intérêts : bel arbre de parcs en Californie.

TRANSVERSE SECTION.

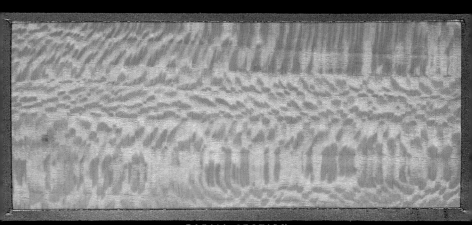

RADIAL SECTION.

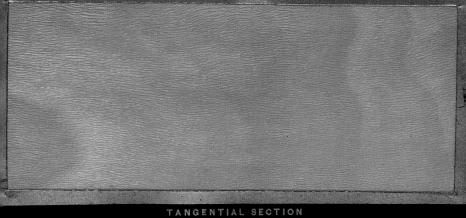

TANGENTIAL SECTION

P | PLANE-TREE FAMILY
Western Plane, American Plane, Buttonwood,
Button-ball Tree

FAMILIE DER PLATANENGEWÄCHSE
Abendländische Platane

FAMILLE DU PLATANE
Platane d'Amérique, Boule de boutons,
Platane d'Occident

PLATANACEAE
Platanus occidentalis L.

Description: 165 to 180 ft. (50 to 60 m) high and 10 to 13 ft. (3 to 4 m) in diameter. Bark separating in irregular plates, with a color range of yellowish white, yellowish green, or brownish gray-green.

Habitat: In eastern North America mainly rich lowland and riverbanks, grows best in the basin of the lower Ohio.

Wood: Heavy, hard, and tough; difficult to split.

Use: Boxes, crates, chopping blocks, etc. When suitably cut, is also used, by virtue of its coloring, in interior construction and for furniture.

A splendid specimen tree, but also a street tree that withstands smoky conditions. Together with the Tulip Tree and the Swamp Cypress, the largest deciduous tree of North America.

Beschreibung: 50 bis 60 m hoch bei 3 bis 4 m Durchmesser. Borke löst sich in unregelmäßigen Platten ab. Gelblich weißes, gelblich grünes und bräunlich graugrünes Farbenspiel.

Vorkommen: Im östlichen Nordamerika vorzugsweise auf reichem Tiefland und an Flussufern, optimal im Becken des unteren Ohio.

Holz: Schwer, hart und zäh; schwer zu spalten.

Nutzung: Kisten, Kästen, Hauklötze usw. Bei geeignetem Zuschnitt aufgrund der Färbung auch im Innenausbau und für Möbel.

Prachtvoller Solitär, aber auch rauchfester Straßenbaum. Mit dem Tulpenbaum und der Sumpfzypresse der größte laubabwerfende Baum Nordamerikas.

Description: 50 à 60 m de hauteu sur un diamètre de 3 à 4 m. Ecor d'un gris jaunâtre s'exfoliant en plaques irrégulières. Belles teinte changeantes, crème, vert jaune o brun gris.

Habitat: préfère les plaines riches les bords de cours d'eau dans l'e de l'Amérique du Nord; optimur de développement dans le bassin du bas Ohio.

Bois: lourd, dur et tenace; difficil à fendre.

Intérêts: employé en caisserie, pour la fabrication de billots etc. Sa teinte permet de l'utiliser en menuiserie intérieure et pour l'ameublement.

Bel arbre solitaire, mais également arbre de rues, résistant au feu. Le plus grand arbre à feuilles caduques du continent nord-amé ricain avec le tulipier et le taxode chauve.

Sycamore, Buttonwood, Button-ball, Plane-Tree.

TRANSVERSE SECTION.

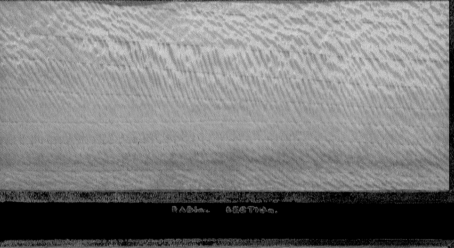

RADIAL SECTION.

TANGENTIAL SECTION.

Ger. Amerikanische Platane. Sp. Platano de America.

Fr. Platane d'occident.

P | PROTEA FAMILY
Silky Oak

FAMILIE DER PROTEUSGEWÄCHSE
Grevillea, Australische Seideneiche

FAMILLE DU PROTEA
Grevillea, Chêne soyeux, Rince-bouteille

PROTEACEAE
Grevillea robusta A. Cunn. ex R. Br.

Description: A tree up to 100 ft. (30 m) high in its native country. Leaves with 11 to 21 leaflets. Leaflets deeply lobed, hairless on the upper side but covered with silvery hairs beneath. Flowers pale orange, in one-sided clusters on bare branches. The fruit is crowned by the long, persistent style.

Habitat: Its native country is Australia, from where it was introduced into the south-western USA. It has escaped from cultivation and has become naturalized around several towns.

Wood: Light, soft, flexible, easy to split, and durable. It contains numerous distinct medullary rays, and has a light reddish heartwood surrounded by even paler sapwood.

Use: The tree was introduced into North America as a park and street tree. In its native country the wood is used in the manufacture of small crates.

Beschreibung: In seiner Heimat ein bis 30 m hoher Baum. Blätter mit 11 bis 21 Blattfiedern. Fiedern tief gelappt mit glatter Ober- und silbrig behaarter Unterseite. Blüten hellorange in einseitswendigen Trauben auf laubblattfreien Ästen. Die Frucht wird von dem langen, ausdauernden Griffel gekrönt.

Vorkommen: Seine Heimat ist Australien, von wo er in den Südwesten der USA eingeführt wurde und in der Nähe einiger Städte verwilderte.

Holz: Leicht, weich, elastisch, einfach spaltbar und dauerhaft. Es enthält zahlreiche deutliche Markstrahlen und besitzt ein hell rötliches Kernholz, das von noch hellerem Saftholz umgeben ist.

Nutzung: Der Baum wurde nach Nordamerika als Park- und Straßenbaum eingeführt. Das Holz wird in seiner Heimat zur Herstellung kleiner Kisten verwendet.

Description : arbre pouvant attein 30 m de hauteur dans son aire naturelle. Feuilles composées de à 21 folioles, profondément lobé à face supérieure glabre et face inférieure couverte d'une pubescence argentée. Les fleurs orange clair forment des grappes dont les ramifications sont émises d'u seul côté et apparaissent sur les branches dénudées. Fruit surmo d'un long style persistant.

Habitat : originaire d'Australie ; introduit dans le sud-ouest des Etats-Unis et devenu subspontan à proximité de quelques villes.

Bois : léger, tendre, élastique, fac à fendre et durable. Bois de cœu rougeâtre clair ; aubier plus clair. Nombreux rayons médullaires distincts.

Intérêts : espèce introduite en Amérique du Nord comme arbr de parcs et d'alignement. Le bois est employé dans son pays d'origine pour la fabrication de petite caisses.

Silky Oak, Grevillea.

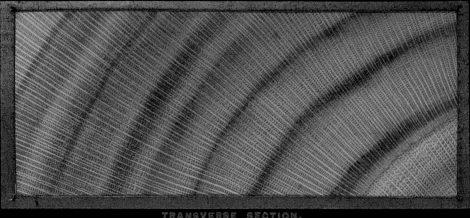

TRANSVERSE SECTION.

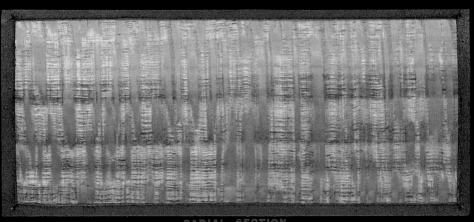

RADIAL SECTION.

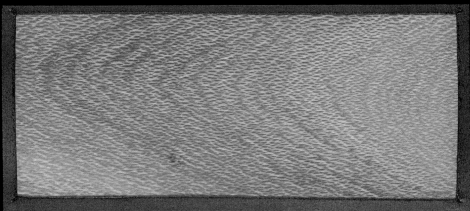

TANGENTIAL SECTION.

Ger. Starke Grevillea.   *Fr.* Grevillea robuste.

Sp. Grevillea robusta.

Q | QUASSIA FAMILY
Paradise Tree, Bitter Wood, Florida Bitter

FAMILIE DER BITTERESCHENGEWÄCHSE
Simaruba

FAMILLE DE L'AILANTE
Simarouba glauque, Bois amer, Quinquina d'Europe

SIMAROUBACEAE
Simarouba glauca DC.

Description: 50 ft. high (15 m), with a straight trunk, 18 to 20 in. (0.45 to 0.5 m) in diameter, and slender, spreading branches that form a rounded crown. Bark moderately thick, light red-brown, broken into broad, thick plates. Leaves divided into six pairs of leaflets. Leaflets ovate to oblong, glabrous, thin and dark red at first, later leathery and dark green, leaf stalks fleshy. Flowers in very loose panicles. Petals pale yellow. Fruit ovoid, dark purple when ripe.

Habitat: Rich soils on hills. Ranges from Cuba to Central America and as far as Brazil. In North America it is confined to Florida, but not in abundance.

Wood: Light, soft, weak, and close-grained. Heartwood light brown; sapwood thick, darker. Many open pores, and thin, well-separated medullary rays.

Use: Because of its beauty, the tree is known in Florida as the "Paradise Tree," and thanks to its excellent shape, the impressive, shiny foliage, and the richly colored fruits, it is a popular ornamental tree in tropical gardens.

Beschreibung: 15 m hoch mit geradem Stamm von 0,45 bis 0,5 m Dicke und schlanken, ausladenden Ästen, die eine runde Krone formen. Borke mäßig dick, hell rotbraun in breite, dicke Schollen zerbrechend. Blätter in 6 Fiederblattpaare geteilt. Blattfiedern oval bis länglich, unbehaart, Blattstiele fleischig. Anfangs dünn und dunkelrot, später ledrig und dunkelgrün. Blüten in sehr lockeren Rispen. Kronblätter blassgelb. Frucht oval-eiförmig, reif dunkelviolett.

Vorkommen: Reiche Böden auf Hügeln. Von Kuba über Mittelamerika bis nach Brasilien verbreitet. Nordamerika nur in Südflorida erreichend und dort nicht häufig.

Holz: Leicht, weich, schwach und mit feiner Textur. Kern hellbraun; Splint dick, dunkler. Viele offene Gefäße und dünne, weit voneinander entfernte Markstrahlen.

Nutzung: Wegen seiner Schönheit in Florida als „Paradise tree" bekannt. Dank seiner exzellenten Wuchsform, dem leuchtenden, stattlichen Laub und den kräftig gefärbten Früchten ein lohnendes Ziergehölz für tropische Gärten.

Description : arbre de 15 m de ha[u]teur, avec un tronc droit de 0,45 [à] 0,5 m de diamètre et des branch[es] fines, étalées, constituant un hou[p]pier rond. Ecorce moyennement épaisse, brun rouge clair, se craquelant en côtes épaisses et larg[es] Feuilles comptant 6 paires de folioles ovales à allongées, glabre[s] à pétiole charnu, fines et rouge foncé à l'état juvénile puis coriac[es] et vert sombre. Fleurs à pétales jaune pâle, groupées en panicule[s] très légères. Fruits ovales-ovoïdes violet sombre à maturité.

Habitat : sols riches à l'étage colli néen ; répandu de Cuba jusqu'au Brésil en passant par l'Amérique centrale. En Amérique du Nord seulement dans le sud de la Floride où il est peu répandu.

Bois : léger, tendre, faible et à gra[in] fin. Bois de cœur brun clair ; aub[ier] épais, plus sombre. Nombreux va[is]seaux ouverts, rayons médullaire[s] fins et espacés.

Intérêts : le «paradise tree» de Floride, est une espèce ornemen tale recommandée pour les jardi[ns] exotiques car son port est attraya[nt] son feuillage, brillant et imposan[t] et ses fruits, d'une couleur inten[se]

TRANSVERSE SECTION.

RADIAL SECTION.

TANGENTIAL SECTION.

Q | QUASSIA FAMILY

Tree of Heaven

FAMILIE DER BITTERESCHENGEWÄCHSE
Götterbaum

FAMILLE DE L'AILANTE
Vernis du Japon

SIMAROUBACEAE
Ailanthus altissima (Mill.) Swingle
Syn.: *A. glandulosa* Desf.

Description: 85 to 100 ft. (25 to 30 m) high and from 2 to 3 ft. (0.6 to 0.9 m) in diameter. Often multi-stemmed. Bark with conspicuous, pale longitudinal fissures. Ornamental pinnate leaves like palm fronds, up to 2 ft. (0.6 m) long with glandular leaflets. Flowers greenish yellow in attractive panicles, the male flowers with an unpleasant odor. Fruits pale reddish brown, in long hanging clusters, decorative in the early stages. Very quick-growing and easily escaping from cultivation into the wild by means of sucker shoots.

Habitat: China, but planted worldwide and partly naturalized (eastern North America).

Wood: Heavy, fairly hard and with large pores; resembles that of the ash. Heartwood grayish to brownish yellow; sapwood broad yellowish white.

Use: In Asia the foliage provides food for silkworms.

Undemanding as regards soils, withstands smoky conditions.

Beschreibung: 25 bis 30 m hoch bei 0,6 bis 0,9 m Durchmesser. Oft mehrstämmig. Borke mit auffallenden hellen Längsrissen. Ornamentale, wedelartige Fiederblätter bis 0,6 m lang, mit drüsigen Blättchen. Blüten grünlich gelb in ansehnlichen Rispen, die männlichen überriechend. Früchte hell rötlich braun, in langen, hängenden, anfangs zierenden Fruchtständen. Sehr schnellwüchsig und leicht verwildernd (Wurzelbrut).

Vorkommen: Ursprünglich aus China, aber weltweit angepflanzt und im östlichen Nordamerika z. T. eingebürgert.

Holz: Schwer, mittelhart und großporig; eschenähnlich. Kern graugelb bis braungelb; Splint breit, gelblich weiß.

Nutzung: Laub in Asien als Seidenraupenfutter verwendet.

Geringe Bodenansprüche, rauchfest.

Description: arbre s'élevant entre et 30 m, avec un tronc de 0,9 m diamètre, présentant souvent plu sieurs tiges. L'écorce se remarqu pour ses fissures longitudinales claires. Grandes feuilles pennées forme de fronde, très décoratives pouvant atteindre 0,6 m de long munies de folioles glanduleuses. Fleurs jaune verdâtre groupées en belles panicules terminales ; les mâles dégagent une odeur désagréable. Samares allongées brun rougeâtre clair, réunies en grappes pendantes et ornementa en début de maturation. Espèce croissance rapide, très envahissa (drageons).

Habitat: originaire de la Chine, mais cultivé un peu partout dans le monde et même naturalisé da l'est de l'Amérique du Nord.

Bois: lourd, moyennement dur e à gros pores ; semblable à celui d frêne. Bois de cœur d'un jaune gris ou brun ; large aubier blanc jaunâtre.

Intérêts: en Asie, les feuilles serv à nourrir une espèce de ver à so

S'accommode de sols médiocres et résiste au feu.

TRANSVERSE   SECTION.

RADIAL   SECTION.

R | ROSE FAMILY
American Mountain Ash, Roundwood

FAMILIE DER ROSENGEWÄCHSE
Kanada-Eberesche

FAMILLE DE LA ROSE
*Sorbier d'Amérique, Sorbier des montagnes d'Amérique,
Cormier (Canada)*

ROSACEAE
Sorbus scopulina Greene

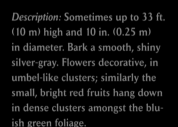

Description: Sometimes up to 33 ft. (10 m) high and 10 in. (0.25 m) in diameter. Bark a smooth, shiny silver-gray. Flowers decorative, in umbel-like clusters; similarly the small, bright red fruits hang down in dense clusters amongst the bluish green foliage.

Habitat: Swamps and banks of streams, and beside mountain springs, chiefly in the far north. Southward extent no further than the northern states of the USA.

Wood: Light, soft, and not much used.

Use: Unknown.

Beschreibung: Manchmal bis 10 m hoch bei 0,25 m Durchmesser. Borke glatt, glänzend silbergrau. Blüten in Scheindolden zierend, ebenso die überhängenden Fruchtstände mit dicht gedrängten, kleinen, hellroten Früchten im blaugrünen Laub.

Vorkommen: Sümpfe, Bachufer und die Ränder von Bergquellen vorwiegend des hohen Nordens. Nach Süden nur bis in die nördlichen Staaten der USA vordringend.

Holz: Leicht, weich und nur wenig genutzt.

Nutzung: Unbekannt.

Description : arbre atteignant parfois 10 m de hauteur sur un diamètre de 0,25 m. Ecorce lisse, d'un gris argenté brillant. Fleurs e corymbes très décoratifs, comme les petits fruits rouge clair réunis en grappes denses et pendantes. Feuillage vert bleu.

Habitat : marécages, bords des tor rents et des sources de montagne en majorité dans le Grand Nord. Vers le sud jusqu'aux Etats du nord des Etats-Unis.

Bois : léger, tendre et peu employe

Intérêts : inconnus.

Elder-leaved Mountain Ash.

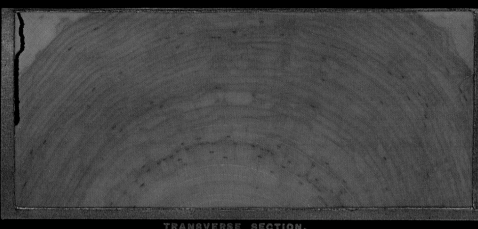

TRANSVERSE SECTION.

RADIAL SECTION.

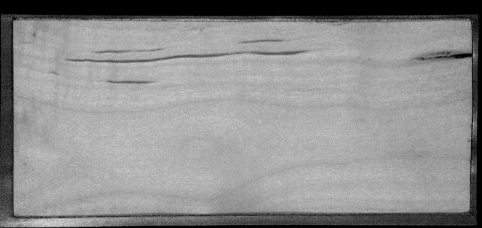

TANGENTIAL SECTION.

Ger. Schlehdorn.   *Fr.* Sorbier.   *Sp.* Mostajo.

R | ROSE FAMILY
American Red Plum, Wild Plum

FAMILIE DER ROSENGEWÄCHSE
Amerikanische Scheinpflaume

FAMILLE DE LA ROSE
Prunier d'Amérique, Guignier,
Prune de la Gallissonnière (Canada)

ROSACEAE
Prunus americana Marshall

Description: 20 to 33 ft. (6 to 10 m) high and rarely 12 in. (0.3 m) in diameter. Trunk densely branched from a height of between 4 and 5 ft. (1.2 to 1.5 m). Lateral branches often drooping at the edges of the crown, producing a broad and compact shape. Small branches partially thorny. Bark thin, dark brown with a red tinge. Leaves ovate or slightly oblong, dark green on the upper surface and pale on the lower. Flowers in two to five-flowered umbels. Sepals entire, hairy on the inner surface. Fruit roundish, at first with a green skin, later turning bright red.

Habitat: Widespread above all in the interior of North America. In the north on the edges of streams and swamps; in the south on swampy ground by rivers, west of the Mississippi on drier chalky soils.

Wood: Heavy, hard, strong, lustrous, and close-grained. Heartwood dark, rich brown, tinged red; sapwood thin, paler. Numerous medullary rays.

Use: The fruits are occasionally used for jam, or eaten raw or cooked. An attractive ornamental tree.

Beschreibung: 6 bis 10 m hoch bei selten 0,3 m Durchmesser. Stamm sich bereits in einer Höhe von 1,2 bis 1,5 m dicht verzweigend. Seitenäste an den Rändern der Krone oft herabhängend, wodurch eine breite geschlossene Gestalt entsteht. Kleine Äste z. T. stachelig. Borke dünn, dunkelbraun mit roter Tönung. Blätter oval oder leicht länglich, dunkelgrün auf der Ober- und blass auf der Unterseite. Blüten in zwei- bis fünfblütigen Dolden. Kelchblätter ganzrandig mit behaarter Innenseite. Frucht rundlich, mit zunächst grüner, später hellroter Haut.

Vorkommen: Weit verbreitet vor allem im Inneren Nordamerikas. Im Norden an den Rändern von Bächen und Sümpfen, im Süden in Sümpfen entlang der Flüsse, westlich des Mississippi auch auf trockenerem Kalkboden.

Holz: Schwer, hart, fest, glänzend und mit feiner Textur. Kern dunkel tiefbraun, rot getönt; Splint dünn, heller. Zahlreiche Markstrahlen.

Nutzung: Die Früchte werden gelegentlich als Marmelade, roh oder gekocht gegessen. Dekoratives Ziergehölz.

Description : arbuste de 6 à 10 m hauteur, avec un tronc atteignan rarement 0,3 m de diamètre et s ramifiant tôt, entre 1,2 et 1,5 m. Branches latérales souvent retom bantes au bord du houppier, ce qui lui donne une forme ample e fermée. Petits rameaux garnis en partie d'aiguillons crochus. Ecorc mince, brun sombre teinté de rouge. Feuilles ovales ou légèrement allongées, vert foncé dessu. pâles dessous. Fleurs en ombelles composées de 2 à 5 fleurs. Sépale entiers, veloutés à la face supérieure. Fruits globuleux, verts pui rouge clair à maturité.

Habitat : répandu surtout dans l'intérieur de l'Amérique du Nor Au nord, en bordure de ruisseau et de marais ; au sud, dans les marais installés le long des rivière à l'ouest du Mississippi, s'accommode aussi de sols calcaires plus secs.

Bois : lourd, dur, résistant, brillan et à grain fin. Bois de cœur brun profond rouge ; aubier mince, cla Nombreux rayons médullaires.

Intérêts : fruits consommés parfois en confitures, aussi crus ou cuits. Espèce ornementale attrayante.

Wild Plum.

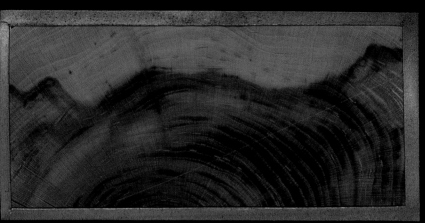

TRANSVERSE SECTION.

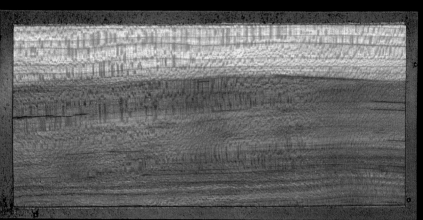

RADIAL SECTION.

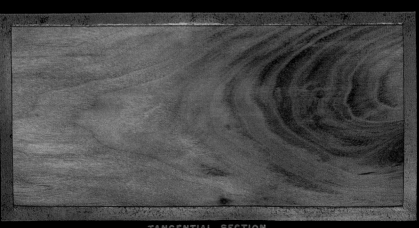

TANGENTIAL SECTION.

Amerikanischer Schlehdorn.   *Fr.* Prunier d'Amerique.

R | ROSE FAMILY

Black Hawthorn, Black Haw

FAMILIE DER ROSENGEWÄCHSE

Douglas-Weißdorn

FAMILLE DE LA ROSE

Aubépine noire, Epine de Douglas,
Aubépine de Douglas (Canada)

ROSACEAE

Crataegus douglasii Lindl.

Description: Bushy small tree with long, pale brown thorns. Leaves broadly ovate, with sharply and often doubly serrated margins, sometimes three-lobed, dark green above, paler beneath. Flowers white, smaller than those of commoner relative *Crataegus coccinea*, in broad inflorescences. Fruit oblong or rounded, small and hardly fleshy, scarlet.

Habitat: Apparently similar to the commoner *Crataegus coccinea*: dense thickets in open, high altitude woodland, on stony slopes or near riverbanks. From the Atlantic coast in the north-east to Winnipeg, and in the west on the eastern slopes of the Cascade Mountains and in the Rocky Mountains.

Wood: Heavy, hard, and close-grained. Heartwood brown with red tinge; sapwood thin and paler. Thin obvious medullary rays (description based on *Crataegus coccinea*).

Use: In Europe as decorative tree (Schmeil, 1994).

Beschreibung: Buschartiger kleiner Baum mit langen, hellbraunen Dornen. Blätter breit und oval, mit scharf und oft doppelt gesägten Blatträndern, manchmal dreilappig, mit dunkelgrüner Ober- und blasser Unterseite. Blüten weiß, kleiner als bei dem häufigeren Verwandten *Crataegus coccinea*, in breiten Blütenständen. Frucht länglich oder rundlich, klein und kaum fleischig, scharlachrot.

Vorkommen: Vermutlich ähnlich der häufigeren *Crataegus coccinea*: dichte Dickichte in offenen Wäldern der Hochlagen, auf steinigen Hängen oder in der Nähe von Flussufern. Von der Atlantik-Küste im Nordosten bis nach Winnipeg sowie im Westen auf den Osthängen des Kaskadengebirges und in den Rocky Mountains.

Holz: Schwer, hart und mit feiner Textur. Kern braun mit roter Tönung; Splint dünn und heller. Dünne deutliche Markstrahlen (Beschreibung bezieht sich auf *Crataegus coccinea*).

Nutzung: In Europa als Ziergehölz (Schmeil 1994).

Description : arbuste formant une touffe assez dense, avec des rameaux armés de longues épine brun clair. Feuilles ovales-larges, aiguës et souvent redentées, quelquefois trilobées, vert sombr dessus, plus pâles dessous. Fleurs blanches, plus petites que celles de *Crataegus coccinea*, espèce voisine plus répandue, et groupée en larges inflorescences. Petits fruits pourpres, allongés ou rond à peine charnus.

Habitat : sans doute similaire à celui de *Crataegus coccinea* : sur le hauteurs, où il forme des fourrés denses dans les forêts ouvertes, s les pentes caillouteuses ou à prox mité des berges. Dans le nord-est, depuis la côte atlantique jusqu'à Winnipeg ainsi que dans l'ouest s le versant est des montagnes Cascades et jusque dans les Rocheus

Bois : lourd, dur et à grain fin. Bois de cœur brun teinté de rouge ; aubier mince et plus clair. Rayons médullaires fins et distinc (description basée sur celle de *Crataegus coccinea*).

Intérêts : plante ornementale en Europe (Schmeil 1994).

Black Thorn, Western Haw, Black Haw.

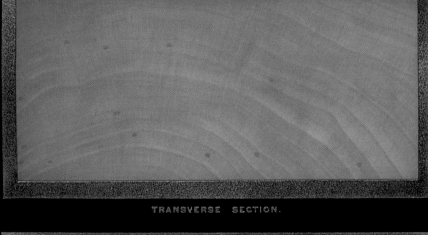

TRANSVERSE SECTION.

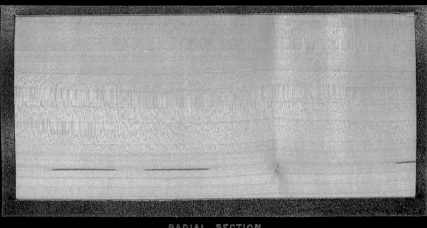

RADIAL SECTION.

TANGENTIAL SECTION.

Ger. Schwarze Hagedorn.   *Fr.* Aubepine noire.

Sp. Espino negro.

R | ROSE FAMILY
Canada Plum

FAMILIE DER ROSENGEWÄCHSE
Kanadische Pflaume

FAMILLE DE LA ROSE
Prunier noir, Prunier canadien

ROSACEAE
Prunus nigra Aiton

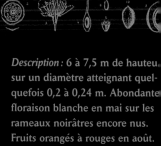

Description: 20 to 25 ft. (6 to 7.5 m) high and as much as 8 to 10 in. (0.2 to 0.24 m) in diameter. The gray branches are still bare in May when the white flowers suddenly appear in abundance. Fruits orange to red in August.

Habitat: Wood margins, clumps of bushes, hedges, in north-eastern North America.

Wood: Heavy, hard, and close-grained.

Use: Flowers extra rich in nectar. The first food plant of the year for bees. Fruits suitable for immediate consumption or for further processing.

Several varieties in cultivation.

Beschreibung: 6 bis 7,5 m hoch bei gelegentlich 0,2 bis 0,24 m Durchmesser. An noch kahlen, schwärzlichen Ästen plötzlich und in Fülle auftretende weiße Blüten im Mai. Früchte orange bis rot im August.

Vorkommen: Waldränder, Gebüsche, Hecken im nordöstlichen Nordamerika.

Holz: Schwer, hart und von sehr feiner Textur.

Nutzung: Blüten überreich an Nektar. Erste Bienenfutterpflanze des Jahres. Früchte zu sofortigem Verzehr oder weiterer Verarbeitung.

Mehrere Varietäten in Kultur.

Description : 6 à 7,5 m de hauteu. sur un diamètre atteignant quel-quefois 0,2 à 0,24 m. Abondante floraison blanche en mai sur les rameaux noirâtres encore nus. Fruits orangés à rouges en août.

Habitat : lisières forestières, fourr haies dans le nord-est de l'Amé-rique du Nord.

Bois : lourd, dur et à grain très fir

Intérêts : fleurs très riches en nectar. Première plante mellifère de l'année. Fruits consommés fra ou préparés sous diverses formes

Plusieurs variétés cultivées.

TRANSVERSE SECTION.

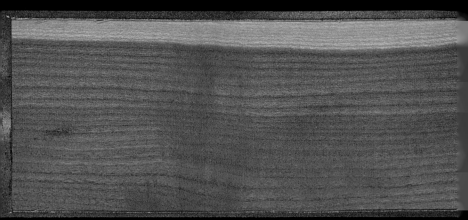

RADIAL SECTION.

R | ROSE FAMILY

Carolina Laurel Cherry, Wild Orange

FAMILIE DER ROSENGEWÄCHSE
Karolinischer Kirschlorbeer

FAMILLE DE LA ROSE
Cerisier du Mississippi

ROSACEAE
Prunus caroliniana (Mill.) Aiton

Description: 30 to 40 ft. (9 to 12 m) high and a maximum of 10 to 12 in. (0.25 to 0.3 m) in diameter. Trunk with small horizontal branches, forming a narrow, oblong crown. Bark thin, smooth, or somewhat roughened, gray in color with large black blotches. Leaves oblong-lanceolate, with a thickened, usually entire margin, and dark green in color, falling only in their second year. Flowers in dense clusters. Sepals rounded. Fruit round, black, with thin dry flesh.

Habitat: On deep, rich, moist, lowland soils in the Gulf states of the USA.

Wood: Heavy, hard, strong, with a silky sheen, and close-grained; polishable. Heartwood light brown, sometimes rich dark brown; sapwood thick, paler.

Use: One of the most popular ornamental trees in the southern USA. As it stands being cut back very well, *Prunus caroliniana* is a good hedge plant.

It contains a considerable amount of cyanide.

Beschreibung: 9 bis 12 m hoch, maximal 0,25 bis 0,3 m Durchmesser. Stamm mit kleinen horizontalen Ästen, bildet eine schmale, längliche Krone. Borke dünn, glatt oder etwas aufgeraut, von grauer Farbe mit großen schwarzen Flecken. Blätter länglich-lanzettlich, mit verdicktem, meist ganzrandigem Blattrand und dunkelgrüner Farbe. Laubfall erst nach 2 Jahren. Blüten in dichten Trauben. Kelchblätter abgerundet. Frucht rund, schwarz, mit dünnem, trockenem Fleisch.

Vorkommen: Auf tiefgründigen, reichen, feuchten Tieflandböden der US-amerikanischen Golfstaaten.

Holz: Schwer, hart, fest, seidig und mit feiner Textur; polierfähig. Kern hellbraun, manchmal tief dunkelbraun; Splint dick, heller.

Nutzung: Eines der beliebtesten Ziergehölze in den südlichen USA. Da sie Rückschnitte gut verträgt, lässt sich *Prunus caroliniana* gut als Hecke ziehen.

Enthält erhebliche Mengen an Zyanid.

Description : arbre de 9 à 12 m de hauteur, avec un tronc ne dépassant pas 0,25 à 0,3 m de diamètre et des petites branches horizontales, formant un houppi étroit et allongé. Ecorce mince, lisse ou un peu rugueuse, brune ponctuée de grandes taches noir Feuilles allongées-lancéolées, ave une marge épaissie, généralemer entières et vert sombre, demeurant sur l'arbre pendant deux an Fleurs en grappes florifères, avec des sépales arrondis. Fruit rond noir, à chair sèche et mince.

Habitat : sols profonds, riches et humides de plaine dans les Etats du Golfe aux Etats-Unis.

Bois : lourd, dur, résistant, satiné à grain fin ; susceptible d'être po Bois de cœur brun clair, parfois brun profond ; aubier épais, plus clair.

Intérêts : une des espèces orneme tales les plus appréciées dans le s des Etats-Unis. Le *Prunus carolini* convient à la plantation de haies car il supporte bien la taille.

La plante contient une grande quantité de cyanur.

WILD ORANGE.   MOCK ORANGE.

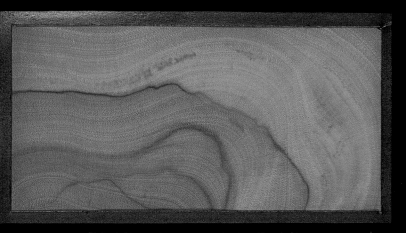

TRANSVERSE SECTION.

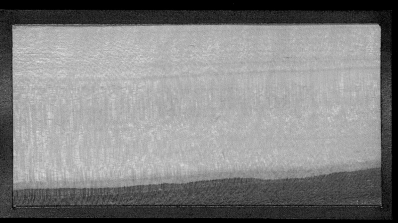

RADIAL SECTION.

TANGENTIAL SECTION.

Gr. Falscher mahahonybaum. Fr. Laurier de Mississippi

R | ROSE FAMILY

Catalina Ironwood

FAMILIE DER ROSENGEWÄCHSE

–

FAMILLE DE LA ROSE

Arbre de Lyon, Bois de fer

ROSACEAE

Lyonothamnus floribundus A. Gray

Description: Rarely to 30 or 40 ft. (10 to 12 m) and if single-trunked 6 to 8 in. (0.15 to 0.20 m) in diameter, otherwise branching from the base, a bushy tree, or shrub. Bark red-brown, with thin papery layers. Dead sections remain as loose strips on the trunk.

Habitat: Dry, rocky slopes, dry soils, walls of deep canyons on Santa Catalina and Santa Cruz Islands off the coast of California.

Wood: Heavy, hard, close-grained. Red, with light orange tinge.

Use: No commercial use known.

Beschreibung: Selten bis 10, 12 m hoch bei 0,15 bis 0,20 m Durchmesser wenn einstämmig, sonst von Grund auf mehrstämmiger buschiger Baum oder Strauch. Borke rotbraun, aus dünnen Papierlagen. Abgestorbene Teile verbleiben als lose Streifen am Stamm.

Vorkommen: Trockene, felsige Hänge, trockene Böden, Wände tiefer Cañons auf Santa Catalina Island und Santa Cruz Island vor der Küste Kaliforniens.

Holz: Schwer, hart, mit feiner Textur. Von roter Farbe, leicht orange getönt.

Nutzung: Keine gewerbliche Nutzung bekannt.

Description : arbre atteignant rare ment 10 ou 12 m de hauteur, av un tronc de 0,15 à 0,20 m de dia mètre lorsqu'il est unique ; sinon divisé dès la base, à port touffu c buissonnant. L'écorce brun rouge forme de minces pellicules rappe lant des feuilles de papier et qui se déchirent en lambeaux restan fixés sur le tronc.

Habitat : pentes rocheuses sèches sols secs, versants de canyons encaissés sur les îles de Santa Ca lina et de Santa Cruz en face de côte californienne.

Bois : lourd, dur et à grain fin. D'un rouge légèrement orangé.

Intérêts : utilisation commerciale non déterminée.

Santa Catalina Iron-wood, Santa Cruz Iron-wood.

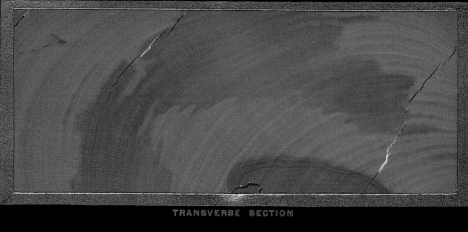

TRANSVERSE SECTION

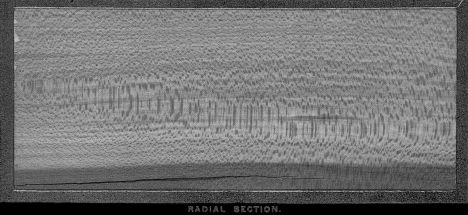

RADIAL SECTION.

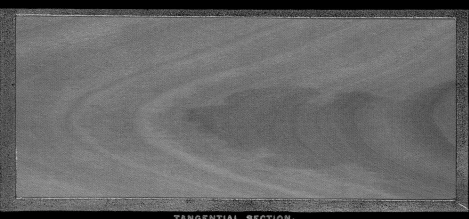

TANGENTIAL SECTION.

Ger. Eisenholtz von St. Catalina. *Fr.* Bois dur de St. Catalina.

Sp. Arbol de Hierro de Santa Catalina.

R | ROSE FAMILY
Cockspur Thorn, New Castle Thorn

FAMILIE DER ROSENGEWÄCHSE
Hahnensporn-Weißdorn

FAMILLE DE LA ROSE
Aubépine ergot de coq, Epine ergot de coq,
Ergot-de-coq (Canada)

ROSACEAE
Crataegus crus-galli L.

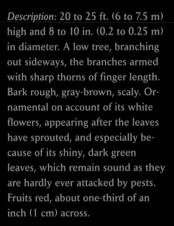

Description: 20 to 25 ft. (6 to 7.5 m) high and 8 to 10 in. (0.2 to 0.25 m) in diameter. A low tree, branching out sideways, the branches armed with sharp thorns of finger length. Bark rough, gray-brown, scaly. Ornamental on account of its white flowers, appearing after the leaves have sprouted, and especially because of its shiny, dark green leaves, which remain sound as they are hardly ever attacked by pests. Fruits red, about one-third of an inch (1 cm) across.

Habitat: Margins of woods, clumps of bushes and hedges in northeastern North America. The name "New Castle Thorn" alludes to its particularly frequent use as a hedge around New Castle, Delaware.

Wood: Heavy, hard, and close-grained.

Use: Suitable for tool handles.

Beschreibung: 6 bis 7,5 m hoch bei 0,2 bis 0,25 m Durchmesser. Niedriger, seitlich sich ausbreitender Baum mit fingerlangen, spitz bedornten Ästen. Borke rau, graubraun, schuppig. Durch seine weißen Blüten nach dem Laubaustrieb und vor allem durch das glänzende Dunkelgrün seiner gesunden, kaum je von Ungeziefer befallenen Blätter sehr dekorativ. Rote, etwa 1 cm dicke Früchte.

Vorkommen: Waldränder, Gebüsche im nordöstlichen Nordamerika. Der Name „New Castle Thorn" spielt auf die besonders häufigen Hahnensporn-Weißdornhecken um New Castle, Delaware, an.

Holz: Schwer, hart, mit sehr feiner Textur.

Nutzung: Für Werkzeuggriffe geeignet.

Description: 6 à 7,5 m de hauteur sur un diamètre de 0,2 à 0,25 m. Petit arbre au port pyramidal inversé dont les rameaux sont ga nis d'épines acérées, de la longue d'un doigt, et arquées. Ecorce rugueuse brun grisâtre, à texture en écailles. Espèce ornementale par ses fleurs blanches qui s'épanouissent lorsque les feuilles son bien développées et surtout par s feuillage d'un vert sombre brillar et sain car il est peu sensible aux attaques d'agents pathogènes. Fruits rouges d'environ 1 cm de diamètre.

Habitat: lisières forestières, fourr haies dans le nord-est de l'Amérique du Nord. Le nom qu'on lui a donné en Amérique, «New Castle Thorn», fait allusion à sa présence particulièrement fréquente dans les environs de Nev Castle, ville du Delaware.

Bois: lourd, dur et à grain très fir

Intérêts: convient bien à la fabrica tion de manches d'outils.

Cock-spur Thorn, Newcastle Thorn.

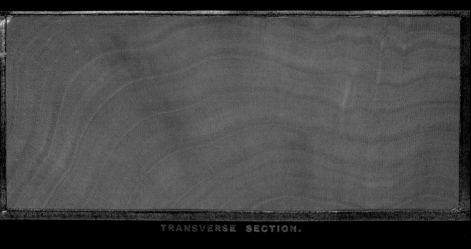

TRANSVERSE SECTION.

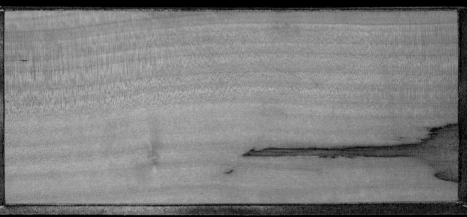

RADIAL SECTION.

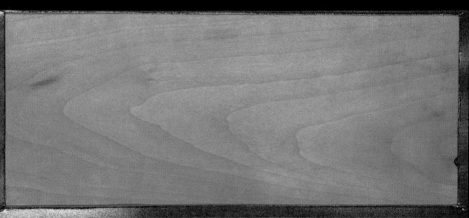

TANGENTIAL SECTION.

Ger. Glänzende Mispel.   **Fr.** Néflier pied de coc.

Sp. Espino del espolon de gallo.

R | ROSE FAMILY
Common Pear

FAMILIE DER ROSENGEWÄCHSE
Gewöhnliche Birne

FAMILLE DE LA ROSE
Poirier commun

ROSACEAE
Pyrus communis L. agg.

Description: In favorable conditions 85 ft. (25 m) high and 2 ft. (0.6 m) in diameter. In the USA, unlike Europe and Asia, the wild form is not *Pyrus communis* s. str. Syn.: *P. pyraster* (L.) Burgsd., *P. achras* Gaertner, which has short thorny shoots and small, yellow to brown fruits, barely edible because of the bitterness of their scanty pulp, but numerous varieties of *P. domestica* Med., the Cultivated Pear, have pleasant tasting, juicy fruit. These forms have become established in America after escaping from cultivation.

Habitat: The wild form prefers riverbanks and the edges of riverside forests, but also ascends high into the mountains.

Wood: Moderately hard and close-grained; polishable, not chipping when cut. Wood of young trees almost white, later reddish brown; often mottled and beautifully grained.

Use: Carving, cabinetwork, turnery. More valuable than the harder wood of the closely related apple tree.

Beschreibung: Unter günstigen Bedingungen 25 m hoch bei 0,6 m Durchmesser. In den USA, anders als in Europa und Asien, nicht die Wildform *Pyrus communis* s. str. Syn.: *Pyrus pyraster* (L.) Burgsd., *Pyrus achras* Gaertner, mit verdornenden Kurztrieben und kleinen, gelben bis braunen und wegen ihres geringen Fruchtfleisches von großer Herbheit kaum genießbaren Früchten. Zahlreiche Varietäten der *P. domestica* Med., einer Kulturbirne, haben hingegen wohlschmeckende, saftreiche Früchte. Diese ist als Kulturflüchtling in Amerika eingebürgert.

Vorkommen: (Wildform) vorzugsweise Flussufer und die Ränder von Auwäldern, im Gebirge hoch aufsteigend.

Holz: Mäßig hart und mit feiner Textur; polierfähig, nicht ausbröckelnd und leicht zu schneiden. Holz junger Bäume fast weiß, bei älteren rötlich braun; oft geflammt und mit schöner Maserung.

Nutzung: Bildhauerarbeiten, Kunstschreinerei, Drechslerarbeiten. Wertvoller als das härtere Holz des nah verwandten Apfelbaums.

Description : peut atteindre 25 m ⌐ hauteur sur un diamètre de 0,6 m Forme distincte du poirier sauvag *Pyrus communis* s. str., originaire d'Europe et d'Asie. Syn. : *Pyrus pyraster* (L.) Burgsd., *Pyrus achras* Gaertner, munis de courts ramea piquants et de petits fruits jaunes bruns, non comestibles (chair très acerbe et peu fournie) ; mais le *P. domestica* Med. ou poirier cultivé a donné naissance à de nombreu cultivars aux fruits juteux et aromatiques. Cette forme hortico échappée des jardins s'est natura sée en Amérique du Nord.

Habitat : à l'état sauvage, il préfèr les berges et les lisières de forêts riveraines, mais il croît aussi à l'étage montagnard.

Bois : moyennement dur et à grai fin ; pouvant se polir, résistant m caniquement et facile à fendre. B quasiment blanc chez les jeunes arbres ; brun rougeâtre, souvent flammé et présentant une belle maillure chez les sujets adultes.

Intérêts : utilisé en sculpture, en ébénisterie et en tournerie. Plus précieux que le bois plus dur de son proche parent, le pommier.

TRANSVERSE SECTION,

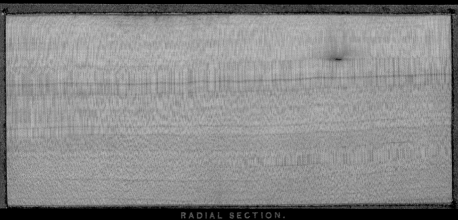

RADIAL SECTION.

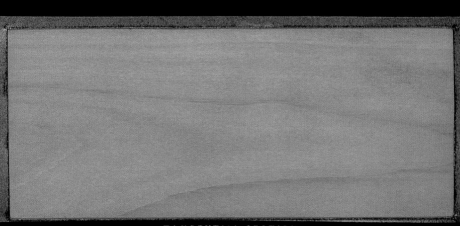

TANGENTIAL SECTION

Ger. Birnbaum.   Sp. Peral.

Fr. Poirier commun.

R | ROSE FAMILY
Cultivated Apple

FAMILIE DER ROSENGEWÄCHSE
Kultur-Apfel

FAMILLE DE LA ROSE
Pommier commun

ROSACEAE
Malus sylvestris Mill.
Syn.: *Pyrus malus* L.

Description: 30 to 50 ft. (10 to 15 m) high and 2 to 3 ft. (0.6 to 0.9 m) in diameter. Beautiful in flower. The fruit is really only the core, the central tissue from which the much-enjoyed "flesh" develops, so the part we eat is in fact a "false fruit."

Habitat: Cultivated ground. Sometimes grows wild in eastern North America.

Wood: Hard and close-grained.

Use: For carving, turnery, and small utensils, but of inferior quality to pear-wood. The fruit is a more valuable product than the wood. It is fermented to make cider, and also used for fresh fruit juice, jelly, as stewed or dried fruit, and in pies.

Has been cultivated outside its native region in western Asia (Alma Ata, the apple town) in all temperate zones since prehistoric times, and new varieties have been continually produced by breeding, with various flavors and storage potential.

Beschreibung: 10 bis 15 m hoch bei 0,6 bis 0,9 m Durchmesser. Schön blühend. Die eigentliche Frucht ist nur das Kerngehäuse; das als Obst genossene „Fleisch" des Apfels geht aus Achsengewebe hervor, daher „Scheinfrucht".

Vorkommen: Kulturland. Bisweilen auch verwildert im östlichen Nordamerika.

Holz: Hart und von feiner Textur.

Nutzung: Für Schnitzereien, Drechslerarbeiten, feine Gerätschaften, dem Birnholz an Qualität etwas untergeordnet. Obstnutzung wertvoller als Holzverarbeitung. Zu Wein vergoren, zu Frischsäften, Gelees und Kompotten, Trockenobst verarbeitet.

Seit vorgeschichtlicher Zeit aus seiner Heimat in Westasien (Alma Ata, die Apfelstadt) überall im gemäßigten Klima in Kultur und dabei fortlaufend züchterisch verändert. Zahlreiche wohlschmeckende Fruchtsorten verschiedener Geschmacksrichtungen und unterschiedlicher Lagerfähigkeit.

Description: 10 à 15 m de hauteur sur un diamètre de 0,6 à 0,9 m. Floraison spectaculaire. Ce que nous appelons «fruit» est le cœur contenant les pépins (réceptacle floral); la chair que nous consommons n'est en fait qu'une excroissance du tissu des rameaux.

Habitat: zones de culture. Dans l'est de l'Amérique du Nord, parfois subspontané sous une forme horticole échappée des jardins.

Bois: dur et à grain fin.

Intérêts: employé en ébénisterie, en sculpture, en tournerie et pour fabriquer des instruments de précision. De qualité légèrement inférieure à celle du poirier. Sa valeur économique est toutefois moindre que celle des fruits qui sont utilisé pour la fabrication de boissons alcoolisées, de jus de fruits, de gelées, de compotes ou séchés.

Cultivé dès le Néolithique dans so aire d'origine, l'Asie occidentale (Alma Ata = ville du pommier), puis dans toutes les régions tempérées. Constamment transformé pa la culture. D'innombrables variété ont été sélectionnées pour leurs fruits.

TRANSVERSE SECTION.

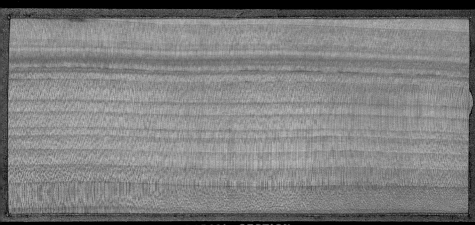

RADIAL SECTION.

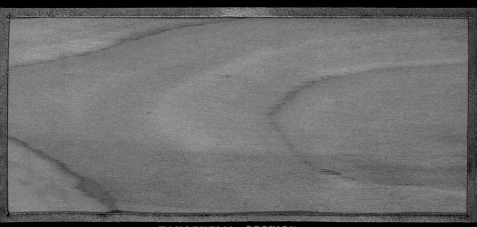

TANGENTIAL SECTION.

Ger. Apfelbaum.   Fr. Pommier.   Sp. Manzana.

R | ROSE FAMILY
Dotted Hawthorn

FAMILIE DER ROSENGEWÄCHSE
Punktierter Weißdorn

FAMILLE DE LA ROSE
Aubépine ponctuée, Aubépine à fruits ponctués,
Pommettier (Canada)

ROSACEAE
Crataegus punctata Jacq.

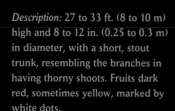

Description: 27 to 33 ft. (8 to 10 m) high and 8 to 12 in. (0.25 to 0.3 m) in diameter, with a short, stout trunk, resembling the branches in having thorny shoots. Fruits dark red, sometimes yellow, marked by white dots.

Habitat: Dry hill-country, but also water-meadows in north-eastern North America. Undemanding.

Wood: Very hard, tough, and very strong. Streaked with brown.

Use: Hammer-handles, threshing-flails, walking-sticks, turnery.

Beautiful in flower, but because of its susceptibility to pests and as a host or intermediate host for fungal leaf-diseases, it is not widely cultivated. A regular nest-tree of the butcher-bird or shrike, which uses the thorns not only to protect itself and its brood, but also to hang its prey until required.

Beschreibung: 8 bis 10 m hoch bei 0,25 bis 0,3 m Durchmesser. Kurzer, kräftiger Stamm, die Äste ähneln verzweigten Dorntrieben. Dunkelrote, manchmal gelbe Früchte, hellgefleckt.

Vorkommen: Trockenes Hügelland, aber auch Flussauen im nordöstlichen Nordamerika. Nicht sehr anspruchsvoll.

Holz: Sehr hart, zäh und sehr fest. Braunaderig.

Nutzung: Hammerstiele, Dreschflegel, Spazierstöcke, Drechslerarbeiten.

Schön blühend, aber wegen seiner Anfälligkeit für Schadinsekten und als Wirt oder Zwischenwirt für Pilzkrankheiten der Blätter mit Zurückhaltung in Kultur. Regelmäßig Nistbaum des Würgers (butcherbird), der seine Brut und sich auf diese Weise schützt und seine Beute auf die Dornen spießt.

Description: arbuste de 8 à 10 m de hauteur, avec un tronc court et puissant de 0,25 à 0,3 m de diamètre, armé comme les branches d'épines droites et piquantes. Frui rouges, quelquefois jaunes, ponctués de taches claires.

Habitat: aussi bien sur sols secs à l'étage collinéen que dans les lits de hautes eaux dans le nord-est de l'Amérique du Nord. Espèce peu exigeante.

Bois: très dur, tenace et très résistant. Veiné de brun.

Intérêts: utilisé pour fabriquer des manches de marteaux, des fléaux et en tournerie.

Belle floraison, mais sa culture est pratiquée avec circonspection en raison de sa réceptivité aux parasites, insectes ou champignon: qui attaquent son feuillage. La piegrièche y élit souvent domicile car ses épines protègent sa couvée des prédateurs et lui servent à

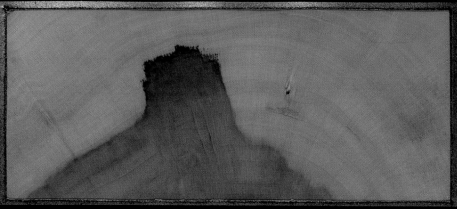

TRANSVERSE SECTION.

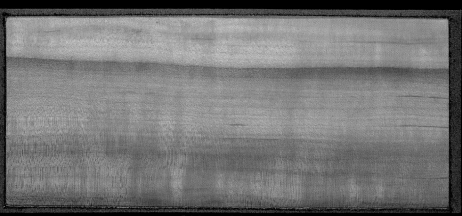

RADIAL SECTION.

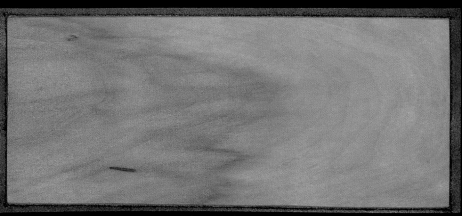

TANGENTIAL SECTION

Ger. Geflecte Mispel.   *Sp.* Espino puntuado.

Fr. Néflier à fruits pointillés.

R | ROSE FAMILY
Hollyleaf Cherry, Islay

FAMILIE DER ROSENGEWÄCHSE
Ilex-Kirschlorbeer

FAMILLE DE LA ROSE
Cerisier à feuilles de houx

ROSACEAE
Prunus ilicifolia (Nutt. ex Hook. & Arn.) D. Dietr.

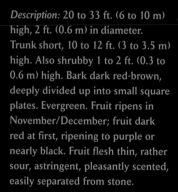

Description: 20 to 33 ft. (6 to 10 m) high, 2 ft. (0.6 m) in diameter. Trunk short, 10 to 12 ft. (3 to 3.5 m) high. Also shrubby 1 to 2 ft. (0.3 to 0.6 m) high. Bark dark red-brown, deeply divided up into small square plates. Evergreen. Fruit ripens in November/December; fruit dark red at first, ripening to purple or nearly black. Fruit flesh thin, rather sour, astringent, pleasantly scented, easily separated from stone.

Habitat: As tree on moist sandy soils near streams on canyon floor. As shrub on dry hills and plateaus. Widespread in the coastal regions from San Francisco to Lower California. Best developed near Santa Barbara, on the islands and in Lower California.

Wood: Heavy, hard, strong, silky-sheened, and close-grained; polishable. Heartwood pale red-brown; sapwood thin and paler.

Use: For cupboards ans well as firewood.

Ornamental and used as hedge plants in California and southern Europe.

Beschreibung: 6 bis 10 m hoch bei 0,6 m Durchmesser. Kurzer Stamm von 3 bis 3,5 m Höhe. Auch als Strauch, 0,3 bis 0,6 m hoch. Borke dunkel rotbraun, tief geteilt in kleine quadratische Platten. Immergrün. Fruchtreife im November/Dezember, Frucht anfangs dunkelrot, bei der Reife purpurn bis nahezu schwarz. Dünnes Fruchtfleisch, leicht säuerlich, adstringierend, mit angenehmem Aroma, leicht vom Stein zu lösen.

Vorkommen: Als Baum auf feuchtem Sandboden in der Nähe von Bächen am Grunde von Cañons. Als Strauch an trockenen Hügeln und auf Tafelland. Verbreitet in den Küstenregionen von San Francisco bis Niederkalifornien. Größte Ausformung nahe Santa Barbara, auf den Inseln und in Niederkalifornien.

Holz: Schwer, hart, fest, seidig und mit feiner Textur; polierfähig. Kern hell rotbraun; Splint dünn und heller.

Nutzung: Für Schränke, auch als Brennholz.

Ornamental und als Hecke in Kalifornien und Südeuropa.

Description : 6 à 10 m de hauteur sur un diamètre de 0,6 m. Tronc court, entre 3 et 3,5 m de haut. Croît aussi sous forme de buisson de 0,3 à 0,6 m de hauteur. Ecorce brun rouge foncé, à fissures profondes et petites plaques carrées. Espèce sempervirente. Fruits mûr. en novembre et décembre ; rouge foncé presque noir à maturité ; chair peu fournie, légèrement acidulée et astringente, aromatique e se détache facilement du noyau

Habitat : dans les sols sablonneux humides près des ruisseaux de canyons ; sous forme buissonnante, sur les collines sèches et le plateaux ; régions côtières de San Francisco à la Basse-Californie. Le plus grands spécimens se trouven près de Santa Barbara, sur les îles et en Basse-Californie.

Bois : lourd, dur, résistant, satiné et à grain fin ; pouvant se polir. Bois de cœur brun rouge clair ; aubier mince et plus clair.

Intérêts : fabrication d'armoires, et bois de chauffage.

Ornemental et planté en haies en Californie et en Europe méridionale.

Islay, Holly-leaved Cherry, Wild or Evergreen Cherry.

TRANSVERSE SECTION.

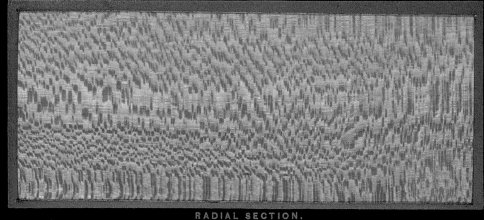

RADIAL SECTION.

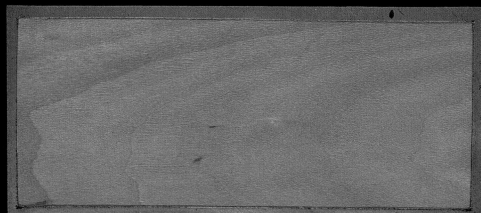

TANGENTIAL SECTION

Ger. Stechpalmenblättrige Kirsche.

Fr. Cerisier à feuilles de Houx. *Sp.* Cerezo de hojas de Aceb

R | ROSE FAMILY
Holmes Hawthorn, Scarlet Hawthorn

FAMILIE DER ROSENGEWÄCHSE
Scharlach-Weißdorn

FAMILLE DE LA ROSE
Aubépine de Holmes

ROSACEAE
Crataegus holmesiana Ashe
Syn.: C. coccinea L.

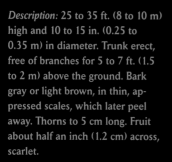

Description: 25 to 35 ft. (8 to 10 m) high and 10 to 15 in. (0.25 to 0.35 m) in diameter. Trunk erect, free of branches for 5 to 7 ft. (1.5 to 2 m) above the ground. Bark gray or light brown, in thin, appressed scales, which later peel away. Thorns to 5 cm long. Fruit about half an inch (1.2 cm) across, scarlet.

Habitat: Fairly common on well-drained slopes and on higher ground, swamp margins, etc., in north-eastern North America.

Wood: Heavy, hard, and very close-grained.

Use: Good for turnery.

Beschreibung: 8 bis 10 m hoch bei 0,25 bis 0,35 m Durchmesser. Richtiger Stamm erst ab 1,5 bis 2 m mit Ästen. Borke grau oder hellbraun, in dünnen, angedrückten Schuppen, die später abblättern. Dornen bis 5 cm lang. Frucht etwa 1,2 cm dick, scharlachrot.

Vorkommen: Weit verbreitet an gut drainierten Hängen und auf höher gelegenem Land, an den Rändern von Sümpfen etc., im nordöstlichen Nordamerika.

Holz: Schwer, hart, mit sehr feiner Textur.

Nutzung: Gut für Drechslerarbeiten.

Description : 8 à 10 m de hauteur sur un diamètre de 0,25 à 0,35 m. Véritable tronc à partir de 1,5 à 2 m avec ramifications. Ecorce grise ou brun clair, développant de minces écailles qui se détachent avec l'âge. Epines jusqu'à 5 cm de long. Fruits écarlates de 1,2 cm environ de diamètre.

Habitat : assez courant sur les pentes bien drainées et sur les terrains situés en hauteur, bords des marais etc. dans le nord-est de l'Amérique du Nord.

Bois : lourd, dur et à grain très fin.

Intérêts : convient bien à la tournerie.

Scarlet Thorn, White Thorn, Red Haw.

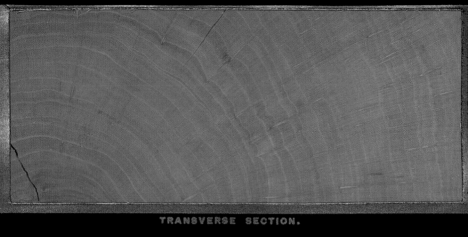

TRANSVERSE SECTION.

RADIAL SECTION.

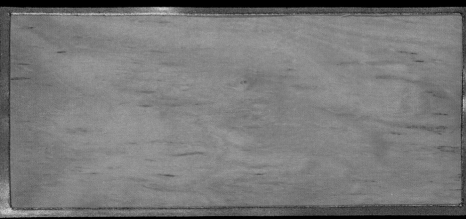

TANGENTIAL SECTION.

Ger. Scharlachfrüchtiger Weissdorn.   Fr. Néflier ecarlate.

Sp. Espino colorado.

R | ROSE FAMILY
Morello Cherry, Sour Cherry, Pie Cherry

FAMILIE DER ROSENGEWÄCHSE
Sauer-Kirsche

FAMILLE DE LA ROSE
Griottier, Cerisier commun

ROSACEAE
Prunus cerasus L.

Description: Rarely more than 20 to 33 ft. (6 to 10 m) high and 8 to 10 in. (0.2 to 0.24 m) in diameter. Crown spreading, generally without a central leading shoot. Bark of young trunks distinctly thin but, unlike the related Sweet Cherry, not remaining that way permanently but breaking up as it grows older and peeling away in thin, curved scales. Underneath is a rough, furrowed layer. Leaves smaller, stiffer, and more erect than those of the Sweet Cherry. Flowers white, approx. 1 inch (2 cm) across, appearing before or at the same time as the leaves, which sprout in small groups from the axillary buds of the previous season. Fruits globular or slightly flattened, red.

Habitat: Forests of northern Iran and Caucasus, but distributed throughout the greater part of Europe, also in India and North Africa. Introduced into North America, where it has since escaped from cultivation and now grows wild.

Wood: Fairly light, hard; liable to chip when cut. Heartwood light brown; sapwood paler.

Use: Of no commercial value in the USA. Many varieties in cultivation.

Beschreibung: Selten höher als 6 bis 10 m bei 0,2 bis 0,24 m Durchmesser. Ausgebreitete Krone im Allgemeinen ohne führenden Mitteltrieb. Borke junger Stämme äußerst dünn, aber im Gegensatz zur verwandten Süß-Kirsche nicht dauerhaft so bleibend, sondern im Alter aufbrechend und in dünnen, gebogenen Schuppen abblätternd. Darunter eine rau-geriefte Schicht. Blätter kleiner, steifer und aufgerichteter als bei der Süß-Kirsche. Blüten weiß, ca. 2 cm groß, erscheinen zu wenigen vor oder mit den austreibenden Blättern aus Achselknospen der vorigen Saison. Früchte kugelig oder wenig abgeflacht, rot.

Vorkommen: Heimat sind die Wälder Nordpersiens und Kaukasiens, aber im größten Teil Europas verbreitet, ebenso in Indien und Nordafrika. Nach Nordamerika eingeführt, als Kulturflüchtling verbreitet.

Holz: Ziemlich leicht, hart; bröckelig. Kern hellbraun; Splint heller.

Nutzung: Trotz guter Eigenschaften ohne Handelswert in den USA. Zahlreiche Sorten in Kultur.

Description : rarement plus de 6 à 10 m de hauteur sur un diamètre de 0,2 à 0,24 m. Houppier ample d'ordinaire sans ramification principale. Chez les jeunes arbres l'écorce est très fine, mais à la différence du merisier, elle change d'aspect avec l'âge, se fragmentant et se desquamant en minces plaquettes recourbées qui font apparaître une couche rugueuse sous-jacente. Feuilles plus petites, plus raides et plus dressées que celles du merisier. Fleurs blanches (2 cm de diamètre environ) avant ou pendant la feuillaison, issues de quelques bourgeons axillaires de saison précédente. Fruits globuleux ou peu aplatis, rouges.

Habitat : Aire d'origine : forêts du nord de l'Iran et du Caucase mais répandu dans presque toute l'Europe ainsi qu'en Inde et en Afrique. Introduit en Amérique, il s'est répandu sous une forme échappée des jardins.

Bois : assez léger, dur, qui tend à se fendre. Bois de cœur brun clair aubier plus clair.

Intérêts : Aucun intérêt économique aux Etats-Unis malgré ses qualités. Nombreuses variétés cultivées.

Sour Cherry. Garden Cherry.

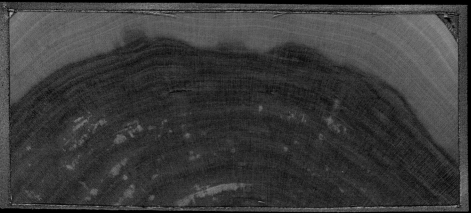

TRANSVERSE SECTION.

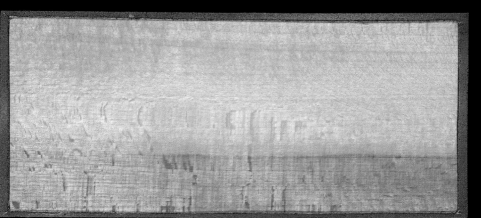

RADIAL SECTION.

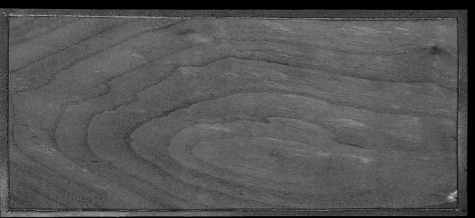

TANGENTIAL SECTION.

Gemeiner Kirschbaum.   Fr. Cerise aigre.   Sp. Cereza agr

R | ROSE FAMILY
Mountain Mahogany

FAMILIE DER ROSENGEWÄCHSE
Berg-Mahagonie

FAMILLE DE LA ROSE
Cercocarpe de montagne

ROSACEAE
Cercocarpus montanus Raf.
Syn.: *C. parvifolius* Nutt.

Description: 25 to 35 ft. (8 to 10 m) high and 8 in. (0.2 m) in diameter. Bushy tree. Bark normally smooth, divided by narrow, shallow furrows and broken up into small, square red-brown scales. Leaves and twigs aromatic. Flowers with long, white-haired, persistent styles.

Habitat: Dry mountain sites in western North America.

Wood: Heavy, hard, and close-grained; difficult to store and work. Heartwood pale red-brown; sapwood thin, pale brown.

Use: Excellent firewood. Sometimes for small-scale turnery.

Food source for cattle, when the grass is desiccated.

Beschreibung: 8 bis 10 m hoch bei 0,2 m Durchmesser. Buschiger Baum. Borke allgemein glatt, geteilt durch enge, flache Risse und in kleine, quadratische, rotbraune Schuppen gebrochen. Blätter und Zweige aromatisch. Blüten mit weiß behaarten, langen, ausdauernden Griffeln.

Vorkommen: Trockene Gebirgszüge im westlichen Nordamerika.

Holz: Schwer, hart und mit feiner Textur; schwierig zu lagern und zu bearbeiten. Kern hell rotbraun; Splint dünn, hellbraun.

Nutzung: Ausgezeichnetes Brennholz. Manchmal für kleine Drechslerarbeiten.

Dient als Nahrungsquelle, Weideplatz für Rinder, wenn die Grasbestände verdorrt sind.

Description: arbre au port touffu de 8 à 10 m de hauteur sur un diamètre de 0,2 m. Ecorce généralement lisse, parcourue de fissures étroites et plates et craquelée en petites écailles carrées, brun roug Feuilles et rameaux odorants. Fleurs aux longs styles duvetés de blanc et persistants.

Habitat: cordillères sèches dans l'ouest de l'Amérique du Nord.

Bois: lourd, dur et à grain fin; di ficile à stocker et à travailler. Bois de cœur brun rouge clair; aubier mince, brun clair.

Intérêts: excellent bois de chauffage. Utilisé quelquefois pour de petits ouvrages de tournerie.

Sert de nourriture et de pacage aux bovins lorsque l'herbe est desséchée.

Mountain Mahogany.

TRANSVERSE SECTION.

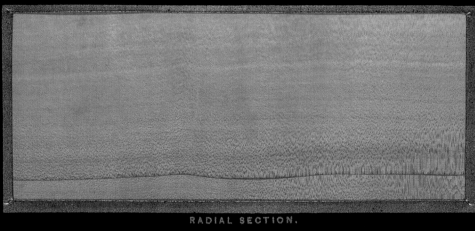

RADIAL SECTION.

TANGENTIAL SECTION

Ger. Gebirgs Mahogany.   *Fr.* Buisson à plumes.

Sp. Caoba de montaña.

R | ROSE FAMILY
Oregon Crab Apple

FAMILIE DER ROSENGEWÄCHSE
Oregon-Holzapfel

FAMILLE DE LA ROSE
Pommier du Pacifique, Pommier de l'Oregon

ROSACEAE
Malus fusca (Raf.) C. K. Schneid.
Syn.: *Pyrus rivularis* Dougl. ex Hook.

Description: 30 to 40 ft. (9 to 12 m) high and 12 to 18 in. (0.3 to 0.45 m) in diameter; often also as many-branched shrub. Bark thin, with thin pale red-brown scales on the surface. Leaves ovate-lanceolate, serrated, often three-lobed; dark green above, hairy beneath. Flowers in clusters on long, thin stalks. Fruit yellow-green or pale yellow with red marks, oblong-ovoid with thin, dry flesh and a pleasant, rather sour flavor.

Habitat: Usually on deep, rich soils near rivers, often forming rather extensive, impenetrable thickets. Coastal states of Pacific North America.

Wood: Heavy, hard, silky-sheened, and very close-grained; polishable. Heartwood pale brown with red tinge; sapwood thick (25 to 30 annual rings), pale. Many obvious medullary rays.

Use: Mallets, tool handles, machine mountings. Fruits collected and consumed by American Indians.

Beschreibung: 9 bis 12 m hoch bei 0,3 bis 0,45 m Durchmesser, oft auch als vielstämmiger Strauch. Borke dünn, an der Oberfläche mit dünnen, hell rotbraunen Schüppchen. Blätter oval-lanzettlich, gesägt, oft dreilappig, Oberseite dunkelgrün, Unterseite behaart. Blüten in Büscheln auf langen dünnen Stielen. Frucht gelbgrün oder hellgelb mit rotem Fleck, oval-länglich mit dünnem, trockenem Fleisch, von angenehm säuerlichem Geschmack.

Vorkommen: Gewöhnlich in tiefgründigen, reichen Böden an Flüssen, oft undurchdringliche Dickichte erheblichen Ausmaßes bildend. Küstenstaaten im pazifischen Nordamerika.

Holz: Schwer, hart, seidig und mit sehr feiner Textur; polierfähig. Kern hellbraun mit roter Tönung; Splint dick (25 bis 30 Jahresringe), heller. Zahlreiche deutliche Markstrahlen.

Nutzung: Holzhämmer, Werkzeuggriffe, Halterungen für Maschinen. Die Früchte werden von den Indianern gesammelt und verzehrt.

Description : 9 à 12 m de hauteur sur un diamètre de 0,3 à 0,45 m, souvent aussi sous forme de buisson à tiges nombreuses. Ecorce mince, recouverte de fines écailles brun rouge clair. Feuilles ovales-elliptiques, dentées, souver trilobées, vert foncé dessus, pubes centes au revers. Fleurs disposées en touffes sur de longs pédoncule Fruits vert jaune ou jaune clair taché de rouge, ovoïdes ; chair sèche, agréablement acide.

Habitat : d'ordinaire dans des sols riches et profonds près des cours d'eau, souvent en vastes taillis impénétrables. Croît dans les Etat sur la côte pacifique de l'Amériqu du Nord.

Bois : lourd, dur, satiné et à grain très fin, susceptible d'être poli. Bois de cœur brun clair, teinté de rouge ; aubier épais (25 à 30 cern annuels), plus clair. Rayons médu laires nombreux et distincts.

Intérêts : utilisé pour la fabrication de maillets, de manches d'outils e de supports de machines. Les frui sont consommés par les Indiens.

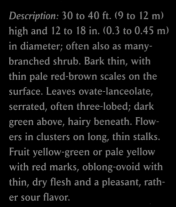

Oregon Crab or Crab Apple.

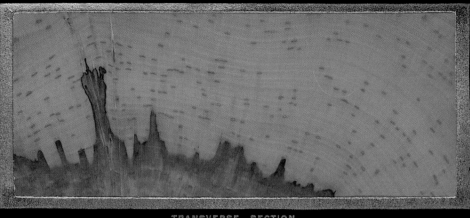

TRANSVERSE SECTION.

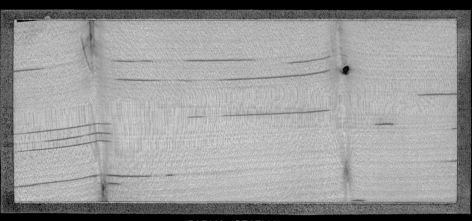

RADIAL SECTION.

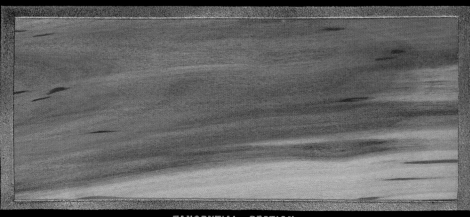

TANGENTIAL SECTION.

Ger. Oregonischer Holzapfel. *Fr.* Pommier sauvage d'Oregon.

Sp. Manzano silvestre de Oregon.

R | ROSE FAMILY
Pin Cherry, Wild Red Cherry, Pigeon Cherry, Bird Cherry

FAMILIE DER ROSENGEWÄCHSE
Feuer-Kirsche

FAMILLE DE LA ROSE
Cerise de Pennsylvanie, Arbre à petites merises,
Petit merisier (Canada)

ROSACEAE
Prunus pensylvanica L.f.

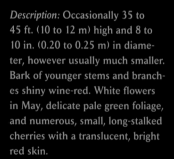

Description: Occasionally 35 to 45 ft. (10 to 12 m) high and 8 to 10 in. (0.20 to 0.25 m) in diameter, however usually much smaller. Bark of younger stems and branches shiny wine-red. White flowers in May, delicate pale green foliage, and numerous, small, long-stalked cherries with a translucent, bright red skin.

Habitat: Dry sandy soils in northeastern North America. In the south of its range primarily along the mountains (Alleghenies). Also extends into western North America in British Columbia on the slopes of the Rocky Mountains.

Wood: Fairly light, soft and closegrained.

Use: Of little commercial use.

Disseminated in vast numbers by birds and also found on tracts of land cleared by forest fires. In these areas it provides shade and shelter for other, more tender species. As these saplings become larger, the pioneer trees die off.

Beschreibung: Gelegentlich 10 bis 12 m hoch bei 0,20 bis 0,25 m Durchmesser, jedoch gewöhnlich viel kleiner. Borke jüngerer Stämme und Äste glänzend weinrot. Weiße Blüten im Mai, zartes, hellgrünes Laub und zahllose durchscheinend hellrote, lang gestielte, kleine Kirschen.

Vorkommen: Trockene Sandböden im nordöstlichen Nordamerika. Im Süden des Verbreitungsgebietes vorwiegend entlang der Gebirge (Alleghenies). In Britisch-Kolumbien an den Hängen der Rocky Mountains auch das westliche Nordamerika erreichend.

Holz: Ziemlich leicht, weich und von sehr feiner Textur.

Nutzung: Von geringem Handelswert.

Durch von Vögeln verstreute Samen selbst auf durch Waldbrände verwüsteten freien Flächen in großer Zahl angesiedelt. Dort als Schatten- und Schutzbaum für andere, zartere, aufkommende Bäume von großem Nutzen. Wenn die Schützlinge größer werden, sterben die Pioniere ab.

Description : atteint parfois 10 à 12 m de hauteur sur un diamètre de 0,20 à 0,25 m, mais d'ordinaire de taille bien plus modeste. Les tiges et les rameaux juvéniles ont une écorce vernissée rouge foncé. Fleurs blanches en mai, feuilles ve clair et innombrables petits fruits rouge clair à chair translucide, por tés par un long pédoncule.

Habitat : sols secs sablonneux du nord-est de l'Amérique du Nord. Au sud de la région de propagatio en majorité le long des montagnes (Alleghanys). En Colombie-Britannique, sur les flancs des Rocheuse et jusqu'à l'ouest de l'Amérique du Nord.

Bois : assez léger, tendre et à grain très fin.

Intérêts : peu de valeur économiqu

Croît en grandes formations grâce aux graines disséminées par les oiseaux et colonise les espaces ouverts, dévastés par le feu. Ici, espèce pionnière, protectrice d'arbres de plus grande utilité. Ma avec la croissance des essences suivantes, les cerisiers de Pennsylvani s'étiolent et finissent par mourir

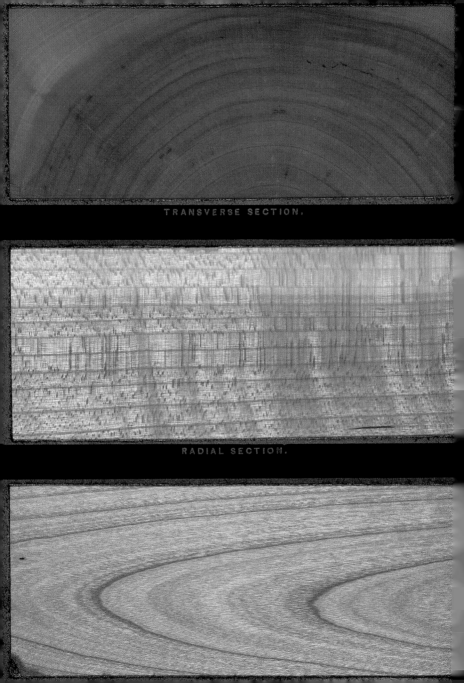

TRANSVERSE SECTION.

RADIAL SECTION.

R | ROSE FAMILY
Serviceberry, Shad Bush, Shad-blow

FAMILIE DER ROSENGEWÄCHSE
Echte Kanadische Felsenbirne

FAMILLE DE LA ROSE
Amélanchier du Canada, Petite poire (Canada)

ROSACEAE
Amelanchier canadensis (L.) Medik.

Description: Occasionally 40 to 50 ft. (12 to 15 m) high and 8 in. to 2 ft. (0.25 to 0.6 m) in diameter. A "white cloud" when in blossom. The name "shad bush" alludes to its coming into blossom at the same time as the shad migrates up the rivers from the sea.

Habitat: Well-drained slopes and uplands in association with American Aspen, Hemlock, White and Red Oak, Sugar Maple, Nettle Tree, etc., in north-eastern North America.

Wood: Heavy, hard, very strong, and close-grained.

Use: Valuable for turnery, for tool handles, and under the name of "lancewood" for fishing rods.

A species often confused in Germany with, for example, the Juneberry or Snowy Mespil (*Amelanchier lamarckii* Schroeder). Many varieties with transitional forms.

Beschreibung: Gelegentlich 12 bis 15 m hoch bei 0,25 bis 0,6 m Durchmesser. Zur Blütezeit eine „weiße Wolke". Blütezeit, wenn der Maifisch (shad) vom Meer her die Flüsse hinaufwandert (daher der engl. Name).

Vorkommen: Gut drainierte Hänge und höhere Lagen in Gesellschaft von Zitter-Pappel, Hemlock, Weiß- und Rot-Eiche, Zucker-Ahorn, Zürgelbaum, u. a. im nordöstlichen Nordamerika.

Holz: Schwer, hart, sehr fest und von feiner Textur.

Nutzung: Wertvoll für Drechslerarbeiten, für Werkzeuggriffe und unter dem Namen „lance-wood" für Angelruten.

Die Art wird in Deutschland oft verwechselt, z. B. mit der Kupfer-Felsenbirne (*Amelanchier lamarckii* Schroeder). Große Formenvielfalt mit Hybriden und Übergangsformen.

Description : arbre atteignant quelquefois 12 à 15 m de hauteur sur un diamètre de 0,25 à 0,6 m. Véritable «nuée blanche» au moment de la floraison. Floraison au moment où l'alose remonte les rivières pour frayer (d'où son nom anglais «shad-bush» = buisson à alose).

Habitat : pentes bien drainées et hauteurs dans le nord-est de l'Amérique du Nord, en association avec le peuplier-tremble, la pruche les chênes blanc et rouge, l'érable à sucre, le micocoulier etc.

Bois : lourd, dur, très résistant et à grain fin.

Intérêts : recherché en tournerie et pour la fabrication de manches d'outils. Connu sous le nom de «lance-wood» pour les cannes à pêche.

Souvent confondu en Allemagne avec une espèce voisine, *Amelanchier lamarckii* Schroeder, l'amélanchier des rochers. Grande diversité d'hybrides et de formes intermédiaires.

June-berry, Service-tree, Shad-bush, Shad-blow.

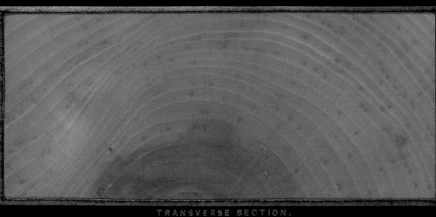

TRANSVERSE SECTION.

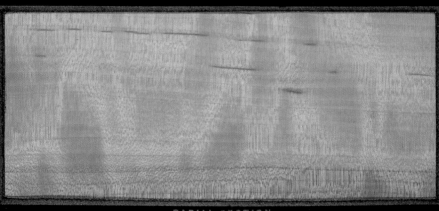

RADIAL SECTION.

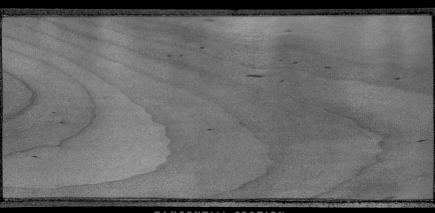

TANGENTIAL SECTION

Ger. Gewohnliche Traubenbirne.   *Sp.* Nispero.

Fr. Grand Amelanchier.

R | ROSE FAMILY
Sweet Cherry, Gean, Wild Cherry, Mazzard

FAMILIE DER ROSENGEWÄCHSE
Süß-Kirsche

FAMILLE DE LA ROSE
Merisier, Cerisier des oiseaux

ROSACEAE
Prunus avium L.

Description: 55 to 85 ft. (16 to 25 m) high and a maximum of 2 to 3 ft. (0.6 to 0.9 m) in diameter. Bark at first with reddish brown horizontal lines, changing to horizontal rows of scales as the tree grows older.

Habitat: From its original home on the Caspian Sea to southern Europe. From there it came with the immigrants first to the east of North America and later spread to the whole of the subcontinent, sometimes growing wild. Extends to uplands, preferring limy soils.

Wood: Quite hard and close-grained; easy to work and similar to other cherry tree species. Reddish brown and often striped or mottled.

Use: For fine cabinet-work and turnery. The naturally exuded gum resin was of importance in earlier times. Many cultivars planted in gardens for their fruit. This fruit is distilled to make kirsch. Many varieties are grown as ornamental trees for their blossom.

The commonest of the cherries introduced and established in North America. Related to the sour Morello Cherry, but much larger.

Beschreibung: 16 bis 25 m hoch bei maximal 0,6 bis 0,9 m Durchmesser. Borke anfangs in rotbraunen Querstreifen, im Alter Übergang zu schuppigen Querreihen.

Vorkommen: Von der Heimat am Kaspischen Meer bis nach Südeuropa. Von dort mit den Einwanderern zunächst in den Osten Nordamerikas und später über den gesamten Subkontinent hin verbreitet, z. T. verwildernd. Gedeiht auch in Berglagen, liebt kalkhaltigen Boden.

Holz: Mittelhart und von feiner Textur; gut zu bearbeiten und mit dem aller Kirschbaumarten übereinstimmend. Rötlich braun und oft gestreift oder geflammt.

Nutzung: Für feine Kunstschreiner- und Drechslerarbeiten. Das Kirschgummiharz war früher von Bedeutung. Meist wegen des Obstes in vielen Kultursorten als Gartenbaum gepflanzt. Das Obst wird zu Kirschwasser gebrannt. Aufgrund der Blüten viele Varietäten als Zierbäume.

Die häufigste der nach Nordamerika eingeführten und dort eingebürgerten Kirschen. Mit der Sauer-Kirsche verwandt, aber viel größer.

Description : 16 à 25 m de hauteur sur un diamètre de 0,6 à 0,9 m maximum. Ecorce à lenticelles formant des stries transversales brun rouge, écailleuses avec l'âge.

Habitat : originaire, de la mer Caspienne, ensuite acclimaté dans le sud de l'Europe. Puis avec l'émigration des populations on le trouve à l'est de l'Amérique du Nord et plus tard sur tout le subcontinent en partie à l'état sauvage. Croit également en zones montagneuses affectionne les sols calcaires.

Bois : moyennement dur et à grain fin ; se travaillant bien, et similaire à celui des autres espèces de cerisier. Brun rougeâtre, souvent rubanné ou flammé.

Intérêts : utilisé en ébénisterie et en tournage. La résine était recherchée autrefois. Aujourd'hui, cultivé sous de multiples formes horticoles essentiellement pour ses fruits utilisés pour le kirsch. Il existe par ailleurs de nombreux cultivars ornementaux aux fleurs attrayantes.

Le plus fréquent des cerisiers naturalisés en Amérique du Nord. Proche parent du grottier, il donne des cerises beaucoup plus grosses

Ox-heart Cherry, English Cherry.

TRANSVERSE SECTION.

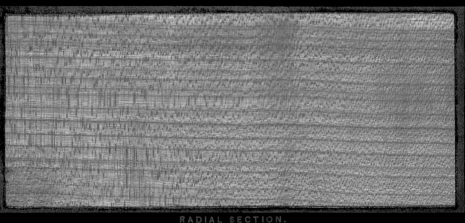

RADIAL SECTION.

TANGENTIAL SECTION

Ger. Susser Kirschbaum. Sp. Cerezo.

Fr. Mérisier.

R | ROSE FAMILY
Sweet Crab Apple, Garland Tree

FAMILIE DER ROSENGEWÄCHSE
Kronen-Apfel

FAMILLE DE LA ROSE
Pommier odorant

ROSACEAE
Malus coronaria (L.) Mill.
Syn.: *Pyrus coronaria* L.

Description: 25 to 33 ft. (8 to 10 m) high and 8 to 12 in. (0.2 to 0.3 m) in diameter. Bark rough, with longitudinal fissures and scales. Flowers pink, very fragrant. Crab apples pale green, also fragrant and decorative.

Habitat: Rich, moist, well-drained soils, often in forest clearings, surrounded by taller trees. Interior of eastern North America

Wood: Heavy and close-grained.

Use: For turnery and tool handles. The fruits are sometimes made into a preserve or used to produce cider and vinegar.

Beschreibung: 8 bis 10 m hoch bei 0,2 bis 0,3 m Durchmesser. Borke rau, längsrissig-schuppig. Sehr stark angenehm duftende rosafarbene Blüten. Blassgrüne Holzäpfel gleichfalls duftend und zierend.

Vorkommen: Reiche, feuchte, gut drainierte Böden, oft auf Waldlichtungen, von höheren Bäumen umgeben. Im Inneren des östlichen Nordamerika.

Holz: Schwer und mit sehr feiner Textur.

Nutzung: Für Drechslerarbeiten und Werkzeuggriffe. Die Früchte werden manchmal zum Einkochen und für Cider und Obstessig verwendet.

Description: 8 à 10 m de hauteur sur un diamètre de 0,2 à 0,3 m. Ecorce rugueuse, parcourue de fissures écailleuses longitudinales. Fleurs roses, très odorantes. Fruit vert pâle, à la fois odorants et décoratifs.

Habitat: sols riches, humides et bien drainés ; souvent dans les clairières, entourés d'arbres plus grands. On le trouve dans les régions intérieures est de l'Amérique du Nord.

Bois: lourd et à grain très fin.

Intérêts: utilisé en tournage et po fabriquer des manches d'outils. L fruits entrent quelquefois dans la fabrication de conserves, de cidre et de vinaigre.

Wild Crab, Sweet-scented Crab.

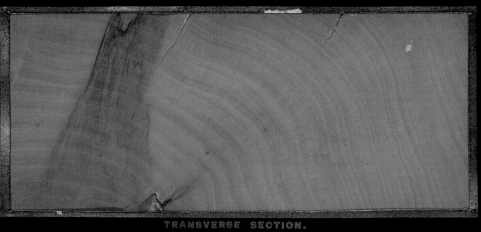

TRANSVERSE SECTION.

RADIAL SECTION.

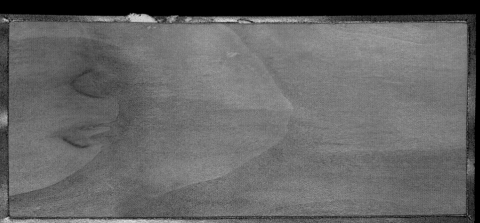

TANGENTIAL SECTION.

Ger. Amerikanischer Kirschapfelbaum.

Fr. Pome sauvage de l'Amerique.   *Sp.* Manzano Americano

R | ROSE FAMILY
Tollon, Toyon

FAMILIE DER ROSENGEWÄCHSE
–

FAMILLE DE LA ROSE
Hétéromèle de Californie

ROSACEAE
Heteromeles arbutifolia (Lindl.) M. Roem.

Description: 33 ft. (10 m) high and 10 to 15 in. (0.25 to 0.36 m) in diameter. Trunk straight, dividing from close to the base into upright trunks, but more commonly small, much-branched shrub. Bark pale gray, smooth. Surface has a reticulate pattern of inconspicuous ridges.

Habitat: Dry hills close to streams, especially on the north sides in California. Clambers up steep, exposed coastal cliffs.

Wood: Very heavy, hard, close-grained. Dark red-brown; sapwood thin and pale.

Use: Twigs, with scarlet fruits and shiny dark green foliage used as Christmas decoration in California.

Beschreibung: 10 m hoch bei 0,24 bis 0,36 m Durchmesser. Gerader Stamm, nahe dem Boden in zahlreiche aufrechte Teilstämme aufgehend, noch häufiger kleiner, vielverzweigter Strauch. Borke hellgrau, allgemein glatt. Oberfläche durch unauffällige Riefen netzförmig zerlegt.

Vorkommen: Trockene Hügel nahe an Bächen, besonders auf Nordseiten in Kalifornien. An der Küste steile Klippen erklimmend, dem Ozean voll ausgesetzt.

Holz: Sehr schwer, hart, mit feiner Textur. Von dunkler rotbrauner Farbe, Splint dünn und heller.

Nutzung: Die mit scharlachroten Früchten und glänzend dunkelgrünem Laub besetzten Zweige dienen in Kalifornien als Weihnachtsschmuck.

Description : petit arbre de 10 m de hauteur, avec un tronc droit d 0,24 à 0,36 m de diamètre, divisé près du sol en de nombreuses tig secondaires droites ; se présente plus souvent encore sous forme de petit buisson très ramifié. Écor gris clair, lisse en général ; surface à texture entrelacée et craquelée.

Habitat : collines sèches à proximi des torrents, en particulier sur les versants d'ubac en Californie. Écueils escarpés, entièrement exp sés à l'action de l'océan.

Bois : très lourd, dur et à grain fin Bois de cœur brun rouge sombre, aubier mince et plus clair.

Intérêts : les rameaux enjolivés de fruits écarlates et de feuilles vert foncé brillant servent de décoratic de Noël en Californie.

TRANSVERSE SECTION.

RADIAL SECTION.

TANGENTIAL SECTION.

R | ROSE FAMILY
Westindian Cherry

FAMILIE DER ROSENGEWÄCHSE
Westindische Kirsche

FAMILLE DE LA ROSE
Cerisier à petites feuilles

ROSACEAE
Prunus myrtifolia (L.) Urban

Description: This slender tree reaches a height of between 35 and 40 ft. (10 to 12 m), and has a trunk 6 in. (0.15 m) in diameter. The bark is very thin, light brown with a reddish tinge. The long-lasting, elliptic leaves are light green on the upperside. The underside is paler. The white flowers are in erect clusters. The roundish fruit is orange-brown in color and has thin flesh.

Habitat: On hills with rich soils and on the banks of small streams and ponds in Florida.

Wood: The light red-brown heartwood is heavy and hard. It is surrounded by only a thin layer of paler sapwood.

Use: The tree does not appear to be used in North America. On the islands of the West Indies the fruits are used in the preparation of a tonic.

Beschreibung: Der schlanke Baum erreicht eine Höhe von 10 bis 12 m und besitzt einen 0,15 m dicken Stamm. Die Borke ist sehr dünn und hellbraun mit rötlicher Tönung. Die festen elliptischen Blätter besitzen eine hellgrüne Oberseite. Die Unterseite ist blasser. Die weißen Blüten erscheinen in aufrechten Trauben. Die dünnfleischige, rundliche Frucht ist orangebraun.

Vorkommen: An Hügeln auf reichen Böden und an den Ufern kleiner Bäche und Teiche Floridas.

Holz: Das hell rotbraune Kernholz ist schwer und hart. Es wird von nur wenig hellerem Saftholz umgeben.

Nutzung: Der Baum scheint in Nordamerika nicht genutzt zu werden. Auf den Westindischen Inseln werden die Früchte zur Zubereitung eines Stärkungsmittels

Description : arbre à port élancé atteignant 10 à 12 m de hauteur e possédant un tronc de 0,15 m de diamètre. L'écorce est très mince, brun clair teinté de rouge. Les feuilles persistantes sont elliptique vert clair à la face supérieure, plus pâles à la face inférieure. Les fleur blanches sont réunies en grappes dressées. Le fruit est globuleux, brun orange et peu charnu.

Habitat : sols riches à l'étage collinéen et rives des ruisseaux et des étangs en Floride.

Bois : lourd et dur. Bois de cœur brun rouge clair ; aubier mince, plus clair.

Intérêts : l'arbre semble ne pas avoir trouvé d'usage en Amérique du Nord. Les fruits entrent dans la composition d'un fortifiant aux Antilles.

TRANSVERSE SECTION.

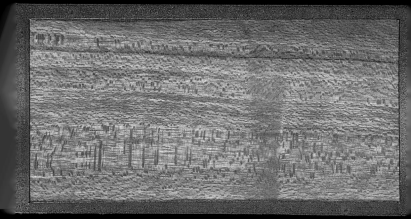

RADIAL SECTION.

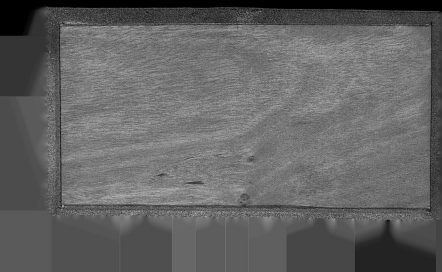

R | ROSE FAMILY
Wild Black Cherry, Rum Cherry

FAMILIE DER ROSENGEWÄCHSE
Späte Traubenkirsche

FAMILLE DE LA ROSE
Cerisier tardif, Cerisier d'automne (Canada)

ROSACEAE
Prunus serotina Ehrh.

Description: 85 to 100 ft. (25 to 30 m) high and 4 to 5 ft. (1.2 to 1.5 m) in diameter. Straight, columnar trunk. One of the most valuable American forest trees. White flowers, appearing later than other species in the genus, form ornamental hanging clusters. The fruits, black when fully ripe, smell pleasantly of wine.

Habitat: Both on dry sandy and gravel soils and on damp soils. A main constituent of the Appalachian forests. Eastern North America.

Wood: Fairly hard, strong, and very close-grained; does not shrink or warp.

Use: One of the most valuable timbers for furniture and veneers. Still exported today as Black Cherry. Fruits used in the production of kirsch and rum. The fragrant bark is used as a tonic and sedative.

Nowadays naturalized on sandy soils in northern Germany and fruiting early there.

Beschreibung: 25 bis 30 m hoch bei 1,2 bis 1,5 m Durchmesser. Gerader, säulenförmiger Stamm. Einer der wertvollsten amerikanischen Waldbäume. Weiße Blüten später als sonst in der Gattung in ornamentalen, hängenden Trauben. Die Früchte, schwarz wenn vollreif, duften angenehm weinartig.

Vorkommen: Sowohl trockene Sand- und Kiesböden als auch feuchte Böden. Ein Hauptelement der Appalachenwälder. Östliches Nordamerika.

Holz: Ziemlich hart, fest und mit sehr feiner Textur; schwindet nicht und wirft sich nicht.

Nutzung: Eines der wertvollsten Möbel- und Furnierhölzer. Noch heute (unter dem Handelsnamen Schwarzkirsche) exportiert. Früchte zur Herstellung von Kirschwasser und Rum. Die wohlriechende Rinde als Tonikum und Sedativum.

—

In Norddeutschland auf Sandböden schon eingebürgert und bereits früh fruchtend.

Description: arbre de 25 à 30 m de hauteur, avec un tronc droit, columnaire de 1,2 à 1,5 m de diamètre. Un des arbres forestiers les plus précieux d'Amérique du Nord. Floraison plus tardive que chez les autres espèces du genre, d'où son nom. Fleurs blanches en grappes pendantes, très décoratives. Fruits ronds, noirs à maturité et exhalan une agréable odeur vineuse.

Habitat: croît indifféremment sur les sols secs sablonneux ou graveleux et les sols humides. Principal essence des forêts appalachiennes On le trouve dans l'est de l'Amérique du Nord.

Bois: assez dur, résistant et à grain très fin; pas de retrait ni de gauchissement au séchage.

Intérêts: un des bois de placage et d'ameublement les plus recherché Exporté encore aujourd'hui sous le nom de «cerisier noir». Fruits utilisés pour parfumer le kirsch et le rhum. L'écorce aromatique est sédative et tonique.

Wild Black Cherry.

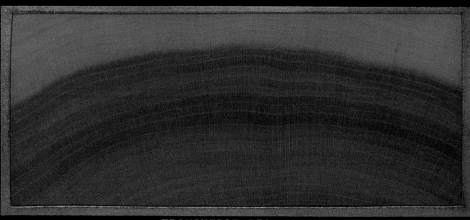

TRANSVERSE SECTION.

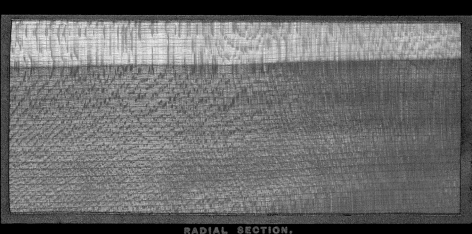

RADIAL SECTION.

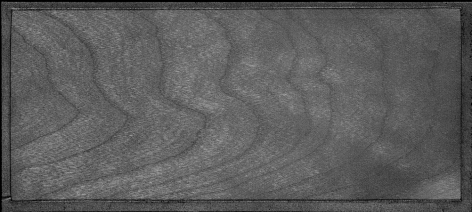

TANGENTIAL SECTION.

Ger. Spätblühende Traubenkirsche.   Fr. Mérisier.

Sp. Cereza silvestre negra.

R | ROSE FAMILY
Wild Cherry

FAMILIE DER ROSENGEWÄCHSE
Steinweichsel

FAMILLE DE LA ROSE
Cerisier amer

ROSACEAE
Prunus emarginata var. mollis (Douglas ex. Hook.) Walp.
Syn.: *P. mollis* Walp.

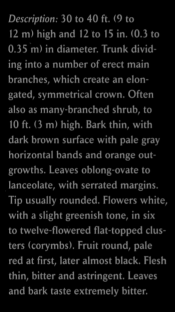

Description: 30 to 40 ft. (9 to 12 m) high and 12 to 15 in. (0.3 to 0.35 m) in diameter. Trunk dividing into a number of erect main branches, which create an elongated, symmetrical crown. Often also as many-branched shrub, to 10 ft. (3 m) high. Bark thin, with dark brown surface with pale gray horizontal bands and orange outgrowths. Leaves oblong-ovate to lanceolate, with serrated margins. Tip usually rounded. Flowers white, with a slight greenish tone, in six to twelve-flowered flat-topped clusters (corymbs). Fruit round, pale red at first, later almost black. Flesh thin, bitter and astringent. Leaves and bark taste extremely bitter.

Habitat: Near river banks on nutrient-poor soils, more rarely on dry slopes, in western North America.

Wood: Soft, brittle, and close-grained. Heartwood brown with green streaks; sapwood paler. Numerous thin medullary rays.

Use: The fruits are eaten by American Indians of the north-western coast.

Beschreibung: 9 bis 12 m hoch bei 0,3 bis 0,35 m Durchmesser. Stamm sich in einige aufrechte Hauptäste spaltend, die eine symmetrische, längliche Krone bilden. Häufig auch als vielstämmiger bis zu 3 m hoher Strauch. Borke dünn, mit dunkelbrauner Oberfläche, auf der sich hellgraue horizontale Bänder und orangefarbene Auswüchse befinden. Blätter länglich oval bis lanzettlich. Blattränder gesägt. Spitze meist abgerundet. Blüten weiß, mit leicht grüner Tönung, in sechs- bis zwölfblütigen Ebensträußen. Frucht rund, zunächst hellrot, später fast schwarz. Fleisch dünn, bitter und adstringierend. Blätter und Borke schmecken außerordentlich bitter.

Vorkommen: In der Nähe von Flussufern auf nährstoffarmen Böden, seltener an trockenen Hängen. Im westlichen Nordamerika.

Holz: Weich, brüchig und mit feiner Textur. Kern braun mit grünen Streifen; Splint blasser. Mit zahlreichen dünnen Markstrahlen.

Nutzung: Die Früchte wurden von den Indianern der Nordwestküste gegessen.

Description : 9 à 12 m de hauteur sur un diamètre de 0,3 à 0,35 m, à tige unique se divisant en quelque branches principales dressées qui forment une couronne allongée et symétrique. Souvent aussi buissor atteignant 3 m de hauteur, à ramifications denses. Ecorce mince, av une surface brun sombre parsemé de bandes horizontales gris clair et d'excroissances orange. Feuilles allongées, ovales à lancéolées, à marge dentée et à apex le plus souvent arrondi. Floraison blanche légèrement verdâtre, en corymbes de six à douze fleurs. Fruits ronds rouge clair devenant presque noir à maturité, peu charnus, amers et astringents. Feuilles et écorce très amères.

Habitat : à proximité des berges dans des sols pauvres en éléments nutritifs, plus rarement sur pentes sèches. Croît dans l'ouest de l'Am rique du Nord.

Bois : tendre, cassant et à grain fin Bois de cœur brun avec des bande vertes ; aubier plus pâle. Rayons médullaires nombreux et fins.

Intérêts : les fruits étaient consommés par les Indiens de la côte pacifique.

Woolly-leaf Cherry, Bitter Cherry.

TRANSVERSE SECTION.

RADIAL SECTION.

TANGENTIAL SECTION.

Ger. Haarigblättrige Kirsche.   *Fr.* Cerisier à feuilles velu

Sp. Cerezo de hojas pelosas.

FAMILIE DER ROSENGEWÄCHSE
Westamerikanische Pflaume

FAMILLE DE LA ROSE
Prunier de Klamath

ROSACEAE
Prunus subcordata Benth.

Description: 20 to 25 ft. (6.0 to 7.5 m) high and to 12 in. (0.3 m) in diameter. Often as shrub with sturdy trunk to 12 ft. (3.6 m) high, or as low many-branched bush. Bark fairly thick, deeply grooved, gray-brown. Leaves broadly ovate to round, with sharply serrated margins. Flowers almost sessile, in small two to four-flowered umbels. Sepals hairy. Fruit ovoid, with dark red to deep purple, sometimes pale yellow skin and juicy, slightly sour flesh with excellent flavor. Stone flattened or swollen, pointed at both ends.

Habitat: Near rivers, often forming extensive thickets. On dry hills and in open woodland, but also on rich, damp soils in Pacific North America.

Wood: Heavy, hard, silky-sheened, and close-grained; polishable.

Beschreibung: 6 bis 7,5 m hoch bei bis zu 0,3 m Durchmesser. Oft auch als Strauch mit kräftigen Stämmchen von bis zu 3,6 m Höhe oder als niedriger, vielstämmiger Busch. Borke mäßig dick, tiefrissig, von graubrauner Farbe. Blätter breit oval bis rundlich. Blattränder scharf gesägt. Blüten fast sitzend in kleinen zwei- bis vierblütigen Dolden. Kelchblätter behaart. Frucht kurz oval mit dunkelroter bis tief violetter, manchmal auch hellgelber Haut und saftigem, leicht saurem Fleisch mit exzellentem Geschmack. Stein abgeflacht oder anschwellend, an beiden Enden zugespitzt.

Vorkommen: In der Nachbarschaft von Flüssen, oft ausgedehnte Dickichte bildend. Auf trockenen Hügeln und offenen Wäldern, aber auch auf reichen, feuchten Böden im pazifischen Nordamerika.

Description : 6,0 à 7,5 m de hauteur sur un diamètre pouvant atteindre 0,3 m. Souvent aussi à port buissonnant, avec un petit tronc vigoureux s'élevant jusqu'à 3,6 m, ou arbrisseau buissonnant. Ecorce brun gris, moyennement épaisse et creusée de fissures profondes. Feuilles ovales-larges, bordées de dents aiguës. Petites ombelles de 2 à 4 fleurs subsessiles; avec des sépales duvetés. Drupes rouge foncé à pourpre profond, parfois aussi jaune clair, ovoïdes, à chair juteuse et acidulée d'un goût délicieux. Noyaux aplatis ou renflés, aux extrémités pointues.

Habitat : près des cours d'eau, souvent en vastes fourrés ; collines sèches et forêts ouvertes, mais aus sols riches et humides du Pacifique de l'Amérique du Nord.

Bois : lourd, dur, satiné et à grain

Pacific Plum, Wild Plum.

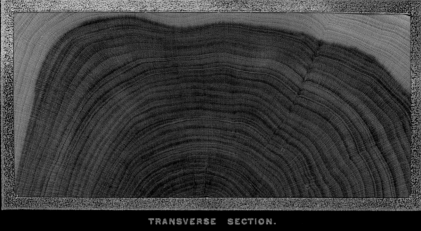

TRANSVERSE SECTION.

RADIAL SECTION.

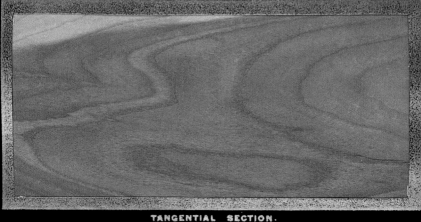

TANGENTIAL SECTION.

Ger. Californische Pflaume. *Fr.* Prune de Californie

Sp. Ciruelo silvestre de California.

S | SAPOTE FAMILY

Ants-wood, Downward Plum, Saffron Plum

FAMILIE DER SAPOTENGEWÄCHSE

–

FAMILLE DU KARITÉ

Acoma faux-célastre, Bumélia à feuilles étroites

SAPOTACEAE

Sideroxylon celastrinum (Kunth) T. D. Pennington
Syn.: *Bumelia angustifolia* Nutt.

Description: 20 ft. (6 m) high and 6 to 8 in. (0.15 to 0.2 m) in diameter. Trunk short, with drooping branches, which form a compact, round crown. In Texas also occasionally a low shrub. Twigs end in a thorn. Bark moderately thick, gray with a red tinge, divided by deep fissures into almost square plates. Leaves leathery, thick, spatulate or linear-lanceolate to broadly ovate, light blue-green on the upper side, paler beneath. Flowers in few or many-flowered clusters on slender stalks. Fruits ovoid-oblong, with thick, sweet flesh, normally only one fruit developing to maturity from each inflorescence.

Habitat: Shores, coastal strips, and riverbanks in Florida, Texas, northern Mexico, and on the Bahamas.

Wood: Heavy, hard, not strong, with a silky sheen, and very close-grained; polishable.

Use: Because of its narrow diameter, probably suitable only as firewood.

Beschreibung: 6 m hoch bei 0,15 bis 0,2 m Durchmesser. Stamm kurz, mit herabhängenden Ästen, die eine kompakte, runde Krone bilden. In Texas gelegentlich auch als niedriger Busch. Zweigenden dornig. Borke mäßig dick, grau, mit roter Tönung, durch tiefe Risse in fast quadratische Felder zerteilt. Blätter ledrig, dick, spatelförmig oder linealisch-lanzettlich bis breit oval, hell blaugrün auf der Oberseite, unterseits blasser. Blüten in wenig- oder vielblütigen Bündeln auf schmalen Stielen. Frucht oval-länglich, mit dickem, süßem Fleisch, normalerweise entwickelt sich nur eine Frucht je Blütenstand zur Reife.

Vorkommen: Strände, Küstenstreifen und Flussufer in Florida, Texas, Nordmexiko und auf den Bahamas.

Holz: Schwer, hart, nicht fest, seidig und von sehr feiner Textur; polierfähig.

Nutzung: Wegen des geringen Durchmessers allenfalls als Brennholz geeignet.

Description : arbuste de 6 m de hauteur, avec un tronc court de 0,15 à 0,2 m de diamètre et des branches retombantes, formant un houppier compact et rond. Sous forme de buisson au Texas. L'extrémité des rameaux est épineuse. Ecorce moyennement épaisse, grise teinté de rouge, à texture fissurée et craquelée en plaques presque carrées Les feuilles coriaces et épaisses sor spatulées ou linéaires-lancéolées à larges-ovales, vert bleu clair dessus plus pâles dessous. Les fleurs sont réunies en groupes multiflores ou pauciflores, sur de fins pédoncules Le fruit est ovale-allongé, avec une pulpe épaisse et douce. D'ordinaire un seul fruit par inflorescence parvient à maturité.

Habitat : plages, franges côtières et rives de cours d'eau de la Floride, du Texas, du nord du Mexique et des Bahamas.

Bois : lourd, dur, non résistant, satiné et à grain très fin ; susceptible d'être poli.

Intérêts : seulement combustible en raison de son faible diamètre.

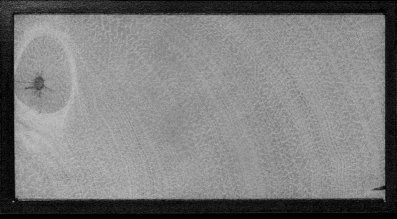

TRANSVERSE SECTION.

RADIAL SECTION.

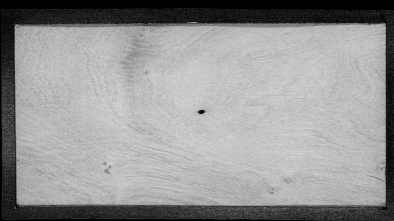

TANGENTIAL SECTION.

Fr. Bois de fee (Martinique)

S | SAPOTE FAMILY

Bustic, Cassada, Sour Wood

FAMILIE DER SAPOTENGEWÄCHSE
Schwarzer Sternapfelbaum

FAMILLE DU KARITÉ
Acomat bastard

SAPOTACEAE
Sideroxylon salicifolium (L.) Lam.
Syn.: *Dipholis salicifolia* (L.) A. DC.

Description: 40 to 50 ft. (12 to 15 m) high and 15 to 18 in. (0.45 to 0.5 m) in diameter. Trunk straight, with slender, erect branches, which form a narrow crown. Bark moderately thick, broken into square scales, brown tinged with red. Leaves oblong-lanceolate or ovate, gradually narrowing into the slender leaf stalk. Flowers in small, dense clusters, shorter than the leaf stalks. Fruits solitary or few together, roundish, black, with thin dry flesh.

Habitat: Shores, rich soils on hills in southern Florida. Most widespread on the Bahamas and other Caribbean islands.

Wood: Very heavy, extremely hard, strong, with a silky sheen, and close-grained; polishable. Heartwood dark brown or red; sapwood thin (four or five annual rings). The wood contains large, open pores and distinct medullary rays.

Use: Especially used for fine cabinet making (Begemann, 1983).

Beschreibung: 12 bis 15 m hoch bei 0,45 bis 0,5 m Durchmesser. Gerader Stamm mit schlanken aufrechten Ästen, die eine schmale Krone bilden. Borke mäßig dick, in quadratische Felder zerfallend, braun mit roter Tönung. Blätter länglich-lanzettlich oder oval, allmählich in den schlanken Blattstiel verschmälert. Dichte kleine Blütenbüschel, die kürzer als die Blattstiele bleiben. Frucht einzeln oder zu wenigen, rundlich, schwarz, mit dünnem, trockenem Fleisch.

Vorkommen: Strände, reiche Böden auf Hügeln in Süd-Florida. Hauptverbeitungsgebiet sind die Bahamas und die Westindischen Inseln.

Holz: Sehr schwer, außerordentlich hart, fest, seidig und von feiner Textur; polierfähig. Kern dunkelbraun oder rot; Splint dünn (4 oder 5 Jahresringe). Es enthält große, offene Gefäße und deutliche Markstrahlen.

Nutzung: Wird besonders in der Kunsttischlerei verwendet (Begemann 1983).

Description : 12 à 15 m de hauteu[r] sur un diamètre de 0,45 à 0,5 m. Tronc droit, sur lequel sont insér[é] des branches fines, dressées, cons[ti]tuant un houppier étroit. Ecorce moyennement épaisse, brune teintée de rouge, à texture craqu[e] lée, en plaques carrées. Feuilles allongées-lancéolées ou ovales, effilées sur le pétiole mince. Fleu[rs] réunies en petites touffes denses, demeurant plus courtes que les pétioles. Fruits ronds et noirs, iso[lés] ou par petits groupes, à chair mi[nce] et sèche.

Habitat : plages, sols riches de col[ ]lines du sud de la Floride. Les régions où il est le plus répandu so[nt] les Bahamas et les îles Caraïbes.

Bois : très lourd, extrêmement du[r,] résistant, satiné et à grain fin ; sus[ ]ceptible d'être poli. Bois de cœur brun sombre ou rouge ; aubier mince (4 à 5 cernes annuels). Grands vaisseaux de type ouvert et rayons médullaires distincts.

Intérêts : utilisé tout particulièrement en ébénisterie (Begemann 1983).

BUSTIC. CASSADA.

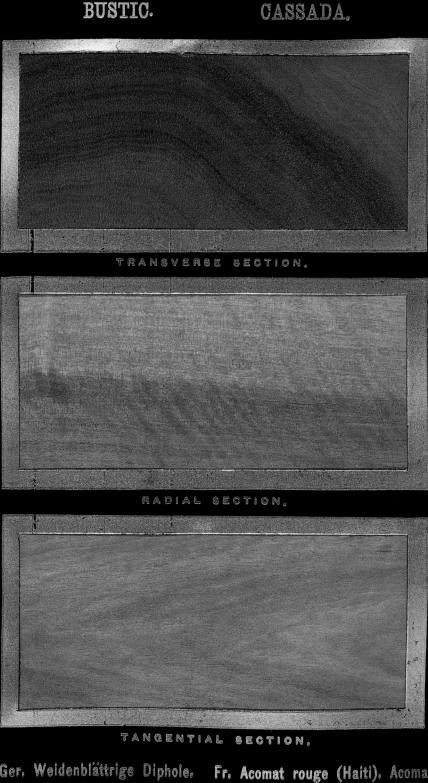

TRANSVERSE SECTION.

RADIAL SECTION.

TANGENTIAL SECTION.

Ger. Weidenblättrige Diphole. Fr. Acomat rouge (Haiti). Acoma bastard (Fr.W.I.). Sp. Tocuma, Almendro sylvestre, Tabloncillo

S | SAPOTE FAMILY
Sapodilla

FAMILIE DER SAPOTENGEWÄCHSE
Sapadill

FAMILLE DU KARITÉ
Balata sapotillier

SAPOTACEAE
Manilkava zapota (L.) van Royen
Syn.: *Sapota achras* Mill.

Description: In North America up to 40 ft. (12 m) high, with a trunk 12 in. (0.3 m) in diameter. The gray-brown bark is deeply fissured, and divided up into small areas. The evergreen leaves are clustered at the ends of the branches. The small, whitish flowers are on short, rust-colored, hairy stalks. The fruits are botanically classified as berries, but they took more like apples. The flesh, hard at first, soon becomes watery and has a pleasant smell and a pear-like taste. The plant contains a whitish sap.

Habitat: The tree is supposed to have originated in the Caribbean. It is cultivated extensively in southern Florida, and is occasionally found growing wild there.

Wood: The rich red-brown heartwood is very heavy, hard, strong, and durable. It is surrounded by paler sapwood.

Use: The fruits are a great favorite. The milky sap was processed in former times to form "gum chicle," a substitute for chewing gum. The wood is used in building.

Beschreibung: In Nordamerika bis zu 12 m hoch, mit einem Stammdurchmesser von 0,3 m. Die graubraune Borke ist tiefrissig und in kleine Felder gegliedert. Die immergrünen Blätter sitzen gebündelt an den Enden der Äste. Die kleinen weißlichen Blüten sitzen auf rostfarben behaarten Stielchen. Die Früchte sind Beeren im botanischen Sinne, ähneln äußerlich aber eher einem Apfel. Das anfangs noch feste Fleisch wird bald wässrig und erhält einen angenehmen Geruch und birnenähnlichen Geschmack. Die Pflanze enthält einen weißlichen Milchsaft.

Vorkommen: Seine Heimat wird in der karibischen Region vermutet. In Süd-Florida in großem Umfang kultiviert und dort gelegentlich auch verwildert anzutreffen.

Holz: Das kräftig rotbraune Kernholz ist sehr schwer, hart, fest und dauerhaft. Es wird von hellerem Saftholz umgeben.

Nutzung: Die Früchte sind als Obst sehr beliebt. Der Milchsaft wurde früher unter dem Namen *gum chicle* als Zusatzstoff für Kaugummis gehandelt. Für Holzbauten verwendet.

Description: arbre atteignant en Amérique du Nord 12 m de hauteur sur un diamètre de 0,3 m. Ecorce brun gris à texture profondément fissurée et craquelée en petites plaques. Feuilles persistantes, disposées en touffes à l'extrémité des rameaux. Petites fleurs blanchâtres portées par des pédoncules recouverts d'une pubescence rouille. Le fruit est un baie, au sens botanique du terme, mais son aspect rappelle celui d'une pomme. La pulpe, ferme au début, devient finalement blette e prenant une odeur agréable et un goût de poire. Présence d'un latex blanc dans toutes les parties de la plante.

Habitat: vraisemblablement originaire des Caraïbes. En grande partie cultivé dans le sud de la Floride, où il est retourné çà et là l'état sauvage.

Bois: très lourd, dur, résistant et durable. Bois de cœur d'un brun rouge intense; aubier plus clair.

Intérêts: la sapote est un fruit très apprécié. Le latex était autrefois employé sous le nom de *gum chicl* comme adjuvant du chewing-gum Le bois est utilisé en construction.

SAPODILLA. CHICLE-TREE.

TRANSVERSE SECTION.

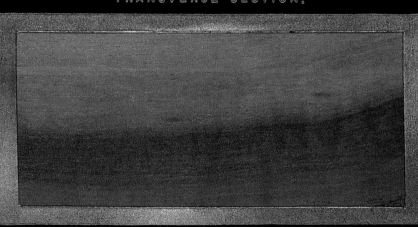

RADIAL SECTION.

TANGENTIAL SECTION.

Ger. Sappadill.   Fr. Sappotillier.

Sp. Nispero (Cent. & So. Amer.).   Chico-zapote. (Mex.).

S | SAPOTE FAMILY
Wild Dilly

FAMILIE DER SAPOTENGEWÄCHSE
–

FAMILLE DU KARITÉ
Balata franc typique, Balata blanc

SAPOTACEAE
Mamilkava jamiqui (C. Wright) Dubard subsp. *emarginata*
L. Cronquist
Syn.: *Mimusops emarginata* L. Britton

Description: Rarely higher than 30 ft. (9 m) or more than 12 to 16 in. (0.3 to 0.4 m) in diameter. Trunk short, gnarled, and mostly hollow, with stout branches, which form a compact, round crown. Bark thin, with deep, irregular fissures, and separating into small pieces. Leaves only at the ends of the branches, elliptic-oblong or slightly ovate, at first bright red, later thick and leathery, bright green, covered with bloom on the lower surface. Flowers at the ends of the branches, petals light yellow, tinged with green. Fruit globular, flattened, or almost ovoid, the flesh containing a whitish, milky juice. As a rule only one seed in each fruit develops to maturity.

Habitat: Most common on the Bahamas, Trinidad, and probably many other islands of the West Indies. In North America confined to the islands off southern Florida.

Wood: Very heavy, hard, strong, and close-grained.

Use: No commercial use known.

Beschreibung: Selten höher als 9 Meter bei 0,3 bis 0,4 m Durchmesser. Kurzer, knorriger, meist hohler Stamm mit kräftigen Ästen, die eine kompakte runde Krone bilden. Borke dünn, von unregelmäßigen tiefen Rissen durchzogen und in kleine Stücke zerfallend. Blätter nur an den Enden der Äste, elliptisch-länglich oder leicht oval. Anfangs hellrot, später dick und ledrig, hellgrün, Unterseite bereift. Blüten an den Enden der Äste, Kronblätter hellgelb mit grüner Tönung. Frucht abgeflacht kugelig oder leicht oval. Fleisch mit weißlichem Milchsaft. Es entwickelt sich i. d. R. nur ein Samen pro Frucht zur Reife.

Vorkommen: Hauptverbreitung auf den Bahamas, Trinidad und wahrscheinlich aufvielen weiteren der Westindischen Inseln. Nordamerika nur auf den Inseln vor Süd-Florida erreichend.

Holz: Sehr schwer, hart, fest und von feiner Textur.

Nutzung: Keine gewerbliche Nutzung bekannt.

Description : petit arbre atteignant rarement plus de 9 m de hauteur sur un diamètre de 0,3 à 0,4 m. Tronc court, noueux, creux généralement, armé de branches vigoureuses, constituant un houppier compact et rond. Ecorce mince, à texture profondément et irrégulièrement fissurée se desquamant e petites plaques. Feuilles terminale uniquement, elliptiques-allongées ou légèrement ovales, rouge clair jeunes et finalement épaisses et coriaces, vert clair, pruineuses au revers. Fleurs terminales avec des pétales jaune clair teintés de vert Fruit globuleux aplati ou légèrement ovale ; chair renfermant un suc laiteux. Une seule graine par fruit parvient à maturité.

Habitat : surtout répandu aux Bahamas, à Trinidad et probablement dans beaucoup d'autres îles des Caraïbes. En Amérique du Nord, seulement sur les îles en face du sud de la Floride.

Bois : très lourd, dur, résistant et à grain fin.

Intérêts : utilisation commerciale non déterminée.

WILD DILLY. WILD SAPODILLA.

TRANSVERSE SECTION.

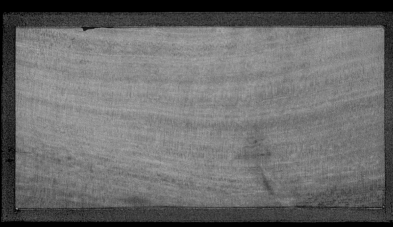

RADIAL SECTION.

TANGENTIAL SECTION.

S | SAPOTE FAMILY

Woolly Bumelia, Gum Elastic, Chittam Wood

FAMILIE DER SAPOTENGEWÄCHSE
Flaumige Bumelie

FAMILLE DU KARITÉ
Acoma laineux typique, Bumélia bois-de-fer laineux

SAPOTACEAE
Sideroxylon lanuginosum Michx. subsp. *lanuginosum*
Syn.: *Bumelia lanuginosa* (Michx.) Pers.

Description: 50 to 60 high ft. (15 to 18 m) and 3 ft. (0.9 m) in diameter. Trunk tall and straight with short, stout branches that form an oblong crown, narrowly rounded at the top. Bark moderately thick, dark gray-brown and deeply furrowed. Leaves oblong-ovate, thick, covered with rust-colored hairs at first, later dark green and shiny above, silky-hairy beneath. Flowers in 16 to 18-flowered fascicles on slender stalks. Fruits more or less ovoid on slender stalks. A clear, sticky gum oozes from damaged plants.

Habitat: East of the Mississippi on dry, sandy soils, more frequent in eastern Texas on the rich soils of the lower reaches of rivers.

Wood: Heavy, rather soft, not strong, and close-grained. Heartwood light brown or yellow; sapwood thick, paler. Numerous distinct medullary rays.

Use: The gum that is exuded in considerable quantities from freshly felled trees is used commercially. The tree is occasionally cultivated in Europe.

Beschreibung: 15 bis 18 m hoch bei 0,9 m Durchmesser. Großer gerader Stamm mit kurzen, kräftigen Ästen, die eine längliche, an der Spitze schmale rundliche Krone bilden. Borke mäßig dick, dunkel graubraun und tiefrissig. Blätter länglich oval, dick, anfangs rostig behaart, später dunkelgrün und blank auf der Oberseite, unterseits seidig behaart. Blüten in 16- bis 18-blütigen Bündeln auf haarfeinen Stengeln. Früchte oval auf schlanken Stielen. Aus verletzten Pflanzen tritt klares, viskoses Gummi aus.

Vorkommen: Östlich des Mississippi auf trockenen, sandigen Böden, häufiger in Ost-Texas auf reichen Böden an den Flussunterläufen.

Holz: Schwer, eher weich, nicht fest und mit feiner Textur. Kern hellbraun oder gelb; Splint dick, heller. Viele deutliche Markstrahlen.

Nutzung: Das in beachtlichen Mengen austretende Gummi frisch gefällter Bäume wird kommerziell verwertet. Der Baum wird in Europa gelegentlich kultiviert.

Description: 15 à 18 m de hauteur sur un diamètre de 0,9 m. Grand tronc droit garni de branches courtes, vigoureuses, constituant un houppier allongé, à cime rond et étroite. Ecorce moyennement épaisse, brun gris sombre et profondément fissurée. Feuilles ovale allongées, épaisses, couvertes de poils roux à l'état juvénile, puis vert foncé, glabres dessus, veloutées dessous. Glomérules de 16 à 18 fleurs portées par des pédoncules filiformes. Fruits ovoïdes-ovles finement pédonculés. Ecoulement de latex blanc très poisseux après traumatisme.

Habitat: à l'ouest du Mississippi, sur les sols sablonneux secs, plus fréquent à l'est du Texas sur les sols riches le long du cours inférieur des fleuves.

Bois: lourd, plutôt tendre, non résistant et à grain fin. Bois de cœur brun clair ou jaune; aubier épais, plus clair. Nombreux rayons médullaires distincts.

Intérêts: exploitation commerciale du latex s'écoulant abondamment des arbres abattus. Cultivé quelquefois en Europe.

TRANSVERSE SECTION.

RADIAL SECTION.

S | SOAPBERRY FAMILY, LYCHEE FAMILY

Florida Soapberry, Wild China Tree

FAMILIE DER SEIFENBAUMGEWÄCHSE
Drummond Seifenbeere

FAMILLE DU SAVONNIER
Savonnier de Drummond

SAPINDACEAE
Sapindus saponaria L. var. *drummondii* (Hook. & Arn.)
L. D. Benson
Syn.: *S. drummondii* Hook. & Arn.

Description: 40 to 50 ft. (12 to 15 m) high and 15 in. (0.45 m) in diameter. Branches stout, usually erect. Bark moderately thick, separating into narrow, reddish brown strips. Leaves pale yellow-green consisting of four to nine pairs of pointed, lanceolate leaflets, glabrous on the upper surface, and covered with short, pale hairs on the lower surface. Leaf stalks usually not winged. Sepals pointed, with finely ciliate margins. Petals white, with a hairy appendage on the inner surface. Fruit ovoid or roundish, with dark orange skin.

Habitat: Prefers moist clay soil, rarely also found on drier, calcareous soil in southern North America.

Wood: Heavy, strong, and close-grained; easy to split. Heartwood light brown tinged with yellow; sapwood paler. Groups of open pores clearly mark the edges of the annual rings, but the medullary rays are thin and indistinct.

Use: Wooden baskets for harvesting cotton, also for the frames of pack-saddles. The fruits are a food for wild and domestic animals. The tree is occasionally cultivated in European and Algerian gardens.

Beschreibung: 12 bis 15 m hoch bei 0,45 m Durchmesser. Äste kräftig, meist aufrecht. Borke mäßig dick, in schmale, rotbraune Streifen zerteilt. Blätter blass gelbgrün, Oberseite kahl, Unterseite mit blasser, kurzer Behaarung, mit 4 bis 9 Fiederpaaren. Blattfiedern lanzettlich, spitz. Blattstiele meist ungeflügelt. Kelchblätter spitz, Ränder fein gewimpert. Kronblätter weiß, auf der Innenseite mit behaartem Anhängsel. Frucht oval oder rundlich, mit dunkelorangener Hülle.

Vorkommen: Bevorzugt auf feuchtem Lehmboden, seltener auch auf trockenerem Kalkboden im südlichen Nordamerika.

Holz: Schwer, fest und mit feiner Textur; leicht spaltbar. Kern hellbraun mit gelber Tönung; Splint heller. Einige Gruppen offener Gefäße, deutlich die Grenzen der Jahresringe markierend, sowie dünne, undeutliche Markstrahlen.

Nutzung: Holzkörbe für die Baumwollernte, auch für die Rahmen von Packsätteln. Früchte Vieh- und Wildfutter. Der Baum wird gelegentlich in europäischen und algerischen Gärten kultiviert.

Description: 12 à 15 m de hauteur, tronc de 0,45 m de diamètre aux ramifications vigoureuses, souvent dressées. Ecorce moyennement épaisse, à fissures brun rouge peu profondes. Feuilles vert jaune clair, glabres dessus, recouvertes d'une courte pubescence pâle au revers, composées de 4 à 9 foliole lancéolées, se terminant en point. Sépales pointus, à marge finement ciliée; pétales blancs, munis d'un languette duvetée à la base de leu face supérieure. Fruit en capsule ronde ou ovale, orange foncé.

Habitat: préfere les sols argileux humides mais croît aussi sur les sols calcaires secs du sud de l'Am rique du Nord.

Bois: lourd, résistant et à grain fin se fendant facilement. Bois de cœ brun clair teinté de jaune; aubier plus clair. Quelques faisceaux vas culaires de type ouvert, délimitan bien les cernes annuels et rayons médullaires fins, indistincts.

Intérêts: bâts et paniers pour la cueillette du coton. Les fruits son consommés par le bétail et les ani maux sauvages. Planté parfois da les jardins en Europe et en Algéri

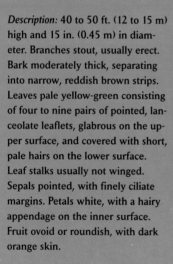

TRANSVERSE SECTION.

RADIAL SECTION.

TANGENTIAL SECTION.

S | SOAPBERRY FAMILY, LYCHEE FAMILY

Inkwood, Ironwood

FAMILIE DER SEIFENBAUMGEWÄCHSE
Tintenholz

FAMILLE DU SAVONNIER
Exothéa paniculé, Bois d'encre

SAPINDACEAE
Exothea paniculata (Juss.) Radlk. ex Durand

Description: 40 to 50 ft. (12 to 15 m) high and 12 to 16 in. (0.3 to 0.9 m) in diameter, with slender, erect branches. Bark thin, with a bright red surface, which separates into large, brown scales. Leaves oblong-ovate. Flowers in panicles.

Habitat: Islands and coastal regions of the Caribbean. In North America confined to southern Florida.

Wood: Heavy, moderately hard, and strong (Bärner, 1962).

Use: Joinery (Bärner, 1962).

Beschreibung: 12 bis 15 m hoch bei 0,3 bis 0,4 m Durchmesser, mit schlanken aufrechten Ästen. Borke dünn mit hellroter Oberfläche, die sich in großen braunen Schuppen abschält. Blätter länglich oval Blüten in Rispen.

Vorkommen: Inseln und Küstengebiete der karibischen Inseln. Nordamerika nur in Süd-Florida erreichend.

Holz: Schwer, mäßig hart und fest (Bärner 1962).

Nutzung: Hauptsächlich für Tischlereiarbeiten (Bärner 1962).

Description : s'élevant entre 12 à 15 m, avec un tronc de 0,3 à 0,4 m de diamètre et des branches fines et dressées. Ecorce mince, rouge clair, s'exfoliant en grandes écailles brunes. Feuilles allongées-ovales. Fleurs groupées en panicules.

Habitat : îles et zones littorales des Caraïbes ; en Amérique du Nord, seulement dans le sud de la Floride.

Bois : lourd, moyennement dur et résistant (Bärner 1962).

Intérêts : utilisé en menuiserie (Bärner 1962).

INK-WOOD.   BUTTER-BOUGH.

TRANSVERSE SECTION.

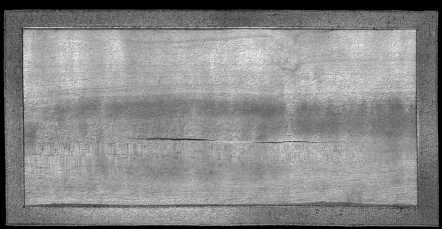

RADIAL SECTION.

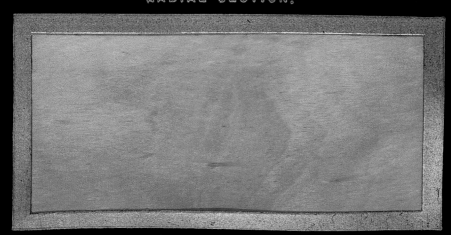

TANGENTIAL SECTION.

Ger. Tinten-holz.   Fr. Bois d'encre.

Sp. Guacarán. Gaita.

S | SOAPBERRY FAMILY, LYCHEE FAMILY

White Ironwood

FAMILIE DER SEIFENBAUMGEWÄCHSE
Weißes Eisenholz

FAMILLE DU SAVONNIER
Gallipeau trifolié

SAPINDACEAE
Hypelate trifoliata Sw.

Description: 35 to 40 ft. (10 to 12 m) high and 15 to 18 in. (0.45 to 0.5) in diameter, but normally smaller. Bark thin and smooth, with many small depressions and lenticels. Leaves divided into three leaflets. Leaflets spatulate, very light green. Flowers very small, in loose panicles. Fruits ripening very sporadically, almost globular to subovoid, with a distinct small tip, and pleasant smell.

Habitat: Islands and coastal strips of Florida, Cuba, and Jamaica.

Wood: Hardly known.

Use: Apparently without any economic importance.

Beschreibung: 10 bis 12 m hoch bei 0,45 bis 0,5 m Durchmesser, normalerweise aber kleiner. Borke dünn und glatt, mit vielen kleinen Einbuchtungen und Atemporen. Blätter in drei Fiedern geteilt. Blattfiedern spatelförmig, sehr hellgrün. Blüten sehr klein, in lockeren Rispen. Früchte sehr zerstreut reifend, fast kugelig, kurz eiförmig, mit deutlicher kleiner Spitze und angenehmem Geruch.

Vorkommen: Inseln und Küstenstreifen von Florida, Kuba und Jamaika.

Holz: Kaum bekannt.

Nutzung: Offenbar ohne ökonomische Bedeutung.

Description : arbre 10 à 12 m de hauteur, avec un tronc de 0,45 à 0,5 m de diamètre, mais d'ordinaire plus petit. L'écorce mince et lisse est couverte de lenticelles et de petites concavités. Les feuilles sont composées de trois folioles spatulées d'un vert très clair. Les fleurs sont très petites et groupées en panicules lâches. Les fruits dont la maturation est très échelonnée sont presque sphériques, courts-ovoïdes avec une petite pointe nette, et dégagent une odeur agréable.

Habitat : îles et franges côtières de la Floride, de Cuba et de la Jamaïque.

Bois : quasiment inconnu.

Intérêts : il n'offre apparemment aucun intérêt économique.

TRANSVERSE SECTION.

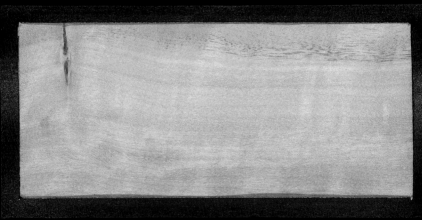

RADIAL SECTION.

TANGENTIAL SECTION.

S | SOAPBERRY FAMILY, LYCHEE FAMILY

Wingleaf Soapberry, Soap Berry

FAMILIE DER SEIFENBAUMGEWÄCHSE
Seifenbaum

FAMILLE DU SAVONNIER
Savonnier des Antilles

SAPINDACEAE
Sapindus saponaria L. var. *saponaria*

Description: 25 to 30 ft. (7.5 to 9.0 m) high and rarely more than 10 to 12 in. (0.25 to 0.3 m) in diameter. Branches erect. Bark moderately thick, light gray, rough from the excrescences. The outer bark breaks away into large scales, exposing an almost black surface. Leaves yellow-green, divided into five to nine leaflets. Leaf stalk and rachis broadly winged. Flowers white, hairy, produced in groups of three and forming elongated panicles. Sepals rounded at the tip. Fruit round, with orange-brown flesh.

Habitat: Shores and islands of the Caribbean. In North America confined to southern Florida.

Wood: Heavy, rather hard, and close-grained. Heartwood light brown with a yellow tinge; sapwood thick, yellow. Numerous thin medullary rays.

Use: The Caribs used the fruits as the basis of a substitute for soup. Cultivated in England since 1697, and introduced into West Africa, where it now grows wild on some of the Cape Verde Islands.

First mentioned by European sources in 1535.

Beschreibung: 7,5 bis 9 m hoch bei selten mehr als 0,25 bis 0,3 m Durchmesser. Äste aufrecht. Borke mäßig dick, hellgrau, durch Auswüchse rau. Die äußere Rinde löst sich in großen Schuppen ab und entblößt eine fast schwarze Oberfläche. Blätter gelbgrün, in 5 bis 9 Blattfiedern geteilt. Blattstiele und Mittelrippe breit geflügelt, Blüten weiß, behaart, in Dreiergruppen auf langgestreckten Rispen. Kelchblätter an der Spitze abgerundet. Frucht rund, mit orangebraunem Fleisch.

Vorkommen: Strände und Inseln der Karibik. Nordamerika nur in Süd-Florida erreichend.

Holz: Schwer, eher hart und von feiner Textur. Kern hellbraun mit gelber Tönung; Splint dick, gelb. Zahlreiche dünne Markstrahlen.

Nutzung: Die Kariben verwendeten die Früchte als Grundlage für einen Suppenersatz. In England seit 1697 kultiviert, eingeführt in Westafrika, und auf einem Teil der Kapverden verwildert.

1535 erste Erwähnung durch europäische Quellen.

Description: arbuste de 7,5 à 9,0 m de hauteur, avec un tronc atteignant rarement plus de 0,25 à 0,30 m de diamètre et armé de branches dressées. Ecorce moyennement épaisse, gris clair et verruqueuse, s'exfoliant en grandes écailles et laissant apparaître une surface presque noire. Feuilles vert jaune, composées de 5 à 9 folioles, pétiole et nervure médiane élargis en aile. Fleurs blanches, duvetées, par groupes de trois fleurs sur des panicules allongées; sépales arrondis au sommet. Fruit rond, à chair brun orange.

Habitat: plages et îles des Caraïbes. En Amérique du Nord seulement dans le sud de la Floride.

Bois: lourd, plutôt dur et à grain fin. Bois de cœur brun clair teinté de jaune; aubier épais, jaune. Nombreux rayons médullaires.

Intérêts: aux Caraïbes, on préparait une soupe avec les fruits. Cultivé en Angleterre dès 1697, introduit en Afrique de l'Ouest et retourné partiellement à l'état sauvage sur les îles du Cap-Vert.

Les premiers documents européens le mentionnant remontent à 1535.

TRANSVERSE SECTION.

RADIAL SECTION.

TANGENTIAL SECTION.

Gr. Seifenbaum. Fr. Savonnier. Sp. Jaboncelle.

S | SPURGE FAMILY
Castor Oil Plant

FAMILIE DER WOLFSMILCHGEWÄCHSE
Rhizinus-Baum, Christpalme

FAMILLE DES EUPHORBES ET DE L'HÉVÉA
Ricin commun, Palma-christi

EUPHORBIACEAE
Ricinus cummunis L.

Description: In regions with a temperate climate throughout the year it grows only as an shrubby annual, but in subtropical and tropical regions it forms a tree up to 25 ft. (7.5 m) high with a trunk 12 in. (0.3 m) in diameter. The smooth, gray bark remains very thin, even in tree-like specimens. Leaves very large, with broad lobes, serrate along the margins. The flowers are clustered, male and female separate, in long inflorescences at the ends of the branches. The fruits are roundish three-celled capsules, containing brown and white mottled, bean-like seeds.

Habitat: The Castor Oil Plant comes from the Old World, but has been introduced into many places in the warm regions of America. In California it prefers rich soils, and has become completely naturalized.

Wood: Very light, soft, and weak. The heartwood is mottled brown in color, and is surrounded by the almost white sapwood.

Use: The Castor Oil Plant is cultivated mainly for the medicinally important oil in its seeds, but it is also a popular ornamental plant.

Beschreibung: In Regionen mit gemäßigtem Jahreszeitenklima nur einjährig und buschartig wachsend, in subtropischen und tropischen Regionen aber auch als bis zu 7,5 m hoher Baum mit 0,3 m dickem Stamm. Die glatte graue Borke bleibt auch bei den baumartigen Exemplaren sehr dünn. Blätter sehr groß, breit gelappt mit gesägten Blatträndern. Die Blüten sind in langen, bereiften Blütenständen am Ende der Äste vereinigt. Die rundlichen Kapselfrüchte enthalten in drei Kammern die bohnenförmigen braun-weiß gescheckten Samen.

Vorkommen: Der Rhizinus stammt aus der Alten Welt und wurde vielfach in die warmen Regionen Amerikas eingeführt. In Kalifornien bevorzugt er reiche Böden und ist vollständig eingebürgert.

Holz: Sehr leicht, weich und schwach. Das von fast weißem Saftholz umgebene Kernholz hat eine fleckig braune Farbe.

Nutzung: Der Rhizinus wird hauptsächlich wegen des medizinisch bedeutsamen Öls seiner Samen angebaut, ist aber auch als Zierpflanze beliebt.

Description : plante annuelle à port buissonnant dans les régions tempérées, mais arborescente dans les régions tropicales et subtropicales où elle peut atteindre 7,5 m de hauteur et 0,3 m de diamètre. L'écorce grise et lisse est très mince, même chez les spécimens arborescents. Grandes feuilles palmées et dentées. Fleurs apétales groupées en grappes de cymes à l'extrémité des rameaux. Fruits en capsules globuleuses, divisées en trois loges contenant des graines en forme de haricot, marbrées de brun et de blanc.

Habitat : originaire du Vieux Monde et introduit dans plusieurs régions chaudes de l'Amérique du Nord. En Californie, il affectionne les sols riches et s'est entièrement naturalisé.

Bois : très léger, tendre et faible. Bois de cœur taché de brun ; aubier presque blanc.

Intérêts : le ricin est une plante médicinale cultivée avant tout pour ses graines, dont on extrait l'huile du même nom, mais il est également très apprécié en ornement.

Castor-bean Tree, Palma Christi.

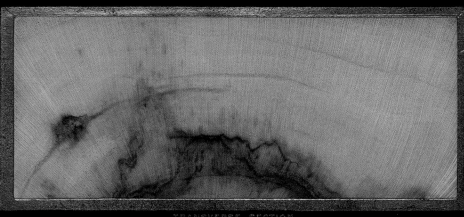

TRANSVERSE SECTION.

RADIAL SECTION.

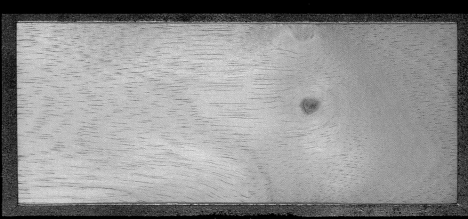

TANGENTIAL SECTION.

Ger. Ricinusbaum,   *Fr.* Arbre de Ricin,

S | SPURGE FAMILY
Chinese Tallow Tree

FAMILIE DER WOLFSMILCHGEWÄCHSE
Chinesischer Talgbaum

FAMILLE DES EUPHORBES ET DE L'HÉVÉA
Arbre à suif chinois

EUPHORBIACEAE
Sapium sebiferum (L.) Roxb.

Description: The tree reaches a height of 70 ft. (20 m) in North America, and the trunk is 10 in. (0.25 m) in diameter. The slender branches are conspicuous for their yellowish brown or gray leaf scars. The smooth bark is red-brown. The very broad, thin leaves are dark green on the upperside, and paler beneath. The dark brown fruit capsules are covered in a thick layer of wax.

Habitat: It is native to eastern Asia. From there it was introduced into the south of the USA, and has since become established in the wild.

Wood: The almost white wood is light, very hard, and strong.

Use: In its native area the wax from the fruit capsules is used in the manufacture of candles and as fuel for lamps. The wood has many uses. A black dye can be obtained from the leaves.

Beschreibung: Der Baum erreicht in Nordamerika eine Höhe von 20 m bei einem Stammdurchmesser von 0,25 m. Die schlanken Äste zeichnen sich durch gelblich braune oder graue Blattnarben aus. Die glatte Borke ist rotbraun. Die Oberseite der sehr breiten, dünnen Blätter ist dunkelgrün. Die Unterseite ist blasser. Die dunkelbraunen Fruchtkapseln werden von einer dicken Wachsschicht bedeckt.

Vorkommen: Seine Heimat ist Ostasien, von wo aus er in den Süden der USA eingeführt wurde und verwilderte.

Holz: Das fast weiße Holz ist leicht, sehr hart und fest.

Nutzung: In seiner Heimat wird das Wachs der Fruchtkapseln zur Herstellung von Kerzen und als Lampenbrennstoff verwendet. Das Holz lässt sich vielseitig nutzen. Aus den Blättern kann ein schwarzer Farbstoff gewonnen werden.

Description: l'arbre atteint en Amérique du Nord une hauteur de 20 m et un diamètre de 0,25 m Les branches fines se remarquent par les cicatrices brun jaunâtre et grises laissées par les feuilles. L'écorce est lisse et brun rouge. Les feuilles très larges et fines sont vert foncé dessus, plus pâles dessous. Les capsules brun sombre sont recouvertes d'une épaisse couche de substance cireuse, un « suif ».

Habitat: originaire de l'Asie orientale ; introduit dans le sud des Etats-Unis, où il est retourné finalement à l'état sauvage.

Bois: presque blanc, léger, très dur et résistant.

Intérêts: le « suif » est utilisé dans son aire d'origine pour fabriquer des bougies et comme combustible de lampes. Le bois a de nombreux usages. On peut extraire des feuilles une teinture noire.

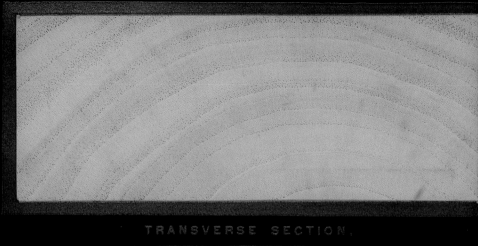

TRANSVERSE SECTION.

RADIAL SECTION.

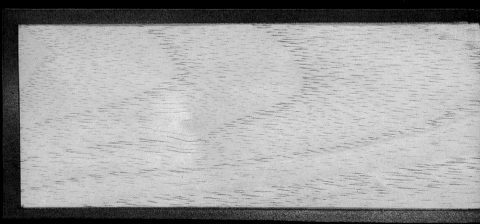

TANGENTIAL SECTION.

Gr. Talgbaum.   Fr. Arbre a suif.

S | SPURGE FAMILY
Mastic, Mastic Bully, Wild Olive

FAMILIE DER WOLFSMILCHGEWÄCHSE
Mastixbaum

FAMILLE DES EUPHORBES ET DE L'HÉVÉA
Acomat franc typique, Acomat

EUPHORBIACEAE
Sideroxylon mastichodendron Jacq.

Description: 60 to 70 ft. (18 to 21 m) high and 3 to 4 ft. (0.9 to 1.2 m) in diameter. Trunk massive, straight. Branches stout, erect, forming a dense irregular crown. Bark moderately thick, dark gray to light brown. Leaves ovate, at first covered with silky hairs on the lower surface, later thin, glabrous, bright green above and yellow-green beneath. Flower parts in fives, in many-flowered clusters that are shorter than the leaf stalks. Corolla united in the lower half. Corolla lobes recurved, and stamens free. From each inflorescence only one fruit develops, and this is oblong, with a yellow skin and thin dry flesh.

Habitat: On hummocks in the Bahamas and other Caribbean islands. In North America it is confined to southern Florida.

Wood: Very heavy, extremely hard, strong, and close-grained.

Use: Not attacked by the shipworm or drill shell (Teredo), so therefore extensively used in ship and boatbuilding. The fruits are an important source of food for animals. The hardest of all the North American woods of comparable size.

Beschreibung: 18 bis 21 m bei 0,9 bis 1,2 m Durchmesser. Stamm massig, gerade, Äste kräftig, aufrecht, eine dichte, unregelmäßige Krone bildend. Borke mäßig dick, dunkelgrau bis hellbraun. Blätter oval, anfangs auf der Unterseite seidig behaart, später dünn, haarlos und hellgrün auf der Ober- und gelbgrün auf der Unterseite. Blüten fünfzählig, in vielblütigen Büscheln, die kürzer als die Blattstiele bleiben. Untere Hälfte der Kronblätter verwachsen. Kronblattlappen zurückgebogen und Staubblätter freistellend. Nur eine reife Frucht je Blütenstand, diese länglich, mit gelber Schale und dünnem, trockenem Fleisch.

Vorkommen: Reiche Böden kleiner Hügel der Bahamas und der Westindischen Inseln. Nordamerika nur in Süd-Florida erreichend.

Holz: Sehr schwer, außerordentlich hart, fest und mit feiner Textur.

Nutzung: Nicht von Schiffsbohrwürmern (*Teredo*) angreifbar, für Schiff- und Bootsbau. Früchte wichtige Nahrungsquelle für Tiere. Schwerstes aller nordamerikanischen Hölzer bei vergleichbarer Größe.

Description: 18 à 21 m de hauteur sur 0,9 à 1,2 m de diamètre. Tronc massif, droit et branches vigoureuses, dressées, constituant un houppier dense et irrégulier. Écorce peu épaisse, gris foncé à brun clair. Feuilles ovales, à face inférieure duvetée jeunes puis fines, glabres, vert clair dessus et vert jaune dessous. Fleurs pentamères, en touffes denses, plus courtes que les pétioles. Corolle à pétales à moitié soudés et lobes recourbés, dégageant les étamines. Un seul fruit par inflorescence parvient à maturité; il est allongé, avec une coque jaune et une chair sèche et mince.

Habitat: sols riches de petites collines des Bahamas et des îles Caraïbes. En Amérique du Nord seulement dans le sud de la Floride.

Bois: très lourd, extrêmement dur, résistant et à grain fin.

Intérêts: inattaqué par les mollusques xylophages (*teredo*), aussi est-il utilisé dans la construction navale. Fruits consommés par les animaux. Bois le plus lourd des essences nord-américaines de taille comparable.

MASTIC.

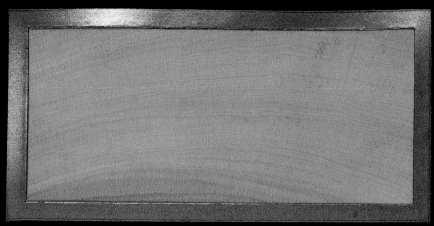

TRANSVERSE SECTION.

RADIAL SECTION.

TANGENTIAL SECTION.

Ger. Mastixbaum. Fr. Acomat (Martinique). Acomat franc (Guadeloupe).

Sp. Tocuma amarillo. Caya.

S | SPURGE FAMILY
Oysterwood, Crabwood

FAMILIE DER WOLFSMILCHGEWÄCHSE
–

FAMILLE DES EUPHORBES ET DE L'HÉVÉA
Gymnanthe vénéneux, Bois droit

EUPHORBIACEAE
Gymnanthes lucida Sw.

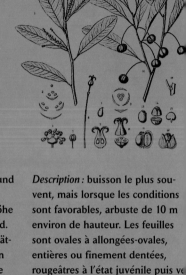

Description: Mostly a shrub, and growing into a small tree some 35 ft. (10 m) high only in favorable conditions. The ovate to oblong-ovate leaves are entire or only finely serrate, and are a reddish color at first, but become dark green later. They remain on the tree until their second summer. The dark brown bark is very thin. The roundish, very dark brown fruits hang on slender stalks.

Habitat: The tree is to be found in abundance on the not too damp valley soils of southern Florida.

Wood: The rich dark brown heartwood is hard, heavy, and polishable. The surrounding sapwood is bright yellow.

Use: The wood is used for the manufacture of small objects, and the larger trunks also for house building. Thin cross sections of the wood are used, under the name Cuban Oyster Wood, in the manufacture of cabinets of the finest quality.

Beschreibung: Meist als Strauch und nur unter günstigen Umständen zu einem kleinen, etwa 10 m Höhe erreichenden Baum aufwachsend. Die ovalen bis länglich ovalen Blätter sind ganzrandig oder nur fein gesägt und besitzen anfangs eine rötliche, später aber dunkelgrüne Farbe. Sie bleiben bis in ihren zweiten Sommer am Baum. Die dunkelbraune Borke ist sehr dünn. Die rundlichen, sehr dunkelbraunen Früchte hängen an schlanken Stielen.

Vorkommen: Der Baum ist häufig auf nicht zu feuchten Niederungsböden Süd-Floridas zu finden.

Holz: Das kräftig dunkelbraune Kernholz ist hart, schwer und schön polierbar. Das umgebende Saftholz ist hellgelb.

Nutzung: Das Holz wird zur Herstellung kleiner Gegenstände verwendet, die größeren Stämme auch für den Hausbau. Dünne Querschnitte des Holzes werden unter dem Namen *Cuban Oyster Wood* zur Herstellung feinster Kabinette gehandelt.

Description : buisson le plus souvent, mais lorsque les conditions sont favorables, arbuste de 10 m environ de hauteur. Les feuilles sont ovales à allongées-ovales, entières ou finement dentées, rougeâtres à l'état juvénile puis ve sombre, et demeurant sur l'arbre jusqu'à leur deuxième été. L'écorce brun foncé est très fine. Les fruits sphériques sont d'un brun très sombre et portés par des pédoncules étroits.

Habitat : fréquent sur les sols pas trop humides des zones basses dans le sud de la Floride.

Bois : dur, lourd ; susceptible de prendre un beau poli. Bois de cœur d'un brun foncé intense ; aubier jaune clair.

Intérêts : utilisé pour fabriquer des petits objets ; les troncs plus gros sont également employés dans la construction de maisons. Les coupes transversales fines, commercialisées sous le nom de *Cuba oyster wood*, servent aux travaux d'ébénisterie les plus fins.

CRABWOOD. POISONWOOD.

RADIAL SECTION.

TANGENTIAL SECTION.

Fr. Bois madre. Sp. Yaiti (Cuba); Palo de tabacco (Santo Domingo)

S | SPURGE FAMILY
White Wood

FAMILIE DER WOLFSMILCHGEWÄCHSE
Florida Pflaumenbaum

FAMILLE DES EUPHORBES ET DE L'HÉVÉA
Drypete à écorce blanche

EUPHORBIACEAE
Drypetes glauca Nutt.
Syn.: *D. keyensis* Urban

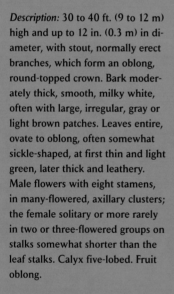

Description: 30 to 40 ft. (9 to 12 m) high and up to 12 in. (0.3 m) in diameter, with stout, normally erect branches, which form an oblong, round-topped crown. Bark moderately thick, smooth, milky white, often with large, irregular, gray or light brown patches. Leaves entire, ovate to oblong, often somewhat sickle-shaped, at first thin and light green, later thick and leathery. Male flowers with eight stamens, in many-flowered, axillary clusters; the female solitary or more rarely in two or three-flowered groups on stalks somewhat shorter than the leaf stalks. Calyx five-lobed. Fruit oblong.

Habitat: Dry sandy soils, together with other shrubs, also as secondary vegetation in deforested regions. In North America, confined to southern Florida, and even there extremely rare.

Wood: Heavy, hard, not strong, brittle, and close-grained.

Use: Because of its small diameter it is used for light structural work, wagon and coach-building, turnery, etc. (Begemann 1983).

Beschreibung: 9 bis 12 m hoch bei bis zu 0,3 m Durchmesser. Kräftige, normalerweise aufrechte Äste, die eine längliche, oben runde Krone bilden. Borke mäßig dick, glatt, milchig weiß, oft mit großen, unregelmäßigen, grauen oder blassbraunen Flecken. Blätter ganzrandig, oval bis länglich, oft etwas sichelförmig, anfangs dünn und hellgrün, später dick und ledrig. Männliche Blüten in achselständigen Büscheln, weiblich einzeln oder seltener in zwei- bis dreiblütigen Gruppen auf Blütenstielen, die etwas kürzer als die Blattstiele bleiben. Kelch fünflappig. Blüte mit acht Staubblättern. Frucht länglich.

Vorkommen: Trockene, sandige Böden, zusammen mit anderen Sträuchern auch als Sekundärvegetation entwaldeter Gebiete. Innerhalb Nordamerikas nur in Süd-Florida, dort einer der seltensten tropischen Bäume.

Holz: Schwer, hart, nicht fest, brüchig und mit feiner Textur.

Nutzung: Wegen des geringen Durchmessers nur eingeschränkt nutzbar, z. B. für leichte Konstruktionen, Wagen- und Karosseriebau, Drechslerei etc. (Begemann 1983).

Description : 9 à 12 m de hauteur sur un diamètre de 0,3 m. Branches vigoureuses, généralement dressées, constituant un houppier allongé et rond sur le dessus. Ecorce moyennement épaisse, lisse, d'un blanc laiteux, souvent ponctuée de grosses taches irrégulières, grises ou brun pâle. Feuilles entières, ovales à allongées, souvent plus ou moins en forme de faucille, fines et vert clair jeunes puis épaisses et coriaces. Fleurs mâles en touffes axillaires, fleurs femelles solitaires ou plus rarement en groupes de 2 à 4 fleurs sur des pédoncules plus courts que les pétioles. Calice pentalobé et 8 étamines. Fruit allongé.

Habitat : sols sablonneux secs ; constitue aussi avec d'autres buissons la végétation secondaire des zones déboisées. En Amérique du Nord, croît uniquement dans le sud de la Floride où il est un des arbres tropicaux les plus rares.

Bois : lourd, dur, non résistant, cassant et à grain fin.

Intérêts : son emploi se limite, en raison de son faible diamètre, aux constructions légères ainsi qu'à la carrosserie, au charronnage et au tournage (Begemann 1983).

FLORIDA WHITEWOOD. FLORIDA PLUM.

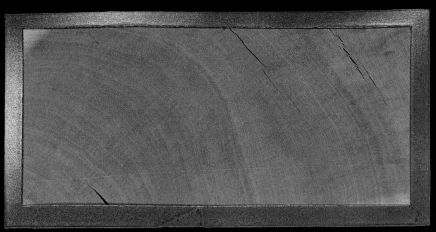

TRANSVERSE SECTION.

RADIAL SECTION.

TANGENTIAL SECTION.

Ger. PFLAUME VON FLORIDA. Fr. PRUNE DE FLORIDA.

Sp. HUESO (Sp. W. I.) VARITAL (Porto Rico)

S | STORAX FAMILY

Silver Bell Tree, Snowdrop Tree, Four-winged Halesia

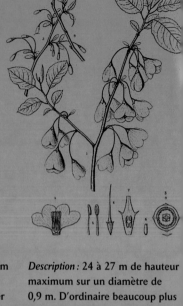

FAMILIE DER STORAXBAUMGEWÄCHSE
Vierflügelige Halesie

FAMILLE DE L'ALIBOUFIER
Halésier de la Caroline, Arbre aux clochettes d'argent

STYRACACEAE
Halesia carolina L.
Syn.: *Mohrodendron carolinum* Britton

Description: A maximum height of 85 to 95 ft. (24 to 27 m) and 3 ft. (0.9 m) in diameter. Normally considerably smaller and frequently a shrub with many stems. Trunk tall and straight with short, stout branches, which form a narrow crown. Bark moderately thick, light reddish brown and marked by mainly longitudinal furrows. The surface separates into small, papery scales. Leaves ovate to ovate-oblong, lustrous pale green, with a hairy lower surface. Flowering begins when about a third of the leaves have unfolded. Petals pure white, slightly lobed, in small fascicles or clusters on long, thin stalks. Fruits four-winged.

Habitat: Densely wooded slopes and riverbanks in south-eastern North America.

Wood: Light, soft, and close-grained.

Use: Since the first half of the 18th century the plant has been grown in parks and gardens of North America, and northern and central Europe, and is valued for the abundance of its strinkingly beautiful flowers.

Beschreibung: Maximal 24 bis 27 m hoch bei 0,9 m Durchmesser. Normalerweise wesentlich kleiner und oft als vielstämmiger Strauch. Gerader, hoher Stamm mit kurzen, kräftigen Ästen, die eine schmale Krone bilden. Borke mäßig dick, hell rotbraun und von breiten, vorwiegend längs orientierten Rissen durchzogen. Die Oberfläche schält sich in papierartigen Schüppchen ab. Blätter oval bis oval-länglich, leuchtend hellgrün mit behaarter Unterseite. Die Blüte beginnt, wenn die Blätter zu etwa einem Drittel entfaltet sind. Blütenkronblätter hellweiß, leicht gelappt, in kleinen Büscheln oder Trauben auf langen, dünnen Stielen. Früchte vierseitig geflügelt.

Vorkommen: Reich bewaldete Hänge und Flussufer im südöstlichen Nordamerika.

Holz: Leicht, weich und mit feiner Textur.

Nutzung: Seit der ersten Hälfte des 18. Jh. in Parks und Gärten Nordamerikas, Nord- und Mitteleuropas eingeführt, wo die Pflanze wegen ihrer in großer Zahl auftretenden, attraktiven Blüten geschätzt wird.

Description: 24 à 27 m de hauteur maximum sur un diamètre de 0,9 m. D'ordinaire beaucoup plus petit et souvent sous forme de buisson. Grand tronc droit garni de rameaux courts et vigoureux, formant un houppier étroit. Ecorce brun rouge clair, moyennement épaisse, à fissures larges, longitudinales pour la plupart, et s'exfoliant en petits feuillets écailleux. Feuilles ovales à ovales-allongées, vert clair brillant, pubescentes à la face inférieure. Les fleurs éclosent lorsqu'un tiers des feuilles environ se sont déployées. Pétales blanc clair, légèrement lobés, en petits bouquets ou en grappes portées par de grêles pédoncules. Fruits piriformes, à quatre ailes.

Habitat: pentes richement boisées et berges dans le sud-est de l'Amérique du Nord.

Bois: léger, tendre et à grain fin.

Intérêts: introduit dans les parcs et les jardins d'Amérique du Nord, d'Europe septentrionale et centrale depuis la première moitié du 18e siècle. Belle et abondante floraison.

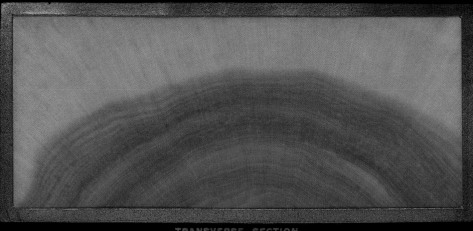

TRANSVERSE SECTION.

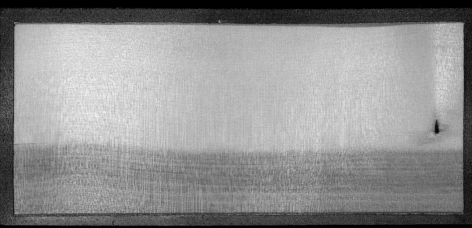

RADIAL SECTION.

TANGENTIAL SECTION.

S | SWEETLEAF FAMILY

Sweet-wood, Horse Sugar, Florida Laurel

FAMILIE DER SAPHIRBEERENGEWÄCHSE
Färber-Rechenblume

FAMILLE DU SYMPLOCOS
Symploque commun, Feuille douce commune

SYMPLOCACEAE
Symplocos tinctoria (L.) L'Hér.

Description: 30 to 35 ft. (9 to 10 m) high and 6 to 8 in. (0.15 to 0.2 m) in diameter, frequently a shrub. Branches slender, erect, forming an open crown. Bark moderately thick, ash-gray with a slightly red tinge, and occasionally with narrow fissures, roughened by warty excrescences. Leaves oblong, mostly distinctly toothed, sweet to the taste. Flowers creamy white, almost sessile, in many-flowered clusters in the axils of the leaves of the previous year. Corolla divided almost to the base into five lobes, stamens in five groups. Fruit a dark orange to brown, ovoid drupe, surrounded by a five-lobed calyx.

Habitat: Moist, rich soils in the shade of dense forests in eastern North America, in the Gulf states also on the edges of cypress swamps.

Wood: Light, soft, and close-grained. Many thin medullary rays. Heartwood light red or brown; sapwood thick, often almost white.

Use: Leaves readily eaten by cattle and horses, also together with the bark the source of a yellow dye.

Beschreibung: 9 bis 10 m hoch bei 0,15 bis 0,2 m Durchmesser. Häufig als Strauch. Äste schlank, aufrecht, eine offene Krone bildend. Borke mäßig dick, aschgrau mit leichter roter Tönung und gelegentlich mit schmalen Rissen, aufgeraut durch warzenartige Auswüchse. Blätter länglich, meist deutlich gezähnt. Süß schmeckend. Blüten cremeweiß, fast sitzend in vielblütigen Büscheln in den Achseln der vorjährigen Blätter. Blütenkrone fast bis zur Basis in 5 Lappen geteilt, Staubblätter in 5 Gruppen. Frucht dunkelorange bis braun, ovale Steinfrucht, außen von einem fünflappigen Kelch umgeben.

Vorkommen: Feuchte, nährstoffreiche Böden im Schatten dichter Wälder im östlichen Nordamerika, in den Golfstaaten auch an den Rändern von Zypressensümpfen.

Holz: Leicht, weich und mit feiner Textur. Viele dünne Markstrahlen. Kern hellrot oder braun; Splint dick, oft fast weiß.

Nutzung: Blätter dienen als Vieh- und Pferdefutter, aus ihnen und der Rinde kann eine gelbe Farbe gewonnen werden.

Description: 9 à 10 m de hauteur sur un diamètre de 0,15 à 0,2 m. Souvent sous forme de buisson. Branches fines, dressées, constituant un houppier ouvert. Ecorce moyennement épaisse, gris cendré légèrement rouge, creusée parfois de fissures étroites et verruqueuses. Feuilles allongées, à marge très dentée en général. Fleurs blanc crème, subsessiles, en grappes multiflores, à l'aisselle des feuilles de l'année précédente. Corolle de pétales soudés à la base, étamines groupées en 5 bouquets. Drupe orange sombre à brune, à noyau ovale et incluse dans le calice pentalobé.

Habitat: sols riches et humides sous le couvert des forêts dans l'es de l'Amérique du Nord, bord des marécages de cyprès dans les Etats du golfe du Mexique.

Bois: léger, tendre et à grain fin. Nombreux rayons médullaires fins Bois de cœur rouge clair ou brun ; aubier épais, souvent presque blanc.

Intérêts: le bétail et les chevaux broutent les feuilles dont on extra ainsi que de l'écorce, une teinture jaune.

TRANSVERSE SECTION.

RADIAL SECTION.

TANGENTIAL SECTION.

T | TEA FAMILY
Loblolly Bay, Black Laurel

FAMILIE DER TEESTRAUCHGEWÄCHSE
Langstielige Gordonie

FAMILLE DU THÉIER
Gordonie à feuilles glabres, Alcée de Floride

THEACEAE
Gordonia lasianthus (L.) Ellis

Description: 70 to 85 ft. (20 to 25 m) high and 14 to 16 in. (0.35 to 0.40 m) in diameter, but usually much smaller, and sometimes shrubby (even in the same locality). Bark thick, dark reddish brown, divided by furrows into conspicuous parallel rounded ridges. Foliage evergreen, the leaves like those of Laurel or Bay.

Habitat: Rich, damp, low-lying land and swamp margins in association with Red Maple, magnolias, avocados, and tupelos. Found with the Cabbage Palm on the poorest sandy soils. (South-)eastern North America.

Wood: Light, soft, not strong, and close-grained; not durable. Heartwood light red; sapwood thick, paler.

Use: For furniture, because of the color of the wood.

Beschreibung: 20 bis 25 m hoch bei 0,35 bis 0,40 m Durchmesser, gewöhnlich aber kleiner, auch strauchig (an der gleichen Lokalität). Dicke, dunkel rotbraune Borke mit auffällig parallel laufenden, abgeflacht gerundeten Riefen. Laub immergrün, Blätter in „Lorbeerform" (Bay).

Vorkommen: Reiche, feuchte Niederungen und Sumpfränder in Gesellschaft von Rot-Ahorn, Magnolien, Avocados, Tupelos. Mit der Palmetto-Palme strauchig auf ärmsten Sandböden. (Süd-) Östliches Nordamerika.

Holz: Leicht, weich, nicht stark und mit feiner Textur; nicht dauerhaft. Kern hellrot; Splint dick, heller.

Nutzung: Für Möbel aufgrund der Holzfarbe.

Description: 20 à 25 m de hauteur sur un diamètre de 0,35 à 0,40 m. Ses dimensions sont en général bien inférieures; on le trouve également sous forme de buisson (dans la même station forestière). Ecorce épaisse brun rouge foncé, se creusant de sillons aux côtes aplaties, nettement parallèles. Feuillage persistant; limbe «en forme d laurier», d'où son nom anglais de «Loblolly Bay» (bay = laurier).

Habitat: sols riches, humides des secteurs déprimés et bords des marais dans le (sud-) est de l'Amérique du Nord, en association avec l'érable rouge, les magnolias, les avocatiers, les nyssas. Constitue de buissons avec le palmetto sur les sols sablonneux les plus pauvres.

Bois: léger, tendre, pas fort et à grain fin; non durable. Bois de cœur rouge clair; aubier épais, plus clair.

Intérêts: pour l'ameublement en raison de la couleur du bois.

TRANSVERSE SECTION.

RADIAL SECTION.

TANGENTIAL SECTION

Ger. Langstielige Gordonie.   Fr. Gordonia à feuilles glabre

Sp. Gordonia.

T | THEOPHRASTA FAMILY
Joewood, Sea Myrtle

FAMILIE DER SCHNECKENSAMENGEWÄCHSE

–

FAMILLE DU THEOPHRASTA
Jacqueminier des Keys

THEOPHRASTACEAE
Jacquinia keyensis Mez

Description: In North America this small tree is not more than 18 to 20 ft. (5 to 6 m) in height, with a trunk 6 to 7 in. (0.15 to 0.18 m) in diameter clothed with a thin, smooth, brownish gray bark. The thick, leathery, yellowish green leaves are oblong-ovate and entire. They are clustered at the ends of the branches. The clusters of pale yellow flowers have a very pleasant scent. The roundish fruit is orange-red in color.

Habitat: On dry coral soils of southwestern Florida. More common in the Caribbean.

Wood: The rich yellow-brown wood is heavy and hard.

Use: The beautifully marked wood would undoubtedly be used in North America if the trees grew to a larger size and were more abundant.

Beschreibung: Der kleine Baum wird in Nordamerika nicht größer als 5 bis 6 m mit einem 0,15 bis 0,18 m dicken Stamm, der von einer dünnen, glatten und braungrauen Borke umkleidet wird. Die ledrigdicken, gelbgrünen Blätter sind länglich oval und ganzrandig. Sie befinden sich gebündelt an den Enden der Äste. Die sehr wohlriechenden Blütentrauben sind blassgelb. Die rundliche Frucht ist orangerot.

Vorkommen: Auf trockenen Korallenböden im Südwesten Floridas. Häufiger in der Karibik.

Holz: Das kräftig gelbbraune Holz ist schwer und hart.

Nutzung: Das schön gemaserte Holz würde sich durchaus für eine Nutzung lohnen, wenn es in Nordamerika nicht so klein bliebe und häufiger vorkäme.

Description : arbuste ne dépassant pas 5 ou 6 m de hauteur en Amérique du Nord, avec un tronc de 0,15 à 0,18 m de diamètre, recouvert d'une écorce gris brun, fine et lisse. Les feuilles épaisses et coriaces sont vert jaune, allongéesovales, entières et terminales. Les grappes de fleurs sont jaune pâle et parfumées. Le fruit globuleux est rouge orange.

Habitat : sols coralliens secs du sud ouest de la Floride. Plus fréquent dans les Caraïbes.

Bois : lourd, dur et d'un brun jaune intense.

Intérêts : le bois présente de belles veinures, ce qui vaudrait la peine de l'utiliser, s'il n'était aussi petit et rare en Amérique du Nord.

TRANSVERSE SECTION.

RADIAL SECTION.

TANGENTIAL SECTION.

T | TRUMPET-CREEPER FAMILY
Black Calabash, Calabash Tree, Melon Tree

FAMILIE DER TROMPETENBAUMGEWÄCHSE
Kalabassenbaum

FAMILLE DU BIGNON
Amphitecna à larges feuilles, Bache marron,
Calebasse zombi

BIGNONIACEAE
Crescentia cujete L.
Syn.: *C. cucurbitina* L.

Description: 18 to 20 ft. (5.5 to 6.0 m) high and 10 to 11 in. (0.1 to 0.13 m) in diameter, with slender, drooping branches covered with wart-like excrescences. Bark thin, light brown with a red tinge, irregularly fissured. Leaves only at the ends of the branches, ovate-oblong, alternate, dark green on the upper surface and pale yellow-green beneath. Flowers solitary, with a strong, unpleasant smell. Corolla-tube thick and leathery, dark purple, the lower surface sometimes creamy white streaked with purple. Calyx two-lobed. Fruit ovoid-oblong, with a hard shell. Seeds large, in two parts.

Habitat: Shores and coastal strips of the Caribbean, nutrient-rich soils on hummocks and riverbanks. In North America confined to southern Florida.

Wood: Heavy, hard, and close-grained. Many scattered, open pores, and thin, scarcely distinguishable medullary rays.

Use: Does not appear to have been used in Florida. In the West Indies, on the other hand, the hard shells of the fruits are used as bowls.

Beschreibung: 5,5 bis 6 m hoch bei 0,1 bis 0,13 m Durchmesser. Mit schlanken, herabhängenden, dicht warzigen Ästen. Borke dünn, hellbraun mit roter Tönung, unregelmäßig rissig. Blätter nur an den Astenden, oval-länglich, wechselständig, dunkelgrün auf der Ober- und blass gelbgrün auf der Unterseite. Blüten einzeln, mit intensivem üblen Geruch. Kronblattröhre dick und ledrig, dunkelviolett, Unterseite mitunter cremeweiß mit violetten Streifen. Kelch zweilappig. Frucht oval-länglich mit harter Schale. Samen groß, zweiteilig.

Vorkommen: Strände und Küstenstreifen der Karibik, nährstoffreiche Böden auf kleinen Hügeln und an Flussufern. Nordamerika nur in Süd-Florida erreichend.

Holz: Schwer, hart und von feiner Textur. Viele, unregelmäßig verteilte offene Gefäße und dünne, kaum unterscheidbare Markstrahlen.

Nutzung: Scheint in Florida nicht genutzt worden zu sein. Im karibischen Raum dagegen werden die harten Fruchtschalen als Schüsseln verwendet.

Description : petit arbre de 5,5 à 6,0 m de hauteur, avec un tronc de 0,1 à 0,13 m de diamètre et des branches fines, retombantes, très verruqueuses. Ecorce mince, brun clair teinté de rouge, à texture irrégulièrement fissurée. Feuilles terminales uniquement, ovales-allongées, alternes, vert foncé dessus d'un vert jaune pâle dessous. Fleurs solitaires, dégageant une odeur désagréable et pénétrante. Corolle tubuleuse épaisse et coriace, violet sombre, parfois blanc crème rayé de violet à la face inférieure ; calice bilabié. Fruit ovale-allongé, entouré d'une coque dure et renfermant de grosses graines biloculaires.

Habitat : plages et franges littorales des Caraïbes, sols riches en éléments nutritifs sur les petites collines et au bord des cours d'eau. En Amérique du Nord, seulement dans le sud de la Floride.

Bois : lourd, dur et à grain fin. Nombreux vaisseaux ouverts, irrégulièrement répartis et rayons médullaires fins, à peine discernables.

Intérêts : semble n'avoir aucun usage en Floride. Aux Caraïbes, en revanche, les coques sont utilisées comme récipients.

BLACK CALABASH.

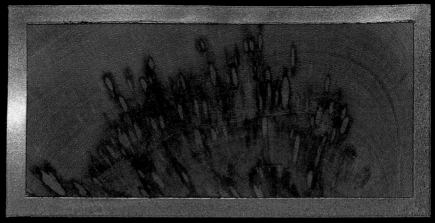

TRANSVERSE SECTION,

RADIAL SECTION,

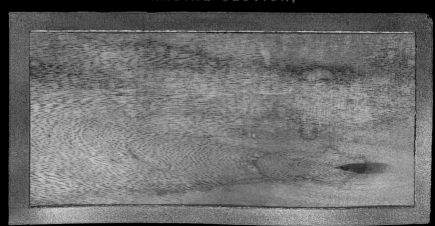

TANGENTIAL SECTION,

Ger. Schwarze Calabasse.   Fr. Colebasse (Fr.W.I.)

Sp. Higuero (Sp.W.I.), Colabazo de playa (Panama), Guautecomate (Mex.)

T | TRUMPET-CREEPER FAMILY

Common Catalpa, Indian Bean, Southern Catalpa

FAMILIE DER TROMPETENBAUMGEWÄCHSE
Gewöhnlicher Tulpenbaum

FAMILLE DU BIGNON
Catalpa commun, Arbre aux haricots,
Catalpa de Caroline

BIGNONIACEAE
Catalpa bignonioides Walt.
Syn.: C. catalpa (L.) H. Karst

Description: Rarely more than 70 ft. (20 m) high and 4 ft. (1.2 m) in diameter. Trunk short and stout, with a few drooping branches and thick, erect branchlets. The leaves, when rubbed, have an unpleasant smell. They turn black and fall immediately after a frost. Leaves large, unfolding relatively late, followed by the highly decorative flowers in erect panicles ("pyramids"). Flowers bell-shaped, flattened, with yellow nectar guides and crowded purple spots on a white ground. Fruits long, thin, tobacco-colored capsules, remaining throughout the winter on the leafless tree.

Habitat: Ornamental park tree in North America and Europe; occasionally found growing wild. Its original home can no longer be determined with certainty. It probably comes from the Gulf of Mexico and its hinterland from eastern Louisiana to western Georgia.

Wood: Soft, coarse-grained, durable in the ground.

Use: For fence posts, balustrades, etc. Bark rich in tannin, bitter.

Beschreibung: Selten höher als 20 m bei 1,2 m Durchmesser. Kurzer, gedrungener Stamm mit wenigen gekrümmten Ästen und dicken, aufrechten Zweigen. Blätter, wenn gerieben, mit unangenehmem Geruch. Nach dem ersten Frost fallen diese schwarz gefärbt sofort ab. Großes Laub, relativ spät entfaltet. Danach die höchst schmucken Blüten in aufrechten Rispen ("Pyramiden"). Die Blüten abgeflacht glockig mit gelben Saftmalen und dicht stehenden purpurnen Flecken auf weißem Grund. Früchte lange, dünne, tabakbraune Kapseln, hängen den Winter über am kahlen Baum.

Vorkommen: In Nordamerika und in Europa überall als ornamentaler Parkbaum gepflanzt, bisweilen verwildert. Das natürliche Heimatgebiet ist nicht mehr sicher auszumachen. Vermutlich ist es die Golfküste und das Hinterland von Ost-Louisiana bis West-Georgia.

Holz: Weich und mit grober Textur, dauerhaft im Boden.

Nutzung: Für Zaunpfähle, Geländer usw. Rinde gerbstoffreich, bitter.

Description : rarement plus de 20 m de hauteur sur un diamètre de 1,2 m. Tronc plus large que haut, armé de quelques grosses branches tortueuses et d'épais rameaux dressés. Grandes feuilles dégageant une odeur désagréable au froissement et tombant sans avoir même changé de couleur, hormis le noir provoqué par le gel. Feuillaison relativement tardive. Fleurs tubuleuses, blanches, tachées de pourpre avec deux raies jaunes à la gorge et s'épanouissant en grosses panicules dressées. Le fruit est une capsule très allongée, pendante, couleur tabac et persistante en hiver sur l'arbre nu.

Habitat : Planté comme arbre d'ornement en Amérique du Nord et en Europe ; parfois à l'état sauvage. La provenance d'origine est difficile à localiser, probablement sur la côte du Golfe et l'arrière-pays de l'est de la Louisiane à l'ouest de la Géorgie.

Bois : tendre et à grain grossier ; durable sur pied.

Intérêts : pieux de clôture, balustrades etc. Ecorce amère, riche en tannins.

TRANSVERSE SECTION.

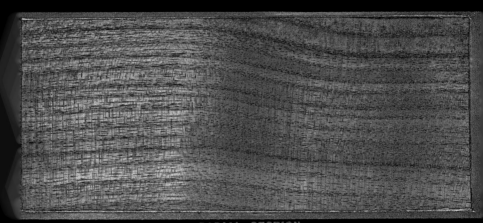

RADIAL SECTION.

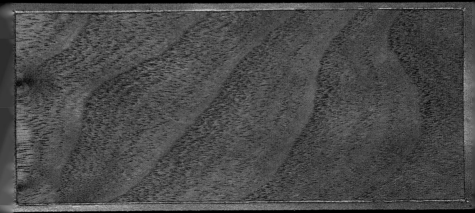

TANGENTIAL SECTION.

Trompetenbaum. Bois Shavanon Sp Catalpa.

T | TRUMPET-CREEPER FAMILY
Desert Willow

FAMILIE DER TROMPETENBAUMGEWÄCHSE
Wüsten-Weide

FAMILLE DU BIGNON
Chilopsis à feuilles linéaires

BIGNONIACEAE
Chilopsis linearis (Car.) Sweet

Description: 24 to 40 ft. (6 to 10 m) high and to 12 in. (up to 0.3 m) in diameter. Trunk slender, more or less inclining, often hollow. Bark dark brown with broad, reticulated ridges and small, thick scales on the surface. Attractive bell-shaped flowers, produced successively over several months; violet-scented at night.

Habitat: Riverbanks, desert valleys, dry, gravely, porous soils in southwestern North America.

Wood: Soft, not strong, close-grained.

Use: Unknown.

Beschreibung: 6 bis 10 m hoch bei bis zu 0,3 m Durchmesser. Schlanker, mehr oder weniger geneigter Stamm, oft hohl. Borke dunkelbraun mit breiten, vernetzten Riefen, an deren Oberfläche kleine, dicke Schuppen. Schöne, glockige Blüten, sukzessive durch mehrere Monate hindurch, nachts mit Veilchenduft.

Vorkommen: Flussufer, Senken in der Wüste, auf trocken-kiesigen, porösen und durchlässigen Böden im südwestlichen Nordamerika.

Holz: Weich, nicht fest und mit feiner Textur.

Nutzung: Unbekannt.

Description : 6 à 10 m de hauteur sur un diamètre allant jusqu'à 0,3 m. Tronc grêle, plus ou moins penché, souvent creux. Ecorce brun sombre parcourue de larges côtes entrelacées et recouvertes de petites écailles épaisses. Belle floraison en clochettes, remontante sur plusieurs mois et exhalant la nuit un parfum de violette.

Habitat : berges, cuvettes dans le désert, sols secs graveleux, poreux et perméables du sud-ouest de l'Amérique du Nord.

Bois : tendre, non résistant et à grain fin.

Intérêts : inconnus.

Desert Willow, Flowering Willow.

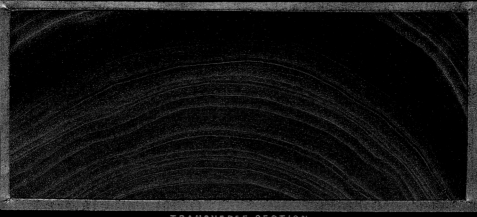

TRANSVERSE SECTION.

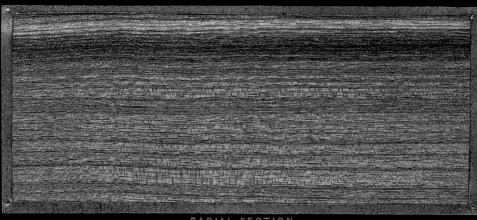

RADIAL SECTION.

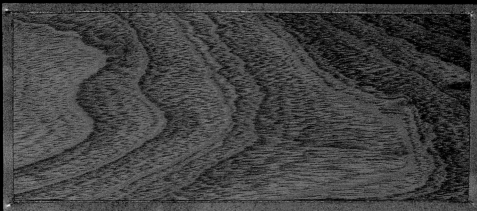

TANGENTIAL SECTION

Ger. Wüste-Weide.　　　Fr. Saule du desert.

Sp. Sauce del desirto.

T | TUPELO FAMILY

Black Gum, Sour Gum, Pepperidge, Black Tupelo

FAMILIE DER TUPELOGEWÄCHSE
Wald-Tupelo

FAMILLE DU TUPELO
Nyssa sylvestre, Gommier jaune, Toulépo (Canada)

NYSSACEAE
Nyssa sylvatica Marshall

Description: 80 to 100 ft. (25 to 30 m) high and 3 to 5 ft. (0.9 to 1.5 m) in diameter, with a round crown. Foliage a fresh green, with red and purple tints in the fall.

Habitat: Swampy regions and wet lowlands, in association with Red Maple, Swamp White Oak, Black Ash, Water Ash, and other water-loving trees in eastern North America. On the slopes of the Allegheny Mountains also found on well-drained soils.

Wood: Fairly light and soft, tough and close-grained; the contorted arrangement of the fibers renders splitting and working rather difficult.

Use: Because of the contorted fibers, it is particularly suitable for hubs of wheels, rollers and wooden cylinders, clogs, etc. Cut thin, used for fruit crates and slatted baskets.

Beschreibung: 25 bis 30 m hoch bei 0,9 bis 1,5 m Durchmesser. Runde Krone. Laub frischgrün, rote und purpurne Herbstfarben.

Vorkommen: Sumpfgebiete und nasses Tiefland, in Gesellschaft von Rot-Ahorn, Sumpf-Weiß-Eiche, Schwarzer Esche, Wasser-Esche und anderen wasserliebenden Bäumen im östlichen Nordamerika. An den Hängen der Alleghenies auch auf gut drainierten Böden anzutreffen.

Holz: Ziemlich leicht und weich, zäh und mit feiner Textur; Drehung der Fasern erschwert die Spaltung und Bearbeitung.

Nutzung: Aufgrund der Faserdrehung besonders geeignet für Radnaben, Walzen und Rollhölzer, Holzschuhe u.a.m. Dünn geschnitten für Fruchtkisten und Lattenkörbe.

Description : arbre atteignant 25 à 30 m de hauteur sur un diamètre de 0,9 à 1,5 m. Couronne arrondie Feuillage d'un vert brillant virant au rouge plus ou moins intense en automne.

Habitat : zones marécageuses et plaines très humides en compagnie de l'érable rouge, du chêne bicolore, du frêne noir, du frêne de la Caroline et d'autres arbres affectionnant les milieux humides de l'est de l'Amérique du Nord. Se rencontre aussi sur les versants des Alleghanys dans des sols bien drainés.

Bois : assez léger et tendre, tenace et à grain fin ; fil de bois sinueux, le rendant difficile à fendre et à travailler.

Intérêts : à cause de la torsion de se fibres convient particulièrement à la fabrication de moyeux, de cylindrages, de rouleaux à pâtisserie, de sabots etc., mais aussi de caisses de fruits et de cageots, lorsqu'il est

Tupelo, Pepperidge, Black or Yellow Gum, Sour Gum.

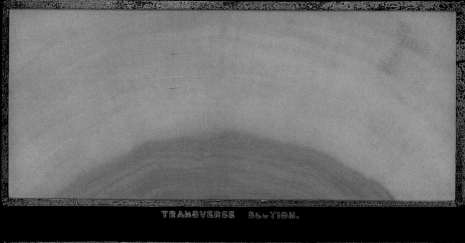

TRANSVERSE SECTION.

RADIAL SECTION.

TANGENTIAL SECTION.

Ger. Sauer Gummibaum.　　　　**Sp.** Tupelo.

Fr. Gommier multiflore.

T | TUPELO FAMILY
Cotton Gum, Tupelo Gum

FAMILIE DER TUPELOGEWÄCHSE
Wasser-Tupelo, Tupelobaum

FAMILLE DU TUPELO
Nyssa aquatique, Tupélo aquatique

NYSSACEAE
Nyssa aquatica L.

Description: 85 to 100 ft. (24 to 30 m) high and 3 to 4 ft. (0.9 to 1.2 m) in diameter. Branches relatively small, spreading, forming a narrowly oblong or pyramidal crown. Bark moderately thick, dark brown, longitudinally furrowed, and roughened on the surface with small scales. Leaves ovate or oblong, pointed, with a few pointed teeth along the margin, dark green above, with appressed hairs beneath. Flowers yellow-green on long, thin stalks in the axils of ciliate, lanceolate bracts. Fruit purple, 1 in. (3 cm) long.

Habitat: Deep and occasionally flooded swamps in south-eastern North America, and in some regions, one of the commonest and largest semi-aquatic trees.

Wood: Light, soft, not strong, and close-grained; not easy to split. Numerous thin medullary rays. Heartwood light brown, often almost white; sapwood thick, sometimes composed of more than 100 annual rings.

Use: Wooden tableware, clogs, broom handles, crates. The wood of the roots is used for the floats of nets.

Beschreibung: 24 bis 30 m hoch bei 0,9 bis 1,2 m Durchmesser. Äste relativ klein, ausladendend, eine schmal-längliche oder pyramidale Krone bildend. Borke mäßig dick, dunkelbraun, längsrissig, an der Oberfläche durch kleine Schüppchen aufgeraut. Blätter oval oder länglich, spitz, an den Rändern mit einzelnen spitzen Zähnen. Oberseite dunkelgrün, Unterseite flach behaart. Blüten gelbgrün auf dünnen, langen Stielen in den Achseln bewimperter und lanzettlicher Tragblätter. Frucht purpurn, 3 cm lang.

Vorkommen: Tiefgründige, zeitweise überschwemmte Sümpfe im südöstlichen Nordamerika, in einigen Regionen einer der häufigsten und größten semiaquatischen Bäume.

Holz: Leicht, weich, nicht fest und mit feiner Textur; schlecht spaltbar. Zahlreiche dünne Markstrahlen. Kern hellbraun, oft fast weiß; Splint dick, manchmal mehr als 100 Jahresringe.

Nutzung: Holzgeschirr, Holzschuhe, Besenstiele, Obst- und Gemüsekisten. Wurzelholz als Ersatz für Schwimmer an Netzen verwendet.

Description : de 24 à 30 m de hauteur, tronc de 0,9 à 1,2 m de diamètre. Branches assez petites, étalées, formant un houppier étroit et allongé ou pyramidal. Ecorce moyennement épaisse, brun sombre, creusée de fissures longitudinales et rendue rugueuse par de petites écailles. Feuilles ovales ou allongées, terminées en pointe et bordées de dents aiguës, vert foncé dessus, à pubescence plate dessous. Fleurs vert jaune sur de longs et fins pédoncules, insérées à l'aisselle de bractées ciliées et lancéolées. Fruits pourpres, de 3 cm de long.

Habitat : marécages profonds du sud-est de l'Amérique du Nord, soumis à des inondations ; un des arbres semi-aquatiques les plus grands et les plus fréquents dans certaines régions.

Bois : léger, tendre, non résistant et à grain fin ; difficile à fendre. Nombreux rayons médullaires. Bois de cœur brun clair, souvent presque blanc ; aubier épais, formé quelquefois de plus de 100 cernes annuels.

Intérêts : vaisselle, sabots, manches à balais, cageots. Les racines offrent un bois de substitution pour les flotteurs de filets.

TRANSVERSE SECTION.

RADIAL SECTION.

TANGENTIAL SECTION.

T | TUPELO FAMILY
Ogeechee Lime, Sour Tupelo

FAMILIE DER TUPELOGEWÄCHSE
Weißlicher Tupelo

FAMILLE DU TUPELO
Nyssa d'Ogeechee

NYSSACEAE
Nyssa ogeche Bartr. ex Marsh.

Description: 40 to 50 ft. (12 to 15 m) high and 2 ft. (0.6 m) in diameter. A shrubby tree, often several-stemmed from the base. Bark irregularly fissured, the dark brown surface being broken up into thick, appressed, plate-like scales. Fruits red to scarlet, attractive on the bare wood.

Habitat: Flooded water meadows, rare and only local along coasts in south-eastern North America between South Carolina and Florida.

Wood: Light, soft, tough, and not strong; difficult to split.

Use: The fruits are made into a pleasant, slightly sour jam ("Ogeechee Limes") in Georgia and South Carolina.

Beschreibung: 12 bis 15 m hoch bei 0,6 m Durchmesser. Buschiger Baum, oft vom Boden aufwärts mehrstämmig. Borke unregelmäßig rissig, die dunkelbraune Oberfläche in dicken, angepressten, plattenartigen Schuppen. Früchte rot bis scharlachrot, zierend am kahlen Holz.

Vorkommen: Überschwemmte Flussauen, selten und nur lokal längs der Küsten im südöstlichen Nordamerika zwischen Südkarolina und Florida.

Holz: Leicht, weich, zäh und nicht fest; schwer zu spalten.

Nutzung: Die Früchte werden zu einer angenehm säuerlichen Marmelade, „Ogeechee Limes", verarbeitet, nur in Georgia und Südkarolina.

Description : 12 à 15 m de hauteur sur un diamètre de 0,6 m. Arbre touffu, dont le tronc est souvent multiple à partir de la base. Ecorce creusée de fissures irrégulières ; la surface brun foncé se craquelle en grosses écailles aplaties. Fruits rouges à écarlates, attrayants sur le bois nu.

Habitat : lits de hautes eaux inondés ; rare et très localisé le long des côtes du sud-est de l'Amérique du Nord entre la Caroline du Sud et la Floride.

Bois : léger, tendre, tenace et pas résistant ; difficile à fendre.

Intérêts : les fruits servent en Géorgie et en Caroline du Sud à faire une confiture agréablement acidulée, appelée «Ogeechee Limes».

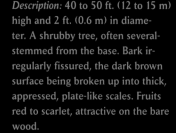

Ogeechee Lime, Sour Tupelo, Gopher Plum.

TRANSVERSE SECTION.

RADIAL SECTION.

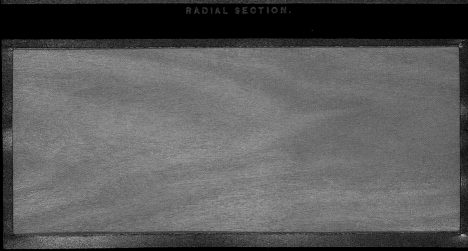

TANGENTIAL SECTION

Ger. Weisslicher Tupelobaum.   Fr. Tupelo blanchâtre.

Sp. Lima de Ogeechee.

V | VERVAIN FAMILY
Fiddlewood

FAMILIE DER EISENKRAUTGEWÄCHSE
Geigenholz

FAMILLE DE LA VERVEINE ET DU TECK
Bois-guitare épineux

VERBENACEAE
Citharexylum fruticosum L

Description: In North America the tree rarely grows more than 25 ft. (7 m) high, and its trunk is usually not more than 6 in. (0.15 m) in diameter. The slender, erect branches form an irregular crown. They are hairy at first and pale yellow. The bark of the trunk is light red-brown and separates into small, flat scales. The flowers form pendent clusters near the ends of the branches. The roundish fruits are reddish brown. Their thin, juicy, and sweet-tasting flesh surrounds the light brown seeds.

Habitat: On sandy soils in Florida.

Wood: The reddish heartwood is very hard, heavy, and strong. It is surrounded by a thin ring of paler sapwood.

Use: The wood is used for many purposes.

Beschreibung: In Nordamerika erreicht der Baum selten mehr als 7 m mit einem meist nicht mehr als 0,15 m dicken Stamm. Die schlanken, aufrechten Äste formen eine unregelmäßige Krone. Sie sind anfangs behaart und hellgelb. Die Borke am Stamm ist hell rotbraun und schuppt sich in Form sehr kleiner, flacher Plättchen. Die Blüten bilden nahe den Astenden herabhängende Trauben. Die rundlichen Früchte sind rotbraun. Ihr dünnes, saftiges und süßes Fruchtfleisch umhüllt die hellbraunen Samen.

Vorkommen: Auf sandigen Böden Floridas.

Holz: Das rötliche Kernholz ist sehr hart, schwer und fest. Es wird von einem dünnen Ring des helleren Saftholzes umgeben.

Nutzung: Das Holz ist vielseitig verwendbar.

Description: l'arbre atteint en Amérique du Nord rarement plus de 7 m de hauteur, avec un tronc ne dépassant pas 0,15 m de diamètre en général. Les branches fines et dressées qui constituent un houppier irrégulier, sont jaune clair et pubescentes à l'état juvénile. L'écorce du tronc est brun rouge clair et se desquame en plaquettes très petites et fines. Les fleurs sont réunies en grappes pendantes, subterminales. Les fruits sont globuleux et brun rouge, à pulpe juteuse, fine et douce enveloppant les graines brun clair.

Habitat: sols sablonneux de la Floride.

Bois: très dur, lourd et résistant. Bois de cœur rougeâtre; aubier mince, plus clair.

Intérêts: le bois se prête à de multiples usages.

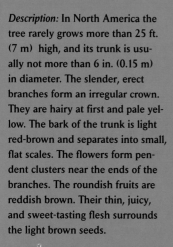

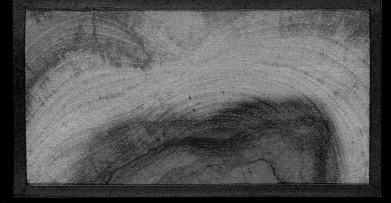

TRANSVERSE SECTION.

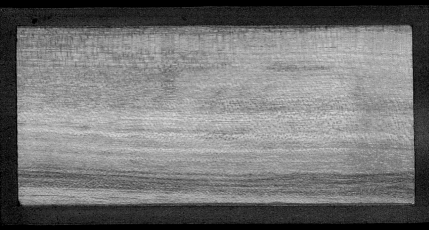

RADIAL SECTION.

TANGENTIAL SECTION.

Fr. Bois de guitare.   Sp. Higuerillo (Porto Rico).

W | WALNUT FAMILY

Big Shellbark Hickory, Bottom Shellbark Hickory, King-nut

FAMILIE DER WALNUSSGEWÄCHSE
Königsnuss

FAMILLE DU NOYER
Caryer lacinié, Hickory écailleux

JUGLANDACEAE
Carya laciniosa (Michx. f.) G. Don
Syn.: Hicoria laciniosa (Michx. f.) Sarg.

Description: 130 ft. high (40 m)and 3 to 4 ft. (0.9 to 1.2 m) in diameter. Similar to the Shagbark Hickory in having gray bark divided into plates, but in this species the edges of the plates do not curl upwards before they detach themselves. The seeds are pleasant in taste, but not quite so good as those of the Shagbark Hickory.

Habitat: Low-lying ground with the Eastern Cottonwood, Nettle Tree, Red Elm, Black Gum, Sweet Gum, Swamp White Oak and Burr Oak, and Black and Red Maple in the interior of (north-)eastern North America.

Wood: Like that of the other hickories and often confused with the wood of the Shagbark Hickory.

Use: As for the other hickories.

Cultivated in England since 1804.

Beschreibung: 40 m hoch bei 0,9 bis 1,2 m Durchmesser. Schindelartige, graue Borke jener der Schindelborkigen Hickory ähnlich, sich aber nicht mit nach außen gebogenen Enden ablösend. Samen frisch wohlschmeckend, aber nicht ganz so gut wie die der Schindelborkigen Hickory.

Vorkommen: Flachland mit Kanadischer Schwarz-Pappel, Zürgelbaum, Rot-Ulme, Wald-Tupelo, Amberbaum, Sumpf-Weiß-Eiche und Klettenfrüchtiger Eiche, Schwarz- und Rot-Ahorn im Inneren des (nord-)östlichen Nordamerika.

Holz: Wie bei den anderen Hickories und oft mit dem der Schindelborkigen Hickory verwechselt.

Nutzung: Wie bei den anderen Hickories.

In England seit 1804 kultiviert.

Description : 40 m de hauteur sur un diamètre de 0,9 à 1,2 m. Ecorce grise avec des lanières se desquamant comme chez le noyer blanc d'Amérique mais sans se relever au bout. La chair des graines fraîches est douce, moins bonne cependant que celle du noyer blanc.

Habitat : sols de plaine dans les régions intérieures du (nord-) est de l'Amérique du Nord, en association avec le peuplier noir, le micocoulier, l'orme rouge, le nyssa sylvestre, le liquidambar, les chênes bicolores et à gros fruits, les érables noir et rouge.

Bois : identique à celui des autres hickorys et souvent confondu avec celui du noyer blanc.

Intérêts : identiques à ceux des autres caryers.

Cultivé en Angleterre depuis 1804.

RADIAL SECTION.

TANGENTIAL SECTION

Ger. Gefurchte Hickory.   Sp. Nogal surcado.

Fr. Noyer grand d'Amerique.

W | WALNUT FAMILY
Bitternut, Swamp Hickory

FAMILIE DER WALNUSSGEWÄCHSE
Bitter-Nuss

FAMILLE DU NOYER
Caryer cordiforme, Hickory amer, Noyer amer (Canada)

JUGLANDACEAE
Carya cordiformis (Wangenh.) K. Koch
Syn.: *C. amara* Nutt., *Hicoria minima* (Marshall) Britton

Description: 100 ft. (30 m) high and 2 to 3 ft. (0.6 to 0.9 m) in diameter. Trunk straight, columnar, with gray-brown, unmistakable reticulate bark. Kernels of the nuts very bitter.

Habitat: Less demanding than all the other hickories and therefore more widespread, it occupies low-lying, damp soils in association with Red and Silver Maple, Black Ash, elms, etc., but also high, rolling uplands in the interior of eastern North America. Extends furthest to the north and west.

Wood: Heavy, very hard, tough, strong, and close-grained. Heartwood dark brown; sapwood light brown to almost white.

Use: For tool handles, agricultural implements, yokes for oxen, hoops, excellent as firewood.

Beschreibung: 30 m hoch bei 0,6 bis 0,9 m Durchmesser. Gerader, säulenförmiger Stamm mit graubrauner, unverkennbar netzförmiger Borke. Kerne der Nüsse sehr bitter.

Vorkommen: Bewohnt, weniger anspruchsvoll als alle anderen Hickories und daher wohl verbreiteter, tiefliegende, feuchte Böden in Gesellschaft von Rot- und Silber-Ahorn, Schwarz-Esche, Ulmen u. a., aber auch Bodenwellen in höheren Lagen im Inneren des östlichen Nordamerika. Geht am weitesten nach Norden und Westen.

Holz: Schwer, sehr hart, zäh, fest und mit feiner Textur. Kern dunkelbraun; Splint hellbraun bis nahezu weiß.

Nutzung: Für Handgriffe an Werkzeug, Ackergerät, Ochsenjoche, Reifen, ausgezeichnetes Brennholz.

Description: 30 m de hauteur sur un diamètre de 0,6 à 0,9 m. Tronc droit, columnaire, recouvert d'une écorce brun gris réticulée, très caractéristique. Fruits globuleux à chair très amère.

Habitat : moins exigeant que toutes les espèces voisines et donc plus répandu ; occupe aussi bien les plaines alluviales, en association avec l'érable rouge et l'érable argenté, le frêne noir, les ormes etc., que les plis de terrains plus élevés dans les régions intérieures est de l'Amérique du Nord. Espèce se rencontrant le plus au nord et à l'ouest.

Bois : lourd, très dur, tenace, résistant et à grain fin. Bois de cœur brun foncé ; aubier brun clair à quasiment blanc.

Intérêts : utilisé pour fabriquer des manches d'outils, des instruments aratoires, des jougs, des roues. Excellent bois de chauffage.

Bitter-nut Hickory,   Swamp Hickory.

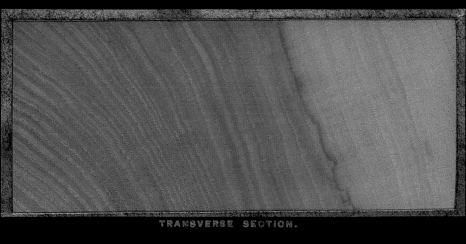

TRANSVERSE SECTION.

RADIAL SECTION.

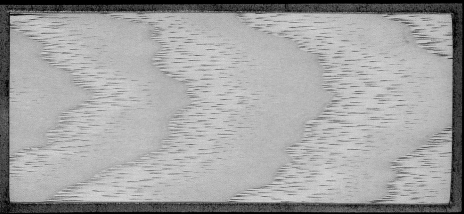

TANGENTIAL SECTION.

Ger. Bitterfrüchtige Hickory.   *Fr.* Noyer amer.

Sp. Nogal amargo..

W | WALNUT FAMILY
Black Walnut

FAMILIE DER WALNUSSGEWÄCHSE
Schwarze Walnuss

FAMILLE DU NOYER
Noyer noir, Noyer noir d'Amérique

JUGLANDACEAE
Juglans nigra L.

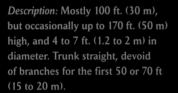

Description: Mostly 100 ft. (30 m), but occasionally up to 170 ft. (50 m) high, and 4 to 7 ft. (1.2 to 2 m) in diameter. Trunk straight, devoid of branches for the first 50 or 70 ft (15 to 20 m).

Habitat: Rich lowlands, fertile hillsides mainly in the interior of eastern North America. Less frequent in the Atlantic states.

Wood: Heavy, hard, strong, with a silky sheen, and rather coarse-grained; polishable, very durable in the ground, and easy to work. Heartwood dark brown; sapwood lighter.

Use: For furniture, interior construction, rifle stocks, ramrods (over-exploited in the Civil War), coffins, boat and shipbuilding. Mentioned as a commercial article as early as 1610. The nuts were used for food by forest-dwelling Indians but these days they are gathered and sold only locally.

Recently replanted. A park tree in the USA and Europe. Susceptible to frost.

Beschreibung: Meist 30 m, gelegentlich bis zu 50 m hoch bei 1,2 bis 2 m Durchmesser. Gerader Stamm, bis 15 oder 20 m ohne Äste.

Vorkommen: Reiches Flachland, fruchtbare Hügel vorwiegend im Innern des östlichen Nordamerika. Weniger häufig in den Atlantikstaaten.

Holz: Schwer, hart, fest, seidig und von ziemlich grober Textur; polierfähig, sehr haltbar im Boden und gut zu bearbeiten. Kern dunkelbraun; Splint hell.

Nutzung: Für Möbel, Innenausbau, Gewehrschäfte, Ladestöcke (Übernutzung im Bürgerkrieg), Särge, Boots- und Schiffbau. Bereits 1610 als Handelsware erwähnt. Die Nüsse waren ein Nahrungsmittel der Waldindianer. Heutzutage nur noch örtlich gesammelt und verkauft.

Seit Kurzem nachgepflanzt. Parkbaum in den USA und in Europa. Frostempfindlich.

Description: généralement 30 m de hauteur, parfois jusqu'à 50 m, sur un diamètre de 1,2 à 2 m. Tronc droit, dénudé jusqu'à 15 ou 20 m de haut.

Habitat: sols riches de plaine, collines fertiles en majorité dans les régions intérieures est de l'Amérique du Nord. Se rencontre moins souvent dans les Etats du littoral atlantique.

Bois: lourd, dur, résistant, satiné et à grain assez grossier; pouvant se polir, très durable sur pied et se travaillant bien. Bois de cœur brun foncé; aubier clair.

Intérêts: construction, meubles, crosses de fusils, refouloirs (exploitation excessive pendant la guerre de Sécession) et de cercueils, construction navale. Mentionné de 1610 comme article de commerce. Les noix, autrefois consommées par les Indiens des forêts, ne sont plus aujourd'hui récoltées et vendues que localement.

Espèce de reboisement depuis peu. Arbre de parcs aux Etats-Unis et en Europe. Sensible au gel.

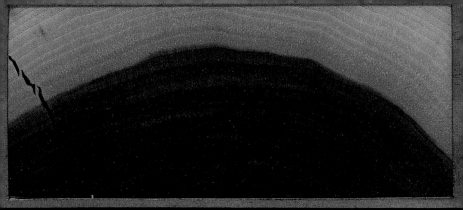

TRANSVERSE SECTION.

RADIAL SECTION.

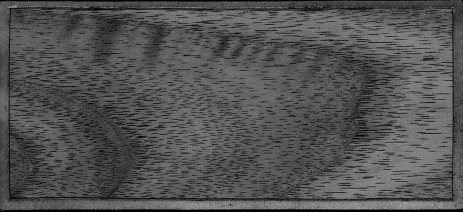

TANGENTIAL SECTION.

W | WALNUT FAMILY

Butternut, White Walnut

FAMILIE DER WALNUSSGEWÄCHSE
Butternussbaum

FAMILLE DU NOYER
Noyer cendré, Noyer à beurre, Arbre à noix longues (Canada)

JUGLANDACEAE
Juglans cinerea L.

Description: 70 to 85 (20 to 25 m), rarely to 100 ft. (30 m) high, and 3 to 4 ft. (0.9 to 1.2 m) in diameter. Without branches for up to half its height, but frequently divided from about 10 ft. (3 m) above the ground.

Habitat: Rich soils along rivers and at the foot of low hills, usually together with Beech, Yellow Birch, maples, elms, Red Spruce, etc., in (north-)eastern North America.

Wood: Light, not strong, with a silky surface and fairly coarse grain; polishable and easy to work.

Use: Interior construction, cupboards. Has a sugary sap, which can be thickened to form syrup. The nuts, from which a fine oil is obtained, are a delicious food. The husks of the fruit are used as a dye.

Beschreibung: 20 bis 25, selten bis 30 m hoch bei 0,9 bis 1,2 m Durchmesser. Bis zur halben Höhe astlos. Häufig jedoch schon bei etwa 3 m geteilt.

Vorkommen: Reiche Böden längs der Flüsse und am Fuß niedriger Hügel, gewöhnlich zusammen mit Buche, Gelb-Birke, Ahornen, Ulmen, Rot-Fichte u. a. im (nord-)östlichen Nordamerika.

Holz: Leicht, nicht fest, mit seidiger Oberfläche und ziemlich grober Textur; polierfähig und leicht zu bearbeiten.

Nutzung: Innenausbau, Schränke. Führt Zuckersaft, der sich zu Sirup eindicken lässt. Die ölhaltigen Nüsse dienen als delikates Nahrungsmittel. Die Nussschalen eignen sich zum Färben.

Description: arbre de 20 à 25 m de hauteur, atteignant rarement 30 m avec un tronc de 0,9 à 1,2 m de diamètre, non ramifié jusqu'à mi-hauteur mais se divisant à environ 3 m de haut.

Habitat: sols riches le long des cours d'eau et au pied de petites collines, généralement en association avec le hêtre, le bouleau jaune, les érables, les ormes, l'épicéa rouge etc. dans le (nord-) est de l'Amérique du Nord.

Bois: léger et peu résistant, à surface satinée et grain assez grossier, facile à travailler et susceptible d'u beau poli.

Intérêts: utilisé en menuiserie intérieure et pour fabriquer des armoires. Donne une sève sucrée qu'on peut épaissir en sirop. Noix comestibles dont on extrait une huile. Brou utilisé pour la teinture.

Butternut, **White-Walnut,** **Oil-Nut.**

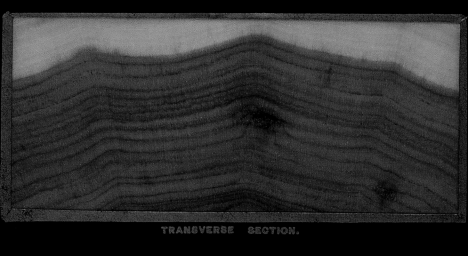

TRANSVERSE SECTION.

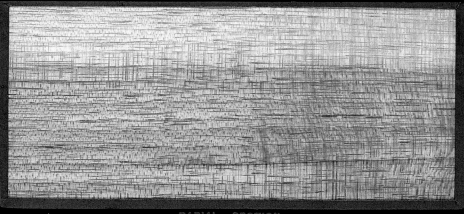

RADIAL SECTION.

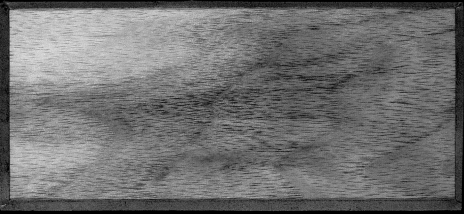

TANGENTIAL SECTION.

Ger. Aschgrauer Wallnuszbaum. *Sp.* Nogal gris.
Fr. Noyer cendré.

W | WALNUT FAMILY

California Walnut

FAMILIE DER WALNUSSGEWÄCHSE
Kalifornische Walnuss

FAMILLE DU NOYER
Noyer de Californie

JUGLANDACEAE
Juglans californica S. Watson

Description: 45 to 70 ft. (13 to 20 m) high and 14 to 16 in. (0.36 to 0.4 m) in diameter. Often smaller and shrubby. Bark dark brown, almost black. Deeply divided by broad, irregular ridges. Surface with layer of thin, appressed scales. Young branches smooth and pale. Large sweet seeds, which may be stored for many months.

Habitat: Beside streams and in low-lying areas in California.

Wood: Heavy, hard, silky-sheened, and rather coarse-grained; polishable. Heartwood dark brown, often attractively speckled and veined; sapwood thick and pale.

Use: Shade tree in California.

Beschreibung: 13 bis 20 m hoch bei 0,36 bis 0,4 m Durchmesser. Oft erheblich kleiner, auch strauchig. Borke dunkelbraun, fast schwarz. Tief geteilt in breite, unregelmäßige Riefen. Oberfläche aus dünnen, anliegenden Schuppen geschichtet. Junge Stämme glatt und fahl. Großer, süßer Samen, mehrere Monate lang haltbar.

Vorkommen: Bachufer und Niederungen Kaliforniens.

Holz: Schwer, hart, seidig und mit ziemlich grober Textur; polierfähig. Kern dunkelbraun, oft hübsch gefleckt und geädert; Splint dick und blass.

Nutzung: Schattenbaum in Kalifornien.

Description: 13 à 20 m de hauteur sur un diamètre de 0,35 à 0,4 m. Souvent bien plus petit ou buissonnant. Ecorce d'un brun sombre presque noir, se fissurant en larges crêtes irrégulières aux fines écailles superposées et adhérentes; lisse et terne sur les jeunes sujets. Grosses noix à chair douce, se conservant plusieurs mois.

Habitat: rives de ruisseaux et zones basses de la Californie.

Bois: lourd, dur, satiné et à grain assez grossier; pouvant se polir. Bois de cœur brun sombre, présentant souvent des taches et des veines d'un bel effet; aubier épais et pâle.

Intérêts: arbre d'ombrage en Californie.

California Walnut.

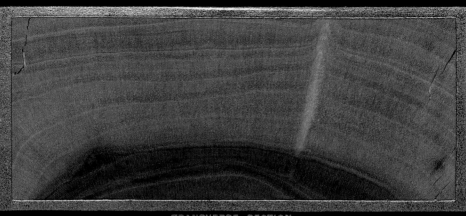

TRANSVERSE SECTION.

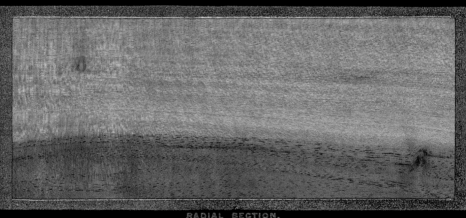

RADIAL SECTION.

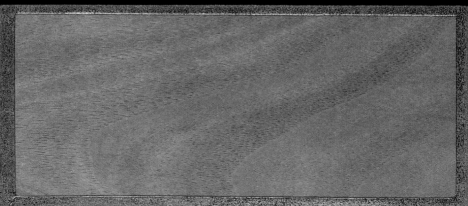

TANGENTIAL SECTION.

Ger. Californische Wallnuszbaum. *Fr.* Noyer de Californie.
Sp. Nogal de California.

W | WALNUT FAMILY

Little Walnut

FAMILIE DER WALNUSSGEWÄCHSE
Felsen-Walnuss

FAMILLE DU NOYER
Noyer à petits fruits, Noyer du Texas

JUGLANDACEAE
Juglans microcarpa Berl. var. *microcarpa*
Syn.: *J. rupestris* Engelm. ex Torr.

Description: Up to 50 ft. (15 m) high and 4 to 5 ft. (1.2 to 1.5 m) in diameter. Trunk short, dividing just above the ground. Branches short, almost rising vertically, forming a narrow crown. On moist land the outer branches hang down, forming a rounder crown. Bark thick, with deep fissures, gray to almost white on branches and on young trees. Leaves divided into 9 to 23 lanceolate to ovate-lanceolate leaflets, reddish at first, later dark yellow-green. Male flowers in long, pendent catkins; female in small, inflorescences with few flowers. Fruits and nuts globular, the latter deeply grooved.

Habitat: In the neighborhood of rivers and riverbeds, always near a continuous supply of water for the roots. South-western North America.

Wood: Heavy, hard, not very strong, with a silky sheen, and close-grained; polishable. Heartwood rich dark brown; sapwood thick, almost white.

Use: The seeds are edible, but of little economic importance, as they are small and their shells are very thick and hard.

Beschreibung: Bis 15 m hoch bei 1,2 bis 1,5 m Durchmesser. Stamm kurz, sich dicht über dem Boden teilend. Äste kurz, fast senkrecht aufstrebend, eine schmale Krone bildend. Auf feuchtem Boden dagegen äußere Äste herabhängend und eine rundere Krone bildend. Borke dick mit tiefen Rissen, an jungen Bäumen und Ästen grau bis fast weiß. Blätter in 9 bis 23 Blättchen geteilt, Fiedern lanzettlich bis oval-lanzettlich, anfangs rötlich, später dunkel gelbgrün. Männliche Blüten in lang herabhängenden Kätzchen, weibliche zu wenigen in kleinen Blütenständen. Frucht und Nuss kugelig, Letztere tief gefurcht.

Vorkommen: In der Nähe von Flüssen oder in deren Betten, stets an Stellen, wo für die Wurzeln eine kontinuierliche Wasserversorgung gewährleistet bleibt. Südwestliches Nordamerika.

Holz: Schwer, hart, nicht sehr fest, seidig, mit feiner Textur; polierfähig. Kern tief dunkelbraun; Splint dick, fast weiß.

Nutzung: Die Samen sind essbar, aber von geringer ökonomischer Bedeutung, da sie klein und die Schalen sehr dick und hart sind.

Description: arbre pouvant atteindre 15 m de hauteur sur un diamètre de 1,2 à 1,5 m. Tronc court, divisé dès la base et armé de branches courtes, presque verticales, formant un houppier étroit. Sur les sols humides, les branches extérieures sont retombantes et forment un houppier rond. Ecorce épaisse, profondément fissurée, grise à presque blanc sur les jeunes arbres et les rameaux. Feuilles composées de 9 à 23 folioles, lancéolées à ovales-lancéolées, rougeâtres puis vert jaune sombre. Fleurs mâles en longs chatons pendants, les femelles en petites inflorescences rares. Fruit et noix globuleux, celle-ci fortement sillonnée.

Habitat: près des cours d'eau ou dans leurs lits, toujours dans les stations où les racines ont une alimentation constante en eau. Dans le sud-ouest de l'Amérique du Nord.

Bois: lourd, dur, résistant, satiné, à grain fin; susceptible d'être poli. Bois de cœur brun profond; aubier épais, presque blanc.

Intérêts: les graines sont comestibles mais d'un rôle économique mineur à cause de leur petite taille et de leur coque dure et épaisse.

Mexican Walnut, Arizona Walnut.

TRANSVERSE SECTION.

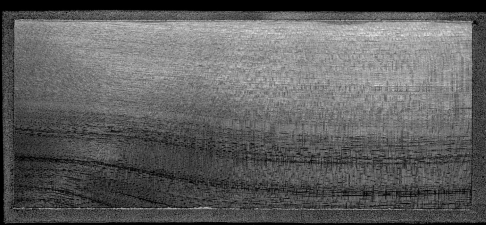

RADIAL SECTION.

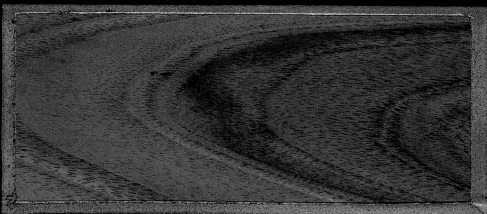

TANGENTIAL SECTION.

Ger. Arizonische Wallnuszbaum. *Fr.* Noyer d'Arizona.
Sp. Nogal de Arizona.

W | WALNUT FAMILY
Mockernut Hickory

FAMILIE DER WALNUSSGEWÄCHSE
Spottnuss

FAMILLE DU NOYER
Caryer blanc, Carya tomenteux, Noyer à noix douces (Canada)

JUGLANDACEAE
Carya alba (L.) Koch
Syn.: *C. tomentosa* (Poir.) Nutt.

Description: 100 to 110 ft. (30 to 33 m) high and 3 ft. (0.9 m) in diameter. Bark gray, rough from the shallow, scaly fissures. Fruit with a thick shell and a sweet seed.

Habitat: More likely to be found on hills and high ridges than the King-nut, which prefers low-lying ground near rivers that is susceptible to flooding. Widespread in eastern North America. In the northern part of its range it occurs with various oaks, Eastern Red Cedar, Sassafras, Sweet Birch, Sweet Gum, and Tulip Tree, mainly in the coastal region. Further to the south it becomes more common and more widely distributed.

Wood: Heavy, hard, tough, and flexible.

Use: Wood similar to that of the Shagbark Hickory and easily confused with it.

It was the first species in the genus to become known to Europeans because of its profusion on the coasts of Virginia.

Beschreibung: 30 bis 33 m hoch bei 0,9 m Durchmesser. Borke grau, rau durch undeutliche, schuppige Riefen. Frucht dickschalig mit süßem Samen.

Vorkommen: Bewohnt eher die Hügel und Höhenrücken als die Königsnuss, die ihrerseits Flussniederungen mit Überschwemmungen bevorzugt. Verbreitet im östlichen Nordamerika. Im nördlichen Teil ihres Gebietes tritt sie mit verschiedenen Eichen, Rotzeder, Sassafras, Zucker-Birke, Amberbaum und Tulpenbaum hauptsächlich in der Küstenregion auf. Weiter nach Süden wird sie häufiger und allgemeiner verbreitet.

Holz: Schwer, fest, zäh und biegsam.

Nutzung: Holz wie bei dem der Schindelborkigen Hickory und damit leicht zu verwechseln.

Durch ihre Häufigkeit an den Stränden Virginias ist sie den Europäern als Erste ihrer Gattung bekannt geworden.

Description : 30 à 33 m de hauteur sur un diamètre de 0,9 m. Ecorce grise, creusée de grossiers sillons écailleux. Fruit à brou épais et à chair douce.

Habitat : occupe plus volontiers les collines et les croupes de montagnes que le caryer lacinié, qui préfère les plaines alluviales inondables. Très répandu dans l'est de l'Amérique du Nord. Associé, dans la partie septentrionale de son aire, principalement dans les régions côtières, à différents chênes, au sassafras, au bouleau flexible, au cèdre rouge, au liquidambar et au tulipier de Virginie. Il est plus fréquent et plus largement réparti vers le sud.

Bois : lourd, résistant, tenace et flexible.

Intérêts : bois identique à celui du caryer lacinié et donc facilement confondu avec celui-ci.

Première espèce de son genre à avoir été découverte par les Européens sur le littoral de la Virginie, où il se rencontre fréquemment.

Moker-nut Hickory.

TRANSVERSE SECTION.

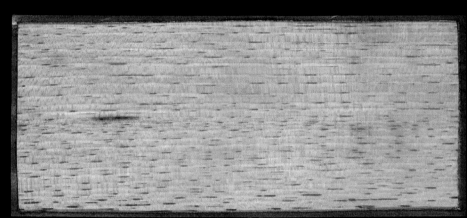

RADIAL SECTION.

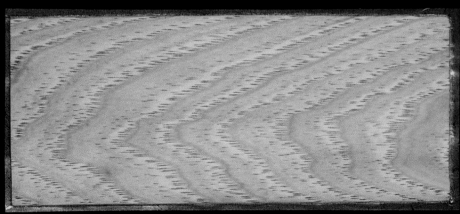

TANGENTIAL SECTION.

Ger. Weichhaarige Hickory. *Fr.* Noyer laineux. *Sp.* Nogal ve

W | WALNUT FAMILY

Pale-leaf Hickory

FAMILIE DER WALNUSSGEWÄCHSE

–

FAMILLE DU NOYER

Caryer pâle

JUGLANDACEAE
Carya texana Buckley
Syn.: *Hicoria villosa* (Sarg.) Ashe

Description: 18 to 20 (40 to 50) ft. (5.5 to 6.0 (12 to 15) m) high and 12 to 18 in. (0.3 to 0.45 m) in diameter. Branches small, the upper ones forming a narrowly oblong crown, the lower ones drooping. Bark moderately thick, light gray or gray-brown and irregularly deeply furrowed. Leaves divided into five to nine leaflets, hairy beneath and covered with small, silvery scales when young, later dark green above. Leaflets serrate in the upper half. Male flowers in slender, pendent inflorescences; the female solitary, oblong with a rectangular outline. Fruit roundish to pear-shaped. Nut small, keeled, thick-shelled.

Habitat: Sandy flatlands or arid, rocky cliffs in eastern North America.

Wood: Hard, strong, rather brittle.

Use: Presumably similar to that of the other hickories.

Beschreibung: 5,5 bis 6 (12 bis 15) m hoch bei 0,3 bis 0,45 m Durchmesser. Äste klein, die oberen eine schmale längliche Krone bildend, die unteren herabhängend. Borke mäßig dick, hellgrau oder graubraun und unregelmäßig tiefrissig. Blätter in 5 bis 9 Fiedern geteilt, unterseits behaart und jung von silbrigen Schüppchen bedeckt, später oberseits dunkelgrün. Fiedern in der oberen Hälfte gesägt. Männliche Blüten in dünnen, herabhängenden Blütenständen, weibliche einzeln, länglich mit vierkantigem Grundriss. Frucht rundlich bis birnenförmig. Nuss klein, gekielt, dickschalig.

Vorkommen: Sandige Ebenen oder sterile steinige Klippen im östlichen Nordamerika.

Holz: Hart, stark, eher brüchig.

Nutzung: Wahrscheinlich ähnlich wie bei anderen Hickories.

Description : arbuste de 5,5 à 6,0 m de hauteur (12 à 15 m max.), avec un tronc de 0,3 à 0,45 m de diamètre, portant de petites branches, les plus hautes formant une couronne étroite et allongée, les plus basses retombantes. Ecorce moyennement épaisse, gris clair ou brun gris, creusée de fissures irrégulières. Feuilles composées de 5 à 9 folioles, pubescentes dessous, couvertes de petites écailles argentées jeunes, puis devenant vert foncé dessus, dentées dans la moitié supérieure. Fleurs mâles en inflorescences fines, pendantes, fleurs femelles solitaires, allongées et à plan quadrangulaire. Fruit sphérique à piriforme. Petite noix carénée, enfermée dans une coque épaisse.

Habitat : plaines sablonneuses ou écueils rocheux, stériles de l'est de l'Amérique du Nord.

Bois : dur, fort, plutôt cassant.

Pale-leaf Hickory.

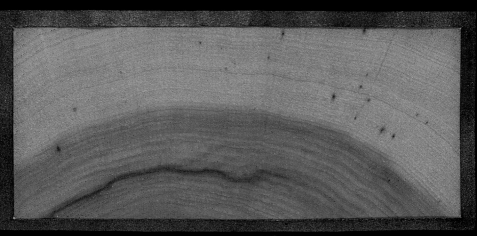

TRANSVERSE SECTION

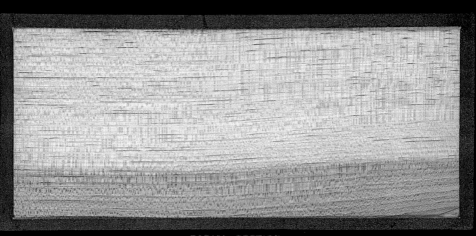

RADIAL SECTION.

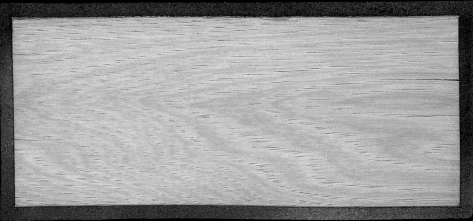

TANGENTIAL SECTION.

Ger. Zottige Hickory.   *Fr.* Noyer villeaux.

Sp. Nogal velludo.

W | WALNUT FAMILY
Pecan

FAMILIE DER WALNUSSGEWÄCHSE
Pekan-Nuss

FAMILLE DU NOYER
Caryer pacanier, Hickory de l'Illinois

JUGLANDACEAE
Carya illinoinensis (Wangenh.) K. Koch
Syn.: *Hicoria pecan* Britton

Description: 100 to 140 ft. (30 to 50 m) high, with a massive trunk up to 6 ft. (1.8 m) in diameter above the enlarged base. Branches stout, only slightly spreading, forming a narrow, inversely pyramidal crown. Only developing a broad, rounded crown where there is ample space. Bark thick, deeply furrowed, brown tinged red. Leaves divided into 9 to 11 leaflets, pale green when young, later dark yellow-green. Male flowers in long, pendant inflorescences, the female solitary, sessile. Fruit winged, nut ovoid-oblong, cylindrical, with a thin shell and a sweet-tasting seed.

Habitat: Low-lying, rich soils near streams and rivers in the interior of eastern North America, the western Gulf states, and the mountain regions of Mexico.

Wood: Heavy, hard, not very strong, brittle, and close-grained.

Use: For firewood, and occasionally in the manufacture of wagons and agricultural implements. The nuts are of great economic importance as a foodstuff. It has been in European gardens since 1766.

Beschreibung: 30 bis 50 m hoch, mit massigem Stamm von maximal 1,8 m Durchmesser oberhalb der verdickten Stammbasis. Äste kräftig, nur wenig ausgreifend, eine schmale, umgekehrt pyramidale Krone bildend. Nur bei viel Platz eine breite, kugelige Krone formend. Borke dick, tiefrissig, braun mit roter Tönung. Blätter in 9 bis 11 Fiedern geteilt, jung hellgrün, später dunkel gelbgrün. Männliche Blüten in lang herabhängenden Blütenständen, weibliche einzeln sitzend. Frucht mit Flügeln, Nuss oval-länglich, zylindrisch, dünnschalig mit süßem Samen.

Vorkommen: Flachgründige, reiche Böden in der Nähe von Bächen und Flüssen im Inneren des östlichen Nordamerika, den westlichen Golfstaaten und in den Bergländern Mexikos.

Holz: Schwer, hart, nicht sehr fest, brüchig und mit feiner Textur.

Nutzung: Feuerholz, gelegentlich im Waggonbau und für landwirtschaftliche Geräte. Nüsse als Nahrungsmittel von großer ökonomischer Bedeutung. In europäischen Gärten seit 1766.

Description: arbre de 30 à 50 m de hauteur, avec un tronc massif atteignant 1,8 m de diamètre au-dessus de la base épaissie. Branches vigoureuses, peu étalées, constituant un houppier étroit, s'inscrivant dans une pyramide renversée, mais ample et globuleux quand l'arbre a de la place. Ecorce épaisse, profondément fissurée, brune teintée de rouge. Feuilles composées de 9 à 11 folioles, vert clair à leur naissance, devenant vert jaune sombre avec l'âge. Fleurs mâles en longues inflorescences pendantes, fleurs femelles solitaires. Fruits ailés, noix ovales-allongées, cylindriques, à brou mince et chair douce.

Habitat: sols riches et peu profonds près des cours d'eau dans l'intérieur à l'est de l'Amérique du Nord, des Etats du Golfe et les régions montagneuses du Mexique.

Bois: lourd, dur, pas très résistant, cassant et à grain fin.

Intérêts: bois de feu, parfois wagons et outils agricoles. La noix est une denrée alimentaire d'une grande valeur économique. Cultivé dans les jardins européens depuis 1766.

TRANSVERSE SECTION

RADIAL SECTION.

W | WALNUT FAMILY
Pignut Hickory

FAMILIE DER WALNUSSGEWÄCHSE
Ferkelnuss

FAMILLE DU NOYER
Caryer glabre, Carya des pourceaux,
Noyer à cochons (Canada)

JUGLANDACEAE
Carya glabra (Mill.) Sweet
Syn.: *Hicoria glabra* (Mill.) Britton

Description: 85 to 100 ft. (25 to 30 m) high and 3 to 4 ft. (0.9 to 1.2 m) in diameter. Trunk slender, often forked a little above the ground. Bark gray with narrow grooves and furrows. Seeds very variable, some bitter and astringent, others with a pleasant scent.

Habitat: In large numbers on the higher ground and on ridges of hills. Apparently found at greater altitude than all other hickories, also on drier soils in eastern North America.

Wood: Heavy, hard, very tough, and strong.

Use: For handles of various kinds, agricultural implements.

Beschreibung: 25 bis 30 m hoch bei 0,9 bis 1,2 m Durchmesser. Schlanker, oft schon in geringer Höhe gegabelter Stamm. Borke grau mit engen Rillen und Riefen. Samen von großer Variationsbreite, manche herb zusammenziehend, andere mit angenehmem Aroma.

Vorkommen: Höher gelegenes Land und Höhenrücken. Angeblich in höheren Lagen als alle anderen Hickories, auch auf trockeneren Böden im östlichen Nordamerika.

Holz: Schwer, hart, sehr zäh und fest.

Nutzung: Für Stiele, Handgriffe und Ackerbaugeräte.

Description: 25 à 30 m de hauteur sur un diamètre de 0,9 à 1,2 m. Tronc élancé, souvent bas branchu. Ecorce grise, parcourue de fins sillons écailleux. Fruit très variable, à chair astringente ou aromatique.

Habitat: en grandes colonies sur les reliefs et les croupes de montagnes. Serait le plus « montagnard » des caryers, mais ne craint pas non plus les sols plus secs de l'est de l'Amérique du Nord.

Bois: lourd, dur, très tenace et résistant.

Intérêts: utilisé pour fabriquer des manches d'outils et des instruments aratoires.

Pig-nut Hickory, Brown Hickory, Black Hickory.

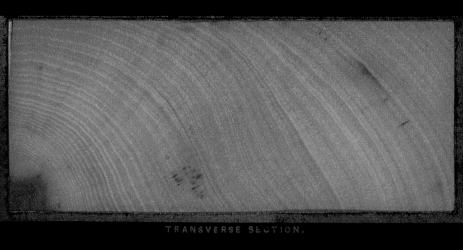

TRANSVERSE SECTION.

RADIAL SECTION.

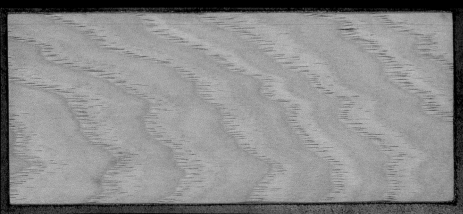

TANGENTIAL SECTION

Ger. Ferkelnusz. *Sp.* Nogal de puercos.

Fr. Noyer de cochon.

W | WALNUT FAMILY
Shagbark Hickory, Little Shellbark Hickory

FAMILIE DER WALNUSSGEWÄCHSE
Schuppenrindige Hickory

FAMILLE DU NOYER
Caryer ovale, Arbre à noix piquées (Canada)

JUGLANDACEAE
Carya ovata (Mill.) K. Koch
Syn.: *Carya alba* Nutt. (L.) Nutt. ex Elliot

Description: 85 to 100 (25 to 30 m), occasionally 135 ft. (40 m) high, and 3 to 4 ft. (0.9 to 1.2 m) in diameter. Trunk straight, columnar, with a curious bark that peels off as gray plates like roof shingles, sometimes curling upwards. The nuts are gathered in the wild for their tasty, aromatic kernels, and are sold commercially as hickory nuts.

Habitat: On low ground at the foot of slopes and on riverbanks together with American Lime, maples, Cottonwood, oaks, other hickories, etc., in the interior of eastern North America.

Wood: Heavy, very hard, tough, strong, flexible, and close-grained. Heartwood pale brown; sapwood almost white.

Use: For agricultural implements, carts, wagons, ax handles, baskets, and firewood.

Grown as a park tree in Europe since the 17th century mainly because of its striking silhouette in winter.

Beschreibung: 25 bis 30 m, gelegentlich 40 m hoch bei 0,9 bis 1,2 Durchmesser. Gerader, säulenförmiger Stamm mit bemerkenswerter grauer, in abwärts gerichteten, z. T. aufgebogenen, schindelartigen Streifen sich ablösender Borke. Die Nüsse werden wegen ihrer wohlriechenden und -schmeckenden Kerne wild gesammelt und sind als Hickorynüsse im Handel erhältlich.

Vorkommen: Tiefliegende Hangfüße und Flussufer gemeinsam mit der Amerikanischen Linde, Ahornen, Kalifornischer Pappel, Eichen, anderen Hickories u. a. im Innern des östlichen Nordamerika.

Holz: Schwer, sehr hart, zäh, fest, biegsam und mit feiner Textur. Kern hellbraun; Splint fast weiß.

Nutzung: Für Ackergerät, Karren, Wagen, Axtstiele, Körbe und Brennholz.

Schon seit dem 17. Jh. vor allem wegen der aparten Schönheit ihrer winterlichen Silhouette als Parkbaum in Europa.

Description: 25 à 30 m de hauteur, quelquefois jusqu'à 40 m, sur un diamètre de 0,9 à 1,2 m. Tronc droit, columnaire. Ecorce grise remarquable, à plaques allongées rigides mais se desquamant à la base. Les noix sont récoltées dans la nature pour leurs «noyaux» à chair douce et aromatique et vendues dans le commerce sous le nom de «noix d'hickory». On le trouve dans les régions intérieures est de l'Amérique du Nord.

Habitat: bas de pentes basses et bords des cours d'eau en association avec le tilleul d'Amérique, les érables, le peuplier deltoïde, les chênes et d'autres caryers.

Bois: lourd, très dur, tenace, résistant, se courbant bien et à grain fin. Bois de cœur brun clair; aubier quasiment blanc.

Intérêts: utilisé pour la fabrication d'instruments aratoires, de manches de haches, de paniers, en charronnage, en carrosserie, et comme combustible.

Arbre de parcs en Europe depuis le 17e siècle en raison de sa belle silhouette en hiver.

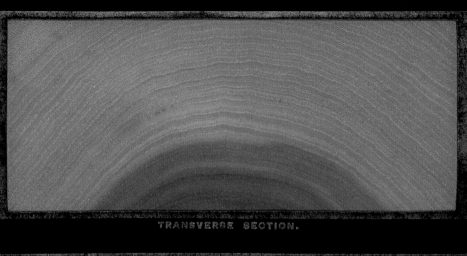

TRANSVERSE SECTION.

RADIAL SECTION.

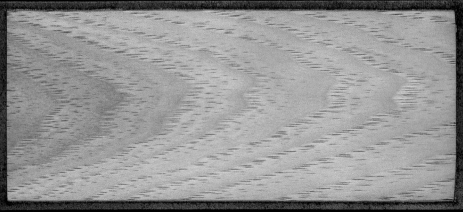

TANGENTIAL SECTION.

Ges. Rindenschälende Hickory.   Fr. Noyer tendre.

Sp. Nogal de America.

W | WALNUT FAMILY
Small-fruited Hickory

FAMILIE DER WALNUSSGEWÄCHSE
Kleinfrüchtige Hickory

FAMILLE DU NOYER
Caryer à petits fruits

JUGLANDACEAE
Carya glabra (Mill.) Sweet
Syn.: *Hicoria microcarpa* (Nutt.) Britton

Description: 85 to 100 ft. (25 to 30 m) high and 2 to 3 ft. (0.6 to 0.9 m) in diameter. Trunk well-developed with rough, gray bark, which separates into narrow scales. Fruits thin-shelled with a sweet seed.

Habitat: Mainly on well-drained slopes and hills in association with Pignut Hickory, Shagbark Hickory, various oaks, Eastern Red Cedar, Dogwood, Sassafras, etc., in eastern North America.

Wood: Strong, hard, and tough.

Use: For agricultural implements, tool handles, etc. Excellent firewood. The pleasant-tasting and aromatic seeds are not sold in great numbers commercially (presumably because of their small size, barely one-third of an inch (1 cm)).

Beschreibung: 25 bis 30 m hoch bei 0,6 bis 0,9 m Durchmesser. Wohlgeformter Stamm mit rauer, grauer Borke, die in schmalen Schuppen abblättert. Früchte kleiner als die der Normalform, dünnschalig mit süßem Samen.

Vorkommen: Hauptsächlich gut drainierte Hänge und Hügel in Gesellschaft von Ferkelnuss-Hickory, Schindelborkiger Hickory, verschiedenen Eichen, Rotzeder, Hartriegel, Sassafras u. a. im östlichen Nordamerika.

Holz: Fest, stark und zäh.

Nutzung: Für Ackerbaugerät, Werkzeuggriffe usw. Ausgezeichnetes Brennholz. Die wohlschmeckenden und duftenden Samen sind im Handel ohne Bedeutung (wahrscheinlich aufgrund ihrer geringen Größe, kaum 1 cm).

Description : 25 à 30 m de hauteur sur un diamètre de 0,6 à 0,9 m. Port élancé, majestueux. Ecorce grise et rugueuse, s'exfoliant en minces écailles. Fruits à brou mince et à chair douce.

Habitat : occupe principalement les pentes et les collines bien drainées de l'est de l'Amérique du Nord, en association avec le caryer glabre, le caryer ovale, différents chênes, le cèdre rouge, le cornouiller, le sassafras etc.

Bois : résistant, fort et tenace.

Intérêts : utilisé pour fabriquer des instruments aratoires, des manches d'outils etc. Excellent bois de chauffage. Les graines aromatiques et comestibles n'ont, probablement en raison de leur petite taille (à peine 1 cm), aucune valeur économique.

Small-fruited Hickory.

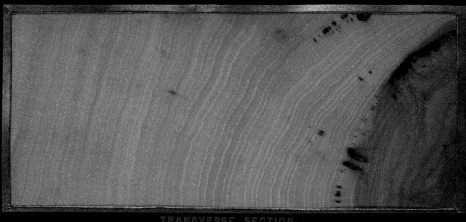

TRANSVERSE SECTION.

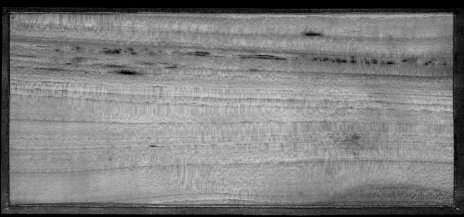

RADIAL SECTION.

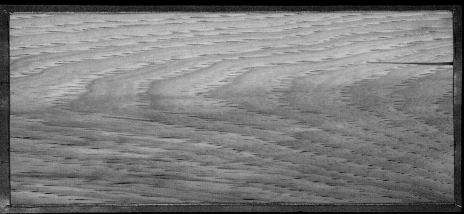

TANGENTIAL SECTION.

Ger. **Kleinfruchtige Hickory.**   *Fr.* Noyer a petit fruit.

Sp. Nogal de fruto pequeno.

W | WALNUT FAMILY
Water Hickory

FAMILIE DER WALNUSSGEWÄCHSE
Wasser-Hickory

FAMILLE DU NOYER
Caryer aquatique, Pécanier, Pécanier sauvage

JUGLANDACEAE
Carya aquatica (Michx. f.) Nutt.

Description: Usually 50 to 75 ft. (15 to 23 m), but in the woods of the lower Mississippi 85 to 120 ft. (25 to 35 m) high and 2 ft. (0.6 to 0.7 m) in diameter. Trunk straight and columnar. Nuts normally dry and bitter, occasionally (south-east Arkansas) sweet and tasty.

Habitat: Usually flat, swampy sites, flooded for much of the year, together with other wetland trees such as Planer Tree, Swamp Privet, Water Locust, Water Ash and Pumpkin Ash, Cotton Gum, Buckwheat Tree, Red Maple, and Swamp Cypress in (south-)eastern North America.

Wood: Heavy, hard, and inflexible.

Use: Mainly as firewood. Has been known to bend steel on cutting. The least commercially valuable of the hickories.

Beschreibung: Gewöhnlich 15 bis 23 m, aber in den Wäldern am unteren Mississippi 25 bis 35 m hoch bei 0,6 bis 0,7 m Durchmesser. Gerader, säulenförmiger Stamm. Nüsse meist herb und bitter, vereinzelt (Südost-Arkansas) süß und schmackhaft.

Vorkommen: Meist flache, längere Zeit des Jahres überschwemmte Sumpfgebiete, zusammen mit anderen wasserliebenden Bäumen wie Wasser-Ulme, Sumpf-Liguster, Wasser-Gleditschie, Wasser-Esche und Kürbis-Esche, Wasser-Tupelo, Cliftonie, Rot-Ahorn, Sumpfzypresse u. a. im (süd-) östlichen Nordamerika.

Holz: Schwer, hart und spröde.

Nutzung: Hauptsächlich als Brennholz. Kann beim Schneiden selbst Stahl verbiegen. Wirtschaftlich die unbedeutendste der Hickories.

Description : d'ordinaire 15 à 23 m de hauteur, mais jusqu'à 25 à 35 m dans le bas Mississippi, avec un tronc droit, columnaire de 0,6 à 0,7 m de diamètre. Les noix sont généralement âpres et amères mais elles peuvent être douces et savoureuses dans certaines régions (sud-est de l'Arkansas).

Habitat : préfère les zones marécageuses plates et inondées une bonne partie de l'année du (sud-) est de l'Amérique du Nord, en association avec des espèces aimant les sols humides : le planéra des marécages, le troène des marais, le févier aquatique, le frêne de la Caroline, le nyssa aquatique, la cliftonie, l'érable rouge, le taxode chauve etc.

Bois : lourd, dur et cassant.

Intérêts : utilisé surtout comme bois de chauffage. Assez dur pour déformer l'acier en le fendant. Le moins important sur le plan économique de tous les caryers.

Water Hickory, Swamp Hickory, Bitter Pecan.

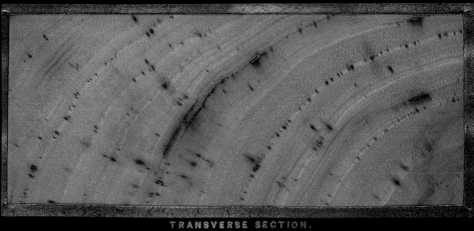

TRANSVERSE SECTION.

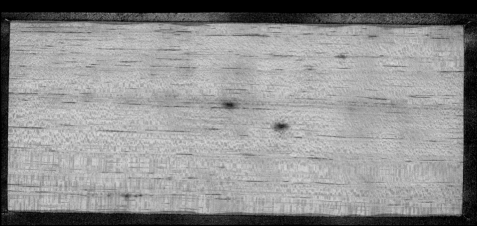

RADIAL SECTION.

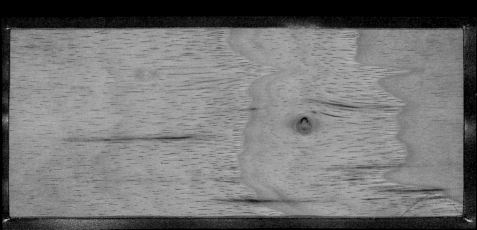

W | WILLOW FAMILY
Almond-leaf Willow, Peach-leaf Willow, French Willow

FAMILIE DER WEIDENGEWÄCHSE
Mandel-Weide

FAMILLE DU SAULE
Saule à trois étamines, Saule à feuilles de pêcher, Saule amandier

SALICACEAE
Salix amygdaloides Anders.

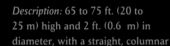

Description: 65 to 75 ft. (20 to 25 m) high and 2 ft. (0.6 m) in diameter, with a straight, columnar trunk. Long, pendent leaves, which resemble those of the Peach or the Almond.

Habitat: In association with Black Willow on riverbanks and lake shores over a vast area, which extends over most of North America with the exception of Mexico, the south and south-east of the USA, Alaska, and the northern part of Canada. While in the east the Black Willow extends over a much wider area than the Almond-leaved Willow; in the west the reverse is the case.

Wood: Light, soft, and not strong.

Use: Mainly for charcoal and as firewood.

Very closely related to the Black Willow, with which it hybridizes.

Beschreibung: 20 bis 23 m hoch bei 0,6 m Durchmesser. Gerader, säulenförmiger Stamm. Lange, hängende Blätter, die denen des Pfirsichs oder der Mandel ähneln.

Vorkommen: In Gesellschaft der Schwarz-Weide an Flussufern und an den Ufern von Seen in einem gewaltigen Areal, das mit Ausnahme von Mexiko, des Südens und Südostens der USA sowie der nördlichen Teile Kanadas und Alaskas den größten Teil Nordamerikas umfasst. Während die Schwarz-Weide im Osten über das Gebiet der Mandel-Weide weit hinausgeht, gilt das Umgekehrte für die Mandel-Weide im Westen.

Holz: Leicht, weich und nicht fest.

Nutzung: Hauptsächlich für Holzkohle und als Brennholz.

Nächstverwandt mit der Schwarz-Weide, mit der sie Bastarde bildet.

Description: arbre de 20 à 23 m de hauteur, avec un tronc droit, columnaire de 0,6 m de diamètre. Longues feuilles pendantes, rappelant celles du pêcher ou de l'amandier.

Habitat: en association avec le saule noir sur les rives des cours d'eau et des lacs, avec une très vaste répartition. On le trouve dans la plupart des régions de l'Amérique du Nord à l'exception du Mexique, du sud et sud-est des Etats-Unis ainsi que dans la partie nord du Canada et de l'Alaska. Alors que le saule noir dépasse largement à l'est les limites d'implantation du saule à trois étamines, celui-ci se rencontre à l'ouest bien au-delà de l'aire de répartition du saule noir.

Bois: léger, tendre et peu résistant.

Intérêts: charbon de bois et bois de chauffage essentiellement.

Le plus proche parent du saule

TRANSVERSE SECTION.

RADIAL SECTION.

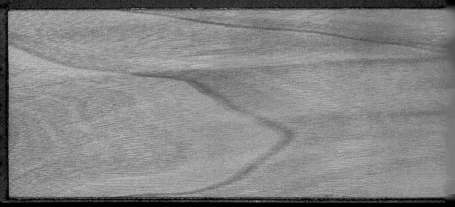

TANGENTIAL SECTION

W | WILLOW FAMILY
American Aspen, Quaking Aspen, Trembling Aspen, Popple

FAMILIE DER WEIDENGEWÄCHSE
Amerikanische Zitter-Pappel

FAMILLE DU SAULE
Peuplier faux tremble, Peuplier d'Athènes (Canada)

SALICACEAE
Populus tremuloides Michx.

Description: In favorable situations over 100 ft. (30 m) high and 3 ft. (0.9 m) in diameter, otherwise considerably smaller. Bark of young trunks and branches yellowish green to almost white and fairly smooth, later darker, scaly, and fissured. The principal characteristic is the quivering of the leaves in the slightest breath of wind. Leaf-stalks flattened above.

Habitat: Grows best in the damp, cold lowlands of the north; in the west along the mountain ranges, however, southwards as far as California and New Mexico. In forests and amongst clumps of bushes near rivers. A pioneer tree on open areas created by forest fires, where it provides protection for the slower-growing seedlings of other trees, which gradually displace it.

Wood: Light and soft.

Use: Primarily for paper pulp and matches.

Beschreibung: Günstigenfalls über 30 m hoch bei 0,9 m Durchmesser, sonst sogar erheblich kleiner. Borke junger Stämme und Äste gelblich grün bis fast weiß und ziemlich glatt, später dunkler und schuppig-rissig. Hauptmerkmal: Zittern der langgestielten Blätter beim geringsten Lufthauch. Blattstiele hochkant abgeflacht.

Vorkommen: Vorzugsweise im feuchten, kalten Flachland des Nordens, im Westen längs der Gebirgsketten jedoch auch noch bis nach Kalifornien und Neu-Mexiko in den Süden vordringend Wälder und Gebüsche in Flussnähe. Pionier auf frischen Freiflächen nach Waldbränden. Dort Schutzbaum für langsamer aufwachsende Keimpflanzen anderer Bäume, die ihn nach und nach verdrängen.

Holz: Leicht und weich.

Nutzung: Vor allem zu Papier-Pulpe und Streichhölzern.

Description : arbre atteignant plus de 30 m de hauteur sur un diamètre de 0,9 m dans des conditions favorables, mais ses dimensions peuvent être bien inférieures. L'écorce des troncs et des branches, vert jaunâtre à presque blanc et assez lisse à l'état juvénile, devient plus sombre, fissurée et écailleuse avec l'âge. Caractère spécifique : le tremblement de ses feuilles longuement pétiolées au moindre souffle d'air. Pétiole aplati près du limbe.

Habitat : préfère les plaines humides et froides du nord. À l'ouest le long des chaînes montagneuses et dans le sud jusqu'en Californie et au Nouveau-Mexique. Forêts et buissons à proximité des cours d'eau. Espèce pionnière sur les zones brûlées et protectrice des semis d'arbres à croissance lente, qui le dominent peu à peu.

Bois : léger et tendre.

Intérêts : utilisé avant tout pour fabriquer de la pâte à papier et

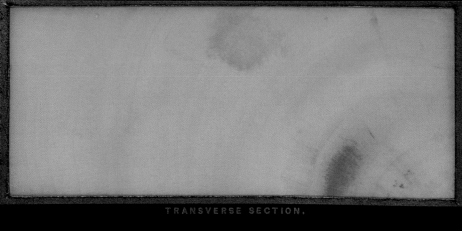

TRANSVERSE SECTION.

RADIAL SECTION.

TANGENTIAL SECTION

Ger. Amerikanische Zitter-espe. *Sp.* Alamo tremblon.

Fr. Le Tremble d'Amerique.

W | WILLOW FAMILY
Balsam Poplar, Hackmatack, Tacamahac

FAMILIE DER WEIDENGEWÄCHSE
Balsam-Pappel

FAMILLE DU SAULE
Peuplier baumier, Tacamahaca, Baumier (Canada)

SALICACEAE
Populus balsamifera L.

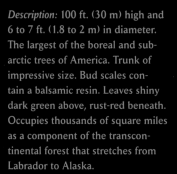

Description: 100 ft. (30 m) high and 6 to 7 ft. (1.8 to 2 m) in diameter. The largest of the boreal and subarctic trees of America. Trunk of impressive size. Bud scales contain a balsamic resin. Leaves shiny dark green above, rust-red beneath. Occupies thousands of square miles as a component of the transcontinental forest that stretches from Labrador to Alaska.

Habitat: Low-lying, often flooded land by rivers and swamps in association with other trees of the transcontinental forest in northern North America.

Wood: Light, soft, not strong, close-grained. Heartwood light brown; sapwood thick, almost white.

Use: Buckets, packing boxes and small crates, paper pulp. On the Great Lakes the bark is used instead of cork for fishing-net floats.

Beschreibung: 30 m hoch bei 1,8 bis 2 m Durchmesser. Größter der borealen und subarktischen Bäume Amerikas. Stamm stattlich. Knospenschuppen balsamisch-harzig. Blätter oberseits glänzend dunkelgrün, unterseits rostrot. Als Mitglied des transkontinentalen Waldes von Labrador bis Alaska Tausende von Quadratkilometern bestockend.

Vorkommen: Tieflagen, oft überschwemmte Böden an Flüssen und Sümpfen in Gesellschaft von anderen Bäumen des transkontinentalen Waldes im nördlichen Nordamerika.

Holz: Leicht, weich, nicht fest und feine Textur. Kern hellbraun; Splint dick, fast weiß.

Nutzung: Eimer, Packschachteln und kleine Kisten, Papier-Pulpe. Die dicke Borke wird an den Großen Seen als Korkersatz für Schwimmer von Fischernetzen verwendet.

Description : 30 m de hauteur sur un diamètre de 1,8 à 2 m. Le plus grand arbre de l'Amérique boréale et subarctique. Tronc imposant. Les écailles des bourgeons sont couvertes d'une résine jaune dégageant une odeur balsamique. Les feuilles sont d'un vert sombre et brillant à l'avers et présentent des reflets roussâtres au revers. Elément de la forêt transcontinentale qui s'étend sur des dizaines de milliers de kilomètres carrés du Labrador à l'Alaska.

Habitat : zones basses, sols souvent inondés au bord des cours d'eau et des marécages dans le nord de l'Amérique du Nord, en association avec d'autres essences de la forêt transcontinentale.

Bois : léger, tendre, non résistant et à grain fin. Bois de cœur brun clair ; aubier épais, presque blanc.

Intérêts : utilisé pour fabriquer des seaux, des emballages, des petites caisses et de la pâte à papier. Dans la région des Grands Lacs, l'écorce épaisse remplace les flotteurs en liège des filets de pêche.

Balsam Poplar,   Tacamahac.

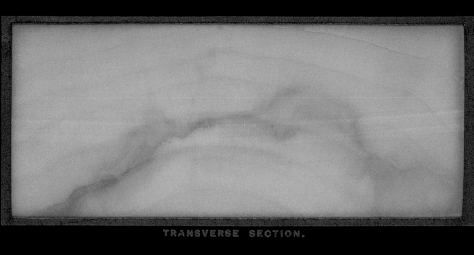

TRANSVERSE SECTION.

RADIAL SECTION.

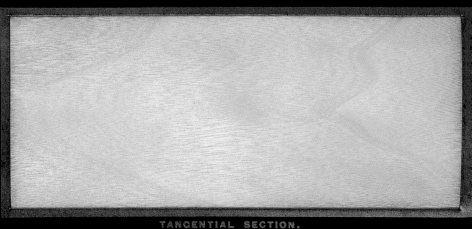

TANGENTIAL SECTION.

Ger. Balsampappel.   *Fr.* Peuplier baumier.

Sp. Alamo balsámico.

W | WILLOW FAMILY
Big-toothed Aspen, Canadian Aspen

FAMILIE DER WEIDENGEWÄCHSE
Großzähnige Espe

FAMILLE DU SAULE
Peuplier à grandes dents, Grand tremble (Canada)

SALICACEAE
Populus grandidentata Michx.

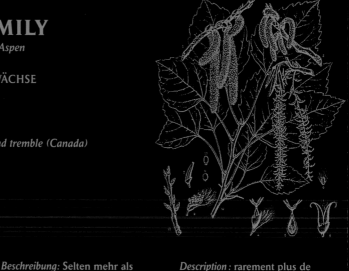

Description: Rarely more than 85 ft. (25 m) high and 2 ft. (0.6 m) in diameter. The leaves quiver in the slightest breath of wind, as is the case with the Aspen, owing to the shape of the leaf stalks.

Habitat: Sandy slopes and riverbanks, together with Hemlock, pines, oaks, maples, Serviceberry, Red Spruce, etc., in north-eastern North America.

Wood: Light, soft, not strong, and close-grained. Heartwood pale brown; sapwood almost white.

Use: Mainly for pulp, occasionally for building, and in turnery.

A rapid early colonizer of wasteland and areas burnt by forest fires, as well as a protective tree for other emerging tree species.

Beschreibung: Selten mehr als 25 m hoch bei 0,6 m Durchmesser. Aufgrund der Anatomie der Blattstiele rascheln die Blätter wie bei der Zitter-Pappel beim leisesten Windhauch.

Vorkommen: Sandige Hänge und Flussufer zusammen mit Hemlock, Kiefern, Eichen, Ahornen, Felsenbirne, Rot-Fichte u. a. im nordöstlichen Nordamerika.

Holz: Leicht, weich, nicht stark, und mit feiner Textur. Kern hellbraun; Splint fast weiß.

Nutzung: Hauptsächlich für Pulpe, gelegentlich Bauholz, Drechslerarbeiten.

Schneller Erstbesiedler auf Öd- und Waldbrandflächen, zugleich Schutzbaum für aufkommende andere Baumarten.

Description: rarement plus de 25 m de hauteur sur un diamètre de 0,6 m. Le «tremblement» des feuilles au moindre souffle d'air est dû, comme chez le peuplier-tremble, à la forme du pétiole.

Habitat: pentes sablonneuses et bords des cours d'eau en compagnie de la pruche, des pins, des chênes, des érables, de l'amélanchier, de l'épicéa rouge etc. On le trouve dans le nord-est de l'Amérique du Nord.

Bois: léger, tendre, pas fort et à grain fin. Bois de cœur brun clair; aubier presque blanc.

Intérêts: utilisé surtout pour la pâte à papier, de temps à autre en construction et en tournage.

Premier arbre à coloniser les terrains ingrats ou victimes d'incendies et protecteur des espèces futures.

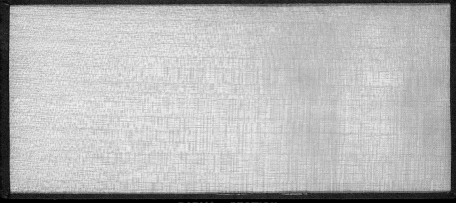

TRANSVERSE SECTION.

RADIAL SECTION.

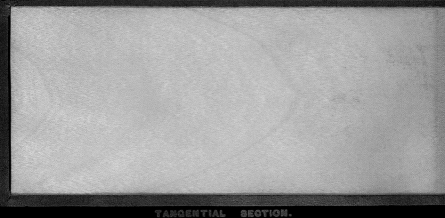

TANGENTIAL SECTION.

Ger. Groszgezänte Espe. Sp. Alamo de diente grande.

W | WILLOW FAMILY

Black Cottonwood, Balsam Cottonwood, Cottonwood,
Balsam Poplar

FAMILIE DER WEIDENGEWÄCHSE
Balsam-Pappel

FAMILLE DU SAULE
Peuplier baumier de l'Ouest, Baumier de l'Occident,
Peuplier de l'Ouest (Canada)

SALICACEAE
Populus balsamifera L. subsp. trichocarpa
(Torr. & A. Gray ex Hook.) Brayshaw
Syn.: P. trichocarpa (Torr. & A. Gray ex Hook.) Brayshaw

Description: 200 ft. (60 m) high and 7 to 8 ft. (2.1 to 2.4 m) in diameter. Crown broad and open. Bark ashgray and deeply furrowed. Leaves broadly ovate, pointed, wide at the base, dark green above, pale reddish or silvery beneath. Dioecious flowers in drooping catkins, appearing before the leaves.

Habitat: Open stands on riverbanks chiefly in the coastal region of Pacific North America.

Wood: Light and soft. Pale brown, thin, almost white sapwood.

Use: Barrels, butter churns, bowls, tableware. Extensive use has resulted in few large trees remaining. Trunks used by American Indians for canoes, and the roots for hats and baskets.

Beschreibung: 60 m hoch bei 2,1 bis 2,4 m Durchmesser. Krone breit und offen. Borke aschgrau und tiefrissig. Blätter breit oval, zugespitzt, mit breiter Basis, Oberseite dunkelgrün, Unterseite blass rötlich oder silbrig. Zweihäusig, Blüten vor den Blättern in herabhängenden Ähren erscheinend.

Vorkommen: Lockere Bestände an Flussufern, bevorzugt in der Küstenregion des pazifischen Nordamerika.

Holz: Leicht und weich. Hellbraun, dünner, fast weißer Bast.

Nutzung: Fässer, Butterfässer, Schüsseln, Geschirr. Wegen des hohen Nutzungsdruckes sind kaum noch große Bäume dieser Art vorhanden. Die Stämme verwendeten die Indianer zur Herstellung von Kanus, die Wurzeln wurden zu Hüten und Körben verarbeitet.

Description : arbre de 60 m de hauteur, avec un tronc de 2,1 à 2,4 m de diamètre et un houppier ample et ouvert. Ecorce gris cendré, à texture profondément fissurée. Feuilles ovales-larges, à apex aigu et élargies à la base, vert foncé dessus, rougeâtre pâle ou argentées dessous. Espèce dioïque à fleurs en chatons pendants, avant la feuillaison.

Habitat : en formations claires sur les berges particulièrement dans les régions de la côte pacifique de l'Amérique du Nord.

Bois : léger et tendre. Brun clair ; liber plus clair, presque blanc.

Intérêts : employé en tonnellerie, pour fabriquer des barattes et de la vaisselle. En raison de son utilisation intensive, il ne subsiste pratiquement plus de spécimens de grande taille. Les troncs étaient utilisés par les Indiens pour fabriquer des canots, les racines pour confectionner des chapeaux et des paniers.

TRANSVERSE SECTION.

RADIAL SECTION

TANGENTIAL SECTION.

Ger. Schwarze Pappel.   *Fr.* Peuplier noir.

W | WILLOW FAMILY
Black Willow

FAMILIE DER WEIDENGEWÄCHSE
Schwarz-Weide

FAMILLE DU SAULE
Saule noir, Saule noir d'Amérique

SALICACEAE
Salix nigra Marshall

Description: Sometimes as much as 140 ft. (40 m) high and 3 to 4 ft. (0.9 to 1.2 m) in diameter, but only under very favorable conditions, usually less than half as large. Commonly with several trunks.

Habitat: Riverbanks and lakesides in southern and eastern North America with a variety of other species.

Wood: Light, soft, weak, and close-grained; shrinks as it dries. Heartwood light reddish brown; sapwood thin, almost white.

Use: The bark is a domestic remedy for fever.

Beschreibung: Manchmal bis zu 40 m hoch bei 0,9 bis 1,2 m Durchmesser, jedoch nur unter sehr günstigen Bedingungen. Meist weniger als halb so groß. Gemeinhin mehrstämmig.

Vorkommen: Fluss- und Seeufer im südlichen und östlichen Nordamerika, in wechselnder Gesellschaft.

Holz: Leicht, weich, schwach und mit feiner Textur; schrumpft beim Trocknen. Kern helles Rotbraun; Splint dünn, fast weiß.

Nutzung: Rinde als Hausmittel gegen Fieber.

Description : peut atteindre, lorsque les conditions sont très favorables, 40 m de hauteur sur un diamètre de 0,9 à 1,2 m, mais ses dimensions sont d'ordinaire deux fois plus petites. Tronc multiple le plus souvent.

Habitat : berges des rivières et des lacs dans le sud et l'est de l'Amérique du Nord, en association avec des espèces différentes selon les conditions.

Bois : léger, tendre, faible et à grain fin ; retrait au séchage. Bois de cœur brun rouge clair ; aubier mince et presque blanc.

Intérêts : l'écorce a des propriétés fébrifuges utilisées en pharmacopé domestique.

Black Willow.

TRANSVERSE SECTION.

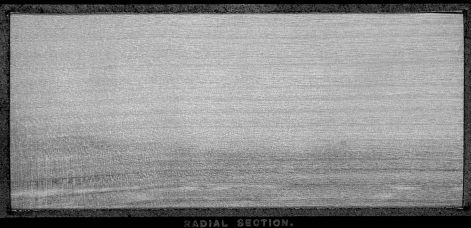

RADIAL SECTION.

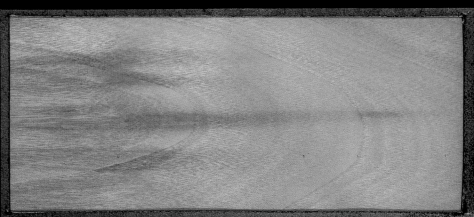

TANGENTIAL SECTION.

Ger. Schwarze Weide. Fr. Saule noir. Sp. Sáuce negro

W | WILLOW FAMILY
Cottonwood

FAMILIE DER WEIDENGEWÄCHSE
Kalifornische Pappel

FAMILLE DU SAULE
Peuplier de Frémont

SALICACEAE
Populus fremontii S. Watson

Description: To 100 ft. (30 m) high and 5 to 6 ft. (1.5 to 1.8 m) in diameter. Trunk short and stumpy, with spreading branches and twigs. Bark on old trunks dark brown, tinged red, deeply divided by irregularly broad, connected, rounded ridges. These covered in closely appressed scales. Deep red inside. Young bark pale gray-brown, thinner and only lightly grooved. Produces large amounts of fluffy white seed-hairs.

Habitat: Beside streams and in river valleys in south-western North America. Optimum either side of the border between USA and Mexico.

Wood: Light, soft, not strong, close-grained; does not store well and warps as it dries. Heartwood pale brown; sapwood thin, almost white.

Use: Pollarded trees sprout quickly, producing abundant firewood. Inner bark used to make petticoats by American Indian women. Splendid street tree in northern Mexico. Indicates presence of water from afar to travelers in the dry Mexican plateau.

Beschreibung: Gelegentlich 30 m hoch bei 1,5 bis 1,8 m Durchmesser. Kurzer, gedrungener Stamm mit ausgebreiteten Ästen und Zweigen. Borke alter Stämme dunkelbraun, rot getönt, tief in unregelmäßig breite, verbundene und abgerundete Riefen geteilt. Diese sind mit dicht anliegenden Schuppen besetzt. Innen kräftig rot. Junge Borke hell graubraun, dünner und nur leicht gerissen. Weiße Samenhaare in großer Menge.

Vorkommen: Bachufer und Flusstäler im südwestlichen Nordamerika. Optimum beiderseits der Staatsgrenze zwischen USA und Mexiko.

Holz: Leicht, weich, nicht fest und mit feiner Textur; schwer zu lagern und wirft sich beim Trocknen. Kern hellbraun. Splint dünn, fast weiß.

Nutzung: Von gekröpften Bäumen schnell und viel Brennholz. Innere Rinde bei den Indianerfrauen des Südwestens zu Petticoats verarbeitet. Großartiger Straßenbaum in Nord-Mexiko. Zeigt im freien Gelände der ariden mexikanischen Hochebene Reisenden von fern Wassernähe an.

Description : peut atteindre 30 m de hauteur, avec un tronc court et trapu de 1,5 à 1,8 m de diamètre, armé de branches et de rameaux étalés. L'écorce des vieux arbres est brun foncé teinté de rouge, à texture entrelacée formant des fissures et de larges crêtes irrégulières et arrondies aux écailles adhérentes ; écorce interne rouge intense. À l'état juvénile, elle est brun gris clair, plus fine et peu fissurée. Abondante bourre blanche formée par les poils des graines.

Habitat : bords des ruisseaux et plaines alluviales du sud-ouest de l'Amérique du Nord. Optimum de développement des deux côtés de la frontière américano-mexicaine.

Bois : léger, tendre, non résistant et à grain fin ; stockage difficile et gauchissement au séchage. Bois de cœur brun clair ; aubier mince, presque blanc.

Intérêts : le bouturage permet d'obtenir du bois de chauffage, rapidement et en grande quantité. Les Indiennes confectionnaient des jupons avec l'écorce interne. Arbre de routes dans le nord du Mexique. Signale de loin les points d'eau du haut-plateau mexicain.

White Cottonwood, Fremont Cottonwood.

TRANSVERSE SECTION.

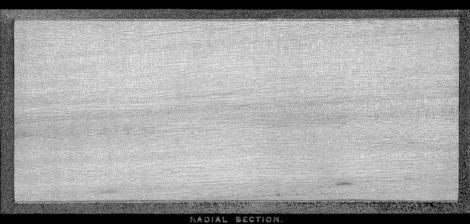

RADIAL SECTION.

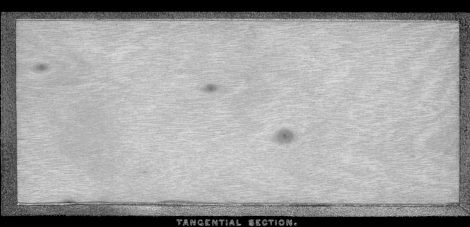

TANGENTIAL SECTION.

Ger. Pappel von Fremont. *Fr.* Peuplier de Fremont.

Sp. Alamo de Fremont.

W | WILLOW FAMILY

Eastern Cottonwood, Necklace Poplar, Carolina Poplar

FAMILIE DER WEIDENGEWÄCHSE
Kanadische Pappel

FAMILLE DU SAULE
Peuplier deltoïde, Liard (Canada)

SALICACEAE
Populus deltoides Bart. ex Marshall ssp. *monilifera* (Aiton)
Eckenwalder
Syn.: *P. monilifera* Aiton

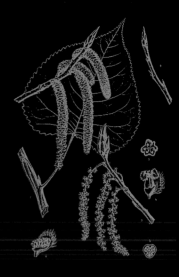

Description: A good 100 ft. (30 m) high and 7 to 8 ft. (2 to 2.5 m) in diameter. The largest representative of the genus. Bark in deep, longitudinal fissures. Seeds hairy.

Habitat: Rich, moist soils in low-lying areas near rivers. A characteristic associate of water, especially in the middle of the subcontinent. Widespread in North America east of the Rocky Mountains through to the Atlantic coast.

Wood: Light, soft, not strong, and close-grained; warps as it dries and is extremely difficult to season. Heartwood dark brown; sapwood thick, almost white.

Use: For paper pulp, cheap packaging, and firewood. The first building material for the settlers on the prairies of the west.

A remarkably quick-growing park tree in Europe since the 18th century. Numerous hybrids important to forestry.

Beschreibung: Gut 30 m hoch bei 2 bis 2,5 m Durchmesser. Größter Vertreter der Gattung. Borke in tiefen Längsrunzeln. Samenhaare.

Vorkommen: Reiche, feuchte Böden der Flussniederungen. Besonders in der Mitte des Subkontinents charakteristischer Begleiter der Gewässer. In weiten Teilen Nordamerikas östlich der Rocky Mountains bis zur Antlantikküste verbreitet.

Holz: Leicht, weich, nicht fest und feine Textur; wirft sich beim Trocknen und extrem schwer zu lagern. Kern dunkelbraun; Splint dick, fast weiß.

Nutzung: Für Papier-Pulpe, billige Verpackungen und Brennholz. Erstes Baumaterial der Präriesiedler im Westen.

Bemerkenswert schnellwüchsiger Parkbaum in Europa seit dem 18. Jh. Zahlreiche forstlich bedeutsame Kreuzungen.

Description : au minimum 30 m de hauteur sur un diamètre de 2 à 2,5 m. Le plus grand représentant de son genre. Ecorce creusée de profondes fissures longitudinales. Les fruits libèrent un «coton» formé par les poils des graines.

Habitat : sols riches, humides des lits de hautes eaux. Elément caractéristique des forêts riveraines, en particulier dans les régions centrales du sous-continent. On le trouve dans des régions étendues de l'Amérique du Nord à l'est des Rocheuses et jusqu'à l'Atlantique.

Bois : léger, tendre, pas fort et à grain fin ; gauchit au séchage et est extrêmement difficile à stocker. Bois de cœur brun foncé ; aubier épais, presque blanc.

Intérêts : utilisé pour la pâte à papier, les emballages bon marché, et comme bois de chauffage. Premier matériau de construction des colons des Grandes Prairies.

Arbre de parcs, à croissance particulièrement rapide, en Europe depuis le 18ᵉ siècle. Espèce forestière importante avec de nombreux hybrides producteurs de bois.

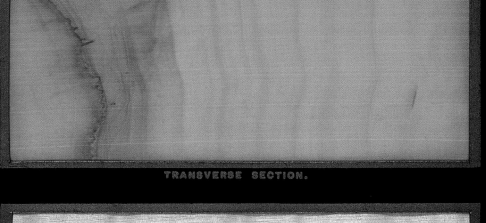

TRANSVERSE SECTION.

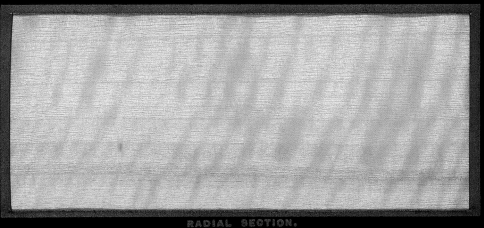

RADIAL SECTION.

TANGENTIAL SECTION.

W | WILLOW FAMILY
Golden Willow

FAMILIE DER WEIDENGEWÄCHSE
Silber-Weide

FAMILLE DU SAULE
Saule jaune

SALICACEAE
Salix alba L. var. *vitellina* (L.) Stokes
Syn.: *S. alba* L. var. *vitellina* Koch.

Description: 75 to 85 ft. (20 to 25 m) high and 3 to 4, sometimes 5 ft. (0.9 to 1.2 (1.5) m) in diameter, with a short, thick trunk, often dividing just above the ground. Bark dark gray with long, deep fissures. Twigs yellow or red instead of green. Quick-growing. Roots bind the soil.

Habitat: Riverbanks, low-lying, damp soils up to alpine peaks and found both in the Arctic and Antarctic. In North America found from Canada through the central and southern states of the USA. Southwards through the West Indies to the Chilean Andes.

Wood: Very light, soft, and tough. Heartwood pale, rarely dark to reddish brown; sapwood pale to white.

Use: Baskets, especially for the transport of stones; cooperage, charcoal, gunpowder. The classic timber for cricket bats. Leaves used as food for domestic animals. Bark medicinal.

Introduced from the Old World and established and naturalized long ago.

Beschreibung: 20 bis 25 m hoch bei 0,9 bis 1,2 (1,5) m Durchmesser. Kurzer, dicker Stamm, oft schon dicht über dem Boden mehrgeteilt. Borke dunkelgrau mit langen, tiefen Runzeln. Zweige gelb oder rot statt grün. Schnellwüchsig. Wurzeln bodenbindend.

Vorkommen: Flussufer, tiefgründige, feuchte Böden bis hinauf zu alpinen Gipfeln und sowohl in der Arktis als auch in der Antarktis vertreten. In Nordamerika von Kanada über die mittleren und östlichen Staaten der USA verbreitet. Südwärts über die Westindischen Inseln bis in die chilenischen Anden.

Holz: Sehr leicht, weich und zäh. Kern hell, selten dunkel- bis rotbraun; Splint blass bis weiß.

Nutzung: Körbe, besonders zum Transport von Steinen, für Böttcherei, Holzkohle, Schwarzpulver. Vor allem Holzarten für Kricketschläger. Blätter als Haustierfutter. Rinde medizinisch.

Aus der Alten Welt eingeschleppt, längst eingewöhnt und eingebürgert.

Description: 20 à 25 m de hauteur sur un diamètre de 0,9 à 1,2 m (1,5 m). Tronc court et large, souvent multiple dès la base. Ecorce gris foncé avec profondes fissures longitudinales. Rameaux jaunes ou rouges. Croissance rapide. Racines fixatrices des sols.

Habitat: berges, sols profonds et humides jusqu'aux sommets alpins et aussi bien dans l'Arctique que l'Antarctique. Se rencontre dans le nord du Canada et dans les Etats du centre et de l'est des Etats-Unis ainsi que dans les Andes chiliennes, en passant par les Antilles.

Bois: très léger, tendre et tenace. Bois de cœur clair, rarement brun rougeâtre à brun foncé; aubier terne à blanc.

Intérêts: paniers pour le transport des pierres en particulier, tonnellerie, charbon de bois et poudre noire. Préféré aux autres bois pour les battes de cricket. Feuilles consommées par les animaux. Usages pharmaceutiques de l'écorce.

Cette essence, originaire du continent européen, s'est depuis longtemps acclimatée et naturalisée en Amérique du Nord.

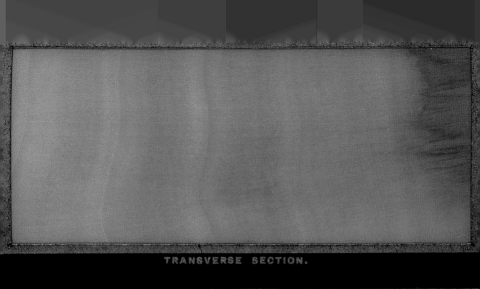

TRANSVERSE SECTION.

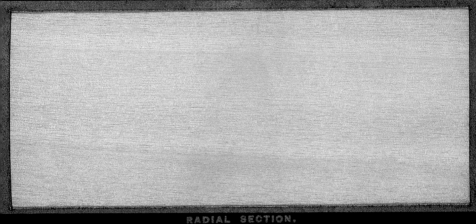

RADIAL SECTION.

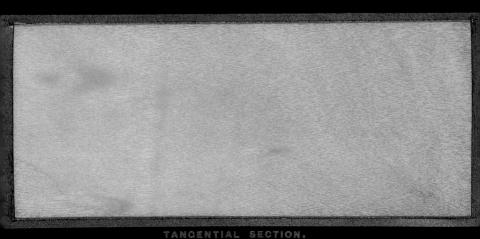

TANGENTIAL SECTION.

Ger. Dotterweide.   Fr. Saule jaune.   Sp. Sauce amarillo

W | WILLOW FAMILY
Lombardy Poplar, Italian Poplar

FAMILIE DER WEIDENGEWÄCHSE
Säulen-Pappel, Pyramiden-Pappel

FAMILLE DU SAULE
Peuplier d'Italie ou de Lombardie

SALICACEAE
Populus nigra L. "Italica"
Syn.: *P. dilatata* Aiton var. *fastigiata* Desf.

Description: 100 ft. (30 m) high and 6 to 8 ft. (1.8 to 2.4 m) in diameter. Bark dark with deep grooves and fissures. The tree has massive buttresses and true prop-roots. Narrow in shape owing to sharply ascending branches. Quick-growing.

Habitat: Planted in almost every part of North America. Thrives especially in water meadows.

Wood: Like that of other poplars.

Use: Extensively used for matches, cardboard and paper pulp, otherwise of little value.

Origin unclear. Is supposed to have appeared as a forest tree in Afghanistan. A second possible homeland is the plain of the river Po in Lombardy. Supposedly favored by Napoleon I for marking military roads across flat country. Not long-lived, but essentially everlasting to the extent of being a weed as a result of the shoots which arise from the roots. An ornamental tree, introduced into America by the settlers, and only male trees are known.

Beschreibung: 30 m hoch bei 1,8 bis 2,4 m Durchmesser. Borke dunkel mit tiefen Rillen und Riefen. Mächtige Wurzelanläufe und wahre Stützwurzeln. Schmale Form durch hoch aufstrebende Äste. Schnellwüchsig.

Vorkommen: In nahezu allen Teilen Nordamerikas angepflanzt. Gedeiht besonders gut in Flussauen.

Holz: Wie das anderer Pappeln.

Nutzung: Kommt außer für Streichhölzer, Pappen und Papier-Pulpe kaum in Betracht.

Herkunft ungeklärt. Soll als Waldbaum in Afghanistan aufgetreten sein. Eine Art zweiter Heimat in der Po-Ebene (Lombardei). Von Napoleon I zur Markierung von Heerstraßen im Flachland als Straßenbaum angeblich gefördert. Nicht langlebig, aber durch Wurzelbrut prinzipiell unsterblich bis zur Landplage. Ornamentaler Baum, mit den Siedlern in Amerika eingeführt und nur in männlichen Exemplaren vorhanden.

Description : 30 m de hauteur sur un diamètre de 1,8 à 2,4 m. Ecorce sombre, parcourue de fissures et de sillons profonds. Base du tronc munie de puissants contreforts et véritables racines d'ancrage dans le sol. Port étroit par ses branches dressées. Croissance rapide.

Habitat : planté dans presque toutes les parties de l'Amérique du Nord. Affectionne les lits de hautes eaux.

Bois : identique à celui des autres peupliers.

Intérêts : peu de valeur économique, mais large utilisation pour fabriquer des allumettes, du carton et de la pâte à papier.

L'origine de cette variété est incertaine : il s'agirait d'une essence forestière d'Afghanistan et qu'elle a été introduite dans la plaine du Pô (Lombardie). Napoléon 1er aurait développé sa culture pour border les routes de sa «Grande Armée». Arbre non longévif, mais son drageonnement lui donne une durée de vie théoriquement infinie et le rend très envahissant. Introduit en Amérique par les colons et très ornemental, le véritable peuplier d'Italie est un arbre mâle.

Lombardy Poplar.

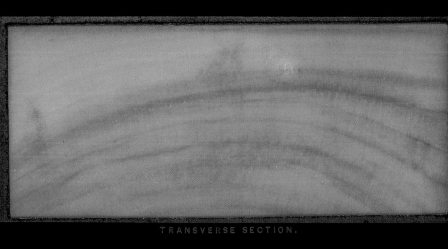

TRANSVERSE SECTION.

RADIAL SECTION.

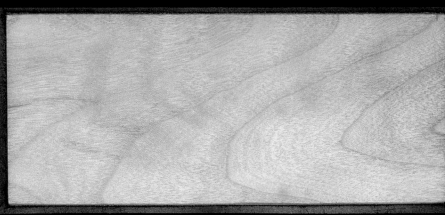

TANGENTIAL SECTION

Ger. Pyramiden-Pappel.   Sp. Alamo de Italia.

W | WILLOW FAMILY
Long-stalk Willow

FAMILIE DER WEIDENGEWÄCHSE
Langstielige Weide

FAMILLE DU SAULE
Saule de la Caroline

SALICACEAE
Salix caroliniana Michx.
Syn.: *S. longipes* Shuttlw. ex Andersson

Description: 20 to 30 ft. (6 to 9 m) high and 1 ft. in diameter. Bark dark brown to almost black, deeply furrowed. Young branches with a fine covering of hair. Leaves lanceolate, pointed, pale green and lustrous above, often silvery white beneath, glabrous or hairy. Flowers in catkins, appearing at the same time as the leaves unfold.

Habitat: Riverbanks, moist soils by springs and streams, even in the deserts of southern Nevada. Most common in the Caribbean, in North America along the Gulf coast to northern Mexico, New Mexico, Arizona and California.

Wood: Probably similar to that of the closely related *Salix nigra.*

Use: Cultivated as a shade tree.

Beschreibung: 6 bis 9 m hoch bei 0,3 m Durchmesser. Borke dunkelbraun bis fast schwarz, tiefrissig. Junge Äste mit feinem Haarbesatz. Blätter lanzettlich, zugespitzt, Oberseite blass hellgrün glänzend, Unterseite oft silbrig weiß, nackt oder behaart. Blüten zur Zeit des Laubaustriebes in Kätzchen.

Vorkommen: Flussufer, feuchte Böden an Quellen und Flüssen, selbst in den Wüsten Süd-Nevadas. Hauptverbreitung in der Karibik, in Nordamerika entlang der Golfstaaten durch Nord-Mexiko, Arizona, Neu-Mexiko und Kalifornien verbreitet.

Holz: Wahrscheinlich ähnlich dem der nahe verwandten *Salix nigra.*

Nutzung: Als Schattenspender kultiviert.

Description: arbuste de 6 à 9 m de hauteur avec un tronc de 0,3 m de diamètre. Ecorce brun sombre à presque noire, profondément fissurée. Jeunes branches pubescentes. Feuilles lancéolées, terminées en pointe, vert pâle et lustrées dessus, souvent blanc argenté, glabres ou tomenteuses dessous. Fleurs en chatons pendant la feuillaison.

Habitat: berges, sols humides près des sources et des rivières, mais se rencontre jusque dans les déserts du sud du Nevada. Surtout répandu dans les Caraïbes, en Amérique du Nord le long des Etats bordant le Golfe en passant par le nord du Mexique, l'Arizona, le Nouveau-Mexique et la Californie.

Bois: probablement similaire à celui de son proche parent *Salix nigra.*

Intérêts: cultivé comme arbre d'ombrage.

Long-stalk Willow.   Ward Willow.

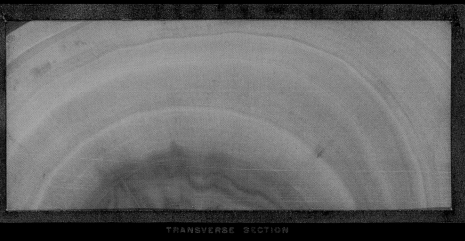

TRANSVERSE SECTION.

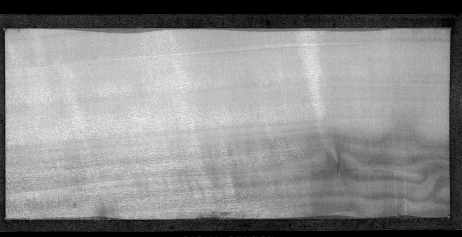

RADIAL SECTION.

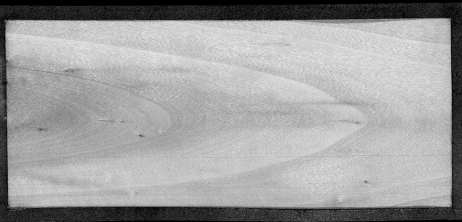

TANGENTIAL SECTION.

Ger. Langstengel-Weide.   *Fr.* Saule à tige long.

Sp. Sauce de tallo largo.

W | WILLOW FAMILY
Missouri Willow

FAMILIE DER WEIDENGEWÄCHSE
Missouri Weide

FAMILLE DU SAULE
Saule à tête laineuse, Saule rigide

SALICACEAE
Salix eriocephala Michx.
Syn.: *S. missouriensis* Bebb

Description: 40 to 50 ft. (12 to 15 m) high and 10 to 12 (18) in. (0.25 to 0.3 (-0.45) m) in diameter. Trunk thick, straight, with a narrow, open crown. Young branches light green, glaucous later red-brown to brown, with small orange lenticels. Bark thin, smooth, light gray with a red tinge and small, closely appressed scales. Leaves lanceolate to bluntly lanceolate, gradually tapering towards the tip, the upper side at first hairy, soon becoming glabrous with the exception of the midrib, the lower side pale and often with silvery hairs. Flowers in erect catkins, appearing before the leaves unfold.

Habitat: Low-lying, alluvial sandy soils, very local distribution in the interior of eastern North America.

Wood: Very resistant to decay. Heartwood dark red-brown; sapwood pale.

Use: Fence posts.

Beschreibung: 12 bis 15 m hoch bei 0,25 bis 0,3 (bis 0,45) m Durchmesser. Stamm dick, gerade, mit schmaler, offener Krone. Äste jung hellgrün, aber von dickem Reif überdeckt, später rotbraun bis braun, mit kleinen, orangen Atemporen. Borke dünn, glatt, hellgrau mit rötlicher Tönung und kleinen, fest anliegenden Schuppen. Blätter lanzettlich bis stumpf lanzettlich, lang zugespitzt, Oberseite anfangs behaart, bald verkahlend mit Ausnahme der Mittelrippe, Unterseite blass und oft silbrig behaart. Blüte vor dem Blattaustrieb, Blüten in aufrechten Kätzchen.

Vorkommen: Tiefgründige, abgelagerte Sandböden, sehr lokal im Inneren des östlichen Nordamerika.

Holz: Sehr fäulnisresistent. Kern dunkel rotbraun; Splint blass.

Nutzung: Zaunpfähle.

Description: 12 à 15 m de hauteur sur un diamètre de 0,25 à 0,3 m (0,45 m max.). Tronc épais, droit, avec une couronne étroite et ouverte. Branches vert clair mais couvertes de pruine blanche jeunes, devenant brun rouge à brunes, avec de petites lenticelles orange à maturité. Ecorce mince, lisse, gris clair teinté de rouge, recouverte de petites écailles adhérentes. Feuilles lancéolées à obtuses-lancéolées, longuement acuminées ; face supérieure garnie de poils soyeux mais devenant bientôt glabre, sauf sur la nervure médiane, face inférieure pâle et souvent couverte de pubescence argentée. Floraison avant la feuillaison, en chatons dressés.

Habitat : alluvions sableux profonds. On le trouve de façon locale à l'intérieur de l'Amérique du Nord.

Bois : très résistant à la décomposition. Bois de cœur brun rouge sombre ; aubier clair.

Intérêts : utilisé pour la fabrication de pieux de clôtures.

Missouri Willow.

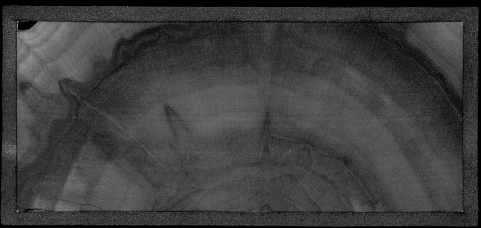

TRANSVERSE SECTION

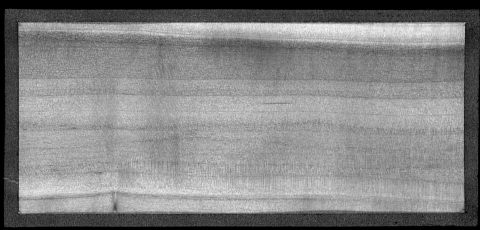

RADIAL SECTION.

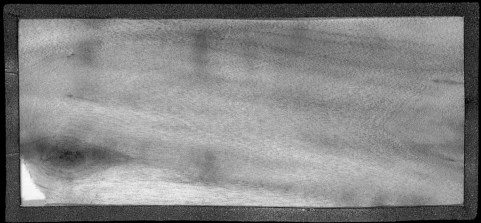

TANGENTIAL SECTION.

Ger. Missouri Weide.   *Fr.* Saule de Missouri.

Sp. Sauce de Missouri.

W | WILLOW FAMILY
Nuttall Willow, Black Willow

FAMILIE DER WEIDENGEWÄCHSE
Nuttall-Weide

FAMILLE DU SAULE
Saule de Scouler

SALICACEAE
Salix scouleriana Barrat ex Hook.
Syn.: *S. nuttallii* Sarg.

Description: Up to 35 ft. (10 m) high and barely 12 in. (0.3 m) in diameter. Trunk short. Bark thin, dark brown, with pale red tinge, in broad, low ridges.

Habitat: Sides of high mountain streams and riverbanks in Pacific North America.

Wood: Light, soft, not strong, close-grained. Heartwood pale brown with red tinge; sapwood thick and almost white.

Use: No commercial use known.

Beschreibung: Gelegentlich 10 m hoch bei kaum 0,3 m Durchmesser. Kurzer Stamm. Borke dünn, dunkelbraun, leicht rot getönt, in breiten, flachen Riefen.

Vorkommen: Ufer der Gebirgsbäche in großen Höhen und Flussufer im pazifischen Nordamerika.

Holz: Leicht, weich, nicht fest und mit feiner Textur. Kern hellbraun mit Rot; Splint dick und fast weiß.

Nutzung: Keine gewerbliche Nutzung bekannt.

Description: arbre atteignant parfois 10 m de hauteur, avec un tronc court d'à peine 0,3 m de diamètre. Ecorce fine, d'un brun sombre légèrement teinté de rouge et creusée de larges fissures plates.

Habitat: bords des torrents de haute altitude et des rivières du Pacifique en Amérique du Nord.

Bois: léger, tendre, non résistant et à grain fin. Bois de cœur brun clair mêlé de rouge; aubier épais et presque blanc.

Intérêts: utilisation commerciale non déterminée.

Nuttall Willow.

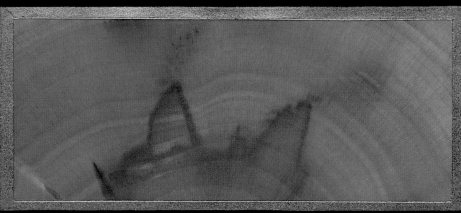

TRANSVERSE SECTION

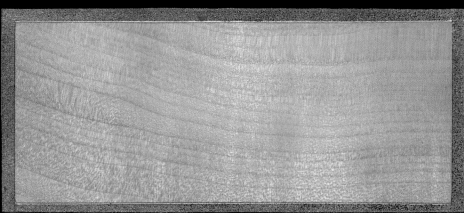

RADIAL SECTION.

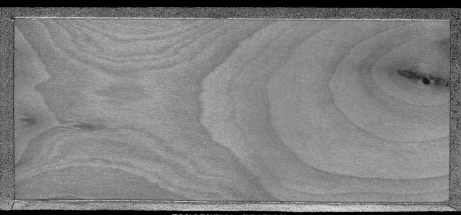

TANGENTIAL SECTION.

Ger. Weide von Nuttall.   *Fr.* Saule de Nuttall.

Sp. Sauce de Nuttall.

W | WILLOW FAMILY
Pacific Willow, Black Willow

FAMILIE DER WEIDENGEWÄCHSE
Zottige Weide

FAMILLE DU SAULE
Saule brillant du Pacifique, Saule laurier de l'Ouest (Canada)

SALICACEAE
Salix lucida Muhl subsp. *lasiandra* (Benth.) E. Murray
Syn.: *S. lasiandra* Benth.

Description: Up to 60 ft. (18 m) high and 2 to 3 ft. (0.6 to 0.9 m) in diameter. In the south of its range and in the interior of the continent it is distinctly smaller and may only be of shrubby growth. Trunk straight, with an open, irregular crown, the ends of the branches usually losing their leaves. Bark thick, dark brown tinged red, divided by fine grooves into broad, flat areas. Young branches short, reddish brown or yellow, towards the end of their first growth period dark purple, bright reddish brown, or pale orange. Leaves lanceolate to ovate-lanceolate, dark green above, often pale or glaucous beneath, the leaf margins finely toothed. Flowers appearing with the leaves in erect catkins.

Habitat: Abundant on riverbanks and lakesides in western North America, and in other moist places.

Wood: Light, soft, and brittle.

Use: Regarded as the most beautiful American willow, also planted in Europe as an ornamental tree (Fitschen, 1994).

Beschreibung: Bis 18 m hoch bei 0,6 bis 0,9 m Durchmesser. Im Süden des Verbreitungsgebietes und im Inneren des Kontinents deutlich kleiner bis hin zu strauchförmigem Wuchs. Stamm gerade, mit offener, unregelmäßiger Krone, Zweigenden meist entlaubt. Borke dick, dunkelbraun mit rotem Farbton, durch feine Risse in breite und flache Schollen zergliedert. Junge Äste kurz, rotbraun oder gelb, gegen Ende der ersten Vegetationsperiode dunkel purpurn, leuchtend rotbraun oder hellorange. Blätter lanzettlich bis eiförmig-lanzettlich, Oberseite dunkelgrün, Unterseite oft blass oder bereift, Blattrand fein gezähnelt. Blüten erscheinen zur Zeit des Laubaustriebes in aufrechten Kätzchen.

Vorkommen: Häufig an Fluss- und Seeufern im westlichen Nordamerika und an anderen feuchten Stellen.

Holz: Leicht, weich und brüchig.

Nutzung: Gilt als eine der schönsten Weiden Amerikas, auch in Europa als Ziergehölz angepflanzt (Fitschen 1994).

Description: jusqu'à 18 m de hauteur sur un diamètre de 0,6 à 0,9 m. Nettement plus petit ou même à port buissonnant dans le sud de son aire et à l'intérieur du continent. Tronc droit; couronne irrégulière, ouverte. Extrémité des rameaux généralement libre. Ecorce épaisse, brun foncé teinté de rouge, finement fissurée et fragmentée en côtes larges et plates. Jeunes rameaux courts, brun rouge ou jaune, devenant pourpre sombre, brun rouge brillant ou orange clair vers la fin de la première saison de végétation. Feuilles lancéolées à ovoïdes-lancéolées, finement dentées, vert sombre dessus, souvent pâles ou blanchâtres dessous. Fleurs en chatons dressés au moment de la feuillaison.

Habitat: affectionne les rives de cours d'eau et de lacs de l'ouest de l'Amérique du Nord, mais aussi dans d'autres lieux humides.

Bois: léger, tendre et cassant.

Intérêts: considéré comme l'un des plus beaux saules américains; cultivé aussi en Europe pour l'ornement (Fitschen 1994).

Western Black Willow.

TRANSVERSE SECTION.

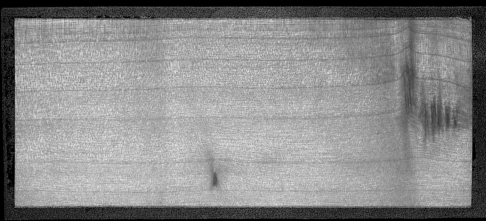

RADIAL SECTION.

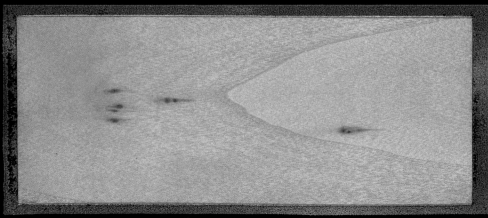

TANGENTIAL SECTION.

Ger. Westliche Schwarzweide.　　*Fr.* Saule noir occidental.
Sp. Sauce negro occidental.

W | WILLOW FAMILY
Red Willow

FAMILIE DER WEIDENGEWÄCHSE
Glattblättrige Weide

FAMILLE DU SAULE
Saule lustré

SALICACEAE
Salix laevigata Bebb.

Description: 45 to 50 ft. (13 to 15 m) high and occasionally 2 ft. (0.6 m), but usually only 10 to 12 in. (0.25 to 0.3 m), in diameter. Trunk straight. Bark dark brown, with a tinge of red, deeply divided into irregularly fused shallow ridges flaking into thick, appressed scales on the surface.

Habitat: Banks of streams in southwestern North America.

Wood: Light, soft, splintering, close-grained. Heartwood pale brown, with reddish tinge; sapwood thick, almost white.

Use: Many willows are used for cane weaving and basketry, etc. Bark yields a traditional medicine against fever.

Beschreibung: 13 bis 15 m hoch bei gelegentlich 0,6 m Durchmesser, meist aber nur 0,25 bis 0,3 m Durchmesser. Gerader Stamm. Borke dunkelbraun, leicht rot getönt, tief geteilt in unregelmäßig verbundene, flache Riefen, an der Oberfläche in dicke, anliegende Schuppen gebrochen.

Vorkommen: Ufer von Bächen im südwestlichen Nordamerika.

Holz: Leicht, weich, splittert leicht, mit feiner Textur. Kern hellbraun rötlich getönt; Splint dick, fast weiß.

Nutzung: Viele Weiden werden für Flechtwerk, Körbe etc. genutzt. Die Rinde wird in der Volksmedizin als Fiebermittel eingesetzt.

Description : arbre de 13 à 15 m de hauteur, avec un tronc droit de 0,25 à 0,3 m de diamètre, pouvant atteindre quelquefois 0,6 m. Ecorce brun sombre, légèrement teintée de rouge, à sillons profonds et à crêtes plates, entrelacées, se craquelant en écailles épaisses et appliquées.

Habitat : berges de ruisseaux du sud-ouest de l'Amérique du Nord.

Bois : léger, tendre, tendant à se fendre et à grain fin. Bois de cœur brun clair teinté de rouge ; aubier épais, presque blanc.

Intérêts : de nombreux saules sont employés localement en vannerie. L'écorce est utilisée en médecine populaire comme fébrifuge.

California Black Willow.

TRANSVERSE SECTION.

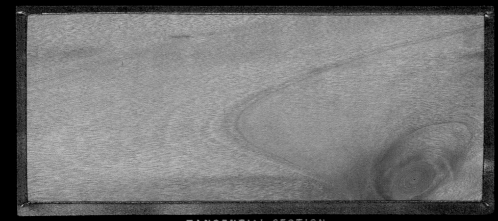

RADIAL SECTION.

TANGENTIAL SECTION

Ger. Californianische Schwartze Weide. *Fr.* Saule noir de Califor
Sp. Sauce negro de California.

W | WILLOW FAMILY
Sand Bar Willow

FAMILIE DER WEIDENGEWÄCHSE
Langblättrige Weide

FAMILLE DU SAULE
Saule riverain

SALICACEAE
Salix melanopsis Nutt.
Syn.: *S. fluviatilis* Nutt.

Description: 20 to 70 ft. (6 to 21 m) high and 4 in. to 2 ft.(0.10 to 0.6 m) in diameter. Mostly only a shrub 5 to 6 ft. (1.5 to 1.8 m) in height, its root suckers forming extensive thickets. Bark thin, smooth, dark brown with a slightly reddish tinge. Leaves linear-lanceolate, green above and beneath, margins finely toothed. Flowers in erect catkins, appearing after the leaves have sprouted.

Habitat: A woody pioneer plant on the sandy and gravely banks of rivers in large areas throughout North America.

Wood: Light, soft, and close-grained. Heartwood light brown tinged with red; sapwood thin, light brown. Numerous distinct medullary rays.

Use: Of no commercial use, but the shrubs stabilize new sand-banks and so prepare the land for colonization by poplars, for example.

Beschreibung: 6 bis 21 m hoch bei 0,10 bis 0,6 m Durchmesser. Meist nur als 1,5 bis 1,8 m hoher Strauch. Durch Wurzelbrut ausgedehnte Dickichte bildend. Borke dünn, glatt, dunkelbraun mit leicht rötlichem Farbton. Blätter linear-lanzettlich, Ober- wie Unterseite grün, Ränder gezähnelt. Blüten nach dem Laubaustrieb in aufrechten Kätzchen.

Vorkommen: Pioniergehölz auf Sand- und Kiesbänken von Flussufern über weite Teile des gesamten Subkontinents.

Holz: Leicht, weich und mit sehr feiner Textur, Kern hellbraun, rot getönt; Splint dünn, hellbraun. Zahlreiche deutliche Markstrahlen.

Nutzung: Keine gewerbliche Nutzung, die Gebüsche stabilisieren junge Sandbänke und bereiten damit den Boden für die Besiedlung z.B. durch Pappeln.

Description: 6 à 21 m de hauteur sur un diamètre de 0,10 à 0,6 m. Généralement sous forme de buisson de 1,5 à 1,8 m de haut et constituant de vastes fourrés avec ses nombreux rejets. Ecorce mince, lisse, brun sombre légèrement teinté de rouge. Feuilles linéaires-lancéolées, vertes, dentées. Fleurs en chatons dressés après la feuillaison.

Habitat: essence pionnière sur les bancs de sable et de gravier des berges dans des grandes régions de l'ensemble de l'Amérique du Nord.

Bois: léger, tendre et à grain très fin. Bois de cœur brun clair teinté de rouge; aubier mince, brun clair. Rayons médullaires nombreux et distincts.

Intérêts: aucun rôle économique; les buissons stabilisent les jeunes bancs de sable et préparent ainsi leur colonisation par d'autres essences, les peupliers par exemple.

Sand-bar Willow. Long-leaf Willow.

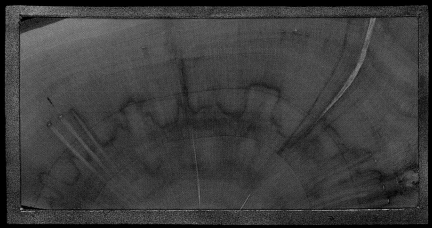

TRANSVERSE SECTION.

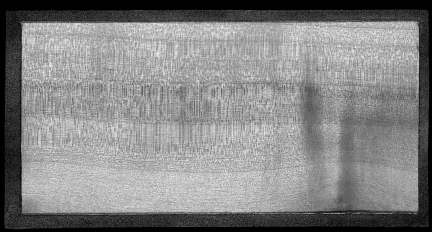

RADIAL SECTION.

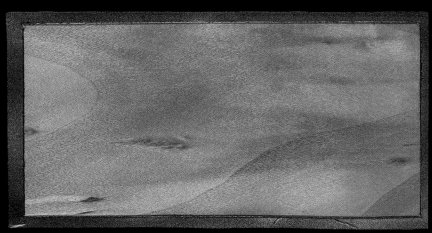

TANGENTIAL SECTION

Ger. Langblättrige Weide.   *Fr.* Saule de longefueille.

Sp. Sauce de hojas largas.

W | WILLOW FAMILY
Sitka Willow

FAMILIE DER WEIDENGEWÄCHSE
Sitka-Weide

FAMILLE DU SAULE
Saule de Sitka

SALICACEAE
Salix sitchensis Sanson ex Bong.

Description: 24 to 30 ft. (7 to 9 m) high and a maximum of 12 in. 0.3 m) in diameter. More frequently only a shrub, 6 to 15 ft. (1.8 to 4.5 m) in height, much-branched, with a short stem. Branches slender. Bark dark brown tinged with red, thin, and divided into irregular scales. Leaves narrowly ovate to broadly lanceolate, entire or finely toothed, pointed, the underside densely covered with woolly hairs. Stipules large and persistent. The flowers appear when the leaves begin to sprout, and are arranged in long, narrow, erect catkins.

Habitat: Riverbanks and other moderately damp places in the coastal region of Pacific North America.

Wood: Light, soft, and close-grained. Heartwood pale red; sapwood thick, almost white.

Use: The wood is used by the Indians of the Alaskan coast for smoking fish. The bark helps in the healing of cuts and wounds. Occasionally cultivated in European gardens.

One of the most beautiful of North American willows.

Beschreibung: 7 bis 9 m hoch bei maximal 0,3 m Durchmesser. Häufiger nur ein 1,8 bis 4,5 m hoher Strauch, stark verzweigt, mit kurzem Stamm. Äste schlank. Borke dunkelbraun mit rotem Farbton, dünn und in unregelmäßige Schuppen zergliedert. Blätter lang oval bis kurz lanzettlich, ganzrandig oder fein gezähnelt, zugespitzt, Unterseite dicht wollig behaart. Nebenblätter groß und ausdauernd. Blüten erscheinen zu Beginn des Laubaustriebes in langen und schmalen aufrechten Kätzchen.

Vorkommen: Flussufer und andere mäßig feuchte Standorte in der Küstenregion des pazifischen Nordamerika.

Holz: Leicht, weich und mit feiner Textur. Kern hellrot; Splint dick, fast weiß.

Nutzung: Das Holz wird von den Indianern der alaskischen Küste zum Räuchern von Fisch verwendet. Die Rinde hilft bei der Heilung von Schnitten und Wunden. Gelegentlich in europäischen Gärten kultiviert.

Eine der schönsten nordamerikanischen Weiden.

Description: 7 à 9 m de hauteur sur un diamètre de 0,3 m maximum. Plus souvent sous la forme d'un buisson de 1,8 à 4,5 m de haut, très ramifié sur un tronc court, avec des branches élancées. Ecorce brun foncé teinté de rouge, mince et à texture craquelée en écailles irrégulières. Feuilles longues-ovoïdes à courtes-lancéolées, entières ou finement dentées, terminées en pointe, glabres dessus, blanc cotonneux dessous. Grandes stipules persistantes. Floraison en début de feuillaison, en longs chatons fins, dressés.

Habitat: berges et autres lieux moyennement humides du littoral pacifique de l'Amérique du Nord.

Bois: léger, tendre et à grain fin. Bois de cœur rouge clair ; aubier épais, presque blanc.

Intérêts: le bois est utilisé par les Indiens des régions côtières de l'Alaska pour fumer le poisson. L'écorce a des propriétés cicatrisantes. Planté quelquefois dans les jardins européens.

L'un des plus beaux saules de l'Amérique du Nord.

Silky Willow, Sitka Willow.

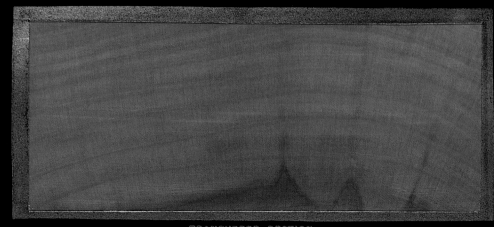

TRANSVERSE SECTION.

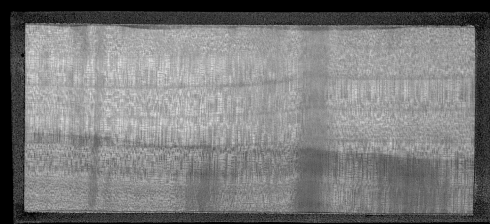

RADIAL SECTION.

TANGENTIAL SECTION.

W | WILLOW FAMILY
Swamp Poplar, Swamp Cottonwood

FAMILIE DER WEIDENGEWÄCHSE
Verschiedenblättrige Pappel

FAMILLE DU SAULE
Peuplier des marais, Peuplier argenté

SALICACEAE
Populus heterophylla L.

Description: In the most favorable conditions (in the Mississippi Valley) 85 to 100 ft. (25 to 30 m) high and 2 to 3 ft. (0.6 to 0.9 m) in diameter, with a straight, columnar trunk. Bark grayish brown with raised plates and deep grooves.

Habitat: The edges of ponds and more or less flooded margins of swamps in association with Honey and Water Locust, Mississippi Hackberry, Swamp White Oak, Red and Drummond Maple, Sweet Gum and tupelos, etc., in eastern North America; however, there are large areas where it does not occur (e. g. the Alleghenies). In the northern Atlantic states, where it is found only along a narrow coastal strip, it is smaller and rarer than in the southern Mississippi basin.

Wood: Light. Heartwood grayish brown; sapwood paler.

Use: As a commercial wood it is known as Black Poplar and is used for interior construction.

Beschreibung: Unter besten Bedingungen (im unteren Mississippi-Tal) 25 bis 30 m hoch bei 0,6 bis 0,9 m Durchmesser. Gerader, säulenförmiger Stamm. Borke graubraun mit hohen Riefen und tiefen Rillen.

Vorkommen: Teichufer und mehr oder weniger überschwemmte Ränder von Sümpfen in Gesellschaft von Amerikanischer und Wasser-Gleditschie, Mississippi-Zürgelbaum, Sumpf-Weiß-Eiche, Rot- und Drummond-Ahorn, Amberbaum und Tupelos u. a. Im östlichen Nordamerika jedoch mit großen Verbreitunglücken (z. B. Alleghenies). In den nördlichen Atlantikstaaten, wo er nur einen schmalen Küstenstreifen besiedelt, kleiner und seltener als im Süden des Mississippi-Beckens.

Holz: Leicht. Kern graubraun; Splint heller.

Nutzung: Als Nutzholz unter dem Namen Schwarz-Pappel zum Innenausbau.

Description : 25 à 30 m de hauteur dans des conditions optimales (vallée inférieure du Mississippi), sur un diamètre de 0,6 à 0,9 m. Tronc droit, columnaire. Ecorce brun gris à sillons profonds et à grosses crêtes.

Habitat : bords des étangs et berges plus ou moins inondées des marais en association avec le févier à trois épines et le févier aquatique, le micocoulier de Virginie, le chêne bicolore, l'érable rouge et l'érable de Drummond, le liquidambar, les nyssas etc. Croît dans l'est de l'Amérique du Nord à l'exception de quelques régions (par ex. les Alleghanys). Plus petit et plus rare dans le nord des Etats de la façade atlantique, où on ne le rencontre que sur une bande étroite le long de la côte, que dans le sud du bassin du Mississippi.

Bois : léger. Bois de cœur brun-gris ; aubier plus clair.

Intérêts : bois utile commercialisé

Swamp Poplar, Downy Poplar, River Cottonwood.

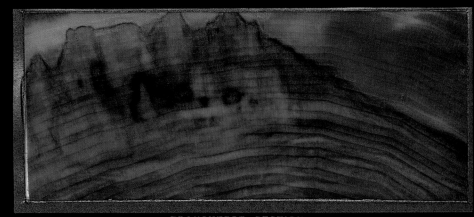

TRANSVERSE SECTION.

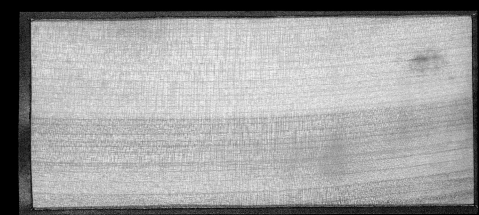

RADIAL SECTION.

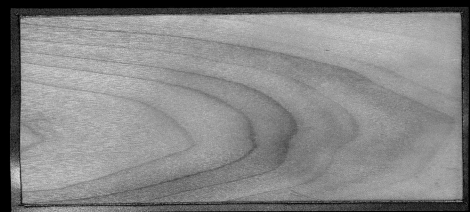

TANGENTIAL SECTION.

Ger. Sumpf-Pappel. Fr. Peuplier marécageaux. Sp. Alamo pant

W | WILLOW FAMILY
White Poplar, Abele, Silver-leaf Poplar

FAMILIE DER WEIDENGEWÄCHSE
Silber-Pappel

FAMILLE DU SAULE
Peuplier blanc

SALICACEAE
Populus alba L.

Description: A good 100 ft. (30 m) high and 3 to 4 ft. (0.9 to 1.2 m) in diameter. Bark greenish gray to whitish on branches and the upper part of the trunk. The older stems and bases of trunks have deeply fissured, dark bark. Leaves dark green above, velvety white beneath. Leaves flutter in the wind. Ornamental. It is quickly growing, and with far-spreading roots and root-suckers.

Habitat: Native to southern Europe, where it occurs in water meadows. It was taken to North America as an ornamental tree by the white settlers, and has become naturalized in some parts of the northeast.

Wood: Light, soft, and tough. Heartwood reddish yellow; sapwood almost white.

Use: Generally only for paper pulp, though locally for small wooden objects.

Beschreibung: Gut 30 m hoch bei 0,9 bis 1,2 m Durchmesser. Borke grünlich grau bis weißlich an den Ästen und am oberen Teil des Stammes. Ältere Stämme und Stammbasen mit tief zerklüfteter dunkler Borke. Blätter oberseits dunkelgrün, unterseits samtig weiß. Bei Windbewegung Wechselspiel. Ornamental. Schnellwüchsig und mit weitstreichenden Wurzeln und Wurzelbrut.

Vorkommen: Seine Heimat liegt in Südeuropa, wo er in Flussauen vorkommt. Von den weißen Siedlern als Schmuckbaum nach Nordamerika eingeführt und im Nordosten teilweise eingebürgert.

Holz: Leicht, weich und zäh. Kern rötlich gelb; Splint fast weiß.

Nutzung: Wohl nur für Papier-Pulpe und regional für kleine Gegenstände aus Holz.

Description: bien 30 m de hauteur sur un diamètre de 0,9 à 1,2 m. Ecorce gris verdâtre à blanchâtre sur les branches et la partie supérieure du tronc. Elle noircit et se crevasse profondément sur les vieux troncs et à la base. Feuilles vert foncé dessus, tomenteuses et blanc argenté dessous. Bel effet de couleurs au moindre souffle de vent. Espèce ornementale. Croissance rapide avec de longues racines et un fort drageonnement.

Habitat: provient du sud de l'Europe, où il croît sur les sols des lits de hautes eaux. À été introduit en Amérique du Nord comme arbre d'ornement par les colons blancs et planté dans certaines régions du nord-est.

Bois: léger, tendre et tenace. Bois de cœur jaune rougeâtre; aubier presque blanc.

Intérêts: utilisé uniquement pour la pâte à papier et, dans certaines régions, pour faire des petits objets en bois.

White Poplar, Abele.

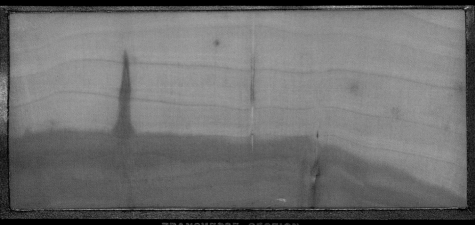

TRANSVERSE SECTION.

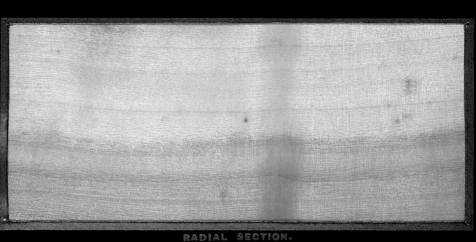

RADIAL SECTION.

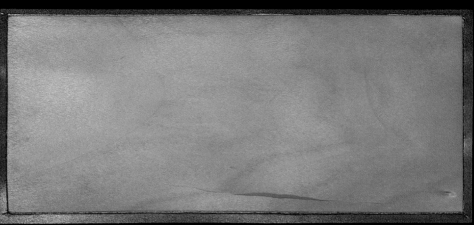

TANGENTIAL SECTION.

Ger. Alber.   Fr. Peuplier blanc.   Sp. Alamo blanco.

W | WILLOW FAMILY
White Willow

FAMILIE DER WEIDENGEWÄCHSE
Bigelow-Weide

FAMILLE DU SAULE
Saule arroyo

SALICACEAE
Salix lasiolepis Benth.

Description: 20 to 35 (occasionally 55) ft. (6 to 10 (16) m) high, 10 to 12 in. (0.24 to 0.36 m) in diameter. Small shrub towards the north and at high altitudes. Older bark dark, roughened by small lenticels and broken up by broad, low, irregular ridges. Younger bark thin, smooth, and pale gray-brown.

Habitat: Commonest and most variable of the Californian willows; banks of streams and in shallow, damp valleys. Apart from California, also in southern Arizona.

Wood: Light, soft, close-grained.

Use: In southern California as firewood.

Beschreibung: 6 bis 10 m (gelegentlich 16 m) bei 0,24 bis 0,36 m Durchmesser. Nach Norden und in großen Höhen ein kleiner Strauch. Borke alter Stämme dunkel, aufgeraut von kleinen Lentizellen und in breite, flache, unregelmäßig verbundene Riefen aufgebrochen. An jungen Stämmen dünn, glatt und hell graubraun.

Vorkommen: Als gemeinste und veränderlichste der kalifornischen Weiden an Bachufern und in flachen, feuchten Senken. Außer in Kalifornien auch in Süd-Arizona.

Holz: Leicht, weich und mit feiner Textur.

Nutzung: In Süd-Kalifornien als Brennholz.

Description: 6 à 10 m de hauteur (parfois 16 m) sur un diamètre de 0,24 à 0,36 m. Petit buisson vers le nord et à haute altitude. Ecorce sombre, rendue rugueuse par de petites lenticelles et parcourue d'un réseau irrégulier de larges fissures plates sur les vieux troncs ; fine, lisse et d'un brun gris clair sur les jeunes arbres.

Habitat : le plus commun et le plus variable des saules américains ; bord des ruisseaux et dépressions plates et humides. En dehors de la Californie, croît aussi dans le sud de l'Arizona.

Bois : léger, tendre et à grain fin.

Intérêts : bois de chauffage dans le sud de la Californie.

California White Willow, Bigelow Willow.

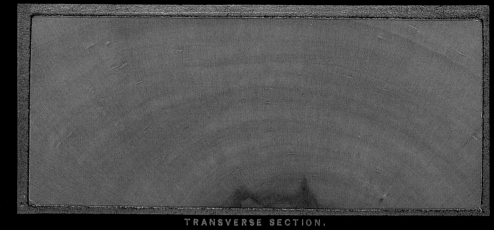

TRANSVERSE SECTION.

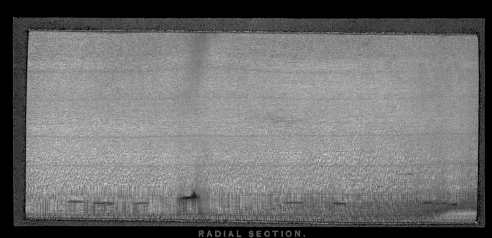

RADIAL SECTION.

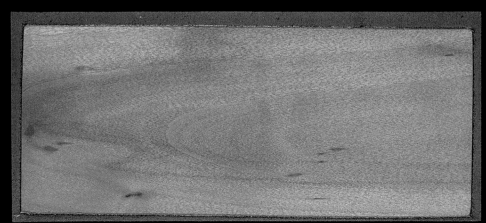

TANGENTIAL SECTION

Ger. Californische Weisze Weide. *Fr.* Saule blanc de Californie
Sp. Sauce blanco de California.

W | WITCH HAZEL FAMILY
Sweet Gum, American Red Gum, Bilsted

FAMILIE DER ZAUBERNUSSGEWÄCHSE
Amerikanischer Amberbaum

FAMILLE DE L'HAMAMÉLIS
Liquidambar d'Amérique

HAMAMELIDACEAE
Liquidambar styraciflua L.

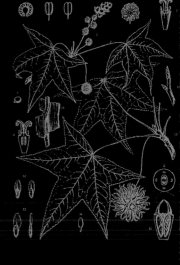

Description: 100 to 130 ft. (30 to 40 m) high and 4 to 5 ft. (1.2 to 1.5 m) in diameter, with a straight, columnar trunk. Young branches with corky wings, not always present. Leaves stellate, five to seven-lobed, bright green, colored red to purple in the fall. Fragrant when bruised.

Habitat: Rich, lowland soils together with Red and Black Maple, Black Gum, Water Tupelo, Laurel and Water Oak, various species of ash, etc., in eastern North America.

Wood: Moderately heavy, fairly soft, and close-grained.

Use: For wooden structures, wood paving, wooden objects, etc., sometimes listed commercially under absurd names such as Satin Walnut, California Red Gum, and so on.

Beschreibung: 30 bis 40 m hoch bei 1,2 bis 1,5 m Durchmesser. Gerader, säulenförmiger Stamm. Junge Zweige mit Korkflügeln, die bisweilen fehlen. Fünf- bis sieben-strahlige, „sternförmige" Blätter von kräftigem Grün und mit roter bis purpurner Herbstfärbung. Duftend, wenn gerieben.

Vorkommen: Reiche Böden im Flachland gemeinsam mit Rot- und Schwarz-Ahorn, Wald- und Wasser-Tupelos, Lorbeer und Wasser-Eiche, verschiedenen Eschen u. a. im östlichen Nordamerika.

Holz: Mäßig schwer, ziemlich weich und mit feiner Textur.

Nutzung: Für Holzkonstruktionen, hölzerne Pflasterung, Holzarbeiten usw., manchmal unter absurden Namen wie Satin Walnut, California Red Gum usw. im Handel.

Description : arbre atteignant 30 à 40 m de hauteur, avec un tronc droit, columnaire de 1,2 à 1,5 m de diamètre. Jeunes rameaux garnis d'ailes liégeuses, absentes cependant quelquefois. Feuilles à 5 ou 7 lobes dentés, avec une nervation palmée, d'un vert intense virant au rouge ou au pourpre en automne et exhalant une agréable odeur balsamique au froissement.

Habitat : sols riches de plaine dans l'est de l'Amérique du Nord, en compagnie de l'érable rouge et de l'érable noir, du nyssa sylvestre, du nyssa aquatique, du chêne noir, du chêne à feuilles de laurier et de divers hêtres.

Bois : moyennement lourd, assez tendre et à grain fin.

Intérêts : bois de gros œuvre ; utilisé aussi pour fabriquer des lambris et divers objets. Parfois commercialisé sous des noms surprenants : Satin Walnut, California Red Gum etc.

TRANSVERSE SECTION.

RADIAL SECTION.

TANGENTIAL SECTION

Ger. Storaxbaum.   Sp. Liquidambar.

Fr. Copalm.

W | WITCH HAZEL FAMILY
Witch-hazel, Snapping Hazel, Spotted Alder, Winter Bloom

FAMILIE DER ZAUBERNUSSGEWÄCHSE
Zaubernuss, Hexenhasel, Virginische Zaubernuss

FAMILLE DE L'HAMAMÉLIS
*Hamamélis de Virginie, Noisetier des sorcières,
Café du diable (Canada)*

HAMAMELIDACEAE
Hamamelis virginiana L.

Description: 25 to 30 ft. (7.5 to 9.0 m) high and 12 to 18 in. (0.3 to 0.35 m) in diameter. Trunk short, with spreading branches, which form a broad, open crown, more frequently however a many-stemmed shrub. Bark thin and smooth, light brown, with small, thin scales, which when they fall allow the dark reddish purple inner bark to become visible. Leaves oblong or ovate, with a very asymmetric base, at first covered with rust-colored hairs, later glabrous and dark green. Flowers in autumn, bright yellow with long, narrow petals. From each inflorescence only two fruits develop to maturity.

Habitat: One of the commonest shrubs in eastern North America, on forest margins on poor soils and on the rocky banks of rivers.

Wood: Heavy, hard, and very close-grained. Heartwood light brown, tinged red; sapwood thick (30 to 40 annual rings). Numerous thin, distinct medullary rays.

Use: The leaves and bark are astringent and are widely used in homeopathy for extracts and decoctions. Formerly used for divining rods.

Beschreibung: 7,5 bis 9 m hoch bei 0,3 bis 0,35 m Durchmesser. Stamm kurz, mit ausladenden Ästen, die eine breite, offene Krone bilden, öfter aber als vielstämmiger Strauch. Borke dünn und glatt, hellbraun, mit dünnen Schüppchen, nach deren Verlust die dunkle, rötlich violette, innere Rinde sichtbar wird. Blätter länglich oder oval mit sehr asymmetrischer Basis, anfangs mit rostfarbener Behaarung, später nackt und dunkelgrün. Blüte im Herbst hellgelb mit langen schmalen Kronblättern. Aus jedem Blütenstand entwickeln sich nur zwei Früchte bis zur Reife.

Vorkommen: Einer der häufigsten Sträucher im östlichen Nordamerika, an Waldrändern auf armen Böden und an steinigen Flussufern.

Holz: Schwer, hart und mit sehr feiner Textur. Kern hellbraun, rot getönt; Splint dick (30 bis 40 Jahresringe). Zahlreiche dünne, deutliche Markstrahlen.

Nutzung: Blätter und Borke wirken adstringierend und werden in großem Umfang für Extrakte und als abgekochter Sud in der Homöopathie verwendet. Früher für Wünschelruten genutzt.

Description : arbuste de 7,5 m à 9,0 m de hauteur, avec un tronc court de 0,3 à 0,35 m de diamètre et des branches étalées, constituant un houppier ample et ouvert. Plus souvent sous forme de buisson à tronc multiple. Ecorce mince et lisse, brun clair, à texture en écailles, qui en se détachant, révèlent une écorce interne pourpre sombre. Feuilles allongées ou ovales, à base très asymétrique, couvertes d'une pubescence rouille jeunes puis devenant glabres et vert foncé. Floraison en automne ; fleurs jaune clair, avec des pétales longs et étroits. Deux fruits seulement par inflorescence jusqu'à maturité.

Habitat : un des buissons les plus fréquents dans l'est de l'Amérique du Nord, en lisière forestière sur des sols pauvres et des berges caillouteuses.

Bois : lourd, dur et à grain très fin. Bois de cœur brun clair teinté de rouge ; aubier épais (30 à 40 cernes annuels). Nombreux rayons médullaires fins et distincts.

Intérêts : les feuilles et l'écorce ont des propriétés astringentes. Utilisé autrefois pour confectionner des baguettes de sourcier.

Witch-hazel.

TRANSVERSE SECTION.

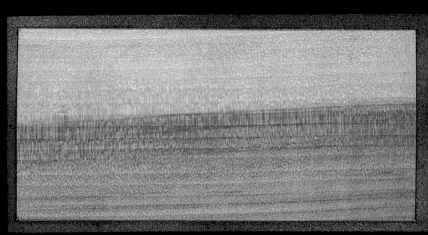

RADIAL SECTION.

Y | YEW FAMILY

Californian Nutmeg

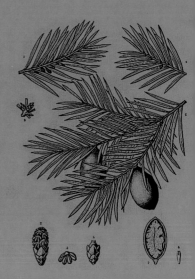

FAMILIE DER EIBENGEWÄCHSE

Muskatnuss-Torreye

FAMILLE DES IFS

Torreya de Californie, Muscadier de Californie,
Torreya muscadier

TAXACEAE

Torreya californica (Torr.)
Syn.: *Tumion californicum* (Torr.) Greene

Description: 50 to 75 (100) ft. (15 to 23 (30) m) high, 1 to 2 (4) ft. (0.3 to 0.6 (1.2) m) in diameter. Sprouts well from stump. Bark gray-brown, with an orange tinge. Irregularly and deeply sculpted into broad furrows with narrow ridges. Has elongated, rather loose scales. The name "nutmeg" comes from the similarity of the inner seed structure to that of a nutmeg. Leaves and twigs very pungent.

Habitat: Banks of mountain streams in California, nowhere common, but widespread. Grows best in the north.

Wood: Light, soft, not strong, silky-sheened, and close-grained; polishable, very durable in the ground.

Use: Occasionally for fence posts.

Rarely cultivated, as park tree (California und western Europe).

Beschreibung: 15 bis 23 (30) m hoch bei 0,3 bis 0,6 (1,2) m Durchmesser. Starkes Ausschlagvermögen des Stubbens. Borke graubraun, orangefarben getönt. Tief und unregelmäßig in breite Rillen und schmale Riefen geteilt. Mit länglichen, lose anliegenden Schuppen. Der Name „Nutmeg" verweist auf eine anatomische Besonderheit im Inneren des Samens ähnlich derjenigen der Muskatnuss. Blätter und Zweige stechend-aromatisch.

Vorkommen: Ufer von Bergbächen in Kalifornien, nirgends häufig, aber weit verbreitet. Am besten im Norden.

Holz: Leicht, weich, nicht fest, seidig und mit feiner Textur; polierfähig, sehr dauerhaft im Boden.

Nutzung: Gelegentlich für Zaunpfähle.

Selten als Parkbaum in Kultur (Kalifornien und Westeuropa).

Description: arbre de 15 à 23 m (30 m) de hauteur, avec un tronc de 0,3 à 0,6 m (1,2 m) de diamètre, rejetant vigoureusement de souche. Ecorce brun gris teinté d'orange à texture irrégulière, creusée de sillons larges aux côtes étroites s'exfoliant en écailles allongées. Il doit son nom de muscadier à sa graine entièrement recouverte, comme la muscade, par l'arille charnu. Ses feuilles et ses rameaux dégagent une odeur pénétrante au froissement.

Habitat: rives de torrents de Californie, nulle part très fréquent mais occupant une aire étendue. Prospère dans le nord.

Bois: léger, tendre, pas résistant, satiné et à grain fin; pouvant se polir et durable sur pied.

Intérêts: utilisé de façon occasionnelle pour fabriquer des piquets de clôtures.

Rarement cultivé comme arbre de parcs (Californie et Europe de l'Ouest).

California Nutmeg.

TRANSVERSE SECTION.

RADIAL SECTION.

TANGENTIAL SECTION

Ger. Californianische Muskatennusz.   Fr. Muskade de Calif

Sp. Nuez moscada de California.

Y | YEW FAMILY

Californian Yew

FAMILIE DER EIBENGEWÄCHSE
Kurzblättrige Eibe

FAMILLE DES IFS
If de l'Ouest, If de Californie, If occidental (Canada)

TAXACEAE
Taxus brevifolia Nutt.

Description: 42 to 50 (75 to 88) ft. (13 to 16 (23 to 26) m) high, 1 to 2 (4) ft. (0.3 to 0.6 (1.2) m) in diameter. Trunk tall and straight, often elliptical in section and irregularly furrowed. Bark with small, thin, dark red to purple scales, pale cinnamon red beneath.	*Beschreibung:* 13 bis 16 (23 bis 26) m hoch bei 0,3 bis 0,6 (1,2) m Durchmesser. Großer, gerader Stamm, häufig mit elliptischem Querschnitt und unregelmäßig gebuchtet. Borke aus kleinen, dünnen, dunkelroten bis purpurfarbenen Schuppen, darunter hell zimtrot.	*Description:* 13 à 16 m (23 à 26 m) de hauteur sur un diamètre de 0,3 à 0,6 m (1,2 m). Tronc droit, élancé, à coupe souvent elliptique et présentant des crevasses irrégulières. Ecorce recouverte d'écailles rouge foncé ou pourpre, petites et fines; couche sous-jacente cannelle clair.
Habitat: Shady banks of mountain streams, deep gorges with high humidity, usually beneath tall conifers. Not very numerous, but widespread, in Pacific North America.	*Vorkommen:* Schattige Gebirgsbachufer, tiefe Schluchten mit hoher Luftfeuchtigkeit, meist unter höheren Koniferen. Nicht sehr häufig, aber weit verbreitet im pazifischen Nordamerika.	*Habitat:* berges de torrents ombragées, gorges profondes à forte humidité atmosphérique, le plus souvent à l'ombre de grands conifères. Peu fréquent mais occupant une aire étendue de la façade pacifique de l'Amérique du Nord.
Wood: Heavy, hard, strong, brittle, and close-grained; polishable, very durable in the ground. Heartwood pale red with darker summer wood; sapwood thin and pale yellow.	*Holz:* Schwer, hart, fest, spröde und mit feiner Textur; polierfähig, sehr haltbar im Boden. Kern hellrot mit dunklem Band des Spätholzes; Splint dünn und hellgelb.	*Bois:* lourd, dur, résistant, cassant et à grain fin; pouvant se polir et durable sur pied. Bois de cœur rouge clair, avec une bande sombre dans le bois final; aubier mince et jaune clair.
Use: Fence posts, used for paddles by American Indians of the northwest, spear shafts, bows, fishing rods.	*Nutzung:* Für Zaunpfähle, Paddel der Indianer des Nordwestens, Speerschäfte, Bogen, Fischangeln.	*Intérêts:* pieux de clôtures, pagaies des Indiens du Nord-Ouest, bois de javelots, arcs, cannes à pêche.
Discovered by Douglas in 1825 on the lower Columbia River. Cultivated in Europe since 1854.	1825 von Douglas am unteren Columbia River entdeckt. Seit 1854 in Europa in Kultur.	Découvert par Douglas en 1825 le long du cours inférieur de la Columbia River. Cultivé en Europe depuis 1854.

Pacific Yew, California or Oregon Yew.

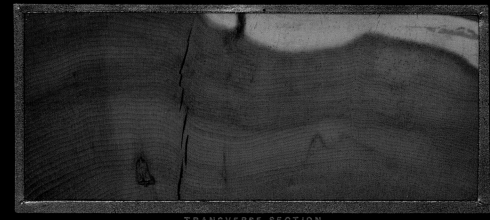

TRANSVERSE SECTION.

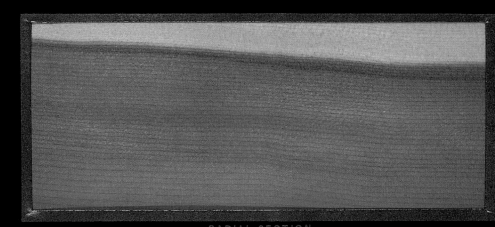

RADIAL SECTION.

TANGENTIAL SECTION

Ger. Californianischer Eibenbaum. *Fr.* If de Californi

Sp. Tejo de California.

Y | YEW FAMILY

Stinking Cedar, Torreya

FAMILIE DER EIBENGEWÄCHSE
Nuss-Eibe, Stinkeibe

FAMILLE DES IFS
Torreya de Californie

TAXACEAE
Torreya taxifolia Arn.
Syn.: *Tumion taxifolium* (Arn.) Greene

Description: Up to 40 ft. (12 m) high and 1 to 2 ft. (0.3 to 0.6 m) in diameter. Usually much smaller. Trunk short, slow-growing. Sometimes shoots vigorously from stump or roots. Bark brown with pale orange tinge. Irregular, broad, flat furrows and broad, flat, rather rounded ridges. Covered with thin, appressed scales, which reveal yellow inner bark when shed. Foliage and so-called nut (seed covered by seed coat) decorative. All parts foul-smelling.

Habitat: Calcareous soils, steep river banks, sides of gorges, and edges of marshes in south-eastern North America from north-west Florida to Tennessee.

Wood: Light, hard, strong, brittle, silky-sheened and close-grained, polishable, and durable in the ground.

Use: Fence posts, overexploited.

Beschreibung: Um 12 m hoch bei 0,3 bis 0,6 m Durchmesser. Gewöhnlich viel kleiner. Kurzer Stamm, langsam wachsend. Manchmal aus Stubben und Wurzeln kräftig austreibend. Borke braun, schwach orange getönt. Unregelmäßige breite, flache Rillen und breite, flache Riefen mit etwas gerundeten Rücken. Bedeckt mit dünnen, anliegenden Schuppen. Wenn diese fallen, wird eine gelbe innere Rinde sichtbar. Die Belaubung und sog. „Nuss" (ein von einem Samenmantel umhüllter Samen) wirken zierend. Alle Teile übelriechend.

Vorkommen: Kalkböden, steile Flussufer, Hänge von Schluchten und Sumpfränder im südöstlichen Nordamerika von Nordwest-Florida bis Tennessee.

Holz: Leicht, hart, fest, spröde, seidig und mit feiner Textur; polierfähig und dauerhaft im Boden.

Nutzung: Für Zaunpfähle, stark übernutzt.

Description : environ 12 m de hauteur sur un diamètre de 0,3 à 0,6 m, mais beaucoup plus petit en général. Tronc court, à croissance lente qui rejette parfois de souche et drageonne vigoureusement. Ecorce d'un brun lavé d'orangé, à larges sillons plats et inégaux et à grandes crêtes plates, légèrement arrondies au sommet. Elle est recouverte de minces écailles plaquées qui en se détachant laissent voir l'écorce interne jaune. Le feuillage et la « noix » (une coque verte enveloppant toute la graine) sont attrayants. Toutes les parties dégagent une odeur désagréable.

Habitat : sols calcaires, berges escarpées, versants de ravins, bords des marais dans le sud-est de l'Amérique du Nord depuis le nord-ouest de la Floride jusqu'au Tennessee.

Bois : léger, dur, résistant, cassant, satiné et à grain fin; pouvant se polir et durable sur pied.

Intérêts : utilisé pour fabriquer des pieux de clôture; exploitation excessive.

Yew-leaved Torreya, Stinking Cedar, Savin.

TRANSVERSE SECTION.

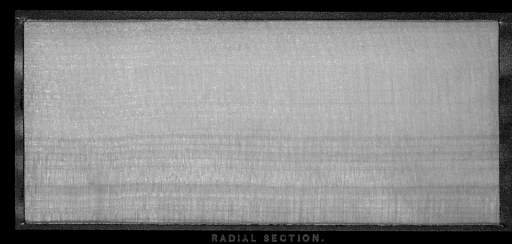

RADIAL SECTION.

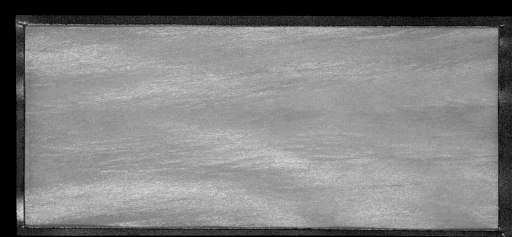

TANGENTIAL SECTION

Ger. Stink-Ceder.   *Fr.* Torreya à feuilles d'If.   *Sp.* Cedro feti

INDEX

The figures in parentheses refer to the plate number in Hough's original edition, the figures in bold type following to the page number in the present book.

Die Ziffern in Klammern verweisen auf die Tafelnummer in Houghs Originalausgabe, die fetten Ziffern dahinter auf die Seitenzahl im vorliegenden Buch.

Les chiffres entre parenthèses renvoient au numéro de la planche dans l'édition originale de Hough, les chiffres en gras qui les suivent au numéro de page du présent ouvrage.

743

BARNER, J.: Die Nutzhölzer der Welt, vol. 1–4. Wiesbaden, 1961/62

BEGEMANN, H.: Das große Lexikon der Nutzhölzer, vol. 1–12, Gernsbach, 1981–1994

BONAP (The Biota of North America Program): A synonymized checklist of the Vascular Flora of the United States, Canada, and Greenland. http:\\mip.berkeley.edu\bonap.

FITSCHEN, J.: Gehölzflora, Heidelberg, 10th edition, 1994

HARRAR, E.: Hough's Encyclopaedia of American Woods, vol. 1–8, New York, 1957–1981

HOUGH, R.B.: The American Woods – Exhibited by actual specimens and with copious explanatory texts, 14 vols, Lowville, NY, 1888–1913, 1928

HOUGH, R.B.: Report upon Forestry, Commissioner of Agriculture (ed.), vol. I–II, Washington, 1877–1880

KALM, P.: Des Herrn Peter Kalm als Beschreibung der Reise, die er nach dem nördlichen Amerika auf den Befehl gedachter Akademie und öffentliche Kosten unternommen hat, vol. I–III, Göttingen, 1754–64

MAYR, H.: Die Waldungen von Nordamerika, München, 1890

MAYR, H.: Fremdländische Wald- und Parkbäume für Europa, Berlin, 1906

MEDICUS, F.C.: Über nordamerikanische Bäume und Sträucher als Gegenstand der deutschen Forstwirtschaft und der schönen Gartenkunst, Mannheim, 1792

NÖRDLINGER, H.: Querschnitte von hundert Holzarten, vol. I–XI, Stuttgart & Tübingen, 1852–1888

PINCHOT, G.: The Fight for Conservation, New York, 1910

SARGENT, C.S.: The Silva of North America, (Reprint 1947, Peter Smith, New York), vol.1–14, 1889–1901

SARGENT, C.S.: Manual of the Trees of North America, vol. 1–2, 2nd edition, New York, 1965

SUPAN, A. (ed.): Dr. A. Petermanns Mitteilungen aus Justus Perthes' Geographischer Anstalt, vol. 32, Gotha, 1886

ACKNOWLEDGMENTS

We should like to thank Andrew McRobb, Kew's photographer, John Harris, Kew's Publications Marketing Officer and the employees at Kew Gardens for their helpful support.
The publication of this book is due very largely to the enormous efforts of Prof. Dr. Klaus Ulrich Leistikow (d. 2002), and we regret deeply that he is no longer able to witness this event.

Wir danken dem Fotografen Andrew McRobb, John Harris, dem Leiter der Publikationsabteilung, und allen anderen Mitarbeitern von Kew Gardens für die freundliche Unterstützung der Publikation. Dem großen Engagement von Prof. Dr. Klaus Ulrich Leistikow († 2002) für das Publikationsvorhaben ist es zu verdanken, dass dieses Buch vorliegt und wir bedauern außerordentlich, dass er diesen Moment nicht mehr erleben konnte.

Nous remercions le photographe Andrew McRobb, John Harris, responsable des publications et tout le personnel de Kew Gardens pour leur aimable soutien.
C'est à l'enthousiasme du Prof. Dr. Klaus Ulrich Leistikow († 2002), qui s'est profondément investi dans ce projet de publication, que l'on doit l'existence de cet ouvrage, et nous sommes extrêmement peinés qu'il n'ait pu vivre lui-même ce moment

IMPRINT

EACH AND EVERY TASCHEN BOOK PLANTS A SEED!
TASCHEN is a carbon neutral publisher. Each year, we offset our annual carbon emissions
with carbon credits at the Instituto Terra, a reforestation program in Minas Gerais, Brazil,
founded by Lélia and Sebastião Salgado. To find out more about this ecological partnership,
please check: www.taschen.com/zerocarbon
Inspiration: unlimited. Carbon footprint: zero

To stay informed about TASCHEN and our upcoming titles, please subscribe to
our free magazine at www.taschen.com/magazine, follow us on Twitter, Instagram,
and Facebook, or e-mail your questions to contact@taschen.com.

Original edition: © 2002 TASCHEN GmbH
© Photo p. 15: Albert Renger-Patzsch Archiv/Ann & Jürgen Wilde/VG Bild-Kunst, Bonn 2019
Photos for R. B. Hough's and C. S. Sargent's plates:
Andrew McRobb, Kew's photographer, Royal Botanic Gardens, Kew, London
Texts: Klaus Ulrich Leistikow (plates 1–200) and Holger Thüs (plates 201–350)
Project management: Petra Lamers-Schütze, Cologne
Editorial coordination: Thierry Nebois and Leslie Weissgerber, Cologne
English translation: Martin Walters and Clive King, Cambridge
French translation: Annie Berthold, Düsseldorf, and
Michèle Schreyer, Cologne (introduction)
Design: Claudia Frey, Cologne
Production: Thomas Grell, Cologne
Cover design: Sense/Net Art Direction,
Andy Disl and Birgit Eichwede, Cologne

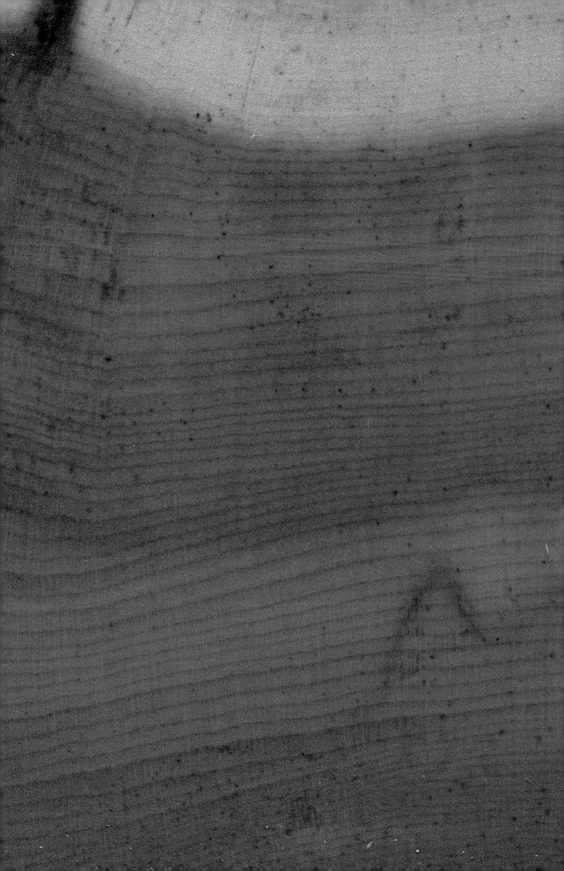